Reproduction in Reef Fishes

Reproduction in Reef Fishes

BY DR. R.E. THRESHER

ISBN 0-87666-808-2

Distributed in the UNITED STATES by T.F.H. Publications, Inc., 211 West Sylvania Avenue, Neptune City, NJ 07753; in CANADA by H & L Pet Supplies Inc., 27 Kingston Crescent, Kitchener, Ontario N2B 2T6; Rolf C. Hagen Ltd., 3225 Sartelon Street, Montreal 382 Quebec; in ENGLAND by T.F.H. Publications Limited, 4 Kier Park, Ascot, Berkshire SL5 7DS; in AUSTRALIA AND THE SOUTH PACIFIC by T.F.H. (Australia) Pty. Ltd., Box 149, Brookvale 2100 N.S.W., Australia; in NEW ZEALAND by Ross Haines & Son, Ltd., 18 Monmouth Street, Grey Lynn, Auckland 2 New Zealand; in SINGAPORE AND MALAYSIA by MPH Distributors Pte., 71-77 Stamford Road, Singapore 0617; in the PHILIPPINES by Bio-Research, 5 Lippay Street, San Lorenzo Village, Makati Rizal; in SOUTH AFRICA by Multipet Pty. Ltd., 30 Turners Avenue, Durban 4001. Published by T.F.H. Publications Inc., Ltd. the British Crown Colony of Hong Kong.

TABLE OF CONTENTS

To
Ann H.G.
and the
"famous father of the author"

Introduction

Reproduction is the central feature of an animal's life history, with all other aspects, from spacing mechanisms to growth rate, reflecting and resulting from selection for maximum reproductive success. It is therefore of utmost importance in studying a species that the investigator be at least cognizant of, if not intimately familiar with, the courtship, spawning, and, if present, patterns of care of eggs and young in that species. For most terrestrial animals this requirement is not generally a burdensome one, since at least general patterns of reproduction are well known for most such animals. Moreover, the investigator can not only fall back upon a substantial literature for details of reproductive biology, but he is also likely, with relative ease, to observe courtship, etc., directly. Unfortunately, this is not always the case with marine animals. Until recently the difficulties of working in an "alien" environment, combined with the use of bulky and often expensive diving gear and the special training required to use this gear safely, have resulted in a dearth of information on spawning by such animals. In fact, observation of a single spawning sequence by some marine species is of such significance as to alone warrant a scientific paper.

This is especially true for the fishes that inhabit the world's coral reefs. Reef fishes are phenomenally diverse and often strikingly beautiful, which has aroused widespread interest in them by both biologists and marine aquarists. Despite this interest, reproduction by these fishes remains poorly known. In the last comprehensive review of spawning in fishes, Breder & Rosen's monumental *Modes of Reproduction in Fishes*, either nothing or at best a few scattered anecdotes about possible courtship in aquaria were available for almost every family of reef-associated fishes. Only a few, such as the damselfishes (Pomacentridae) and wrasses (Labridae), were known in any detail, and even this knowledge was based on studies of only a handful of the numerous species in each family. Since the publication of Breder & Rosen's book in 1966, however, there has been a tremendous surge of interest in reef fishes and increases in the number of biologists working with them, in the ease with which studies can be carried out on the reef, and in the size of the marine aquarium industry and the number of aquarists. As a result, Breder & Rosen's book is now largely out of date with respect to the reproductive biology of reef fishes. While it is still true that for a few families details of courtship and spawning remain sparse, for most groups sufficient information is now available that at least the basic reproductive biology of the group can be blocked out.

Nonetheless, it is still widely held that little is known about reproduction by reef fishes, in part because Breder & Rosen, the standard reference on such matters, says so and in part for two other reasons. First, the average casual diver is not likely to see spawning on the reef. Spawning typically occurs only during very specific and brief periods, usually at times divers are rarely in the water (such as the crack of dawn or just prior to dark) and often only at specific, limited sites to which the fish migrate. Unless one is literally in the right place at the right time, one will never see even some of the most common fishes spawn. Second, the literature on spawning by reef fishes is so widely scattered that unless one digs deeply for it, one inevitably concludes that little such information is available. Relatively few papers on the spawning of reef fishes reach "main-line" journals such as *Copeia, Marine Biology*, or *Animal Behavior*; far more can be found in aquaculture journals, publications of various universities, museums, and public aquaria, and, perhaps most difficult of all to obtain, many small, often short-lived, popular aquarium magazines.

Few biologists have the time or resources to successfully dig through this dispersed and often obscure literature to ferret out information on spawning by reef fishes. One purpose of this book, then, is to review this literature for the professional biologist, summarizing it on a family-by-family basis, and in so doing pointing out both salient features of spawning in each group and also areas of likely profitable research.

This book, however, is also designed for the marine aquarist who desires to spawn and rear reef fishes in small aquaria. In addition to the basic information in each chapter, there is a final section that reviews successful spawnings and rearings to date (if such have been accomplished) and provides details of the conditions of both the fish and the aquaria when such spawnings occurred. For those fishes that have not yet been spawned in captivity, the prospects for such spawnings are discussed, along with suggestions of techniques that might help. Finally one chapter deals specifically with techniques of rearing the larvae of reef fishes and is written primarily for the aquarist. Although professional biologists may well find the information and references in this chapter useful, it is not meant to be either comprehensive or "state of the art." Rather, it deals with basic, relatively uncomplicated procedures that are best suited for small closed-system aquarium rearing of larvae.

Because the book is designed to serve two audiences with markedly different backgrounds and different approaches to reef fishes, the language used in the text is a compromise—one that is not so technical that a reasonably intelligent and motivated aquarist cannot follow it, or, on the other hand, so simple-minded that biologists will find it trivial and imprecise. If anything, the text hopefully errs on the side of simplicity on the assumption that professionals can, if they so desire, seek further details in the literature cited. Finally, the reader is assumed to be reasonably familiar with the external anatomy of fishes. Simple descriptive terms, such as names of fins, are not explained, and readers unfamiliar with them are referred to any of a number of good basic texts on ichthyology or comparative vertebrate anatomy. Technical terms pertinent to reproduction, however, are on occasion expanded upon in the text and are defined in a glossary at the end of the book.

ORGANIZATION AND SCOPE OF THE BOOK

Unlike the fish-by-fish approach used by Breder & Rosen, I have opted for a more synthetic approach, treating each family or group of closely related fishes as a whole and using species to exemplify each point. Except for those chapters covering rearing techniques

and general patterns of reproduction, each chapter deals either with a single family of fishes, with a subfamily if the family is so large as to be unwieldy to cover in a single chapter, or in a few cases with groups of closely related families whose systematic positions are not well spelled out. I have generally followed Nelson (1976) with respect to fish classification. I deviate from Nelson's arrangement only where the reproductive biology of the fishes clearly suggests some changes in the classification are necessary, as in, for example, the pseudochromoids. Each chapter consists of seven sections (Introduction, Sexual Dimorphism, Spawning Season, Reproductive Behavior, Eggs and Larvae, Reproduction in Captivity, and Literature Cited) unless so little information is known about the group in question that there wasn't any point to dividing it up in that fashion. In toto, the book covers some 53 families (depending upon where you want to draw family lines). To a large extent the selection of families to include was an arbitrary one. While I clearly endeavored to include all those families that are closely and regularly associated with the world's coral reefs, I had to draw the line somewhere with respect to the numerous families that are found only occasionally on the reef or that have members on the reef but are best represented in other habitats. In the end a good part of the decision on whether or not to include a group was made on the basis of their importance to aquarists, as well as biologists, and my interest in them.

In a similar fashion, this book is also likely to be criticized for the literature I've missed. While I've made every effort over the past several years to track down every bit of information, even that about which I may have just heard rumors, it is inevitable in a book of this scope that I've missed things (I'm especially worried about the European and Japanese popular literature, which I was only able to obtain on a fragmentary basis). Hopefully I haven't missed too much material or any major studies, but if so, I'd be more than happy to hear about them. Along the same lines, I sincerely hope that this book stimulates readers to publish observations they've made on the courtship and spawning of coral reef fishes. One of the most frustrating aspects of this work was writing "nothing is known about . . .," when I'm certain that somebody knows but hasn't published.

ACKNOWLEDGMENTS

During the several years over which this book has been written, a tremendous number of people have helped me. Of these, three are by far the most important. Without the help of Ann Gronell, Pat Colin, and Jack Moyer, this book would neither have been finished nor have been as comprehensive as it hopefully is. I deeply appreciate their comments, encouragement, and free sharing of information. I also thank Walter Heiligenberg and Peter Sale for sponsoring me as a postdoctoral fellow at, respectively, Scripps Institution of Oceanography and the University of Sydney, where much of the work on this book was carried out. I thank Jan Aldenhoven, Charles Arneson, Paul Atkins, Johann Bell, Lori Bell, Bruce Carlson, Peter Doherty, William Douglas, Hans Fricke, Barry Goldman, Janet Gomon, Martin Gomon, Bob Johannes, Alex Kerstitch, Peter Klimley, Rudy Kuiter, Jeff Leis, Phil Lobel, Brian Luckhurst, Richard Rosenblatt, Barry Russell, Mike Schmale, Patti Schmidt, Greg Stroud, Mike Sutton, Hugh Sweatman, Yutaka Yogo, and Martha Zaiser for assistance (often of a major nature) with various groups. Peter Doherty, Ann Gronell, Grahaeme Pyke, Peter Sale, and Richard Shine commented on various parts of the review chapter. I especially thank several kind souls who volunteered to review several chapters—Jerry Allen, Jack Briggs, Bruce Carlson, Jack Moyer, Ernie Reese, Doug Shapiro, and David Smith suggested many useful changes, made me aware of literature I would otherwise have missed, and in many cases allowed me to use some of their unpublished material. I also thank Charles Arneson, Pat Colin, Ann Gronell, Alex Kerstitch, Jack Moyer, Mike Sutton, Roger Steene, Katsumi Suzuki, and Noriaki Yamamoto for providing many of the fantastic photographs used in this book. June Jeffrey drew the beautiful figures for the text. Finally, I was supported during the production of this book by a National Needs Postdoctoral Fellowship from the National Science Foundation, a Queen Elizabeth II Fellowship in Marine Science, Scripps Institution of Oceanography, the University of Sydney, a grant from the Australian Research Grants Committee, and a grant from the Hildebrand Foundation, all of which are most gratefully acknowledged.

Literature Cited

Breder, C.M., Jr. and D.E. Rosen. 1966. *Modes of Reproduction in Fishes*. Natural History Press, Garden City, N.Y. 941 pp. (Reprinted by T.F.H. Publications, Neptune, N.J.)

Nelson, J.S. 1976. *Fishes of the World*. Wiley-Interscience, N.Y. 416 pp.

Eels
(Anguilliformes)

Eels are extremely elongate "snake-like" fishes that have successfully radiated into almost every aquatic habitat from small freshwater streams and ponds to the abyssal plains of the deep ocean. Anguilliformes, the order of true eels (as opposed to, for example, the freshwater spiny eels, family Mastacembelidae), contains approximately 600 species in 22 families, of which five are found on or around the reef: the conger and garden eels (Congridae), the spaghetti eels (Moringuidae), the morays (Muraenidae), the snake and worm eels (Ophichthidae), and the false morays (Xenocongridae). As a group, the eels are extremely common on the reef, much more so than they are usually given credit for. Except for the garden eels, which are planktivores and form immense and often conspicuous colonies in sandy areas, the reef-dwelling eels are a retiring group, foraging well back in cracks and crevices or burrowing silently through sand or mud. When approached by an intruder, all withdraw to disappear into cover. Some are so secretive that they have yet to be seen alive, having been collected only in poison stations. Except for the garden eels, the reef-dwelling species are also usually considered nocturnal fishes, foraging over the reef at night with the cardinalfishes, squirrelfishes, and other night-active animals. Many are also active by day, and it is likely that eels, like other predators, hunt opportunistically over long periods.

Reproduction of eels was in the last century one of the great mysteries of ichthyology. Naturalists had long observed that the common freshwater eels, *Anguilla* species, each year left streams and ponds in a mass migration to the sea. Months later tiny eels returned to laboriously work their way upstream, even overland through wet grass, to the waters occupied by the adults. Spawning clearly took place at sea, but where and when could only be speculated upon. Finally, in the early 1900's Johannes Schmidt (in a series of papers reviewed by Breder & Rosen, 1966) followed back along a trail of increasingly younger larvae from the mouths of European rivers to the Sargasso Sea, the center of an immense oceanic gyre located south of Bermuda. Since then, several other workers (e.g., Bruun, 1963; Smith, 1968) have confirmed that both the American and European eels spawn in this area and that the larvae migrate from there back to their respective shores.

Such spawning migrations are a general, though not universal, characteristic of eel reproduction. Another is their distinctive "leaf-shaped" larva known as a leptocephalus. These extremely long, flattened, and transparent creatures were commonly collected in plankton tows well before their relationship to the eels was known and were placed by naturalists in their own genus, *Leptocephalus*. In the late 1800's, Delage (1886) reared the leptocephali of the European eel *Conger conger* until transformation into a recognizable juvenile eel. Since then considerable effort has been made by ichthyologists to link each leptocephalus with its adult form. Many such linkages have been established, aided immeasurably by the discovery that the number of myomeres (muscle bands) in the larvae corresponds to the number of vertebrae in the adult. Nonetheless, there are still many larvae for which the adults are unknown and which in some cases cannot even be identified at the family level (e.g., Smith, 1979); similarly, the larvae of many species have not been determined. With some effort, however, it is now possible to identify, at least to family level, the most common leptocephali.

Leptocephalus larvae are also characteristic of two orders clearly related to the true eels: Elopiformes (tarpon and their relatives) and Notacanthiformes (a small group of deep oceanic eels). Their larvae can be readily distinguished from those of the true eels by the shape of the caudal fin (Smith, 1979). The caudal fin of the true eels, if present, is broadly rounded, often with the dorsal and anal fins merging into it;

elopiform larvae have a conspicuously forked caudal fin, while notacanthiform larvae lack a caudal fin altogether and instead have a single long, trailing filament.

SEXUAL DIMORPHISM

The degree and type of sexual dimorphism vary widely among the different families of eels, depending in part on whether or not they undertake a spawning migration. Migrators typically change dramatically at maturity, usually developing a bicolored pattern (dark above and pale below), enlarged fins, and enlarged eyes. This change is often most pronounced in the males, resulting in conspicuous sexual dimorphism in which males have proportionately larger eyes or longer fins. Such morphological changes with maturity have been reported for ophichthids by Böhlke & Chaplin (1968) and Cohen & Dean (1970) and for some congrids by various authors, e.g., Bigelow & Schroeder (1953). The most striking changes, however, occur in a small family (probably only one genus and less than a dozen species) of inshore burrowing eels known as spaghetti eels, family Moringuidae. Sexually immature individuals have small eyes and very reduced fins and look basically eel-like. At sexual maturity the fish become dark above and silvery below and the fins and eyes enlarge. The males' fins develop more than those of females, with the dorsal and anal fins becoming very conspicuous and the caudal fin becoming almost forked (e.g., Gosline & Strasburg, 1956; Böhlke & Chaplin, 1968; Castle, 1968; Castle & Böhlke, 1976). Morphological differences between immatures and adults and between males and females are so extreme that different stages of these fishes have in the past been placed in four families and nine genera (Castle & Böhlke, 1976).

Sexual dimorphism in most nonmigratory eels is far less conspicuous, and many are probably monomorphic. In the garden eel *Gorgasia sillneri* (Heterocongrinae, Congridae) males have larger jaws and a more pointed caudal tip than do females (Fricke, 1970). In the xenocongrid *Kaupichthys hyoproroides* (= *atlanticus*) males have shorter pectoral fins and possibly shorter snouts than the females and differ in dentition; other members of the family, however, are apparently monomorphic (Böhlke, 1956).

Brock (1972) found no definitive external sexual dimorphism in the moray *Gymnothorax javanicus*, and, based on general observations, none (with a single exception) seems likely in the family. The single exception concerns a recent study by Shen, et al. (1979), who were stimulated by the taxonomic difficulties of distinguishing between the two nominal species of *Rhinomuraena*, *R. ambonensis* and *R. quaesita*. *R. ambonensis* is black, *R. quaesita* is blue, but otherwise the two species are nearly identical in morphology. By rearing the fish in captivity, Shen, et al. found that with growth black *R. ambonensis* transform into blue *R. quaesita* and then ultimately develop a bright yellow pattern. Histological examination of the gonads of different sized individuals demonstrated that the single species, *R. quaesita*, is a protandrous hermaphrodite which in the course of its maturation changes from a black juvenile to a blue male and ultimately to a yellow female. Transition sizes between the three phases are, respectively, about 65 cm and about 99 to 120 cm. Hermaphroditism was confirmed by observations on blue-and-yellow individuals, all of which had varying amounts of testicular and ovarian tissues in their gonads.

Despite the occurrence of protandry in *R. quaesita*, it is not likely that such hermaphroditism is widespread in eels. In at least some groups sex appears to be set early in life and may be genetically determined; that is, the animals are not sequential hermaphrodites despite significant size differences between the sexes. Sexual differences in vertebral counts are reported for pelagic larvae of the western Atlantic spaghetti eel *Moringua edwardsi* by Castle & Böhlke (1976), indicating that sex is determined at least that early. Similarly, larvae in several families (e.g., Xenocongridae, Nettastomatidae) have been collected already containing ova (Castle & Raju, 1975; Castle, 1978). This may be an adaptation to the long planktonic stages of these fishes and may indicate only a brief existence as an adult, benthic fish.

Along with the features discussed above, there are often significant differences in the sizes of the sexes at maturity. The pattern of this difference seems to vary with the presence or absence of a spawning migration. In all migrators thus far examined the female reaches a significantly larger size than the male (e.g., Cohen & Dean, 1970; Castle & Böhlke, 1976), sometimes twice his size. Based on the little data yet available, in the nonmigrators males are often larger than females. Male *Gorgasia sillneri* are approximately a third larger than females (Fricke, 1970); similarly, male *Gymnothorax javanicus* average 8.4 kg, while females average only 5.5 kg (Brock, 1972). In contrast, Böhlke (1956) found that males of the xenocongrid *Kaupichthys hyoproroides* average smaller than the females, a point that will be returned to later. The factors that select for size differences between the sexes are not known.

SPAWNING SEASON

Only scattered information is available regarding spawning seasons of eels, most of it based on seasonal patterns of appearance of eggs and larvae. Again, there seems to be a consistent difference between migrators and nonmigrators. Muraenids in general seem to spawn during the warmer months of the year —in late summer for Australian species (Castle, 1965a), late June to early July for *Gymnothorax javanicus* at Hawaii (Brock, 1972), and possibly year-round with a late spring - early summer peak in Florida e.g., *G. nigromarginatus* (Eldred, 1969a). Early summer spawnings are also characteristic of heterocongrins, based on observations of spawning (Fricke, 1970) and the occurrence of larvae (Raju, 1974). In contrast, reported spawning seasons for migrating species are usually late fall to early spring, during the cooler months of the year. Such reports include congrids (Castle & Robertson, 1974), moringuids (Castle, 1979), and ophichthids (Cohen & Dean, 1970). Why migrators and nonmigrators should differ in this regard is not known.

There are no reports of lunar periodicity in the spawning of eels. Castle (1979), however, suggests a monthly interval to repeated spawnings of the migrating western Atlantic species *Moringua edwardsi*. *G. javanicus* females, in contrast, appear to spawn once annually.

REPRODUCTIVE BEHAVIOR

There is a general tendency among the eels to migrate offshore and spawn pelagically, as is well documented for the genus *Anguilla*. Such migrations are apparently necessitated by the planktonic larval stage characteristic of the order. Leptocephali have little or no ability to regulate their salt and water balance (Hulet, et al., 1972) and so are restricted to areas of constant salinity, that is, the open ocean (D. Smith, pers. comm.). Offshore spawning migrations have been reported for members of a number of inshore families including congrids (e.g., Schmidt, 1931; Castle & Robertson, 1974), moringuids (e.g., Castle, 1979), and ophichthids (e.g., Cohen & Dean, 1970). P. Klimley (pers. comm.) also reports observing an immense aggregation of larger ophichthids (probably *Myrichthys* sp.) near the surface well off the Atlantic coast of Panama.

In these offshore areas the eels are commonly seen swimming singly or in small groups near the surface. They are often collected at night by dip-netting around a light. Spawning by such species has not yet been observed in either the field or the aquarium, however. Apparent prespawning behavior has been observed in aquaria for the Indian Ocean ophichthid *Leiuranus semicinctus* by Deraniyagala (1930). Three such eels were collected at the surface in 18 m of water and were immediately placed in a water-filled basin. One, which subsequently proved to be a male, seized another (later proved to be a female) with his jaws dorsally just behind the head and ahead of the gill openings. Repeated attempts to break his hold only resulted in him repeatedly seizing the female in the same position. The following morning the exhausted eels were still in the same position, but when broken apart again the male seized the underside of the female such that he was belly upward under her. When disturbed, the male released a cloud of sperm, dropped off the female, and thereafter showed no interest in her. The female did not release any eggs during this interaction, but they could be forced out of her by means of slight pressure on her abdomen, indicating her ripeness. A generally similar nuzzling of the female by the male in the vicinity of her head has recently been observed in hormone-injected *Anguilla rostrata* (D. Smith, pers. comm.).

At least two groups of eels do not leave the reef to spawn, apparently because they are already in areas of oceanic water (though the same is also true for various ophichthids which do migrate). Spawning by the garden eel *Gorgasia sillneri* begins early in the morning (Fricke, 1970; Clark, 1972). Such eels are found in variable-sized colonies with each eel occupying a burrow it has dug and reaching into the water column to pick out the plankton on which it feeds. Prior to spawning the male and female move their burrows close together; whether the male or the female moves is not known. The male then defends the female from other males, and prolonged and vigorous inter-male combat is common. On the day of spawning the male extends out of his burrow toward the female. Both spread their dorsal fins broadly and the male rubs his head over the female's body. If not ready, she withdraws into her burrow; if ready, she extends toward the male and wraps her body once or twice around his, bringing her urogenital opening close to his. The pair remain in this position for up to nine and a half hours. Actual spawning was not seen, but pairs collected while in this entwined position are clearly in the process of spawning. The female has large eggs in her urogenital opening, while only slight pressure on his sides is enough to cause the male to release sperm.

Spawning by morays is less well documented and, until very recently, had not been definitely witnessed at all. Apparent spawning behavior had been observed on only three occasions: in *Gymnothorax javanicus* by Brock (1972), in *Echidna zebra* by Johannes (cited in Brock, 1972), and in *G. kidako* by Yamamoto (pers. comm.). The reported behavior was similar in all three species. In *G. javanicus*, a pair was observed lying on the bottom under a large head of *Acropora* sp., entwined around one another. While in this position several waves of motion passed down the bodies of the eels, but no visible gamete release was apparent. After a minute or so the eels separated and departed, possibly because of the close approach of the diver. The pair of *E. zebra* were similarly entwined, but were observed to do so while hovering, head up, in the water column over the reef. Finally, in an apparent prespawning sequence in *G. kidako*, a pair of eels was observed at midday to approach one another on a sand patch between reefs. The fish slowly raised their heads off the bottom, facing one another with jaws agape. When fully erect, they suddenly opened their jaws even wider and raised their dorsal fins. The fish held this pose for 10 to 15 seconds, then reached toward one another such that their bodies crossed, and one turned to wrap around the other. At this point the approaching diver-photographer appeared to frighten the fish, and they fled in opposite directions.

Since the original draft of this chapter was written, confirmed spawnings by morays, ending in clear and conspicuous gamete release, have been observed twice, in *Gymnothorax kikado* off southern Japan by Moyer and Zaiser (1982) and in a species tentatively identified as *G. brunneus* (C. Ferrari, in prep.). In the first case a pair of eels, similar in size and appearance, were discovered (at a depth of 12 meters two hours prior to sunset) lying on the bottom pressed tightly together with the posterior third of their bodies entwined. As the divers approached, the pair pressed their abdomens together and then turned sharply apart and separated, leaving behind a cloud of gametes. The eggs were large and floated in the water column where they were inspected by the divers. Moyer and Zaiser speculate that the spawning they witnessed terminated a sequence the beginning and middle of which were observed, respectively, by Yamamoto and Brock. In the first step the eels approach one another and, with bodies partially raised, entwine themselves. Falling back to the bottom they lie entwined, and then ultimately spawn in the

fashion described.

Spawning by *G. brunneus* is quite different, however. In this case the divers chanced upon a group of five small individuals entwined about one another and about a single larger and obviously heavier individual in the middle. Several of the smaller fish (males?) were biting the larger one just behind the head in a fashion very reminiscent of the earlier described observations on migrating eels. While being observed, the entire writhing mass of eels suddenly dashed upward toward the surface and then returned to the bottom, leaving behind a conspicuous cloud of floating eggs. After spawning, the mass broke apart and the eels dispersed. Spawning in this species occurred at dusk. Moyer and Zaiser (1982) also witnessed a mass of dark-colored morays at dusk moving across the bottom with their heads close together and their bodies rapidly waving. Within a few minutes the group dispersed, with no obvious spawning taking place.

Spawning by xenocongrids has not yet been observed, possibly because of their extreme shyness. Alternatively, indirect evidence suggests that the fishes in fact undergo at least a brief spawning migration such that spawning does not occur on the reef. First, unlike known nonmigrators such as the muraenids, in which the male is the larger sex, the xenocongrid *Kaupichthys hyoproroides* is similar to known migrators in that the female is much larger than the male (Böhlke, 1956). Second, although neither sex undergoes dramatic changes in morphology at maturation (D. Smith, pers. comm.), the permanent differences between the sexes are similar, if less pronounced, to those developed by eels that do change; that is, the sexes differ in the relative length of the pectoral fins and in the diameter of the eyes. Third, the sexes of *K. hyoproroides* differ in dentition, with the male's teeth longer and more recurved than the female's. In the only report of prespawning behavior in a pelagic spawning eel, Deraniyagala noted that the male clamped his jaws onto the female and held this position until spawning, a matter of several hours. Differences in dentition between the sexes, in exactly the manner exhibited by *K. hyoproroides*, would be expected (a sexual difference in dentition is, in fact, mentioned briefly by Deraniyagala, 1930, in his description of apparent spawning behavior of *Leiuranus*). The apparent lack of morphological changes at maturity in xenocongrids and the lack of reports of observations of them near the surface may only indicate that their migration is a short one. Spawning migrations by such secretive fishes may be a means of bringing the sexes together for spawning. Such synchronized activity may not be required in the more active and conspicuous garden eels and morays.

In this regard, it is interesting that the xenocongrid *Powellichthys ventriosus* was described by J.L.B. Smith (1966) on the basis of a single specimen collected by a fisherman in a small boat off the Cook Islands. The specimen, a female greatly swollen with eggs, was one of a large number of such eels seen swimming that night near the surface. The species has never been collected again.

In the only estimate of fecundity yet available for a reef-associated eel, Brock (1972) found 200,000 to 300,000 ripe eggs in each of four 5.0 to 6.8 kg *Gymnothorax javanicus*. Each female is thought to spawn once annually.

EGGS AND LARVAE

Eel eggs are spherical and pelagic, with a wide periviteline space, usually no oil droplets, and, in some species, a densely reticulated yolk. They are also quite large, ranging in diameter from 1.8 mm in an unidentified muraenid off India (Bensam, 1966) to 4.0 mm in *Gymnothorax nigromarginatus* off Florida (Eldred, 1969a). One hundred forty-five eel eggs collected off Hawaii by Watson & Leis (1974) ranged in diameter from 2.4 to 3.8 mm, with a peak at 3.0 mm. The eggs of at least one temperate species are larger than those yet reported from the tropics; eggs of *Muraena helena* in the Mediterranean Sea reach 5.5 mm in diameter (D'Ancona, 1931). Eggs of the various families of reef-associated eels are similar— those of muraenids are described by Schmidt (1913), D'Ancona (1931), Bensam (1966), and Eldred (1968a, 1969a); congrid eggs are described by Delsman (1933), Thomopoulos (1956), and Castle (1969a, 1974); and those of ophichthids are described by Thomopoulos (1956). Egg development by *Gymnothorax nigromarginatus* is depicted by Eldred (1969a). Hatching of an unidentified 1.8 mm muraenid egg took approximately 100 hours (Bensam, 1966) and is likely to be at least that long for other, larger eggs.

Newly hatched leptocephali range in size from 5 to approximately 10 mm and have pigmented eyes, well developed jaws with extremely elongate teeth (because of these teeth, the jaws are probably not functional), and a small yolk sac. The literature concerning eel larvae is an extensive one, and descriptions of lep-

Leptocephalus larva of *Gymnothorax moringa* collected at 30-55 meters depth. Photo by Dr. David G. Smith.

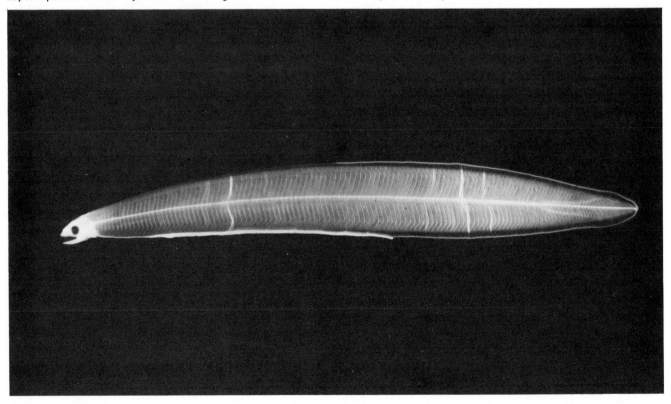

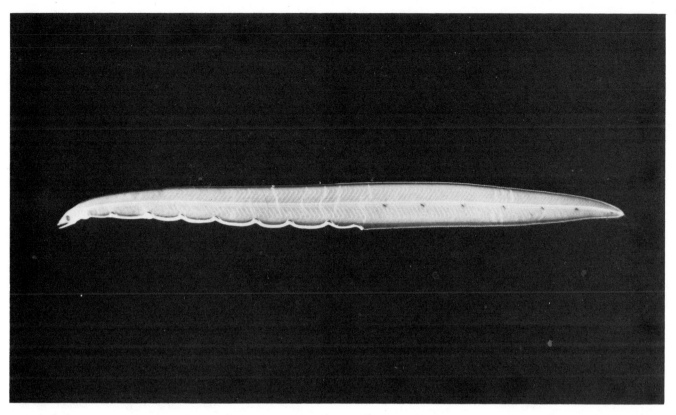

Leptocephalus larva of unidentified ophichthid eel collected off Veracruz, Mexico, at 35-48 meters depth. Photo by Dr. David G. Smith.

tocephali in various families include the following: muraenids - Schmidt (1913), D'Ancona (1931), Nair (1947), Patalu & Jones (1954), Castle (1965a), Bensam (1966), Eldred, in a series of papers (1968a & b, 1969a & b, 1970a-c), and Blache (1971); congrids, including garden eels - D'Ancona (1928), Sparta (1938), Castle (1964, 1969b, 1977), Castle & Robertson (1974), Raju (1974), and, in great detail, Hulet (1978); moringuids -Castle (1965b, 1979), Eldred (1967), and Smith & Castle (1972); ophichthids - Castle (1965c), Eldred (1966), Dean (1968), Richardson (1974), and Leiby (1979); and xenocongrids - Smith (1969) and Blache (1972). Despite its length, this list is far from complete; more detailed bibliographies are available in Castle (1969c) and Smith (1979).

Identification of leptocephali at even the family level is often difficult, and at the genus and species levels it is nearly impossible, if not completely so, for anyone but a specialist. An excellent review of larval characteristics of fishes in the various families is given by Smith (1979), which is highly recommended to anyone interested in the field. In brief, most leptocephali can be identified at the family level by means of a variety of features. All are elongate, flattened, and transparent, with a small head, large eyes and teeth, and a rounded caudal area (as opposed to the larvae of tarpon and their allies, which have forked caudal fins, and those of notacanthiform eels, which have a caudal streamer). Among the five reef-associated families, larval muraenids have a smooth, uninterrupted gut (that is, no conspicuous thickenings or arches) and extremely small, almost indistinguishable, pectoral fins; xenocongrids are similar but have larger pectoral fins and a gut that is less than half the length of the larva; moringuid larvae have large pectoral fins and a conspicuous arch in the gut just ahead of the anus; ophichthid larvae have three or more swellings in the gut, spaced along its length, and may lack a caudal fin; and finally, congrid larvae are similar to those of xenocongrids but have guts that extend well beyond half the length of the larva. Identifications at the genus and species levels involve such characters as distribution and size of pigment (if any), location of the anus, and, especially, the number of muscle bands (myomeres).

Leptocephali reach a length of up to 200 mm before settlement and metamorphosis. Duration of the planktonic stage can be as long as three years in some eels (e.g., *Anguilla* spp.) but is only on the order of three to five months for moringuids (Castle, 1979), six to ten months for muraenids (Eldred, 1969a; Castle, 1965a), and about ten months for at least some congrids (Castle & Robertson, 1974). As noted earlier, leptocephali are incapable of controlling their salt and water balance, and so are restricted to oceanic waters with constant salinity (Hulet, et al., 1972). There is also some question about their nutrition, since many have jaws that do not appear functional (because of the long teeth) and guts that appear to be blocked. Hulet (1978) suggested that at least some leptocephali absorb organic materials directly from sea water.

REPRODUCTION IN CAPTIVITY

There are no reports of spawning by reef-associated eels in captivity, and for most families none seem likely, either because the fish engage in long spawning migrations or because of their large size at maturity. Garden eels, however, may ultimately prove spawnable in large aquaria. To do so will require a tank with a deep bed of gravel in which the eels can burrow, a steady current into which the fish can face and feed, a diet that includes heavy feedings of live foods, and enough space for probably five or six eels. Larval eels have not yet been reared in captivity; such a feat seems unlikely in the near future due to the very long planktonic larval stage characteristic of the order.

Literature Cited

Bensam, P. 1966. The eggs and early development of a muraenid eel. *J. Mar. Biol. Assoc. India*, 8:181-187.

Bigelow, H.B. and W.C. Schroeder. 1953. Fishes of the Gulf of Maine. *Fish. Bull. U.S. Fish. Wildl. Serv.*, (74):577 pp.

Blache, J. 1971. Larves leptocephales des poissons Anguilliformes dans le Golfe de Guinee (Zone Sud). 1re note: Larves de Muraenidae. *Cahiers O.R.S.T.O.M., Ser. Oceanogr.*, 9:203-246.

Blache, J. 1972. Larves leptocephales des poissons Anguilliformes dans le Golfe de Guinee (Zone Sud). 2^e note: Les especes adultes de Xenocongridae et leurs larves. *Cahiers O.R.S.T.O.M., Ser. Oceanogr.*, 10:219-241.

Böhlke, J.E. 1956. A synopsis of the eels of the family Xenocongridae (including the Chlopsidae and Chilorhinidae). *Proc. Acad. Natur. Sci. Phila.*, 108:61-95.

Böhlke, J.E. and C.C.G. Chaplin. 1968. *Fishes of the Bahamas and adjacent tropical waters*. Livingston Publ. Co., Wynnewood, Pa. 771 pp.

Brock, R.E. 1972. A contribution to the biology of *Gymnothorax javanicus* (Bleeker). Master's Thesis., U. Hawaii, Honolulu, Hawaii.

Bruun, A.F. 1963. The breeding of the North Atlantic fresh-water eels. *Adv. Mar. Biol.*, 1:137-169.

Castle, P.H.J. 1964. Congrid leptocephali in Australian waters with descriptions of *Conger wilsoni* (Bl. and Schn.) and *C. verreauxii* Kaup. *Zool. Publ., Vict. Univ., Wellington*, 37:45 pp.

Castle, P.H.J. 1965a. Muraenid leptocephali in Australian waters. *Trans. Roy. Soc. N.Z.*, 7:57-84.

Castle, P.H.J. 1965b. Moringuid leptocephali in Australian waters. *Trans. Roy. Soc. N.Z.*, 7:125-133.

Castle, P.H.J. 1965c. Ophichthid leptocephali in Australian waters. *Trans. Roy. Soc. N.Z.*, 7:97-123.

Castle, P.H.J. 1968. A contribution to a revision of the moringuid eels. *Spec. Publ. Dept. Ichthyol., Rhodes Univ.*, (3):1-29.

Castle, P.H.J. 1969a. Eggs and early larvae of the congrid eel *Gnathophis capensis* off southern Africa. *Spec. Publ. Dept. Ichthyol., Rhodes Univ.*, (5):5 pp.

Castle, P.H.J. 1969b. The eel genera *Congrina* and *Coloconger* off southern Mozambique and their larval forms. *Spec. Publ. Dept. Ichthyol., Rhodes Univ.*, (6):10 pp.

Castle, P.H.J. 1969c. An index and bibliography of eel larvae. *Spec. Publ. Dept. Ichthyol., Rhodes Univ.*, (7):121 pp.

Castle, P.H.J. 1977. Leptocephalus of the muraenesocid eel *Gavialiceps taeniola*. *Copeia*, 1977:488-492.

Castle, P.H.J. 1978. Ovigerous larvae of the nettastomid eel genus *Facciolella*. *Copeia*, 1978:29-33.

Castle, P.H.J. 1979. Early life history of the eel *Moringua edwardsi* (Pisces, Moringuidae) in the western north Atlantic. *Bull. Mar. Sci.*, 29:1-18.

Castle, P.H.J. and J.E. Böhlke. 1976. Sexual dimorphism in size and vertebral number in the western Atlantic eel *Moringua edwardsi* (Anguilliformes, Moringuidae). *Bull. Mar. Sci.*, 26:615-619.

Castle, P.H.J. and S.N. Raju. 1975. Some rare leptocephali from the Atlantic and Indo-Pacific. *Dana Rpt.*, (82).

Castle, P.H.J. and D.A. Robertson. 1974. Early life history of the congrid eels *Gnathophis habenatus* and *G. incognitus* in New Zealand waters. *N.Z. J. Mar. Freshwater Res.*, 8:95-110.

Clark, E. 1972. The Red Sea's Gardens of Eels. *Nat. Geogr.*, 142:724-734.

D'Ancona, U. 1928. Murenoidi (Apodes) del Mar

Rosso e del Golfo di Aden. *Mem. R. Com. Talassogr. Ital.*, 146:146 pp.

D'Ancona, U. 1931. Ordine Apodes. *In*: Uova, larve e stadi giovanili di teleostei. *Fauna e flora del Golfo de Napoli, Monogr.*, (38):94-156.

Dean, D.M. 1968. The metamorphosis of the ophichthid eel *Myrophis egmontis*. Master's Thesis, U. Miami, Fla.

Delage, M.Y. 1886. Sur les relations de parenté du congre et du leptocéphele. *C.R. hebd. Seanc. Acad. Sci.*, 108:698-699.

Delsman, H.C. 1933. Fish eggs and larvae from the Java Sea. 21. Eel eggs. *Treubia*, 14:237-247.

Deraniyagala, P.E.P. 1930. Notes on the breeding habit of the eel *Leiuranus semicinctus*. *Spolia Zeylanica*, 16:107.

Eldred, B. 1966. The early development of the spotted worm eel, *Myrophis punctatus* Lütken (Ophichthidae). *Fla. Bd. Conserv. Mar. Lab., Leafl. Ser.*, 4(3):13 pp.

Eldred, B. 1967. Eel larvae, *Leptocephalus tuberculatus* Castle, 1965, (Moringuidae) in South Atlantic waters. *Fla. Bd. Conserv. Mar. Lab., Leafl. Ser.*, 4(5):6 pp.

Eldred, B. 1968a. The larval development and taxonomy of the pygmy moray eel, *Anarchias yoshiae* Kanazawa 1952. *Fla. Bd. Conserv. Mar. Lab., Leafl Ser.*, 4(10):8 pp.

Eldred, B. 1968b. Larvae of the marbled moray eel, *Uropterygius juliae* (Tommasi, 1960). *Fla. Bd. Conserv. Mar. Lab., Leafl. Ser.*, 4(8):4 pp.

Eldred, B. 1969a. Embryology and larval development of the blackedge moray, *Gymnothorax nigromarginatus* (Girard, 1859). *Fla. Bd. Conserv. Mar. Lab., Leafl. Ser.*, 4(13):16 pp.

Eldred, B. 1969b. The larvae of the redface moray, *Rabula acuta* (Parr, 1930) Böhlke and Chaplin, 1968. *Fla. Bd. Conserv. Mar. Lab., Leafl. Ser.*, 4(11):5 pp.

Eldred, B. 1970a. Larva of the green moray, *Gymnothorax funebris* Ranzani, 1840. *Fla. Bd. Conserv. Mar. Lab., Leafl. Ser.*, 4(16):2 pp.

Eldred, B. 1970b. Larva of the spotted moray, *Gymnothorax moringa* (Cuvier, 1829). *Fla. Bd. Conserv. Mar. Lab., Leafl. Ser.*, 4(15):10 pp.

Eldred, B. 1970c. Larva of the purplemouth moray, *Gymnothorax vicinus* (Castlenau, 1855). *Fla. Bd. Conserv. Mar. Lab., Leafl. Ser.*, 4(14):7 pp.

Fricke, H.W. 1970. Okologische und verhaltensbiologische Beobachtungen an den Rohrenaalen *Gorgasia sillneri* und *Taenioconger hassi* (Pisces,

Apodes, Heterocongridae). *Z. Tierpsychol.*, 27:1076-1099.

Gosline, W. and D.W. Strasburg, 1956. The Hawaiian fishes of the family Moringuidae: another eel problem. *Copeia*, 1956:9-18.

Hulet, W. 1978. Structure and functional development of the eel leptocephalus *Ariosoma balearicum* (De La Roche, 1809). *Phil. Trans. Roy. Soc. Lond.*, B, 282:107-138.

Hulet, W.H., J. Fischer, and B.J. Reitberg. 1972. Electrolyte composition of Anguilliform leptocephali from the Straits of Florida. *Bull. Mar. Sci.*, 22:432-448.

Leiby, M.M. 1979. Morphological development of the eel *Myrophis punctatus* (Ophichthidae) from hatching to metamorphosis, with emphasis on the developing head skeleton. *Bull. Mar. Sci.*, 29:509-521.

Moyer, J.T. and M.J. Zaiser. (1982). Reproductive behavior of moray eels at Miyake-jima, Japan. *Japan. J. Ichthyol.* 28:466-468.

Nair, R.V. 1947. On the metamorphosis of two leptocephali from the Madras plankton. *Proc. Indian Acad. Sci.*, 25:1-14.

Palatu, V.R. and S. Jones. 1954. On some metamorphosing stages of eels (Muraenidae) from the estuary of the Burhabulong River, Orissa State. *Proc. Indian Acad. Sci.*, 39:24.

Raju, S.N. 1974. Distribution, growth and metamorphosis of leptocephali of the garden eels, *Taenioconger* sp. and *Gorgasia* sp. *Copeia*, 1974:494-500.

Richardson, S.L. 1974. Eggs and larvae of the ophichthid eel *Pisodonophis cruentifer*, from the Chesapeake Bight, western North Atlantic. *Chesap. Sci.*, 15:151-154.

Schmidt, J. 1913. On the identification of muraenoid larvae in their early ("preleptocephaline") stages. *Medd. Komm. Havundersogelser.*, 4:1-13.

Schmidt, J. 1925. The breeding places of the eel. *Annu. Rep. Smithson. Inst.*, 1924:279-316.

Schmidt, J. 1931. Eels and conger eels of the north Atlantic. *Nature, Lond.*, 128:602-604.

Shen, S., R. Lin, and F.C. Liu. 1979. Redescription of a protandrous hermaphroditic moray eel (*Rhinomuraena quaesita* Garman). *Bull. Inst. Zool., Academia Sinica*, 18:79-87.

Smith, D.G. 1968. The occurrence of larvae of the American eel, *Anguilla rostrata*, in the Straits of Florida and nearby areas. *Bull. Mar. Sci.*, 18:280-293.

Smith, D.G. 1969. Xenocongrid eel larvae in the western north Atlantic. *Bull. Mar. Sci.*, 19:377-408.

Smith, D.G. 1979. Guide to the leptocephali (Elopiformes, Anguilliformes, and Notacanthiformes). *NOAA Tech. Rpt., NMFS Circ.*, (424):39 pp.

Smith, D.G. and P.H.J. Castle. 1972. The eel genus *Neoconger*: systematics, osteology, and life history. *Bull. Mar. Sci.*, 22:196-249.

Smith, J.L.B. 1966. An interesting new eel of the family Xenocongridae from Cook Island, Pacific. *Ann. Mag. Natur. Hist.*, 8:297-301.

Sparta, A. 1938. Contributo alla conoscenza dello sviluppo embrionale e postembrionale nei Murenoidi. 4- *Ophisoma balearicum* De la Roche. *Mem. R. Com. Talassogr. Ital.*, 252:9 pp.

Thomopoulos, A. 1956. Sur quelques oeufs planctoniques de teleosteens de la Baie de Ville-Franche. II. Peches du mois de Septembre. *Bull. Inst. Oceanogr., Monaco*, (1072):16 pp.

Watson, W. and J.M. Leis. 1974. Ichthyoplankton of Kaneohe Bay, Hawaii. *Sea Grant Tech Rept.*, (TR-75-01).

Lizardfishes
(Synodontidae)

Lizardfishes are long, heavy bodied, and heavily scaled fishes characteristically seen lying on a sandy bottom or propped up on a rock or coral head, angled head-up into the water. All have prominent jaws which readily testify to their activities as voracious piscivores, striking upward from their partially concealed positions at a variety of small (and sometimes not so small) fishes. One large lizardfish was observed to chase a Moorish idol nearly to the surface, taking repeated shots at the fish in attempts to capture it (J. Moyer, pers. comm.). Lizardfishes are relatively primitive animals most closely related to a variety of deep-water families. Synodontids are found in shallow waters circumtropically. There are three genera in the family and about 40 species.

Until recently, little was known about spawning in synodontids, though various aspects of egg and larval development have been known for some time. So far as is known, most species are sexually monomorphic, though occasional observations of pairs of the animals at dusk suggest one sex may be smaller than the other and one (the male?) more slender in several western Atlantic species (pers. obs.). Liu & Tung (1959) reported that the second dorsal fin ray of the male is longer than that of the female in the Indo-West Pacific species *Saurida tumbil*. Such dimorphism has not been noted elsewhere in the family but is similar to that of the presumed ancestral group, the Aulopidae, in which the male has a much larger first dorsal fin than the female, apparently for use in courtship displays (e.g., Coleman, 1974).

Information on lizardfish spawning seasons is equally sparse. Liu & Tang (1959) (subsequently confirmed by Yamada, 1968 and Saishu & Ikemoto, 1970) found that spawning activity of *Saurida tumbil* reaches a peak in April. Similarly, Munro, et al., (1973) collected a ripe female *Synodus intermedius* off Jamaica in April. Spawning by *Synodus ulae* off Japan, a subtropical region, was observed in late summer (Zaiser & Moyer, 1981).

Detailed information on spawning behavior is available for only one species in the family, *Synodus ulae*, by Zaiser & Moyer (1981). They witnessed a pair of fish ascend off the bottom at dusk in a fast and high (about 4 m) ascent, at the peak of which an immense cloud of gametes was released. The male and female were about the same size (about 20 cm TL); spawning occurred at 1804 hours, shortly before sunset, in September, and in an area consisting of sand and volcanic rubble immediately next to a large boulder. Spawning was seen only once, whereas courtship was observed numerous times during the day. Such courtship consists of the male actively pursuing the female across the bottom and vigorously displaying to her whenever she stops. Such displays involve the male sitting with or even lying across the female and swimming around her with his gills fully flared and his hyomandibular processes puffed out. Both male and female make frequent lengthy movements across the bottom, and fights between males over females are regularly seen. Zaiser & Moyer suggested that the social system of *S. ulae* consists of broadly overlapping home ranges, with males competing for ripe females whenever they are encountered. Such males apparently stay with the female, defending her from all rivals, until she is ready to spawn.

Similar paired dusk spawning has also been observed for a western Atlantic lizardfish (P. Colin, in prep.), and Delsman (1938) also suggested lizardfish spawning occurs at "night," based on the time at which ripe eggs appeared in plankton tows. In a number of species one commonly sees pairs of individuals at dusk sitting close together on top of a coral head or prominent rock. Based on Zaiser & Moyer's observations, these may be pairs preparing to spawn.

Lizardfish eggs have been described by numerous authors, including Delsman (1938), Breder (1944), Vijayaraghavan (1957), Kuthalingham (1959), Mito (1966), and Miller, et al., (1979). Such eggs are large (1.1 to 1.46 mm in diameter), spherical, pelagic, lack oil droplets (or have only a few very small ones), and are readily identified by a network of hexagonal meshing on the surface. Such hexagonal patterns occur elsewhere only on the eggs of callionymids and a few flounders. Synodontid eggs develop slowly, hatching in two to three and a half days to produce a long skinny larva. Larval development is described by Delsman (1938), Breder (1944), and Gibbs (1959), among others. Planktonic synodontid larvae are characterized by their extremely elongate shape and by conspicuous black pigment spots along the gut. Spot patterns vary among species, permitting even larvae to be identified at the species level. Settlement to the bottom occurs at a length of about 30 to 35 mm and takes place at night. Duration of the planktonic larval stage is not known.

Literature Cited

Breder, C.M., Jr. 1944. The metamorphosis of *Synodus foetens* (Linnaeus). *Zoologica, N.Y.*, 24:13-15.

Coleman, N. 1974. *Australian Marine Fishes in Color*. Reed, Sydney, Aust. 108 pp.

Delsman, H.C. 1938. Fish eggs and larvae from the Java Sea. 24. Myctophoidea. *Treubia*, 16:415.

Gibbs, R.H., Jr. 1959. A synopsis of the postlarvae of western Atlantic lizard-fishes (Synodontidae). *Copeia*, 1959:232-236.

Kuthalingham, M.D.K. 1959. *Saurida tumbil* (Bloch): development and feeding habits. *J. Zool. Soc. India*, 11:116-124.

Liu, C.K. and I.H. Tung. 1959. The reproduction and the spawning ground of the lizard fish, *Saurida tumbil* (Bloch), of Taiwan Strait. *Rpt. Inst. Fish. Biol. Ministry Econ. Aff. and Natl. Taiwan Univ.*, 1:1-11.

Miller, J.M., W. Watson, and J.M. Leis. 1979. An atlas of common nearshore marine fish larvae of the Hawaiian Islands. *Sea Grant Misc. Rpt.* (INIHI-SeaGrant-MR-80-02).

Mito, S. 1966. Fish eggs and larvae. *Illustrations of the Marine Plankton of Japan*, 7:74 pp.

Munro, J.L., V.C. Gaut, R. Thompson, and P.H. Reeson. 1973. The spawning seasons of Caribbean reef fishes. *J. Fish. Biol.*, 5:69-84.

Saishu, K. and R. Ikemoto. 1970. Reproductive curve of the lizardfish, *Saurida tumbil* (Bloch), of the East China Sea group. *Bull Seikai Reg. Fish. Res. Lab.*, 38:41-59.

Vijayaraghavan, P. 1957. *Studies on Fish-eggs and Larvae of Madras Coast*. Madras Univ., Madras, India. 79 pp.

Yamada, U. 1968. Spawning and maturity of the lizard-fish, *Saurida tumbil* (Bloch) in the East China Sea. *Bull. Seikai Reg. Fish. Res. Lab.*, 36:21-37.

Zaiser, M.J. and J.T Moyer. 1981. Notes on the reproductive behavior of the lizardfish *Synodus ulae* at Miyake-jima, Japan. *Japan. J. Ichthyol.*, 28:95-98.

Cardinalfishes (Apogonidae)

Of the several families of nocturnally active reef fishes, the best known and, relatively speaking, best studied is the family Apogonidae or cardinalfishes. The cardinalfishes are a diverse group, ranging from a few secondarily freshwater species to elongate predators and short chunky planktivores. They are united by three characters: 1) most, if not all, are nocturnally active and spend the day hovering within or near the mouths of caves and crevices; 2) all have two clearly separate dorsal fins; and 3) so far as is known, all are mouthbrooders. The exact limits of the family are at present somewhat vague, and subsequent work may ultimately demonstrate a polyphyletic origin for the family. As the family now stands, most cardinalfishes are found in shallow water circumtropically. The Apogonidae consists of approximately 200 species in about two dozen genera. More than half of these species are in the genus *Apogon* itself (Fraser, 1972). The family reaches its peak in abundance and diversity on Indo-Pacific reefs, with several genera, such as *Foa, Siphamia,* and *Cheilodipterus,* found only in this area.

Despite their abundance, few studies have specifically dealt with the biology of the reef-dwelling cardinalfishes. Most are reddish in color or horizontally lined and somewhat translucent fishes that only occasionally exceed 9 or 10 cm in total length. Aside from a few apparent predators, such as *Cheilodipterus quinquelineata* (W. Douglas, pers. comm.), most are nocturnally active planktivores that disperse from their daytime shelters to forage over the reef at night. Habitat partitioning is based in part on preferred shelter sites, on foraging areas (for example water-column versus close to the bottom) and preferred depths (Livingstone, 1971; Hobson, 1973, 1977), and in part on different times of emergence and return to shelter for sympatric species. This temporal partitioning is apparently based on species-specific responses to different levels of light intensity (Chave, 1978).

Finally, as a result of their use of shelter during the day, a number of species have developed close associations with various, usually noxious or poisonous, invertebrates in which they shelter. Such hosts include sea urchins (Eibl-Eibesfeldt, 1961), the crown-of-thorns starfish (Allen, 1972), anemones (Colin & Heiser, 1973), and a large gastropod (Pagan-Font, 1967).

SEXUAL DIMORPHISM

In most, if not all, species of cardinalfishes, females are noticeably swollen with eggs immediately before spawning. Otherwise, external sexual dimorphism for most species is slight or non-existent (e.g., Böhlke & Randall, 1968; Fraser, 1972). Lachner (1953) noted that females of *Siphamia* are on the average larger than males whereas males tend to have a deeper body and a larger and deeper head; the latter he felt reflected the males' role as mouthbrooders. Females also average larger than males in the Japanese species *Apogon niger* (Mine & Dotsu, 1973). The only report of permanent sexual dichromatism is for *Apogon trimaculatus,* an Indo-West Pacific species in which large males have a dark spot on the bottom of the second dorsal fin that is lacking in both small males and females (Lachner, 1953). Temporary color differences during courtship and spawning occur in several species, usually involving the male lightening in color (e.g., *Apogon imberbis* – Garnaud, 1950a; *Apogon* sp. at One Tree Island – pers. obs.). *Cheilodipterus lineatus* not only lightens in color, but also partially loses the horizontal banding shortly before spawning and develops instead a white bar across the center of the body (Fishelson, 1969).

SPAWNING SEASONS

Data on apogonid spawning seasons are still sketchy. In relatively cool, temperate regions, spawning occurs in the summer, when water temperatures are near their annual maximum (Garnaud, 1950a;

Nakahara, 1962; Mine & Dotsu, 1973). In more tropical areas, data are as yet somewhat contradictory, and it seems likely that different species have quite different seasonal patterns of reproductive activity. Russell, et al. (1977), for example, found late spring-early summer peaks of settlement for juveniles of two species at One Tree Island, Great Barrier Reef, but scattered year-round settlement for four other species. Similarly, Powles (1975) and Luckhurst & Luckhurst (1977) found two peaks of settlement, one in the spring and another in the fall, for apogonids in the southern Caribbean, while Smith, et al. (1977) and Charney (1976) suggested year-round spawning in the Bahamas. The factors underlying these apparent differences are not known.

Allen (1972) found a conspicuous lunar rhythm in the spawning activity of *Sphaeramia orbicularis* at Palau, based on the changing number of males brooding eggs. Peak spawning activity occurred between the first quarter and the full moon, with a second minor peak between the last quarter and the new moon. Allen suggested that this rhythm was a direct function of the tidal cycle; peak high tides occurred at near midnight at these times, and only such tides allowed the fish to remain in their preferred areas throughout the night. J. Moyer (pers. comm.), however, also found a weak bimonthly spawning cycle in *Apogon cyanosoma* in Japan, suggesting that such a rhythm may be widespread in the apogonids.

REPRODUCTIVE BEHAVIOR

Courtship, and in several cases spawning, has been observed for a number of reef-associated apogonids, including *Apogon notatus* (Nakahara, 1962), *A. maculatus* (Coleman, 1966), *Cheilodipterus lineatus* (Fishelson, 1970), *Sphaeramia orbicularis* (Allen, 1975), *Phaeoptyx* sp. (R. Livingstone, pers. comm., described in Thresher, 1980), and *Apogon cyanosoma* (J. Moyer, pers. comm.). It has also been observed in a temperate species, *A. imberbis* (Garnaud, 1950a, b), and a freshwater one, *Glossamia gilli* (Rudel, 1934). Approximately half of these are based on field observations which, in general, agree closely with behavior observed in aquaria.

Spawning by cardinalfishes is generally assumed to occur at night, with several authors noting that fishes lacked eggs at dusk and had them at dawn the next morning. Not suprisingly, however, spawnings at night have not been witnessed. Spawning during the day has been documented in the field for *A. cyanosoma* (approximately midday) and *C. lineatus* (at dusk) and confirmed by several aquarium reports. That cardinalfishes should spawn at night is somewhat unusual for a demersal spawner, most of which spawn in the early morning, and could well be a consequence of the fishes' nocturnal activity periods (early night for an apogonid might be the equivalent of early morning for a diurnally active fish). Conversely, spawning by cardinalfishes may well take place very early in the morning, before first light and before their behavior is readily visible to observers. Either case would be of great interest, and the behavior of the animals in the field should be looked at in greater detail.

In a number of species spawning is associated with conspicuous pair formation which in some species also involves pair defense of a spawning territory. At One Tree Island, for example, such pairing is prominent in *Apogon macrodon, A. cyanosoma, A. leptacanthus*, and an unidentified species of *Apogon. Apogon leptacanthus* hover in large numbers around isolated heads of coral during the day, but do so in conspicuous pairs, with each egg-brooding male shadowed by a non-brooding individual, presumably the female whose eggs he is carrying. Such pairing has been examined in detail by Usuki (1977) for the Indo-West Pacific species *A. notatus*. Through use of tagged individuals, he demonstrated: first, that each pair clearly prefers a specific small site on the bottom; second, that this site is defended from other cardinalfishes with the female the more active defender; and third, that pair members recognize one another as individuals, apparently based on behavioral cues. In *A. notatus* and also *A. imberbis* (Garnaud, 1950a), pairing appears to be temporary, with pair members returning to the large foraging aggregations at the end of the spawning season. What appears to be year-round pairing, however, is reported for *Apogonichthys waikiki* by Chave (1978); the social system of this species has not been examined in detail. More promiscuous spawning systems have been suggested for *Cheilodipterus lineatus* and *Sphaeramia orbicularis*, but without tagging experiments it is difficult to determine how much mate fidelity, if any, occurs in such species. The point is graphically illustrated by the Indo-West Pacific species *Apogon fragilis* at One Tree Island. During the day this fish hovers in groups of up to several hundred around coral heads in shallow water, showing no clear social system and appearing basically promiscuous. Just at dusk, however, when the fish become more active and prepare to leave shelter to forage in the water column during the night, all of the individuals suddenly form conspicuous pairs as they

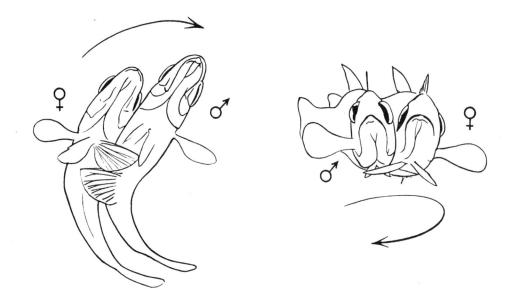

Courtship and spawning in the cardinalfish *Apogon imberbis*. Based on Garnaud (1960).

swim about. There is no evident aggressive interaction between pairs and, once established, no movement of individuals between pairs. In pairs, then, the fish disperse for the night. Observations the following morning indicate no sudden increase in the number of males carrying eggs, indicating that pairing is not simply a response to imminent spawning. Pairing by this species, then, is prominent only for short periods and could easily be overlooked. Similar behavior may occur in other species.

Courtship and spawning in cardinalfishes are always paired rather than group activities and typically involve the female as the aggressive partner. Four different spawning sequences have been described. The first has been reported for three geographically distant species of *Apogon* and is the nearest thing to a common spawning behavior. In *A. imberbis* (Mediterranean), *A. maculatus* (western Atlantic), and *A. cyanosoma* (Indo-West Pacific), courtship consists of prolonged (up to several hours) side-by-side swimming in tight circles by the pair, with the female on the outside and the male on the inside. Such pairs are frequently interrupted, break apart, and then reform into the circling behavior as the female follows the male around and prods him in the general region of his operculum. In all three species this circling leads to the male wrapping his anal fin around the abdomen of the female, covering her urogenital opening. Spawning in at least *A. imberbis* and *A. cyanosoma* is signaled by the pair quivering side-by-side, after which the egg-ball is extruded. The male turns and "inhales" the eggs. Thereafter the pair remain together in the territory they've established, and the female defends the brooding male.

Spawnings by *Cheilodipterus lineatus*, *Apogon notatus*, and *Phaeoptyx* sp. all differ from that above as well as from each other. In *C. lineatus* spawning begins with the male swimming among a group of females while "S" curving or undulating his body and holding his dorsal fin fully erect. Eventually he is followed by a female and the pair swim off into the coral, often moving in large (about 1 m diameter) circles with the male in the lead. Intermittently the pair swim side-by-side, and eventually the male approaches a clear spot on the bottom and skims over it, trembling slightly. The female follows close behind, presses her abdomen against the substrate, and expels a small (2.0 to 2.5 mm) ball of eggs. The male circles behind her and "inhales" the eggs. The pair then return to the main aggregation and the male again begins courting. In *Apogon notatus* spawning is also a paired affair but takes place off the bottom. The female courts a male by arching her body before him in a lateral display, often displaying intermittently for several days before the male begins to respond. When he does respond, spawning follows quickly. The pair swim side-by-side, and then the male rolls around the female, turning himself upside-down such that their vents are in contact. While in this position the female extrudes her eggs, after which the male turns to "inhale" them. In *Phaeoptyx* sp. from the western Atlantic, spawning takes place in a third fashion. The pair orient themselves such that the male is perpendicular to and facing the female. Both begin to quiver as they slowly turn parallel to one another and bring their vents together. The female quickly extrudes her eggs and the male takes them into his mouth. Finally, although spawning has not been observed, in

Sphaeramia orbicularis apparent courtship consisted of alternating chasing and displaying by a pair. The most common display consists of one fish approaching the other while "flicking" its brightly colored pelvic and first dorsal fins. Some mutual biting was also observed, usually directed at the sides or the anal region.

Mouthbrooding (oral incubation) has been reported for a large number of apogonid species by numerous workers, including all of those cited above (for partial reviews of the literature see Breder & Rosen, 1966; Oppenheimer, 1970; Charney, 1976; and Usuki, 1977). Species reported as mouthbrooders include members of the genera *Apogon*, *Cheilodipterus*, *Phaeoptyx*, *Foa*, *Sphaeramia*, *Archamia*, *Astrapogon*, and the freshwater genus *Glossamia*. Oral incubation is clearly the rule for the family and may eventually be documented for all members. In almost all cases the male is the incubating sex (though in many of the earlier reports, such as Fowler & Bean, 1930, this was assumed rather than demonstrated). Female brooding has been conclusively documented only in *Phaeoptyx affinis*, a western Atlantic species in which both sexes have been found carrying eggs (Smith, et al., 1971). An earlier and widely cited report of similar behavior in *Apogon semilineatus* by Ebina (1932) needs to be confirmed. Ebina reported finding 17 males and 24 females carrying eggs out of 500 fish collected, but he does not indicate how the fish were sexed. J. Moyer, who has examined the species in the field, doubts that females carry the eggs in this Japanese species and suggests that Ebina could have been deceived by the external appearance of the tending male (pers. comm.). In *A. cyanosoma*, for example, "females," as evidenced by their distended abdomens, were regularly observed brooding eggs; upon dissection, all proved to be males that had recently swallowed a previous clutch of eggs.

A similar report of female brooding in *Foa madagascariensis* by Petit (1931) should also be checked.

Fecundity in apogonids is not well known, and the only direct estimates are those of Smith & Tyler (1972), who counted the ripe eggs in the ovaries of several western Atlantic species. The number counted ranged from 610 to 3,722, taking into account only those that were nearly ripe. Several females also contained a number of immature eggs, which suggests that each female spawns several times during each spawning season (multiple spawnings are also suggested for *Sphaeramia orbicularis* by Allen, 1975). Estimates of the number of eggs carried by a male vary a great deal (from 75 to 100 eggs for *A. maculatus* by Coleman, 1966, to 21,000 for *Phaeoptyx affinis* by Smith, et al., 1971, and 22,000 for *A. imberbis* by Garnaud, 1950a) but in general are much higher than the estimates of female fecundity, suggesting that males of at least some species routinely carry eggs from several different spawnings. Whether this is true for the apparently pair-forming species such as *Apogon notatus* is not known.

EGGS AND LARVAE

Cardinalfish eggs are bound together in a tight ball by threads that originate from one pole of the egg and, in at least some species, e.g., *Cheilodipterus lineatus* (Fishelson, 1970), by a fine mucous membrane that surrounds them. Such eggs are spherical, contain several oil droplets, and range in diameter (for tropical species) from 0.24 to 0.7 mm. The largest eggs yet reported for a cardinalfish are those of *A. conspersus*, a cold temperate species found off southern Australia; they are approximately 4.5 mm in diameter (Hale, 1947). For the tropical species *Sphaeramia orbicularis* eggs hatch in approximately eight days at 27 to 30°C (Allen, 1975); the eggs of the temperate *A. imberbis* hatch in seven to nine

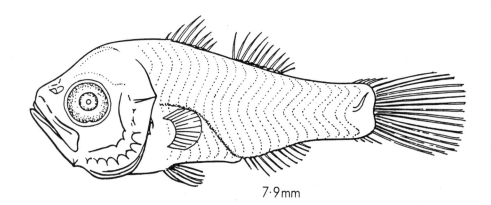

7·9mm

Larval cardinalfish (*Apogon* sp.) Based on Vatanachai (1974).

Postlarval apogonids collected near a submerged light at night near Papua New Guinea. The upper individual is 37 mm total length. Photo by Dr. R. E. Thresher.

days at 17 to 18°C (Garnaud, 1950a).

Newly hatched larvae range in size from 1.0 mm total length (*Phaeoptyx affinis* according to Smith, et al., 1971, although the larvae were still curled and probably hatched prematurely; the next smallest larvae reported are 2.3 mm long) to 3.3 mm (*Sphaeramia orbicularis* according to Allen, 1975). At hatching they are well developed with a small to moderate sized yolk, pigmented eyes, and an open mouth. Each is elongate and robust, with a complete vertical fin fold, an indentation in the fold indicating the limits of the caudal fin, and a large head (e.g., Garnaud, 1960). As they grow, the larvae quickly develop a deeper body and a more pointed snout. Planktonic postlarvae can be as long as 20 mm and are easily recognizable as cardinalfishes (e.g., Fourmanoir, 1976).

The duration of the planktonic larval stage is not known for certain. The mean interval between the first observation of the male brooding eggs and first appearance of newly settled juveniles for three species of *Apogon* and two of *Cheilodipterus* in the One Tree Island lagoon, however, was 58 days and only ranged from 56 to 61 days. The size at settlement ranges from 10 to 25 mm. Allen (1975) reported that juvenile *S. orbicularis* grow as much as 6.4 mm per month. The age at sexual maturity is not known, but given the small size of many species it is probably a year or less.

REPRODUCTION IN CAPTIVITY

Several species of cardinalfishes have been spawned in aquaria, including *Apogon imberbis* (Garnaud, 1950a), *A. notatus* (Nakahara, 1962), and *A. maculatus* (Coleman, 1966). There are few details about aquarium conditions in any of these reports, but it appears that all involved small tanks and abundant cover. The spawning pairs of *A. imberbis* and *A. maculatus* were isolated from other fishes, but those of *A. notatus* were housed in an aquarium with several other pairs, suggesting that isolation is not a prerequisite in at least this species. In the field one sees many pairs of this species courting within 30 cm or so of each other, suggesting that so long as each pair has some minimal space to itself (chases between members of different pairs are common), putting several pairs together is not likely to be a problem.

More details are available regarding aquarium spawning of the widely distributed Indo-Pacific species *Apogon cyanosoma*, recently spawned by A. Gronell (pers. comm.). Two apparent pairs of this species were collected from the reef and transferred to an approximately 40-gallon (150 L) aquarium. The aquarium was sparsely furnished with a gravel bottom (for the undergravel filters) and several pieces of pipe in which the fish sheltered. They were fed twice daily with live shrimp and frozen fish, both of which were taken readily. Courtship was first noted about a week after the fish were collected, with the largest female nuzzling and following the largest male, as described earlier, beginning in early morning. Aside from a few interruptions when one female chased another, courtship continued unabated until early afternoon, when the fish first assumed a side-by-side position. This position was at first held for only a few seconds, after which the pair broke apart and continued courtship. Positioning was attempted several times, until finally the pair held it, began to quiver, and the female extruded a large white ball of eggs. The male immediately dropped back and engulfed the egg mass, shifting it in his mouth and occasionally spitting it out and re-engulfing it, apparently to settle it into the right position in his mouth. After four or five hours of courtship, spawning itself took less than five seconds. Such spawnings occurred at roughly six-day intervals (at 27-28°C) for several weeks. In all cases it followed a lengthy courtship and occurred in late afternoon (roughly between 1430 and 1600).

As yet there are no reported attempts to rear cardinalfish larvae. Given their large size and advanced state of development at hatching, however, they should prove fairly easy to rear. All should take rotifers as a first food.

Literature Cited

Allen, G.R. 1972. Observations on a commensal relationship between *Siphamia fuscolineata* (Apogonidae) and the crown-of-thorns starfish *Acanthaster planki. Copeia*, 1972:595-597.

Allen, G.R. 1975. The biology and taxonomy of the cardinalfish, *Sphaeramia orbicularis* (Pisces; Apogonidae). *J. Roy. Soc. West. Aust.*, 58:86-92.

Böhlke, J.E. and J.E. Randall. 1968. A key to the shallow-water West Atlantic cardinalfishes (Apogonidae), with descriptions of five new species. *Proc. Acad. Nat. Sci., Phila.*, 120:175-208.

Charney, P. 1976. Oral brooding in the cardinalfishes *Phaeoptyx conklini* and *Apogon maculatus* from the Bahamas. *Copeia*, 1976:198-200.

Chave, E.V. 1978. General ecology of six species of Hawaiian cardinalfishes. *Pac. Sci.*, 32:245-269.

Coleman, R.K. 1966. Spawning red cardinals. *Salt Water Aquar.*, 2(6):144-148.

Colin, P.L. and J.B. Heiser. 1973. Associations of two species of cardinalfishes (Apogonidae: Pisces) with sea anemones in the West Indies. *Bull. Mar. Sci.*, 23:521-524.

Ebina, K. 1932. Buccal incubation in the two sexes of a percoid fish, *Apogon semilineatus* T. and S. *J. Imp. Fish. Inst., Tokyo*, 27:19-21.

Eibl-Eibesfeldt, I. 1961. Eine Symbiose zwischen Fischen (*Siphamia versicolor*) und Seeigeln. *Z. Tierpsychol.*, 18:56-59.

Fishelson, L. 1970. Spawning behavior of the cardinal fish, *Cheilodipterus lineatus*, in Eilat (Gulf of Aqaba, Red Sea). *Copeia*, 1970:370-371.

Fourmanoir, P. 1976. Formes post-larvaires et juveniles de poissons cotiers pris au chalut pelagique dans le sudouest Pacifique. *Cahiers du Pacifique*, (19):47-88.

Fowler, H.W. and B.A. Bean. 1930. The fishes of the families Amiidae, Chandidae, Duleidae, and Serranidae, obtained by the United States Bureau of Fisheries steamer "Albatross" in 1907 to 1910, chiefly in the Philippine Islands and adjacent seas. *Bull. U.S. Nat. Mus.*, 100(10).

Fraser, T.H. 1972. Comparative osteology of the shallow water cardinal fishes (Perciformes: Apogonidae) with reference to the systematics and evolution of the family. *Rhodes Univ. Ichth. Bull.*, (34).

Garnaud, J. 1950a. La reproduction et l'incubation branchiale chez *Apogon imberbis* G. et L. *Bull. Inst. Oceanog., Monaco.*, 47(977):1-10.

Garnaud, J. 1950b. Notes partielles sur le reproduction d'*Apogon imberbis. La Terre et la Vie*, (1):39-42.

Garnaud, J. 1960. Ethologie d'un poisson extraordinaire: *Apogon imberbis. Bull. Inst. Oceanog., Monaco, Spec. Iss.* (1D):51-60.

Hale, H.M. 1947. Evidence of the habit of oral gestation in a south Australian marine fish, *Apogon conspersus* Klunzinger. *S. Aust. Natur.*, 24(3).

Hobson, E.S. 1974. Feeding relationships of teleostean fishes on coral reefs in Kona, Hawaii. *Fish. Bull.*, 72:915-1031.

Hobson, E.S. and R. Chess. 1978. Trophic relationships among fishes and plankton in the lagoon at Enewetak Atoll, Marshall Islands. *Fish. Bull. (U.S.)*, 76:133-153.

Lachner, E.A. 1953. Family Apogonidae: Cardinal Fishes. Pp. 412-498. *In*: Fishes of the Marshall and Marianas Islands. (L.P. Schultz, et al., Eds.). *Bull. U.S. National Mus.* (202).

Livingstone, R.J. 1971. Circadian rhythms in the respiration of eight species of cardinalfishes (Pisces: Apogonidae): comparative analysis and adaptive significances. *Mar. Biol.*, 9:253-266.

Luckhurst, B.E. and K. Luckhurst. 1977. Recruitment patterns of coral reef fishes on the fringing reef of Curacao, Netherlands Antilles. *Can. J. Zool.*, 55:681-689.

Mine, K. and Y. Dotsu. 1973. On the mouth breeding habits of the cardinal fish, *Apogon niger. Bull. Fac. Fish., Nagasaki Univ.*, 36:1-6.

Nakahara, K. 1963. On the spawning behavior of a cardinal fish, *Apogon notatus* (Houttuyn). *Mem. Fac. Fish., Kagoshima Univ.*, 11:14-17.

Oppenheimer, J.R. 1970. Mouthbreeding in fishes. *Anim. Behav.*, 18:493-503.

Pagan-Font, F.A. 1967. A study of the commensal relationship between the conchfish, *Astrapogon stellatus* (Cope), and the queen conch, *Strombus gigas* Linnaeus, in southwestern Puerto Rico. Master's Thesis, Univ. Puerto Rico, Mayaguez.

Petit, M.G. 1931. Une espéce nouvelle du genre *Foa* présentant un cas d'incubation bucco-branchiale. *Bull. Mus. Natl. Hist. Nat. Paris, Ser. 1*, 3:91-95.

Powles, H. 1975. Abundance, seasonality, distribution and aspects of the ecology of some larval fishes off Barbados. Ph. D. Diss., McGill Univ., Montreal, Canada.

Rudel, A. 1934. Ein neuer Maulbrüter (*Glossamia gilli*). *Blätt. Aquar.-Terrarienk.*, 45(9):141-143.

Russell, B.C., G.R.V. Anderson, and F.H. Talbot. 1977. Seasonality and recruitment of coral reef fishes. *Aust. J. Mar. Fresh. Res.*, 28:521-528.

Smith, C.L., E.H. Atz, and J.C. Tyler. 1971. Aspects of oral brooding in the cardinalfish *Cheilodipterus affinis* Poey (Apogonidae). *Amer. Mus. Novit.*, (2456):1-11.

Smith, C.L. and J.C. Tyler. 1972. Space resource sharing in a coral reef fish community. Pp. 125-170. *In*: Results of the Tektite program: ecology of coral reef fishes. (B.B. Collette and S.A. Earle, Eds.). *Sci. Bull. Natl. Hist. Mus. Los Angeles County*, (14).

Thresher, R.E. 1980. *Reef Fish*. Palmetto Publ. Co., St. Petersburg, Fla. 172 pp.

Usaki, H. 1977. Underwater observations and experiments on pair formation and related behaviors of the apogonid fish, *Apogon notatus* (Houttuyn). *Publ. Seto Mar. Biol. Lab.*, 24:223-243.

Vatanachi, S. 1974. The identification of fish eggs and larvae obtained from the survey cruises in the South China Sea. *Proc. Indo-Pac. Fish. Council*, 15, Sect. 3:111-130.

Squirrelfishes (Holocentridae)

The squirrelfishes and soldierfishes, family Holocentridae, are the most abundant of the large nocturnally active fishes found on coral reefs. By day they hover in or close to crevices, either as solitary individuals (in at least some species of *Holocentrus* such individuals are territorial — e.g., Winn, et al., 1964) or in small groups. At dusk, however, they slowly emerge from their shelters and disperse across the reef to feed on a variety of small fishes and invertebrates. Like most other nocturnally active fishes, squirrelfishes tend to be wholly or largely red and have large, prominent dark eyes. The family is otherwise characterized by large jaws, prominent spines in the forward section of the dorsal fin and the leading edges of the pelvic and anal fins, and a forked caudal fin. Of the 70 or so species included in the nine genera, few exceed 25 to 30 cm, although at least one species (*Myripristis murdjan*) is reputed to reach 60 cm.

There is little information about reproduction in squirrelfishes other than numerous descriptions of their distinctive planktonic larvae. According to Wyatt (1976), the sex ratio is close to 1:1 for two western Atlantic species, *Holocentrus rufus* and *H. ascensionis*, and the sexes are similar in appearance and size. Males tend to average slightly larger than females, but the difference is slight and the overlap of size range is nearly complete. Spawning by both species apparently occurs year-round off Jamaica, with peaks of activity from January to March and in October (Munro, et al., 1973; Wyatt, 1976). Farther north spawning apparently occurs later in the year, in mid-summer (Winn, et al., 1964). Johannes (1981) suggests that an unidentified species of *Myripristis* off Palau spawns entirely during the first five days of the lunar month (based on gonad samples obtained by native fishermen). Otherwise there is no other infor-

Squirrelfishes are nocturnal, usually have large eyes, and are largely red in color. This is *Holocentrus rufus*. Photo by Wilhelm Hoppe.

mation on lunar cycles of activity, if any.

Possible courtship by a squirrelfish has been described for *Holocentrus xantherythrus (= Adioryx xantherythrus)* in aquaria by Herald & Dempster (1957), who observed two individuals, out of several kept together, to repeatedly swim rapidly toward one another, turn away quickly, and brush their caudal peduncles and tails together. Before separating, each fish backed up until the caudal peduncle of one was above the other and its erected anal spine contacted the other's ventral surface. After assuming this position, the individuals vibrated their tails rapidly and then separated. During the entire sequence the fish made frequent "clicking" sounds. No gamete release was observed. While the behavior described could well have been courtship, given the territoriality characteristic of at least some species of squirrelfishes, it is equally appealing to interpret the behavior as basically agonistic.

Apparent pre-spawning behavior has also been observed in an unidentified species of *Holocentrus* off One Tree Island, Great Barrier Reef. A trio of fish, consisting of one very heavy-bodied individual and two longer but slimmer fish following it, was seen at dusk (A. Gronell, pers. comm.). One of the larger fish changed color briefly while following the heavier fish about. Diver activity drove the trio into hiding before anything further occurred, but it is tempting to speculate that the three fish represented a ripe female and two courting males, and that spawning, at dusk or early night, might have been imminent.

There is also no information on holocentrid eggs, though they are presumably pelagic. Larval development, based on specimens obtained in plankton tows, was detailed by McKenny (1959) and Jones & Kumaran (1964), who also reviewed the large literature describing the late stage larvae. The smallest individuals collected, about 2 to 3 mm long, lack yolk sacs and already have well pigmented eyes and prominent dorsal and opercular serration. This "spininess"

reaches a peak in slightly larger larvae, which not only have a sharply pointed snout but also have large prominent spines on the operculum and at the leading edge of the dorsal fin. By a length of 25 mm the spines have become reduced somewhat in relative size, and by about 32 mm the fish is clearly a juvenile squirrelfish. Settlement of at least *Adioryx* occurs at night (pers. obs.). Sexual maturity in *H. rufus* is reached before a length of 13 to 14 cm (the smallest size group examined), while full maturity in *H. ascensionis* is attained by 18 to 20 cm (Wyatt, 1976).

Holocentrids have not yet been spawned in captivity. The smaller species of *Myripristis*, which are not usually territorial, may well prove spawnable with reasonable effort.

Literature Cited

Herald, E.S. and R.P. Dempster. 1957. Courting activity in the white-lined squirrel fish. *Aquar. J.*, 28(10):366-367.

Johannes, R.E. 1981. *Words of the Lagoon*. Univ. Calif. Press, Los Angeles, Calif. 320 pp.

Jones, S. and M. Kumaran. 1964. Notes on the eggs, larvae and juveniles of fishes from Indian waters. XII. *Myripristis murdjan* (Forskal) and XIII. *Holocentrus* sp. *Ind. J. Fish.*, 18:155-167.

McKenny, T.W. 1959. A contribution to the life history of the squirrelfish *Holocentrus vexillarius* Poey. *Bull. Mar. Sci.*, 9:174-221.

Munro, J.L., V.C. Gaut, R. Thompson, and P.H. Reeson. 1973. The spawning seasons of Caribbean reef fishes. *J. Fish Biol.*, 5:69-84.

Winn, H.E., J.A. Marshall, and B. Hazlett. 1964. Behavior, diel activities and stimuli that elicit sound production and reactions to sounds in the longspine squirrelfish. *Copeia*, 1964:413-425.

Wyatt, J.R. 1976. The biology, ecology, and bionomics of Caribbean reef fishes: Holocentridae (Squirrelfishes). *Res. Rpt. Zool. Dept. Univ. West Indies*, (3):41 pp.

Sweepers
(Pempheridae)

The pempherids are a small family (two genera and about 25 species) of deep-bodied nocturnally active fishes. During the day they are commonly seen in aggregations of up to several hundred individuals dashing about in a milling mass inside caves and under ledges. By night they disperse into the water column to feed on plankton. The fishes are characterized by a short, high dorsal fin, a forked caudal fin, a long, low anal fin, and a strongly laterally compressed body. Like other nocturnally active fishes, they also tend to be reddish in color and have relatively large eyes. The family is found in the western Atlantic, Indian, and Pacific oceans and is entirely tropical. Most species are entirely marine and are found on or near reefs, but a few Australian species occupy brackish waters.

Nothing is known about spawning by pempherids, but it is likely that they spawn in groups, probably at night (dusk?) (a few species have luminescent organs, which might be involved in social behavior at night), and produce pelagic eggs. Newly settled *Pempheris schomburgki* off Florida arrive in late summer and join adult aggregations, though usually hovering near the periphery of the group. Such small juveniles are nearly transparent; otherwise, they resemble the adults.

During the day pempherids are commonly seen in large aggregations in the shelter of the reef. At night they move into the water column to feed. Photo by Walter Deas.

Bigeyes
(Priacanthidae)

Bigeyes are moderate-sized, laterally compressed nocturnal predators characterized by an oblique mouth, usually a bright red color, and, not surprisingly, big eyes. During the day they characteristically hover as solitary fish around small caves and crevices (often so small the fish cannot get into them if threatened); at night they forage either singly or in small groups over the reef, feeding mainly on crustaceans and soft-bodied invertebrates. The family is found worldwide in tropical and temperate seas and consists of three genera (*Priacanthus, Pristigenys* (= *Pseudopriacanthus*), and *Cookeolus*) and about 18 species.

Little is known about reproduction in the priacanthids. Colin & Clavijo (1978) reported observing a large aggregation of *Priacanthus cruentatus* (about 200 individuals) off the Virgin Islands near a reef at which a number of other species were spawning. They suggested this may have been a spawning aggregation of the bigeyes, but saw no courtship or spawning activity. The aggregation of so many of what is normally a solitary fish, however, is quite suggestive. P. Young (in prep.) found that females are usually larger than males in two species of *Priacanthus* from Australia.

Priacanthids presumably produce a pelagic egg, but such eggs have not yet been described. Postlarvae of *Priacanthus hamrur* and *Pristigenys niphonia* collected off the east coast of Africa are briefly described and depicted by Fourmanoir (1976). The only detailed description of larval development is that for the deepwater subtropical species *Pristigenys alta* by Caldwell (1962). According to Caldwell, spawning by this species occurs in the summer. Small larvae, from 2.2 to about 8 mm total length, each have a bony crest on their heads, another smaller one over each eye, and a long prominent spine on the preopercle. The spine is progressively reduced as the fish develops. Prejuveniles range in size from 10 to about 60 mm and are laterally compressed with large eyes, rounded fins, and prominent dorsal fin spines. Settlement occurs at a length of about 65 mm after an unknown time spent in the water column. Postlarval *Priacanthus hamrur*

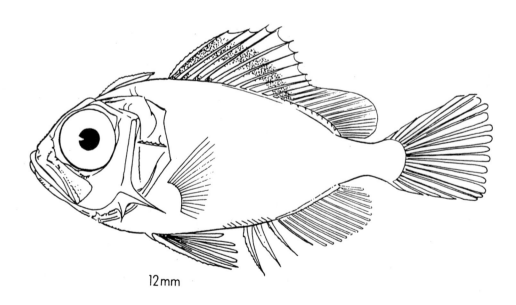

Postlarval priacanthid,
Priacanthus hamrur. Based on
Fourmanoir (1976).

12 mm

A postlarval priacanthid, 40.8 mm total length, collected near a submerged light at night off the coast of Papua New Guinea. Photo by Dr. R. E. Thresher.

described by Fourmanoir are virtually identical to small *Pristigenys alta*. Postlarval *Pristigenys niphonia*, however, are more heavily armored on the head.

Bigeyes have neither been spawned nor reared in captivity.

Literature Cited

Caldwell, D.K. 1962. Development and distribution of the short bigeye, *Pseudopriacanthus altus* (Gill) in the western North Atlantic. *Fish. Bull. (U.S.)*, 62:103-149.

Colin, P.L. and I. Clavijo. 1978. Mass spawning by the spotted goatfish, *Pseudupeneus maculatus* (Bloch) (Pisces: Mullidae). *Bull. Mar. Sci.*, 28:78-82.

Fourmanoir, P. 1976. Formes post-larvaires et juveniles de poissons cotiers pris au chalut pelagique dans le sud-ouest Pacifique. *Cah. du Pacifique*, (19):47-88.

Marine Catfishes
(Plotosidae)

Catfishes are a predominantly freshwater group. A few families, however, have successfully invaded the marine environment, of which the most successful is the Plotosidae. Like most other marine catfishes, they are most common in inshore, usually soft-bottomed, areas, but unlike the other marine catfishes the plotosids are also seen regularly on coral reefs. Adults are generally retiring, dull-colored fishes seen as solitary individuals hiding by day under branching corals or deep in crevices under the reef. Juveniles, however, are horizontally striped with black and white (or black and yellow) and are usually found in shoals of up to several hundred individuals in caves on the reef. These small juveniles are popular aquarium fishes.

The systematics of the plotosids is currently being revised (J. Gomon, pers. comm.). At present the fami-ly consists of about seven genera and 25 to 30 species. About half of these are secondarily freshwater fishes; the rest are primarily estuarine or, if entirely marine, live in near-shore areas. Only one species, *Plotosus lineatus*, is common on reefs.

Spawning behavior of a plotosid has been described thus far only for a secondarily freshwater species, *Tandanus tandanus*, found in northern Australia. Various Australian naturalists have commented on its reproduction (reviewed by Whitley, 1941; Lake, 1959), all agreeing on the basic elements involved. Spawning is preceded by the construction by the male of a mound-like nest that may be as much as 1.8 m in diameter. The nest is constructed by fanning sediment away from the site and by the male carrying to it stones, twigs, and other debris. After the nest is com-

Plotosus lineatus are dully colored as adults but striped as juveniles. Photo by Dr. Herbert R. Axelrod.

plete the male seeks a ready female and goes through an elaborate courtship that apparently involves both threatening and "persuasive" elements. Spawning takes place with the pair in a side-by-side position in the center of the nest, with the large (2.0 to 2.5 mm diameter) eggs scattered into the rubble. After the female departs, the eggs are tended and guarded by the male. Hatching occurs in seven to ten days at a water temperature of about 18°C.

Spawning by a marine plotosid has yet to be described but is presumably similar. In *Plotosus lineatus* (as *P. anguillaris*) off Japan, spawning occurs in late spring and early summer, at which time males construct nests under rocks and other large pieces of debris (Kajikawa, 1973; Moriuchi & Dotsu, 1973). After spawning the male also guards the eggs. Eggs are spherical, non-adhesive, demersal, and roughly 3.12 to 3.5 mm in diameter. Each contains a large bright yellow yolk. Newly hatched larvae are about 6.9 mm long and, like the freshwater *T. tandanus*, carry a large yolk sac. The yolk is absorbed in about ten days and the young rapidly become free-swimming and begin to feed (such young are large enough to take newly hatched brine shrimp immediately). By the time they reach a length of 15 mm, they clearly resemble the adults, though differing in color. Sexual maturity for females can be reached in a year, when they reach a length of about 14 cm, although most fish are mature by three years. The maximum life span is about seven years.

There are no details of spawning plotosids in aquaria, though at least two large public aquaria in Japan have been doing just that for several years (Kajikawa, 1973). Based on their general reproductive biology, spawning these fishes should require little more than a pair of well fed fish and a large aquarium. Due to their large size at hatching, the young should be easy to rear.

Literature Cited

Kajikawa, H. 1973. Determination of age in the marine catfish *Plotosus anguillaris* by the use of otolith. *Sci. Rpt. Shima Marineland*, (2):5-22.

Lake, J.S. 1959. The freshwater fishes of New South Wales. *Res. Bull. St. Fish. N.S.W.*, (5):1-19.

Moriuchi, S. and Y. Dotsu. 1973. The spawning and the larva rearing of the sea catfish, *Plotosus anguillaris*. *Bull. Fac. Fish., Nagasaki Univ.*, 36:7-12.

Whitley, G.P. 1941. The catfish and its kittens. *The Aust. Mus. Mag.*, June 10:306 ff.

Brotulids
(Brotulidae)

Brotulids are small, extremely secretive fishes that are probably common on coral reefs and in other inshore areas but are rarely seen. All are characterized by an elongate robust body that tapers from a relatively massive head to a pointed tail. In some species the dorsal and anal fins meet at this point; in others there is a small but distinct caudal fin. Brotulids occupy one of the widest ranges of habitats of any extant fish families, ranging from inshore, and possibly even intertidal, areas to the ocean depths. Indeed, the deepest collected fish of any kind is a brotulid, *Bassogigas profundus*, taken at a depth of nearly 7 km. There are six genera in the family, and about 150 species have been described. They are close relatives of the Ophidiidae (some workers consider them to be in the same family) and also of the pearlfishes, family Carapidae.

Virtually nothing is known about spawning by brotulids, but their overall reproduction has aroused considerable interest. The family contains both livebearing and egglaying species, separated by Radcliffe (1913) into different subfamilies, a condition that is probably not representative of the animals' phylogeny (see Suarez, 1975). Little is known about spawning by the egglaying (oviparous) species, at least some of which produce gelatinous egg masses similar to those of carapids (Mito, 1962). In part, this lack of information is due to the habitat preferences of the oviparous species; all are found at depths greater than 150 m, and more than half occur deeper than 2,000 m (Wourms & Bayne, 1973). Reproduction of deep-sea fishes, including the brotulids, is reviewed by Mead, et al. (1964).

Reproduction by the livebearing species has been examined in greater detail, though studies to date have focused on the anatomy of reproduction rather than on behavior. Livebearing brotulids inhabit a variety of habitats, including coral reefs. All tend to be small, few reaching a maximum length of more than 7 or 8 cm, and all are cryptic. In at least one species females tend to be larger than males (Suarez, 1975). Otherwise, the sexes can be readily distinguished by the male's intromittent organ, which is located ahead of the anal fin and behind the anus (Turner, 1946; and especially Suarez, 1975). The female lacks the intromittent organ, having in its place a genital pore. Suarez speculated, on the basis of morphology, that copulation involves a flexion of the male's body and a pumping of sperm into the female. Following copulation, sperm penetrates to the ovaries and fertilizes the eggs present. Details of ovarian structure are provided by Lane (1909) and Suarez (1975), and details of embryonic development are provided by the same two authors as well as by Wourms & Bayne (1973). The last authors, working with the widely distributed species *Dinematichthys ilucoeteoides*, indicated that embryos were retained in the ovaries until about 6 mm long, and that right up until parturition they bore large yolk sacs and were probably deriving nutrition from these reserves, making the species ovoviviparous rather than viviparous. Newly born young are apparently planktonic. In contrast, both Lane and Suarez reported young retained well after the yolk was absorbed, noting that in many cases these late-stage embryos were oriented such that their well-developed jaws were around an "ovigerous bulb" that arises from the wall of the ovary. Suarez suggested that the young are, effectively, suckling, i.e., deriving some fluid nutrition from the mother. Newly born young of these species are relatively large (more than 12 mm long) (Suarez, 1975) and are apparently benthic. Longley & Hildebrand (1941) reported that in the field 15 mm young actively swim about with the parent and that their bodies are transparent and colorless except for a broad dusky stripe down each side. Parental care was not reported but might well be expected.

Literature Cited

Lane, H.H. 1909. On the ovary and ova in *Lucifuga* and *Stygicola*, pp. 226-231. *In:* Cave Vertebrates of America (C.H. Eigenmann, Ed.). *Carnegie Inst. Wash. Publ.*, (104):241 pp.

Longley, W.H. and S.F. Hildebrand. 1941. Systematic catalog of the fishes of Tortugas, Florida with observations on color, habits, and local distribution. *Carnegie Inst. Wash. Publ.*, (535):331 pp.

Mead, G.W., E. Bertelsen, and D.M. Cohen. 1964. Reproduction among deep-sea fishes. *Deep-Sea Res.*, 11:569-596.

Mito, S. 1962. Pelagic fish eggs from Japanese waters. V. Callionyma and Ophidina. *Sci. Bull. Fac. Agric., Kyushu Univ.*, 19:377-380.

Radcliffe, L. 1913. Descriptions of seven new genera and 31 new species of fishes of the family Brotulidae and Carapidae from the Philippine Islands and Dutch East Indies. *Proc. U.S. Nat. Mus.*, 44:135-176.

Suarez, S.S. 1975. The reproductive biology of *Ogilbia cayorum*, a viviparous brotulid fish. *Bull. Mar. Sci.*, 25:143-175.

Turner, C.L. 1946. Male secondary sexual characters of *Dinematichthys ilucoeteoides*. *Copeia*, 1946:92-96.

Wourms, J.P. and O. Bayne. 1973. Development of the viviparous brotulid fish, *Dinematichthys ilucoeteoides*. *Copeia*, 1973:32-40.

Pearlfishes
(Carapidae)

Pearlfishes derive their common name from the life style of many species, which involves sheltering during the day inside the visceral cavities of a variety of marine invertebrates, including molluscs and holothurians. After foraging during the night for the small animals that make up at least part of their diet (several species appear to eat parts of the viscera of the holothurians they inhabit), the fish seek a host as dawn approaches and then literally corkscrew their way, tailfirst, into the anal opening (Arnold, 1956; Strasburg, 1961; Smith, 1964). Free-living carapids, of which there are many, are thought to be burrowers in sand and silt. There are about 30 species in about six genera in the family, distributed circumtropically.

Little is known about spawning in carapids. It presumably occurs at night when the fish are active and, in some species at least, occurs in the summer (e.g., Arnold, 1956; Smith, 1964) or from late winter through to late summer (Robertson, 1975). All produce pelagic eggs that are oval and slightly over 1 mm in the longest dimension and which, in at least some species, are imbedded in a gelatinous mass (Emery, 1880; Robertson, 1975). Larval development in carapids is rather spectacular and has been studied since the late 1800's. Recent references include Arnold (1956), Strasburg (1961), Robertson (1975), and, especially, Olney & Markle (1979), who not only described the early stages of western Atlantic *Echiodon* in great detail but also reviewed the pertinent literature. After hatching, prolarval carapids are long and slender with a small, elliptical yolk sac and long, low, and even dorsal and anal fin precursors. As the animals develop, they lengthen greatly and develop an extremely long filament, the vexillum, that arises ahead of the dorsal fin. Such larvae are known as vexillifers and are unique to the carapids. Robertson reported that the vexillifer of one species, *Echiodon rendahli*, also has its gut exterior to the body proper, extending below the body like a ventral filament. This condition has not been reported for the larvae of any other species. As the vexillifer develops, its dorsal filament gradually shrinks until, as it just disappears, the larva ceases its planktonic stage and settles to the bottom as a "tenuis"-stage benthic larva. This stage is characterized by a body that is cylindrical in cross section and by a relatively tiny head; in some species the head is only a few percent of the animal's total length. As the larva develops its body lengthens, but its anal opening remains the same distance from its head so that eventually the gut curls through the body and then must return anteriorly before ending. As maturity is reached the body of the larva deepens, assuming the laterally compressed shape of the adult, and it shrinks, in some cases losing as much as 65% of its total length in transforming into an adult. Such adults are pale, largely transparent, blind, and nearly finless.

Literature Cited

Arnold, D.C. 1956. A systematic revision of the teleost family Carapidae (Percomorphi, Blennioidea), with descriptions of two new species. *Bull. Brit. Mus. Nat. Hist. (Zool.),* 4:245-307.

Emery, C. 1880. Le specie del genere *Fierasfer* nel golfo di Napoli e regione limitrofe. *Fauna u Flora Neapel.,* 2:1-76.

Olney, J.E. and D.F. Markle. 1979. Description and occurrence of vexillifer larvae of *Echiodon* (Pisces: Carapidae) in the western North Atlantic and notes on other carapid vexillifers. *Bull. Mar. Sci.,* 29:365-379.

Robertson, D.A. 1975. Planktonic stages of the teleost family Carapidae in eastern New Zealand waters. *N.Z. J. Mar. Freshwater Res.,* 9:403-409.

Smith, C.L. 1964. Some pearlfishes from Guam, with notes on their ecology. *Copeia,* 1964:34-40.

Strasburg, D.W. 1961. Larval carapid fishes from Hawaii, with remarks on the ecology of the adults. *Copeia,* 1961:479-480.

Anglerfishes
(Antennariidae)

The frogfishes or anglerfishes are lumpy amorphous creatures best known for the small "bait" many wriggle to attract prey. All are voracious carnivores with large oblique mouths lined with many rows of conical teeth. Almost all are cryptically colored, sometimes amazingly so, and patiently sit for hours awaiting the approach of prey (small fishes and crustaceans). Although some lack a special luring apparatus, the family is best known for those that have evolved a "baiting" device with which they attract prey. Located just ahead of the dorsal fin, from which it is derived, the bait, known as the illicium, is angled forward over the mouth and is variously wriggled, jerked, or waved, depending upon the animal it is meant to mimic. Different species of anglerfishes have evolved lures resembling worms, shrimps, and, in one case, even a fish.

Anglerfishes are found in shallow tropical and temperate seas worldwide, including coral reefs, but are usually present only in low numbers. It is rare for a museum collection to contain more than a few of these odd fishes. An exception to this pattern is the sargassum fish, *Histrio histrio*, a small mottled brown species that inhabits floating masses of sargassum weed. Hundreds can often be found in large masses of the weed, each actively stalking both one another and any other suitably sized prey (which makes it, at best, an iffy customer for a community aquarium).

The family Antennariidae contains about 60 known species in two genera (*Histrio* is monotypic; the remaining species are in *Antennarius*). All are characterized by a lumpy shape, large heads (and mouths), foot-like pectoral fins with which they can "walk" across the bottom, and a narrow tube-like gill opening just behind each pectoral. They are characteristic of shallow rock and reef areas where, because of their camouflage, they are often overlooked even by experienced collectors.

Spawning by anglerfishes is a spectacular process involving the production of an enormous jelly-like egg "raft." Because of this and the ease with which at least some species spawn in aquaria, reproduction in a few species has been studied in great detail over the past 60 to 70 years. With one exception, discussed below, no recent work has been done with the animals, however.

SEXUAL DIMORPHISM

For most species there are no conspicuous external differences between the sexes. Cadenat (1937) (cited in Breder & Rosen, 1966) reported that the males of *Antennarius* (formerly *Phrynelox*) *scaber* have more extensively developed cutaneous appendages than the females and differ in color, but this observation has yet to be confirmed for this or any other species. Böhlke & Chaplin (1968), however, noted that all three black-phase individuals of this species they collected were female, while the normal mottled brown phase individuals included both males and females. Other than this possible difference, sexual dimorphism only becomes apparent shortly before spawning, when the female is noticeably distended by the developing egg raft.

SPAWNING SEASON

No detailed study of spawning in the field has been undertaken for an anglerfish, so hard data on spawning seasonality is lacking. Adams (1960) collected postlarvae of the sargassum fish, *Histrio histrio*, year-round, indicating that this species as a whole does not have any extended periods of no reproduction. In aquaria female *Histrio* will produce egg rafts regularly and apparently indefinitely, at all times of the year, again suggesting little or no annual or lunar cycles (Mosher, 1954).

If the species as a whole lacks such cycles, then at least individual females do show a reproductive cycling. Walters (reported in Breder & Rosen, 1966) reported that over 67% of 150 egg masses released by *Histrio* in aquaria were produced between three and five days after the preceding raft. Intervals between successive spawnings, however, ranged from 2 to 39 days. One female studied by Mosher for 20 days produced seven egg rafts at precise three-day intervals.

REPRODUCTIVE BEHAVIOR

Spawning has been reported for *Histrio histrio* by Mosher (1954) and Fujita & Uchida (1959), for *Antennarius nummifer* by Ray (1961), and for *A. zebrinus* by Burgess (1976). Spawning behavior is virtually identical in all three and is probably similar for most members of the family.

Courtship occurs during the day, with spawning itself taking place late in the day or at night. Eight to twelve hours prior to spawning the female's abdomen begins to swell just ahead of the genital pore, a swelling that increases steadily until, immediately prior to spawning, the female begins to look like a mottled brown balloon with fins. During this prespawning period the female sits quietly, either on the bottom or, in the case of the sargassum fish, in *Sargassum* (if it is available). The male, meanwhile, becomes extremely attentive of the female and rarely leaves her immediate vicinity. He may circle around her and repeatedly nudge and nibble at her abdomen or may "feel" her with his hand-like pectoral fins. Mosher reports that at this time the female is more or less normally colored, while the male develops a darker brown or copper color.

Shortly before spawning the female, now so distended that she semi-floats tail-up, begins to move, followed closely by the male. In aquaria she either "walks" on her pelvic fins or swims slowly across the bottom, angled sharply tail-up into the water. The male is close behind, nudging at her abdomen repeatedly. Both sexes quiver sporadically as they move about. "Following" goes on for a few minutes and then, together, the pair dash for the surface. Egg release is almost explosive and occurs just below the surface. Slow motion film analysis indicates that the complete egg raft can be extruded in as little as 0.4 seconds and that the entire spawning sequence, from initial dash upward to the female's return to the bottom, may take as little as 2.9 seconds. During the dash upward the male follows closely behind the female with his mouth close to her venter. Just before reaching the surface he makes a 90° turn, moving away from her, and then vaults ahead of her, breaking the surface before she does. There is some evidence, as yet inconclusive, that the male actually grasps the end of the egg raft as it ex-trudes from the female and pulls it out and surface-ward while shedding his milt on it. After spawning the male descends immediately to the bottom, while the female remains at the surface for several seconds. Then, shaking herself free of the expanding egg raft, she slowly spins to the bottom. She frequently appears dazed and exhausted after spawning, but normally begins feeding within a few hours.

EGGS AND LARVAE

The most striking feature of antennariid spawning is the production by most of the gelatinous floating egg raft, or veil, in which the eggs remain embedded until hatching. The size, shape, and color of the raft vary from species to species. That of *Histrio histrio* is transparent, about 9 cm long without being unrolled, and shaped like a flattened band with rolled ends (it is generally described as a "scroll") (Mosher, 1954; Fujita & Uchida, 1959). *Antennarius multiocellatus, A. tigrinus,* and *A. zebrinus* produce similar though slightly larger rafts (Mosher, 1954; Burgess, 1976). *Antennarius nummifer* produces a colorless, transparent, balloon-like raft which has a pore at one end; the raft is about 5 cm across and about 7 cm high (Ray, 1961). In contrast to these small rafts, those produced by other species are immense. Hornell (1922) reports that an egg raft produced by *Antennarius hispidus* in an aquarium was 9.5 feet long and 6.25 inches wide (2.9 m by 15.9 cm). Egg rafts of the related goosefish, *Lophius americanus* (Lophiidae), a temperate water predator, reach 30 feet (9.2 meters) in length and 3 feet (92 cm) in width (Bigelow & Welsh, 1925). Similar egg rafts are also produced by the batfishes, family Ogcocephalidae (Rasquin, 1958).

Rasquin (1958) examined the structure and production of the gelatinous egg rafts of *Histrio histrio, Antennarius scaber,* and *Ogcocephalus vespertilio* in detail. Each raft replicates the shape and internal structure of the female's ovaries. Thus, the flattened scroll-shaped raft produced by the sargassum fish mimics the animal's bilobed ovaries, the lobes being fused in the center and scrolled at either end. Eggs are produced in follicles that are suspended on tree-like structures known as lamellae in the empty central region of the ovary (the lumen). The inner surface of the lumen is covered with a thin mucoidal material. At ovulation the eggs erupt from the follicles but are immediately held in the mucous capsule that surrounded them. At spawning the entire mucous coat is cast free, with the spaces occupied by the lamellae empty and now forming pores that connect each round egg chamber both with each other and with the outer sea water. Once released, sea water carrying sperm rushes into these pores to fertilize each encapsulated egg. Ray (1961) calculated that the relatively small raft produced by *A. nummifer* contained somewhere around 48,800 eggs. Since rafts can be produced at three- to four-day intervals, the fecundity of this and other species is truly immense.

Following spawning, the eggs develop slowly. *Histrio* eggs are initially oval, 0.7 mm high and 0.6 mm across, but become spherical after a single cleavage (Mosher, 1954). Similar sized and shaped eggs are reported for *Antennarius marmoratus* and *A. nummifer* (Padmanabhan, 1957, and Ray, 1961, respectively). Sargassum fish eggs hatch in four to five days at 21 to 23°C (Mosher, 1954) and two to three days at 30° (Rasquin, 1958). The raft slowly disintegrates until, by the time the eggs are hatching, moderate wave action is enough to release the young. Newly hatched larvae are approximately 1.4 mm long, are only slightly pigmented, and have a large yolk sac.

For the first five to six days the prolarvae drift quietly at the surface. By the end of this time the yolk sac is absorbed, the mouth and anus open, and the eyes are pigmented. Adams (1960) described the development of subsequent larval stages of *H. histrio* based on plankton collections. At a total length of about 2.0 mm the prolarvae enter a postlarval stage characterized by a large round head, fully formed fins, a pigmented double-notched eye, and, most strikingly, a clear, more or less spherical, integumentary envelope in which the larva is suspended. The larva contacts the envelope only at the mouth, the anus, and at the base of the fins, with each fin projecting through it. This inflated integument is thought to help the larva osmoregulate, maintain its buoyancy, and possibly facilitate nutrient transport to different parts of the animal. In the past these odd-looking larvae were considered a separate genus of anglerfishes known as *Kanazawaichthys* (Schultz, 1957; Hubbs, 1958). With development, the postlarva essentially grows into this tough, thin envelope, until by the time it reaches 9 mm TL the envelope has completely fused with the larva's epidermis and is no longer visible. At this stage the animal is effectively a juvenile and begins to develop the adult coloration. By the time it reaches 15 mm in length, it is essentially a miniature adult. Growth and development are extremely fast; a generation can take as little as 21 days.

Adams (1960) reported that postlarval *H. histrio* can be collected year-round. Those larger than 9 mm are invariably found in *Sargassum* patches; smaller fish can be found anywhere in the upper 100 m of the water column.

Very recently, Pietsch & Grobecker (1980) described quite a different mode of larval development and egg care that is, so far as is known, unique among antennariids. While examining museum specimens of various antennariid species, they discovered an 85 mm male *Antennarius caudimaculatus* that had approximately 650 spherical eggs attached to its side. Each egg was about 3.2 to 3.6 mm in diameter and was attached to the surface of the male (and to each other) by hook-like looping threads emerging from each that wove together, and by thin sheets of a transparent acellular material. Close examination of the embryos indicated they were indeed those of an antennariid, similar in fact to those subsequently obtained from the ripe ovaries of female *A. caudimaculatus*. Based on this evidence, Pietsch & Grobecker (1980) suggest that males of *A. caudimaculatus* brood large demersal eggs, which each male carries and presumably defends. They further note that the embryos examined were large compared with those described for other antennariids, and apparently hatch in an advanced stage of development. Subsequent larval development of this unusual species has not been described, nor are there any field observations of the behavior of this Philippine species.

REPRODUCTION IN CAPTIVITY

A number of antennariids, including *Histrio histrio* and several species of *Antennarius*, have spawned in small aquaria. Indeed, even solitary females will routinely produce infertile egg rafts (e.g., Paull, 1969). Spawning these and other small anglerfishes should present few problems to a determined aquarist once a pair have been established in a tank. Unfortunately, there is the problem that anglerfishes are all voracious predators and have no compunctions about eating one another, even if the other is a potential mate. In the field courtship is undoubtedly initiated by the male, who remains near, but not too near, a female until she is ready to spawn, i.e., until she begins to visibly swell with the developing egg raft. In the confines of an aquarium, however, one partner may not last long enough for the other to reach this state.

The simplest and most basic solution to this problem is to feed a pair heavily with live fish. Given the constant presence of prey, hopefully the pair will refrain from eating one another. If cannibalism occurs anyways, it may indicate that the "pair," in fact, consisted of two members of the same sex (as yet, no one has looked at the gonads of antennariids to determine if they are sequential or simultaneous hermaphrodites; such might be expected given the widely dispersed, low density populations of most species). Sufficient cover for both individuals should also be provided. An alternative approach, and one likely to be less costly in terms of fish, is to separate members of a potential pair by a glass partition while feeding both heavily with live fish until one begins to swell. Individuals on one side of the partition could be replaced regularly until a good combination was achieved.

Eggs and prolarvae of several species have done well in aquaria with no special equipment or conditions. No successful rearings of the larvae through metamorphosis have been reported as yet, however (probably because no one has tried), so that what problems, if any, are involved are not known.

Literature Cited

Adams, J.A. 1960. A contribution to the biology and postlarval development of the sargassum fish, *Histrio histrio* (Linnaeus), with a discussion of the *Sargassum* complex. *Bull. Mar. Sci.*, 10:55-82.

Bigelow, H.B. and W.W. Welsh. 1925. Fishes of the Gulf of Maine. *Bull. U.S. Bur. Fish.*, 40:567 pp.

Böhlke, J.E. and C.C.G. Chaplin. 1968. *Fishes of the Bahamas and adjacent tropical waters*. Livingston Publ., Wynnewood, Pa. 771 pp.

Breder, C.M., Jr. and D.E. Rosen. 1966. *Modes of Reproduction in Fishes*. Natural History Press, Garden City, N.Y. 941 pp.

Burgess, W.E. 1976. Salts from the seven seas. *Tropical Fish Hobbyist*, 25:57-64.

Fujita, S. and K. Uchida. 1959. Spawning habits and early development of a sargassum fish, *Pterophryne histrio* (Linne.). *Sci. Bull. Fac. Agric. Kyushu Univ.*, 17:277-282.

Hornell, J. 1922. The Madras marine aquarium. *Bull. Madras Fish. Bur.*, (5):57-96.

Hubbs, C.L. 1958. *Dikellorhynchus* and *Kanazawaichthys*: nominal fish genera interpreted as based on prejuveniles of *Malacanthus* and *Antennarius*, respectively. *Copeia*, 1958:282-285.

Mosher, C. 1954. Observations on the spawning behavior and the early larval development of the sargassum fish, *Histrio histrio* (Linnaeus). *Zoologica, N.Y.*, 39:141-152.

Padmanabhan, K.G. 1957. Early stages in the development of the toad fish, *Antennarius marmoratus* Bleeker. *Bull. Cent. Res. Inst. Univ. Travancore, Trivandrum, Ser. C., Nat. Sci.*, 5:85-92.

Paull, R.C. 1969. Sargassum fish spawns. *Salt Water Aquarium*, 5:77-78.

Pietsch, T.W. and D.B. Grobecker. 1980. Parental care as an alternative reproductive mode in an antennariid Anglerfish. *Copeia*, 1980:551-553.

Rasquin, P. 1958. Ovarian morphology and early embryology of the pediculate fishes *Antennarius* and *Histrio*. *Bull. Amer. Mus. Nat. Hist.*, 114:331-371.

Ray, C. 1961. Spawning behavior and egg raft morphology of the ocellated fringed frogfish, *Antennarius nummifer* (Cuvier). *Copeia*, 1961:230-231.

Schultz, L.P. 1957. The frogfishes of the family Antennariidae. *Proc. U.S. Nat. Mus.*, 107:47-105.

Trumpetfishes (Aulostomidae) and Cornetfishes (Fistulariidae)

The trumpetfishes and cornetfishes are a pair of similar appearing and closely related families, in turn somewhat more distantly related to the pipefishes and seahorses of the family Syngnathidae. Fishes in both families are extremely long and slender and have small dorsal and anal fins directly above one another, pelvic fins located well back on the body, and a long tubular snout. They can be distinguished by the shape of the caudal fin (in trumpetfishes, rounded; in cornetfishes, lunate, with a long trailing central streamer) and body shape (trumpetfishes compressed side-to-side; cornetfishes depressed top-to-bottom). Members of both families are predators, slowly stalking the small fishes that make up the bulk of their diet. Both are also camouflage artists well known for their abilities to change colors to match background objects. Trumpetfishes are also prone to hover quietly parallel to a soft

coral strand or to follow along in a school of grazing fishes, such as parrotfishes, apparently as a ruse to approach their prey. The two families differ in preferred habitats and, to a lesser extent, in distribution. The trumpetfishes are characteristically reef fishes and are distributed in the tropical western Atlantic and Indo-Pacific; cornetfishes are more common inshore, though they are abundant on eastern Pacific reefs, and are found in the tropical Atlantic, Indo-Pacific and eastern Pacific. Finally, each family consists of only one genus (*Aulostomus* and *Fistularia*) and four species.

Until recently, little was known about spawning by members of either family. Munro, et al. (1973) collected ripe females of *A. maculatus* off Jamaica in November and January, indicating that spawning takes place at least during the winter. Delsman (1921)

Trumpetfishes are camouflage artists. This *Aulostomus chinensis* is exhibiting the plain gold pattern that in one observation turned out to be the female. Photo by Walter Deas.

suggested spawning occurs year-round for *Fistularia serrata* in the Java Sea, since he was able to collect eggs and larvae year-round, and also suggested spawning was likely to occur at dusk or at night. Recently, A. Gronell (pers. comm.) witnessed apparent courtship and spawning of *Aulostomus chinensis* off One Tree Island, Great Barrier Reef. On a dusk dive to the outer, windward edge of the reef in late summer, a trio of *Aulostomus* were observed actively chasing one another a few meters off the bottom. Closer investigation revealed one gold-colored individual with a heavily distended abdomen being pursued by two gray individuals in the normal color pattern (pale gray, with faint barring, dark caudal peduncle and fin, and dark rear edges of the dorsal and anal fins). The two gray fish were more slender and slightly smaller than the gold one and were clearly fighting one another while pursuing the apparent female. Eventually one was driven off. The female hung in the water column a few meters up, with her swollen abdomen toward the winning male. The male approached her, developing first a pink wash across his ventral surface and then darkening all over with only a faint trace of barring. Upon reaching the female the pair swam forward with the apparent male weaving back and forth about the female. In the apparent spawning ascent, the female started up with the male weaving behind her, and the two ascended to a height of 5 to 8 m before turning back to the bottom. Such ascents were seen twice, along with a third in which the pair made a smooth arc upward, side-by-side. Gamete release, if it occurred, was not conspicuous. All activity was observed after sunset and about 30 minutes before complete darkness.

Aulostomid eggs and larvae have not yet been described. Fistulariid eggs and larvae, however, have been described for three species: *Fistularia petimba* by Mito (1966), *F.* sp. by Watson & Leis (1974), and *F. serrata* by Delsman (1921). The eggs of *F. petimba* range in size from 1.8 to 2.1 mm in diameter and are spherical, transparent, and pelagic. Hatching occurs in about 170 hours, producing a long, slender yolk-sac larva 7.08 mm long. The eggs of *Fistularia* sp. from Hawaii are similar to those of *F. petimba* but only 1.65 to 1.75 mm in diameter. The eggs of *F. serrata* are smaller (1.5 to 1.7 mm in diameter) and, like those of the other two species, contain no oil droplets. The eggs also have a double egg membrane, the outer being thicker and stronger than the inner. Such eggs hatch in four days to produce a 6 mm long larva characterized by an elongate yolk sac, long pointed jaws, and many branching pigment spots along the underside. The eyes are pigmented within a day of hatching. Two days after hatching the larvae are 6.5 mm long and have a lower jaw that extends in front of the upper jaw. By the time the larvae reach a length of 12 mm the jaws have developed their tube-like shape and the dorsal, anal, and caudal fin rays are developing. The caudal fin streamer begins to develop by the time a length of 20 mm is attained, and it is well developed by 45 mm.

Literature Cited

Delsman, H.C. 1921. Fish eggs and larvae from the Java Sea. 1. *Fistularia serrata. Treubia*, 2:97-108.

Mito, S. 1966. Fish eggs and Larvae. *Illustrations of the Marine Plankton of Japan.*, 7:74 pp.

Munro, J.L., V.C. Gaut, R. Thompson, and P.H. Reeson. 1973. The spawning seasons of Caribbean reef fishes. *J. Fish Biol.*, 5:69-84.

Watson, W. and J.M. Leis. 1974. Ichthyoplankton of Kaneohe Bay, Hawaii. *Sea Grant Tech. Rpt.* (TR-75-01).

Jacks
(Carangidae)

Variously referred to as jacks, pompano, or trevally, carangids are mid-water to surface nektonic predators usually found in schools and moving at high speed. Like other, ecologically similar fishes, such as tuna and mackerel, they tend to be silvery and streamlined, with small scales, swept-back fins, and a deeply forked or lunate caudal fin. The body is laterally compressed and the shape varies from long and slender to extremely deep-bodied. Most species (especially the larger ones) are characteristic of areas other than the reef, but many of the smaller species routinely visit reefs and may be significant predators on smaller reef fishes. There are 24 genera in the family, with about 200 species distributed worldwide in tropical and temperate seas.

Spawning has thus far been described for only three species in the family: *Seriola dorsalis*, a large deeper-water species, by Walford (1937); *Caranx sexfasciatus*, by Breder (1951); and *Caranx ignobilis*, by Von Westernhagen (1974). In the first case, spawning oc-

curred well offshore, a condition that appears to be the case for many carangids (e.g., Johannes, 1981). Roughly 112 km off the coast of California, an aggregation of several hundred individuals was seen swimming just below the surface in tight circles, clearly releasing gametes. Spawning in this, and probably most other, species takes place during the day. In the case of *C. sexfasciatus*, apparent spawning took place in a 75 by 25 feet (approximately 22.5 by 7.6 m.) enclosure in late afternoon. Gamete release was not seen, but two individuals swimming faster than other fish in the school were observed to approach one another closely and to press their ventral surfaces together. One fish turned almost entirely horizontal to do so, holding this position for a long period of time. Finally, confirmed spawning has been observed "in-water" only in *Caranx ignobilis* in the Philippines. An aggregation of the fish was located on a shoal 35 to 45 m deep in a channel between two islands. At the margins of the aggregation three or four males could

Larval carangids, 2.5 and 5.5 mm total length, collected in plankton tows off Miami, Florida. Photo by Dr. R. E. Thresher.

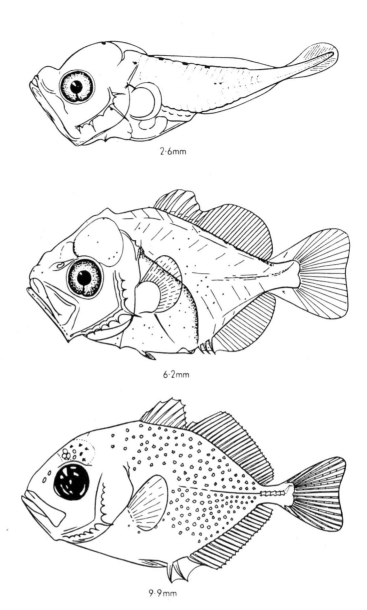

Larval development of the jack
Caranx crysos. Based on McKenney,
et al. (1958), with permission of the
Bulletin of Marine Science.

2·6mm

6·2mm

9·9mm

be seen vigorously pursuing running ripe females. Eventually, only one male remained with each female. The pair then sank to the bottom and, swimming in tight circles, released their gametes. Similar behavior, though not actual gamete release, was observed in the same area in *Alectis indicus.*

Spawning seasons for most species appear to be long (e.g., Munro, et al., 1973; Kwei, 1978), with a peak in activity during the summer. Johannes (1981) suggests that a number of species at Palau spawn only at the full and/or new moons. Though such lunar cycles have not yet been reported elsewhere, both De Jong (1940) and Von Westernhagen (1974) suggested their respective species spawn repeatedly and periodically.

So far as is known all carangids are gonochoristic, and in most, if not all, cases the sexes are about the same size (e.g., Thompson & Munro, 1974). The only species thus far reported to be sexually dichromatic is *Caranx ignobilis*, in which the males are darker than the females (Talbot & Williams, 1956; Von Westernhagen, 1974).

There is an extensive literature on carangid eggs and larvae, in large part because the fishes are often commercially important. Particularly relevant literature includes Delsman (1926), McKenny, et al. (1958), Berry (1959), Demir (1961), Aprieto (1974), and Miller & Sumida (1974). Carangid eggs are pelagic and transparent, ranging in diameter from roughly 0.6 to 1.6 mm. Incubation lasts 12 to 24 hours, apparently increasing with egg diameter. At hatching, larvae have a large yolk sac that contains an anteriorly located oil droplet. Older larvae are large-headed and laterally compressed, with prominent eyes, well developed jaws, and prominent spines on the preopercle. By the time they reach a length of 8-9 mm they are clearly recognizable as small jacks. These juveniles are typically spotted or barred and are commonly seen swimming below drifting objects, including jellyfishes. Larger juveniles are often common inshore but move to deeper water as they mature.

Literature Cited

Aprieto, V.L. 1974. Early development of 5 carangid fishes of the Gulf of Mexico and the south Altantic coast of the U.S. *Fish. Bull. (U.S.),* 72:415-443.

Berry, F.H. 1959. Young jack crevalles (*Caranx* species) off the southeastern Atlantic coast of the United States. *Fish. Bull. (U.S.),* 59:417-535.

Breder, C.M., Jr. 1951. A note on the spawning behavior of *Caranx sexfasciatus. Copeia,* 1951:170.

De Jong, J.K. 1940. A preliminary investigation of the spawning habits of some fishes of the Java Sea. *Treubia,* 17:307-330.

Delsman, H.C. 1926. Fish eggs and larvae from the Java Sea. 5. *Caranx kurra, macrosoma* and *crumenophthalmus. Treubia,* 8:199-218.

Demir, M. 1961. On the eggs and larvae of the *Trachurus trachurus* (L.) and *Trachurus mediterraneus* (Steind.) from the sea of Marmara and the Black Sea. *Rapp. P.-v. Réun. CIESMM,* 16:317-330.

Johannes, R.D. 1981. *Words of the Lagoon.* Univ. California Press, Los Angeles. 320 pp.

Kwei, E.A. 1978. Food and spawning activity of *Caranx hippos* (L.) off the coast of Ghana. *J. Natur. Hist.,* 12:195-215.

McKenney, T.W., E.C. Alexander, and G.L. Voss. 1958. Early development and larval distribution of the carangid fish, *Caranx chrysos* (Mitchell). *Bull. Mar. Sci.,* 8:160-200..

Miller, J.M. and B.Y. Sumida. 1974. Development of eggs and larvae of *Caranx mate* (Carangidae). *Fish. Bull. (U.S.),* 72:497-514.

Munro, J.L., V.C. Gaut, R. Thompson, and P.H. Reeson. 1973. The spawning seasons of Caribbean reef fishes. *J. Fish Biol.,* 5:69-84.

Talbot, F.H. and F. Williams. 1956. Sexual color differences in *Caranx ignobilis* (Forsk.) *Nature, Lond.,* 178:934.

Thompson, R. and J.L. Munro. 1974. The biology, ecology and bionomics of Caribbean reef fishes: Carangidae (Jacks). *Res. Rpt. Zool. Dept., Univ. West Indies,* (3):43 pp.

Von Westernhagen, H. 1974. Observations on the natural spawning of *Alectis indicus* (Rüppell) and *Caranx ignobilis* (Forsk.) (Carangidae). *J. Fish Biol.,* 6:513-516.

Walford, L.A. 1937. *Marine Game Fishes of the Pacific Coast from Alaska to the Equator.* Univ. California Press, Berkeley. 205 pp.

Drums and Jackknife Fishes (Sciaenidae)

The jackknife fishes and their relatives are, for the most part, bottom-oriented predators characterized by a robust body, a single dorsal fin that is almost, but not quite, divided into two, and, in many species, one or more small barbels on the underside of the jaw. Many grow to a large size, some being reported to exceed a weight of 90 kg. Not surprisingly, such fishes are commercially important, which has led to a considerable amount of work being done on their biology. Information on the smaller reef-associated species, which constitute only a small minority in this typically inshore family, is far less complete. The family consists of 28 genera and about 160 species; of these, perhaps four genera and less than a dozen species are found on the reef. The best known of these are the jackknife fishes, genus *Equetus*, a group of striking black-and-white species.

The bulk of the work done on sciaenids has involved description of eggs and larvae and methods of rearing the commercially important species in captivity. Relatively little is known of their spawning behavior. To date, spawning has been described for only one reef-associated species, *Equetus acuminatus*, by Straughan (1968) (for a non-reef species, see Guest & Lasswell, 1978). Spawning by *Equetus* took place in a small aquarium containing three individuals between 12 and 15 cm long. A pair of these began to swim in tight head-to-tail circles, one essentially chasing the other, such circling becoming increasingly faster and the fish increasingly closer together. Finally, the fish quivered simultaneously and released a visible cloud of gametes. Such courtship and spawning were seen several times within a single day. The eggs sank to the bottom, suggesting that they were not fertile.

The larger non-reef associated species appear to characteristically spawn in spring and summer (e.g., Hildebrand & Cable, 1934; Fujita, 1956; Takemura, et al., 1978; Taniguchi & Okada, 1979). No information is available concerning lunar cycles, though such are likely to occur in the family.

Sciaenid eggs range in size from about 0.6 to 1.3 mm in diameter and are spherical, transparent, and pelagic. Early in their development they tend to possess many small oil globules which eventually

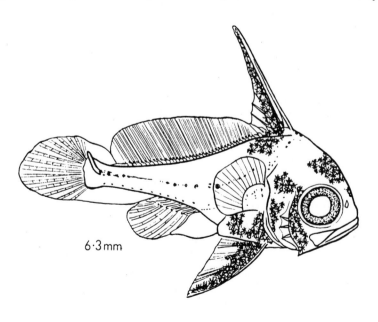

Late stage larva of *Equetus* sp.
Based on Powles & Burgess (1978).

6·3 mm

coalesce into a single large globule. There is a considerable literature on the development of eggs and larvae of non-reef sciaenids (e.g., Welsh & Breder, 1923; Fujita, 1956; Joseph, et al., 1964; Taniguchi, et al., 1979; and see Breder & Rosen, 1966), but little on the reef-associated species. Powles & Burgess (1978) discussed and described the apparently benthic larvae of several western Atlantic species of *Equetus* (= *Pareques*). These larvae, like those of non-reef species, have large heads and eyes and are slightly laterally compressed. By the time they reach a length of 4.4 mm, they already have well developed jaws and the beginnings of fin ray development. By 6.6 mm, such larvae look much like the adults. Powles & Burgess collected the postlarvae from under rocks in shallow areas off the coast of Florida and Colombia, having discovered them swimming in small groups in the protected areas created by the rocks. Larger individuals of *Equetus* are commonly found in this area and are readily identifiable as jackknife fishes. Each resembles a tiny black-and-white dot from which stream four long filaments: the dorsal fin, the caudal fin, and the two pelvic fins. With development, the relative sizes of these streaming fins decreases until adult proportions are reached. Age at maturity is not known.

Larvae of commercially important sciaenids are commonly reared in captivity, and an extensive literature exists on the various techniques used (see Arnold, et al., 1976). Eggs are usually obtained by injecting hormones into the adults and stripping ripe gametes (e.g., Haydock, 1971; Lasker, 1974; Colura, 1974; Middaugh & Yoakum, 1974), though hormone-induced natural spawning has also been reported (e.g., Garza, et al., 1978).

Literature Cited

Arnold, C.R., T.D. Williams, W.A. Fable, Jr., J.L. Lasswell, and W.H. Bailey. 1976. Methods and techniques for spawning and rearing spotted sea trout (*Cynoscion nebulosus*) in the laboratory. *Proc. Ann. Conf. Southeast Assoc. Game Fish. Comm.*, 30:167-178.

Breder, C.M., Jr. and D.E. Rosen. 1966. *Modes of Reproduction in Fishes*. Natural History Press, Garden City, N.Y. 941 pp.

Colura, R.L. 1974. Induced spawning of the spotted seatrout, *Cynoscion nebulosus* (Cuvier). *Fifth Ann. Workshop, World Maricult. Soc., 1974*:319-330.

Fujita, S. 1956. On the development of the egg and prelarval stages of *Nibea argentata* (Houttuyn). *Sci. Bull. Fac. Agric., Kyushu Univ.*, 15:537-540.

Garza, G., W.H. Bailey, and J.L. Lasswell. 1978. Rearing of black drum in fresh water. *Progress. Fish-Cult.*, 40:170.

Guest, W.C. and J.L. Lasswell. 1978. A note on courtship behavior and sound production of red drum. *Copeia*, 1978:337-338.

Haydock, I. 1971. Gonad maturation and hormone-induced spawning of the gulf croaker, *Bairdiella icistia*. *Fish. Bull. (U.S.)*, 69:157-180.

Hildebrand, S.F. and L.E. Cable. 1934. Reproduction and development of whitings or kingfishes, drums, spot, croaker, and weakfishes or sea trouts, family Sciaenidae, of the Atlantic coast of the United States. *Bull. U.S. Bur. Fish.*, 48:41-117.

Joseph, E.B., W.H. Massman, and J.J. Norcross. 1964. Pelagic eggs and early larval stages of the black drum from Chesapeake Bay. *Copeia*, 1964:425-434.

Lasker, R. 1974. Induced maturation and spawning of marine fish at the southwest fisheries center, La Jolla, California. *Fifth Ann. Workshop, World Maricult. Soc., 1974*:313-318.

Middaugh, D.P. and R.L. Yoakum. 1974. The use of chorionic gonadotropin to induce laboratory spawning of the Atlantic croaker, *Micropogon undulatus*, with notes on subsequent embryonic development. *Chesapeake Sci.*, 15:110-114.

Powles, H. and W.E. Burgess. 1978. Observations on benthic larvae of *Pareques* (Pisces: Sciaenidae) from Florida and Colombia. *Copeia*, 1978:169-172.

Straughan, R.P. 1968. First spawning of the high hat. *Salt Water Aquarium*, 4:73-74.

Takemura, A., T. Takita, and K. Mizue. 1978. Studies on the underwater sound. VII. Underwater calls of the Japanese marine drum fishes (Sciaenidae). *Bull. Japan. Soc. Sci. Fish.*, 44:121-123.

Taniguchi, N. and Y. Okada. 1979. Study of the maturation and spawning of the nibe-croaker collected from Tosa Bay. *Rpt. Usa Mar. Biol. Inst., Kochi Univ.*, (1):41-49.

Taniguchi, N., T. Kuga, Y. Okada, and S. Umeda. 1979. Studies on the rearing of artificially fertilized larvae and early developmental stage of the nibe-croaker, *Nibea mitsukurii*. *Rpt. Usa Mar. Biol. Inst., Kochi Univ.*, (1):51-58.

Welsh, W.W. and C.M. Breder, Jr. 1923. Contributions to life histories of Sciaenidae of the eastern United States coast. *Bull. U.S. Bur. Fish.*, 39:141-201.

Mullet
(Mugilidae)

Mullet are heavy-bodied elongate fishes characterized by two widely separated dorsal fins (as in the related barracudas), pectoral fins inserted well above and before the pelvics, a small mouth, and the complete absence of a lateral line. The family consists of ten genera and about 100 species distributed in tropical and temperate seas worldwide. Most of these species inhabit inshore environments, including a few that are completely fresh water, and only a few are routinely seen on the reef. Even these are not characteristic of the reef itself, but rather tend to be found along shore lines and in shallow back-reef areas.

Mullet are commercially important food fishes and have been the subject of considerable mariculture work. Consequently, there is an immense literature on artificially spawning and rearing the animals; for access to this literature, see reviews by Thomson (1963, 1964, 1966), Breder & Rosen (1966), and Donaldson (1977). Most of this work concerns the gray mullet, *Mugil cephalus*, an inshore species distributed circumtropically.

In general, mullet are sexually dimorphic with respect to size at maturity (female larger than male), although to date there is no evidence of sequential hermaphroditism in the family. Spawning occurs at various times of the year but most commonly in spring and summer. Johannes (1981) suggests that two species at Palau characteristically spawn on or near the nights of the full moon. At least some species migrate to specific spawning grounds, though the location of these spawning areas varies widely both within and between species. *Mugil cephalus*, for example, has been reported spawning well offshore by Dehnik (1953), Anderson (1957), Arnold & Thomson (1958), and Fitch (1972), in estuaries by Breder (1940), and even in freshwater rivers by Johnson & McClendon (1970). Offshore spawning involves immense aggregations of fish near the surface. According to Anderson and Arnold & Thomson, spawning takes place late at night (close to midnight) and involves the males following closely after and nudging at individual females. During the nudging the fish would occasionally stop swimming and quiver vigorously, apparently releasing gametes. Generally similar behavior is reported for *M. cephalus* in estuaries (Breder, 1940), although spawning groups consisted of only four males and a female, rather than the dense aggregations reported offshore, and spawning took place during the day, rather than at night.

The only report of spawning by a reef-associated mullet is that of Helfrich & Allen (1975), who observed spawning of *Crenimugil crenilabis* in the lagoon at Enewetak, Marshall Islands. Late at night in midsummer, an aggregation of from 500 to 1,500 individuals was observed swimming rapidly in a counter-clockwise motion close to the surface in shallow water. Collection of a few specimens indicated that the fish were running-ripe and apparently near spawning. Close to midnight the fish suddenly became more active, swimming faster and breaking up into several groups, which subsequently re-merged. Just after midnight the fish suddenly erupted from the water, splashing and thrashing about for one to two seconds. Collection of seawater samples immediately afterward revealed large numbers of newly fertilized pelagic eggs, presumably those of the mullet. For at least the next two hours, when observations ended, the jumping behavior was repeated every five to ten minutes. By dawn the next day the school of fish had disappeared.

Mugilid eggs tend to be small, roughly 0.6 to 0.8 mm in diameter, spherical, transparent, and contain a single large oil droplet. Development of eggs and larvae has been discussed in considerable detail; pertinent references include Anderson (1957, 1958), Ochiai & Umeda (1969), and Houde, et al. (1976). Newly hatched larvae are approximately 2 mm long and have a large yolk sac that contains an oil droplet at its anteriormost point. The yolk sac is absorbed in about three days. Development is rapid, with transformation to a silvery schooling juvenile occurring in about 12 days and at a length of 5.5 to 6.5 mm.

There are no reports of natural spawning by mullet in captivity. Induced spawning, involving injections of hormones and stripping gametes, is widely used to obtain stock (see Yashouv, 1969; Shehadeh & Ellis, 1970; Liao, et al., 1971; Kuo, et al., 1973; Shehadeh, et al., 1973). Several species are routinely reared in captivity; for a recent example of the techniques used see Houde, et al. (1976).

Literature Cited

Anderson, W.W. 1957. Early development, spawning, growth, and occurrence of the silver mullet (*Mugil curema*) along the south Atlantic coast of the United States. *Fish. Bull. (U.S.),* 57:397-414.

Anderson, W.W. 1958. Larval development, growth, and spawning of striped mullet (*Mugil cephalus*) along the south Atlantic coast of the United States. *Fish. Bull. (U.S.),* 58:501-519.

Arnold, E.L., Jr. and J.R. Thomson. 1958. Offshore spawning of the striped mullet, *Mugil cephalus,* in the Gulf of Mexico. *Copeia,* 1958:130-132.

Breder, C.M., Jr. 1940. The spawning of *Mugil cephalus* on the Florida west coast. *Copeia,* 1940:138-139.

Breder, C.M., Jr. and D.E. Rosen. 1966. *Modes of Reproduction in Fishes.* Natural History Press, Garden City, N.Y. 941 pp.

Dehnik, T.V. 1953. Spawning of mullet in the Black Sea. *Dokl. Acad. Nauk SSSR,* 93:201-204.

Donaldson, E.M. 1977. Bibliography of Fish Reproduction 1963-1974. *Fish. & Mar. Serv. Canada, Tech. Rpt.* (732).

Fitch, J.E. 1972. A case for the striped mullet, *Mugil cephalus,* spawning at sea. *Calif. Fish and Game,* 58:246-248.

Helfrich, P. and P.M. Allen. 1975. Observations on the spawning of mullet, *Crenimugil crenilabis* (Forskal), at Enewetak, Marshall Islands. *Micronesica,* 11:219-225.

Houde, E.D., S.A. Berkeley, J.J. Klinovsky and R.C. Schekter. 1976. Culture of larvae of the white mullet, *Mugil curema* Val. *Aquacult.,* 8:365-370.

Johannes, R.E. 1981. *Words of the Lagoon.* Univ. Calif. Press, Los Angeles. 320 pp.

Johnson, D.W. and E.L. McClendon. 1970. Differential distribution of the striped mullet, *Mugil cephalus* Linnaeus. *Calif. Fish and Game,* 56:138-139.

Kuo, C.M., Z.H. Shehadeh, and C.E. Nash. 1973. Induced spawning of captive grey mullet (*Mugil cephalus* L.) females by injection of human chorionic gonadotropin (HCG). *Aquacult.,* 1:429-432.

Liao, I.C., Y.J. Lu, T.L. Huang, and M.C. Lin. 1971. Experiments on induced breeding of the grey mullet, *Mugil cephalus* Linnaeus. *Aquacult.,* 1:15-34.

Ochiai, A. and S. Umeda. 1969. Spawning aspects of the grey mullet, *Mugil cephalus* L. living on the coastal region of Kochi Prefecture. *Jap. J. Ichthyol.,* 16:50-54.

Shehadeh, Z.H. and J.N. Ellis. 1970. Induced spawning of the striped mullet *Mugil cephalus* L. *J. Fish Biol.,* 2:355-360.

Shehadeh, Z.H., C.M. Kuo, and K.K. Milisen. 1973. Induced spawning of grey mullet *Mugil cephalus* L., with fractionated salmon pituitary extract. *J. Fish Biol.,* 5:471-478.

Thomson, J.M. 1963. Synopsis of biological data on the grey mullet *Mugil cephalus* Linnaeus 1758. *C.S.I.R.O. Aust. Fish. Oceanogr. Fish. Synopsis,* (1):66 pp.

Thomson, J.M. 1964. A bibliography of systematic references to the grey mullets (Mugilidae). *C.S.I.R.O. Aust. Fish. Oceanogr. Tech. Pap.,* (16):127 pp.

Thomson, J.M. 1966. The grey mullets. *Ann. Rev. Oceanogr. Mar. Biol.:* 301-335.

Yahouv, A. 1969. Preliminary report on induced spawning of *M. cephalus* (L.) reared in captivity in freshwater ponds. *Bamidgeh,* 21:19-24.

Barracudas
(Sphyraenidae)

Barracudas are the unmistakable top-level predators on a coral reef. Their long sleek bodies and rows of conspicuous sharp teeth radiate an aura of immense ferocity and an unparalleled ability to swiftly decimate unwary prey. Reputed to reach lengths of over 3 m, such fish are an impressive sight on the reef but, despite their fearsome reputations, are relatively harmless so far as divers are concerned, if not to the fishes that entirely make up their diets. Even larger fishes have to beware of barracudas, since the fish are known to strike so powerfully as to cut their prey in half.

The barracudas are closely related to the mullets (Mugilidae), with which they share a long robust body, large scales, and two clearly separated dorsal fins. Only one genus, *Sphyraena*, consisting of about 20 species, is currently recognized in the family (Schultz, 1953; de Sylva, 1963; Kwong-chak Au, 1979). The family is distributed in warm seas circumtropically and occupies a wide variety of habitats.

Thus far not much is known about spawning in barracudas, though the few points that have been established are well substantiated for a number of species. There is no evident external sexual dimorphism in the fishes (e.g., de Sylva, 1963), though males reach sexual maturity at a smaller size than females. Barracudas, in general, migrate to specific spawning areas, often aggregating there in large numbers (e.g., Walford, 1932; de Sylva, 1963; Houde, 1971; Johan-nes, 1981). Such spawning grounds are located at the reef edge or, in the case of the two Florida species, off-shore at the junction between oceanic and continental circulations. Spawning takes place in the warmer months of the year and often covers an extended period of time (e.g., Walford, 1932; Shojima, et al., 1957; de Sylva, 1963; Munro, et al., 1973; Nzioka, 1979) within which individual females spawn repeatedly. Johannes (1981) suggested that *S. bar-racuda* and *S. qenie* off Palau spawn only within a few days of the full moon.

Although little is known about when and how barracudas spawn, their eggs and larvae have been well examined (e.g., Barnhart, 1927; Vialli, 1937; Orton, 1955; Shojima, et al., 1957; de Sylva, 1963; Houde, 1971). Sphyraenid eggs range in diameter from 0.69-0.82 mm (*S. pinguis*) to 1.5 mm (*S. argentea*). Such eggs are pelagic, spherical, and transparent, each containing a single clear or yellow oil droplet. The eggs of *S. pinguis* hatch in 24 to 30 hours at 21.3 to 26°C.

Newly hatched prolarvae range in size from 1.7 to about 2.6 mm and have a large yolk sac that contains a single oil droplet located either anteriorly (*S. pinguis*, *S. borealis*) or posteriorly (*S. argentea*). The larvae begin to feed about three days after hatching, lunging at small copepods with a quick strike from an S-curved position. By the time they reach a length of about 5 mm and an age of five to six days, the teeth

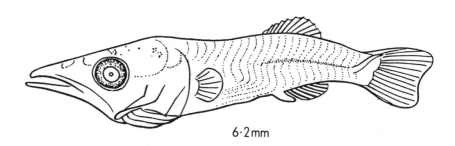

Larval barracuda (*Sphyraena* sp.).
Based on Vatanachai (1974).

6·2mm

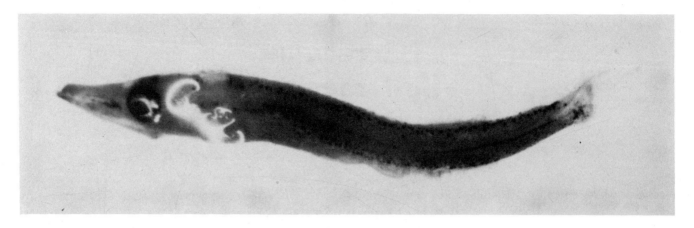

Larval barracuda, 7.0 mm total length, collected in a plankton tow off Miami, Florida. Photo by Dr. R. E. Thresher.

and finfolds begin to develop; fin rays are fully developed at a length of about 13 mm. These larger larvae voraciously attack a number of plankters, including other fish larvae, and are also aggressive toward one another. Settlement to the bottom, usually in areas inshore from the reef, occurs at a length of approximately 18 mm. Newly settled young resemble the adults in general color and shape and are voracious piscivores. The young of at least two species, *S. obtusata* and *S. barracuda*, are thought to mimic blades of sea grass and floating debris (Waite, 1923; Randall & Randall, 1960). At slightly larger sizes young barracuda are often found in schools, while fully mature adults are generally solitary and apparently territorial, except for spawning periods. Maturity in a large species, *S. barracuda*, occurs at an age of two years for the males and at about four years for the females.

Barracudas have not been spawned in captivity. The young of *S. borealis* have been reared by Houde (1971), who used eggs collected from the plankton. Larvae were initially fed zooplankton less than 0.1 mm in length and subsequently both larger zooplankton and fish larvae.

Literature Cited

Barnhart, P.S. 1927. Pelagic fish eggs off La Jolla, California. *Scripps Inst. Oceanog., Bull., Tech. Ser.*, 1:91-92.

de Sylva, D.P. 1963. Systematics and life history of the Great Barracuda, *Sphyraena barracuda* (Walbaum). *Stud. Trop. Oceanog.*, 1:179 pp.

Houde, E.D. 1971. Development and early life history of the Northern Sennett, *Sphyraena borealis* DeKay (Pisces: Sphyraenidae) reared in the laboratory. *Fish. Bull. (U.S.)*, 70:185-196.

Johannes, R.E. 1981. *Words of the Lagoon*. Univ. Calif. Press, Los Angeles. 320 pp.

Kwong-chak Au. 1979. Systematic study on the barracudas (Pisces: Sphyraenidae) from a northern sector of the South China Sea. *J. Nat. Hist.*, 13:619-647.

Munro, J.L., V.C. Gaut, R. Thompson, and P.H. Reeson. 1973. The spawning seasons of Caribbean reef fishes. *J. Fish Biol.*, 5:69-84.

Nzioka, R.M. 1979. Observations on the spawning seasons of East African reef fishes. *J. Fish Biol.*, 14:329-342.

Orton, G.L. 1955. Early developmental stages of the California Barracuda, *Sphyraena argentea* Girard. *Calif. Fish and Game*, 41:167-176.

Randall, J.E. and H.A. Randall. 1960. Examples of mimicry and protective resemblance in tropical marine fishes. *Bull. Mar. Sci. Gulf and Caribb.*, 10:444-480.

Schultz, L.P. 1953. In: Schultz, et al. Fishes of the Marshall and Marianas Islands. *Bull. U.S. Nat. Mus.*, (202):279-287.

Shojima, Y., S. Fujita, and K. Uchida. 1957. On the egg development and prelarval stages of a kind of barracuda, *Sphyraena pinguis* Günther. *Sci. Bull. Fac. Agric., Kyushu Univ.*, 16:313-318.

Waite, E.R. 1923. *The Fishes of South Australia*. R.E. Rogers, Government Printers, Adelaide. 240 pp.

Walford, L.A. 1932. The California barracuda (*Sphyraena argentea*). *Fish. Bull., Calif. Fish and Game*, 37:120 pp.

Vialli, M. 1937. Plate 35. In: Fauna e Flora del Golfo di Napoli. *Publ. Staz. Zool. Napoli, Monogr*, (38).

Tilefishes
(Malacanthidae)

Tilefishes are long slender fishes characterized by a nearly horizontal mouth and long and even dorsal and anal fins. The family is a small one, with only about 25 species in six genera, of which the best known are *Malacanthus* and *Hoplolatilus*. The family is distributed worldwide in tropical and temperate seas, often being found at great depths. The shallow-water species, such as *Malacanthus plumieri* in the western Atlantic, are burrowers, diligently digging a double-ended burrow, the top of which becomes covered with rubble. All appear to be carnivores, taking a variety of benthic animals along with substantial amounts of plankton.

Until recently little has been known about the reproductive behavior of tilefishes. Several species of *Malacanthus* and *Hoplolatilus* are commonly found in pairs (e.g., Clark & Ben-Tuvia, 1973; Thresher, 1980, pers. obs.), suggesting long-term mate fidelity. Details

of the reproductive behavior have recently been obtained for the western Atlantic species *Malacanthus plumieri* by Colin (in prep.). Spawning occurs from December to March off the coast of Puerto Rico. Courtship involves an arching swim by the male up-and-down in the water column in the vicinity of the burrows and ends in a short upward spawning ascent. The pelagic eggs are spheroid, about 0.72 mm in diameter, and contain a single oil globule.

Larval tilefishes have been described by a number of workers, including Smith (1956), Berry (1958), Hubbs (1958), Okiyama (1964), and Fourmanoir (1970, 1976). Initially the larvae were assigned provisionally to a new genus, *Dikellorhynchus*, by Smith (1956), though he also suggested that they might represent a late-stage larva of *Malacanthus*. This question was subsequently treated by Berry (1958), who again considered them a distinct genus of pelagic

A juvenile *Malacanthus latovittatus* has a color pattern much like that of *Labroides dimidiatus*, and it has been suggested that the tilefish mimics the cleaner wrasse. Photo by Dr. G. R. Allen.

tilefishes closely related to *Malacanthus*, and by Hubbs (1958), who demonstrated that, in fact, they were only prejuvenile stages of the common tilefishes. These distinctive looking prejuveniles are long and slender and are characterized by a large head covered with numerous small spines, including an anchor-shaped spine on the snout and several keeled scales.

Following settlement to the bottom, at a size of about 6 cm, *Malacanthus* immediately dig burrows under rocks in shallow water. Araga (1969) suggested that the Indo-Pacific species *Malacanthus latovittatus* may be a mimic of the cleaner wrasse *Labroides dimidiatus*.

Literature Cited

Araga, C. 1969. Young form of the rare coral fish, *Malacanthus latovittatus* (Lacépède) from Tanabe Bay. *Pub. Seto Mar. Biol. Lab.*, 16:405-410.

Berry, F.H. 1958. A new species of fish from the western North Atlantic, *Dikellorhynchus tropidolepis*, and relationships of the genera *Dikellorhynchus* and *Malacanthus*. *Copeia*, 1958:116-125.

Clark, E. and A. Ben-Tuvia. 1973. Red Sea fishes of the family Branchiostegidae with a description of a new genus and species *Assymeturus oreni*. *Bull. Sea Fish. Res. Stat., Haifa*, 60:63-74.

Fourmanoir, P. 1970. Notes ichthyologiques (I). *Cah. ORSTOM Océanogr.*, 8:19-33.

Fourmanoir, P. 1976. Formes post-larvaires et juveniles de poissons cotiers pris au chalut pelagique dans le sud-ouest Pacifique. *Cah. du Pacifique*, (19):47-88.

Hubbs, C.L. 1958. *Dikellorhynchus* and *Kanazawaichthys*: nominal fish genera interpreted as based on prejuveniles of *Malacanthus* and *Antennarius*, respectively. *Copeia*, 1958:282-285.

Okiyama, M. 1964. Early life history of the Japanese Blanquillos, *Branchiostegus japonicus japonicus* (Houttyn). *Bull. Japan. Sea Reg. Fish. Res. Lab.*, (13):1-14.

Smith, J.L.B. 1956. An extraordinary fish from South Africa. *Ann. Mag. Nat. Hist., Ser. 12*, 9:54-57.

Thresher, R.E. 1980. *Reef Fish.* Palmetto Publ. Co., St. Petersburg, Fla. 172 pp.

Chubs
(Kyphosidae)

Sea chubs are oval-shaped, laterally compressed fishes commonly seen on the reef in small to large schools swimming several meters above the bottom. There are about ten species in the family, in three genera, all of which tend to be silvery or pale-colored, usually with some indication of faint horizontal lines. All are herbivorous, feeding on a combination of planktonic and benthic algae. The family is distributed circumtropically and is thought to be closely related to the more temperate family Girellidae.

Not much is known about reproduction in the kyphosids. Moore (1962) collected juveniles year-round for two species in the western Atlantic, suggesting that some spawning occurs year-round. The eggs of *Kyphosus vaigiensis* from off Hawaii are spherical, planktonic, and about 1.0 to 1.1 mm in diameter (Miller, et al., 1979). Virtually nothing is known about larval development in these fishes, but planktonic prejuveniles of several species are commonly collected under floating masses of *Sargassum*. Such juveniles tend to be mottled brown or green and resemble the floating algae (Randall & Randall, 1960).

Literature Cited

Miller, J.M., W. Watson, and J.M. Leis. 1979. An atlas of common nearshore marine fish larvae of the Hawaiian Islands. *SeaGrant Misc. Rpt.* (INIHI SeaGrant-MR-80-02).

Moore, D. 1962. Development, distribution, and comparison of rudder fishes *Kyphosus sectatrix* (Linnaeus) and *K. incisor* (Cuvier) in the western North Atlantic. *Fish. Bull. (U.S.)*, 61:451-480.

Randall, J.E. and H.A. Randall. 1960. Examples of mimicry and protective resemblance in tropical marine fishes. *Bull. Mar. Sci.*, 10:444-480.

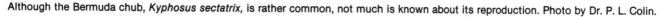

Although the Bermuda chub, *Kyphosus sectatrix,* is rather common, not much is known about its reproduction. Photo by Dr. P. L. Colin.

Scorpionfishes
(Scorpaenidae)

Scorpionfishes are heavy-bodied ambush predators frequently seen sitting quietly on the reef, apparently waiting for small fishes to swim by. Although probably abundant, they are frequently overlooked due to their remarkable camouflage; even closeup one can easily mistake the fish for an algae-covered rock due to the mottled colors and numerous algae-like fleshy appendages that cover the body. Scorpionfishes derive their common name from the poisonous spines present in the dorsal fin. Most species are not deadly but can give a painful wound if stepped on. There are about 60 genera and perhaps 300 species in the family, which is distributed worldwide in the tropics and temperate seas. The dominant reef genus, however, is *Scorpaena*. The pteroids, which are often included in the family Scorpaenidae, are treated separately in the next chapter.

Spawning by the North Pacific *Sebastes* (= *Sebastodes*) has been extremely well studied, in large part because of the commercial value of the fishes. All members of the genus are livebearers (ovoviviparous), and the mechanisms involved have been examined in great detail (e.g., De Lacy, et al., 1964; Moser, 1967; Igarashi, 1968a & b; Shimizu & Yamada, 1980). A discussion on the development of the young and general information about reproduction are provided by Fujita (1957, 1958, 1959), Hitz (1962), Moser (1967, 1972, 1974), Takai & Fukunaga (1971), and Sasaki (1974). The species in *Helicolenus* are apparently also livebearing (Thompson & Anderton, 1921; Graham, 1939).

Species of the genus *Scorpaena* are not livebearers, but instead produce floating, gelatinous egg masses much like those of the closely related pteroids. Spawning has not been observed in any species of the genus, but David (1939) suggested that in *S. guttata* it occurs around midnight. The egg mass produced is thin-walled, hollow, and bilobed, with the eggs imbedded in a single layer (Sparta, 1941; Orton, 1955; Miller, et

al., 1979). Similar egg masses are reported for *Sebastolobus* by Pearcy (1962) and Moser (1974), and for *Inimicus* by Fujita & Nakuhara (1955). Individual eggs range in size from 0.7 to 1.2 mm. Hatching in *S. guttata* occurs in 58 to 72 hours, producing a 1.9-2.0 mm larva. Each larva bears a large yolk sac that extends no farther forward than the leading edge of the eye. The yolk sac is absorbed in about three days. Older larvae are described for *Sebastolobus* by Moser (1974) and for *Scorpaena* by Fourmanoir (1976) and Miller, Watson & Leis (1979). Such larvae have large and usually pigmented pectoral fins and, in early stages, a dermal sac that encloses most of the body. At lengths greater than about 5 mm each larva has a pair of prominent horn-like spines on its head, surrounded by a large number of smaller spines. The duration of the planktonic larval stage is not known.

Orton (1955) reported that *S. guttata* spawned repeatedly in captivity in a large public aquarium. No details were provided.

Nothing is known about reproduction in the related family Synancejidae (the stonefishes).

Literature Cited

David, L.R. 1939. Embryonic and early larval stages of the grunion, *Leuresthes tenuis,* and the sculpin, *Scorpaena guttata. Copeia,* 1939:75-81.

De Lacy, A.C., C.R. Hitz, and R.L. Dryfoos. 1964. Maturation, gestation and birth of rockfish (*Sebastodes*) from Washington and adjacent waters. *Fish. Res. Pap. State Wash.,* 2:51-67.

Fourmanoir, P. 1976. Formes post-larvaires et juveniles de poissons cotiers pris au chalut pelagique dans le sud-ouest Pacifique. *Cah. du Pacifique,* (19):47-88.

Fujita, S. 1957. On the larval stages of a scorpaenid fish, *Sebastodes pachycephalus nigricans* (Schmidt). *Japan. J. Ichthyol.,* 6:91-93.

Fujita, S. 1958. On the egg development and larval

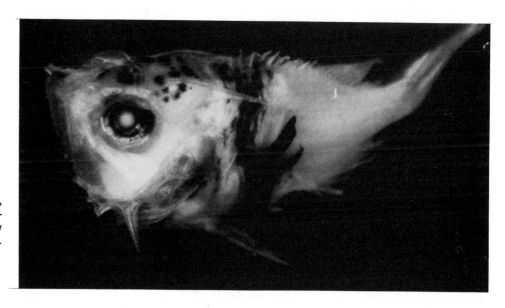

Postlarval scorpionfish, 9.0 mm total length, collected in a plankton tow off Miami, Florida.

stages of a viviparous scorpaenid fish, *Sebastes oblongus* Günther. *Bull. Jap. Soc. Sci. Fish.*, 24:475-479.

Fujita, S. 1959. On the egg development of a viviparous scorpaenid fish, *Sebastes pachycephalus nigricans* (Schmidt). *Jour. Jap. Assoc. Zool. Gard. Aquar.*, 1:42-43.

Fujita, S. and K. Nakuhara. 1955. On the development of the egg and prelarval stages of a scorpaenid fish, *Inimicus japonicus* (Cuvier et Valenciennes). *Sci. Bull. Fac. Agric. Kyushu Univ.*, 15:233-228.

Graham, D.H. 1939. Breeding habits of the fishes of Otago Harbour and adjacent seas. *Trans. Royal Soc. N. Zeal.*, 69:361-372.

Hitz, C.R. 1962. Seasons of birth of rockfish (*Sebastodes* spp.) in Oregon coastal waters. *Trans. Amer. Fish. Soc.*, 91:231-233.

Igarashi, T. 1968a. Ecological studies on a marine ovoviviparous teleost, *Sebastes taczanowskii* (Steindachner): seasonal changes of the testis. *Bull. Fac. Fish. Hokkaido Univ.*, 19:19-26.

Igarashi, T. 1968b. Ecological studies on a marine ovoviviparous teleost, *Sebastes taczanowskii* (Steindachner): about the copulatory organ. *Bull. Fac. Fish. Hokkaido Univ.*, 19:27-31.

Miller, J.M., W. Watson, and J.M. Leis. 1979. An atlas of common nearshore marine fish larvae of the Hawaiian Islands. *Sea Grant Misc. Rept.* (INIHI SeaGrant-MR-80-02).

Moser, H.G. 1967. Reproduction and development of *Sebastodes paucispinis* and comparison with other rockfishes off southern California. *Copeia*, 1967:773-797.

Moser, H.G. 1972. Development and geographic distribution of the rockfish, *Sebastes macdonaldi* (Eigenmann and Beeson, 1893), family Scorpaenidae, off Southern California and Baja California. *Fish. Bull. (U.S.)*, 70:941-958.

Moser, H.G. 1974. Development and distribution of larvae and juveniles of *Sebastolobus* (Pisces; family Scorpaenidae). *Fish. Bull. (U.S.)*, 72:865-884.

Orton, G.L. 1955. Early developmental stages of the California scorpionfish *Scorpaena guttata*. *Copeia*, 1955:210-214.

Pearcy, W.G. 1962. Egg masses and early developmental stages of the scorpaenid fish, *Sebastolobus*. *J. Fish. Res. Bd. Canada*, 19:1169-1173.

Sasaki, T. 1974. On the larvae of three species of rockfish (genus: *Sebastes*) in Hokkaido. *Bull. Fac. Fish. Hokkaido Univ.*, 25:169-173.

Shimizu, M. and J. Yamada. 1980. Ultrastructural aspects of yolk absorption in the vitelline syncytium of the embryonic rockfish, *Sebastes schlegeli*. *Japan. J. Ichthyol.*, 27:56.

Sparta, A. 1941. Contributo alla conoscenza di uova, stadi embrionali e post-embrionali negli Scorpenidi. II. *Scorpaena scrofa* L. III. *Scorpaena ustulate* Lowe. IV. *Scorpaena dactiloptera* Delaroche. *Arch. Oceanogr. Limnol., Rome*, Vol. 1, Fasc. 3, Mem. 292, pp. 203-210.

Takai, T. and T. Fukunaga. 1971. The life history of a ovo-viviparous scorpaenoid fish, *Sebastes longispinis* (Matsubara). I. Egg and larval stages. *J. Shimonoseki Univ. Fish.*, 20:91-95.

Thomson, G.M. and T. Anderton. 1921. History of the Portobello Marine Fish-Hatchery and Biological Station. *Bull. Board Sci. Art, New Zealand, Wellington*, (2), pp. 1-131.

Lionfishes
(Pteroidae)

Lionfishes are the scorpaeniform fishes most closely associated in the popular mind with coral reefs, although, in fact, the less spectacular scorpionfishes proper (Scorpaenidae) are far more abundant on coral reefs, if less conspicuous. Like the scorpionfishes, lionfishes are essentially solitary predators that feed on both small fishes and invertebrates. Unlike the scorpionfishes, the lionfishes are mainly active at night; by day they are found in caves or under ledges, either "sitting" quietly on the roof or hovering in one corner. Whether active or hovering, however, lionfishes are instantly recognizable fishes, combining a robust body with large fan-like pectoral fins, long (often separate) dorsal spines, and a subdued but striking color pattern of lines and spots. Their striking appearance has long made them a favorite with aquarists despite the need for regular feedings with live food and often a tank of their own. There are two genera in the family, both found only in the Indo-West Pacific: *Dendrochirus*, the dwarf lionfishes, and *Pterois*, the lionfishes or turkeyfishes. There are less than a dozen species in the family.

Rather remarkably for such a small family, and especially for one that is nocturnally active, spawning by lionfishes has been described for four species: *Dendrochirus brachypterus* by Fishelson (1975) and Brandt (1976), *D. zebra* by Moyer & Zaiser (1981), *Pterois volitans* by Fishelson (1975), and *P. radiata* by Fishelson (1975) and A. Gronell (pers. comm.). Overall, courtship and spawning are similar in all four species, though there are significant differences between them. As noted, all are retiring by day and do not appear on the reef until late in the afternoon or early in the evening. *Dendrochirus* are often found in small groups during the day; in *D. brachypterus*, at least, these groups often consist of a large male, several females, and several small males. Fishelson interprets this as a harem-like social structure; more detailed work by Moyer & Zaiser, involving individ-

ually recognizable fish, however, suggests instead that the grouping of fishes by day has little or no bearing on their spawning patterns.

Courtship and spawning activities begin 20 to 40 minutes after sunset and occur each day, i.e., available evidence suggests at best a weak semi-lunar cycle of activity (Moyer & Zaiser, 1981). Males are extremely aggressive prior to and during reproductive activities, and fights between males involving various displays, biting, and ramming with the dorsal spines are common. Such males will even threaten divers, which can result in some stimulating moments underwater. Dominant fish are generally darkly colored; subordinate fish remain light-colored or pale when threatened. Each dominant male appears to patrol a large home range from which he drives other males (though such home ranges can overlap, in part accounting for the high frequency of fights) and within which he courts any female encountered. Most species are normally sexually monomorphic; the only permanent dimorphism thus far described in the family occurs in *D. brachypterus*, in which the pectoral fins of the males are longer than those of the females (those of a male reach to the base or middle of the caudal peduncle and have six to ten bands of color, whereas the pectoral fins of a female barely reach the caudal peduncle and have only four to six bands); males also have a more robust head (Fishelson, 1975). All species, however, develop temporary color differences during courtship and spawning. In *D. brachypterus* the male darkens and the female lightens, at the same time developing an oblique bright white line that runs between the eyes and the conspicuous papilla on the lower jaw. In *D. zebra* both sexes develop such lines, but the female also develops a bright silver tip on the papilla and similar tips on the first three dorsal spines. Males of the two species of *Pterois*, like *D. brachypterus*, become darker than females. In all species the females also develop a conspicuously swollen abdomen before spawning, and the abdomen often has a silvery sheen to it.

When a male encounters a female he swims to her and may circle her slowly. Positioning himself beside her, he then swims toward the surface; if she is ready, she follows. In *D. brachypterus* and, apparently, the two species of *Pterois*, such paired ascents may occur several times before spawning occurs. The latter takes place close to the surface (see below) and is signalled by the conspicuous enlargement of the female's genital pore. The male presses his body against that of the female, often pushing her surfaceward. The female quickly extrudes a pair of floating balls of eggs. After spawning, the female descends to the bottom, while the male begins to search for more females. Spawning by *D. zebra* off Japan is similar but is much faster and involves an upward ascent of only about 60 cm. Moyer & Zaiser (1981) speculated that this difference in ascent height and speed is the result of an increased likelihood of the adults being preyed upon by sharks at that locality. Alternatively, it is not clear from Fishelson's description how many spawnings were observed in the field as opposed to the aquarium; if most or all occurred in captivity, then spawning by *D. brachypterus* near the surface may have been an artifact resulting from the relatively shallow waters

available to the fish (J.T. Moyer, pers. comm.). In any case, females are apparently able to spawn repeatedly at roughly three-day intervals.

Spawning in the field has also been observed in *Pterois radiata* by A. Gronell (pers. comm.) at Enewetak Atoll, Marshall Islands. Shortly after sunset (about ten minutes) a pair of individuals were seen to emerge from a crevice in the reef; the male was estimated to be about 12 cm long and the female 8 to 10 cm long. The male, which had been seen in the same area on previous nights, though never watched for any length of time, appeared jet black; illuminated with a dive light, he turned out to actually be a very deep red. The female, in contrast, was extremely pale, almost white. After emerging, the pair swam away from the cave, the female in front of the male, both with their fins fully spread. They moved about a half meter or so, then turned back and swam to a small coral pinnacle near their shelter hole. The female swam slowly up the side of the coral head, followed by the male. About two-thirds of the way up the male moved forward, next to the female, and the pair abruptly swam, quickly but not extremely fast, about a meter upward and away from the coral. At the peak of the ascent, they suddenly turned downward and very rapidly dashed to the bottom. Once there, the male rapid-

Dendrochirus brachypterus females are reported to have shorter pectoral fins with fewer bands. This fish is apparently a female. Photo by H. Hansen, Aquarium Berlin.

ly swam off, while the female settled quietly to the bottom and then, after several minutes, backed into a hole in the reef. Spawning occurred almost exactly 15 minutes after sunset.

All pteroids, so far as is known, produce gelatinous egg balls similar to those of scorpionfishes. Fishelson (1975) discussed in detail the morphology of the egg balls and related it to the internal structure of the ovaries (additional information on ovary structure and oogenesis in *D. brachypterus* was provided in Fishelson, 1977 and 1978). Such balls contain from 2,000 to 15,000 spherical eggs imbedded in a gelatinous matrix which, in some species, may be distasteful (Moyer & Zaiser, 1981). Shortly after extrusion each ball swells to a size of 2.0 to 5.0 cm in diameter. Each egg is about 0.8 mm in diameter and contains a single oil droplet. Its development, and that of the early larvae, was discussed by Mito & Uchida (1958) and Fishelson (1975). Hatching in *D. brachypterus* occurs in 36 hours at 26°C; hatching of *P. lunulata* occurs in 24 hours at 27 to 30°C. The newly hatched prolarvae are, respectively, 1.0 to 1.25 and 1.52 to 1.58 mm long. Each has a large yolk sac that contains a single posteriorly located oil droplet. Feeding begins four days after hatching, at which time the larvae are long and slender, have prominent dark eyes, and have very large pectoral fin rudiments, each with a dark outer margin. Larvae have not yet been reared past this stage, but Fishelson also described a 6 mm individual collected from the water column. The larva was readily identifiable as a pteroid due to its extremely long free pectoral fin rays. Duration of the larval stage is not known, but settlement to the bottom occurs when the juveniles are 10 to 12 mm long. Such juveniles are extremely cryptic.

Spawning by pteroids in aquaria has been reported by Fishelson (1975, 1978) and Brandt (1976). In the latter case, dealing with *Dendrochirus brachypterus*, only a few details are provided: a single pair was present, the fish spawned in a 50-gallon (227 L) aquarium filtered by a sub-sand system, and the water temperature was 74°F (23°C); spawning occurred very late at night and was repeated at three-day intervals for several weeks. Fishelson reported that his fish were kept in 55- to 165-gallon (250 to 750 L) aquaria containing sub-sand filters. The pH was approximately 8.0 and the water temperature varied between 23 and 26°C. The fish were fed small live fishes, shrimps, and live adult brine shrimp daily. According to Fishelson (1977), spawning occurs at dawn (though see field observations by Moyer & Zaiser, 1981) and eight to ten times for each female during the year. Pteroid eggs were hatched in small containers that had two drops of a 1:10 solution of penicillin and streptomycin per liter of water. No larvae lived longer than six days.

Literature Cited

Brandt, C. 1976. Sea Spawnings—Dwarf Lionfishes. *Octopus*, 3:4-7.

Fishelson, L. 1975. Ethology and reproduction of the pteroid fishes found in the Gulf of Aqaba (Red Sea), especially *Dendrochirus brachypterus* (Cuvier) (Pteroidae, Teleostei). *Publ. Staz. Zool. Napoli*, 39 (Suppl.) :635-656.

Fishelson, L. 1977. Ultrastructure of the epithelium of the ovary wall of *Dendrochirus brachypterus* (Pteroidae, Teleostei). *Cell Tissue Res.*, 177:375-381.

Fishelson, L. 1978. Oogenesis and spawn-formation in the Pigmy Lion Fish *Dendrochirus brachypterus* (Pteroidae). *Mar. Biol.*, 46:341-348.

Mito, S. and K. Uchida. 1958. On the egg development and hatched larvae of a scorpaenid fish, *Pterois lunulata* Temminck et Schlegel. *Sci. Bull. Fac. Agric., Kyushu Univ.*, 16:381-385.

Moyer, J.T. and M.J. Zaiser. 1981. Social organization and spawning behavior of the pteroine fish *Dendrochirus zebra* at Miyake-jima, Japan. *Japan. J. Ichthyol.*, 28:52-69.

Hawkfishes
(Cirrhitidae)

Hawkfishes are generally small colorful fishes that are usually seen on the reef perched on top of a coral head or scurrying about on a rock outcrop. Heavy-bodied and distinctly benthic, all are active predators, pouncing upon a variety of small invertebrates and fishes in a sudden swoop from their coral perches. Given their boldness and generally peaceful manner, it is not surprising that many of the more colorful species are regularly imported as aquarium fishes.

Hawkfishes range in size at maturity from less than 10 cm to almost a meter. Robust and large-headed, they are easily distinguished from other benthic fish groups, such as the blenniids, by their generally more bass-like appearance. Specifically, hawkfishes are characterized by a combination of features: thickened and separate lower pectoral fin rays, a continuous dorsal fin with prominent spinous and soft portions, conspicuous scaling, and one or more hair-like cirri projecting from the interspinous membranes near the top of the dorsal spines (these cirri are often conspicuously colored and easily seen). Randall (1963) recognized ten genera and 34 species of cirrhitids, all found in the tropics and all closely associated with coral reefs. Most are Indo-West Pacific; the Atlantic boasts only two species, the common and colorful western Atlantic *Amblycirrhites pinos* and a relatively drab species, *Cirrhitus atlanticus*, thus far reported only from islands off the coast of Africa.

SEXUAL DIMORPHISM

There are few reports of sexual differences in cirrhitids, and most species appear to be externally monomorphic. Takeshita (1975) suggested that males of the longsnout hawkfish, *Oxycirrhites typus*, are smaller than females (9 cm SL versus 10.8 cm), have darker red lower jaws, and have black edges on the pelvic and caudal fins that are lacking in females. These observations have yet to be confirmed. Schultz, et al. (1960) suggested that *Paracirrhites polystictus* and *P. hemistic-*

tus were, respectively, male and female of a single dichromatic species, but Randall (1963) subsequently found both sexes in both color phases.

The apparent nonsexual dichromatism of *P. hemistictus* may be comparable to that in the widely distributed *Cirrhitichthys oxycephalus*, which also has two color phases: one with vivid red spots on a pink background (the so-called spotted hawkfish) and one with brown spots on a white background (the so-called leopard hawkfish). These color differences are not related to sex (indeed, there is no conspicuous external sexual dimorphism in the species) but rather are a function of habitat (pers. obs.). Examination of the gonads of several groups collected from isolated coral heads revealed: first, that there was never more than one male in a group; second, that the male was always the largest individual present; and third, that the males of some groups were smaller than the largest females in other groups. (The groups ranged in size from three to nine individuals.) These results suggest socially controlled protogynous hermaphroditism, perhaps similar to that of some labrids and pomacanthids. Histological examination of hawkfish gonads may prove informative. (In *Cirrhitichthys oxycephalus* the ovaries are paired, white, cylindrical, and granular, while the testes are paired, slender, and flattened.)

SPAWNING SEASON

There is little information on spawning seasons of cirrhitids. In the warm-temperate Gulf of California, *Cirrhitichthys oxycephalus* spawns daily throughout the warmer parts of the year (late fall and early winter). There is no conspicuous lunar rhythm.

REPRODUCTIVE BEHAVIOR

The only literature on cirrhitid spawning concerns the longsnout hawkfish, *Oxycirrhites typus*, a species characteristically found in pairs in relatively deep water (Takeshita, 1975; Savitt, 1976). Lobel (1974) re-

Courtship and spawning in the eastern Pacific hawkfish *Cirrhitichthys oxycephalus*. Prior to spawning, the male moves rapidly from female to female within his territory, nudging each on the abdomen and apparently soliciting spawning. Females ready to spawn move to the top of prominent coral heads and await the return of the male. The male moves beside and slightly above the female, lying against her side (a). Spawning involves an extremely fast ascent into the water column, at the peak of which gametes are shed (b). A second female (c) waits to spawn.

ported that in an aquarium such a pair produced two "nickel-sized" patches of adhesive eggs. Spawning and courtship were not observed, nor were any details given of parental care. Lobel, however, was reasonably sure the eggs observed were produced by the hawkfish since they were the only fish present (Lobel, pers. comm.).

Spawning by *Cirrhitichthys oxycephalus* in the Gulf of California occurs daily at dusk during the warmer half of the year (pers. obs.). Its social system is based on male territoriality and male dominated harems. The size of the harem varies with the size of the territory: on small isolated coral heads only one or two females are present; on larger ones or in areas of continuous coral there may be six or seven females in a harem. Females are not territorial but seem to remain in preferred parts of the male's territory. At dusk, roughly 20 minutes before complete darkness (and well after sunset), the male, who has largely ignored the females all day, begins to show great interest in them and shuttles quickly from one to another, remaining with each for only a minute or so. During this "census" the male may be determining which, if any, females will spawn that night. Just prior to complete darkness each ripe female moves to await the male at the top of a prominent coral knoll in her preferred area (though if only one such area is available in the male's territory, all females will move there). When he arrives he nudges the female's abdomen with his snout and the two frequently hop about the spawning site, with the male close behind the female. She quickly sets herself at the highest spot in the area, angled slightly head-up; the male moves to a spot slightly behind, beside, and above her, and both sit quietly for 10 to 30 seconds. Then together they rapidly dash into the water column, releasing gametes at the peak of the dash, 0.75 to 1.0 m off the bottom. The female then leaves the spawning site and the male moves immediately to another female to repeat the sequence. Spawning occurs simultaneously all across the reef, with all spawning rushes completed within a brief three or four minutes. By the time the last one is completed, it is usually so dark that it is difficult to see the fish.

This spawning behavior is very different from that reported by Lobel and casts doubts on the latter. Though possible, it is unlikely that reproductive biology in a family as small and clearly monophyletic as the cirrhitids would vary so widely as to include both demersal and pelagic spawners. Further, production of pelagic eggs by the cirrhitids fits their apparent phyletic position, i.e., between scorpaenids and serranids, both of which have species that produce pelagic eggs. Finally, Takeshita (1974) reported behavior of *O. typus* in aquaria that suggests they spawn in a manner similar to *C. oxycephalus*. The fish were reported to be most active in the early evening, at which time they "courted" by swimming about each other in tight circles (nudging?) and then suddenly stopping, sitting side-by-side. Spawning was never observed (nor were demersal eggs found), but Takeshita reported that he did not watch the fish late in the evening. Based on the behavior of *C. oxycephalus*, spawning may have occurred at that time or after dark. Similar pairing and increased levels of activity at dusk also occur in the eastern Pacific species *Cirrhitus rivulatus*, though no conspicuous courtship or spawning was observed. These large hawkfish are very shy, however, and the presence of the diver clearly disturbed them. Similar courtship at dusk, though not yet spawning, has also been observed in *Paracirrhites arcatus* at One Tree Island, Great Barrier Reef (pers. obs.).

EGGS AND LARVAE

Neither cirrhitid eggs nor larvae have been described beyond Lobel's brief statement about adhesive eggs in *Oxycirrhites typus*. Apparently ripe ova of *Cirrhitichthys oxycephalus* are spherical and approximately 0.5 mm in diameter (pers. obs.). Their size after fertilization is not known. What appeared to be a postlarval *Amblycirrhites pinos* was observed while nightlighting off Puerto Rico (pers. obs.). It was approximately 4 cm long and slightly laterally compressed, with a slender pointed head, a terminal mouth, and enlarged pectoral fins. The fish was pale with faint red margins on its caudal fin. Similar appearing large postlarvae (30 mm and 31 mm) have been described for *Oxycirrhites* sp. and *Cyprinocirrhites polyactis* by Formanoir (1971, 1973). Duration of the planktonic larval stage is not known, but based on the size of these apparent postlarvae, it may be fairly long. A lengthy pelagic stage is also suggested by the widespread distribution and limited geographic variation of some species. *Cirrhitichthys oxycephalus*, for example, is found from the eastern Pacific to the Red Sea and varies only slightly throughout this range (Fröiland, 1976).

Juveniles of most species resemble the adults. In the few species where there is some color change with maturity (e.g., *Cirrhitus rivulatus*, *Paracirrhites forsteri*), the juveniles are more brightly and more crisply colored than the adults (e.g., Savitt, 1976).

REPRODUCTION IN CAPTIVITY

Aside from Takeshita's observations of apparent courtship in *Oxycirrhites typus* and Lobel's questionable account of demersal eggs in the same species,

Postlarval hawkfish tentatively identified as *Cyprinocirrhites polyactis.* Based on Fourmanoir (1973).

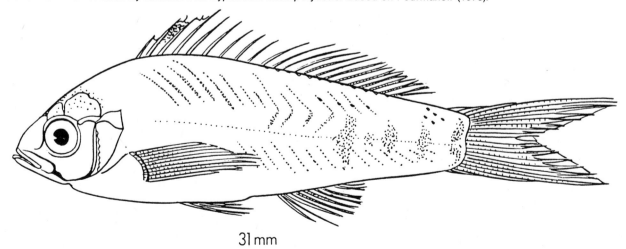

31 mm

there are no reports of spawning by cirrhitids in captivity. The absence of such reports seems strange considering that many hawkfishes, such as *O. typus, Amblycirrhites pinos,* and the colorful *Neocirrhites armatus,* are both small at maturity (less than 10 cm SL) and strongly site-attached to a small portion of the reef (e.g., Carlson, 1975), and so would probably not be unusually stressed by the confines of an aquarium. Indeed, if protogynous hermaphroditism is also characteristic of the family, there might not even be a problem in obtaining a pair. In part, the absence of reports of cirrhitid spawning may be due to its likely occurrence very late in the evening or even after dark. Spawning might be facilitated or at least more readily observed by a gradual reduction of light levels in the evening rather than an abrupt onset of darkness. Otherwise spawning hawkfish should require little more than a reasonably large aquarium (250 liters or larger), heavy feedings of live food, privacy, and good water conditions.

With so little known about eggs and larvae, it is difficult to predict the degree of difficulty of rearing the fishes in captivity. If a long planktonic stage is characteristic of the family, however, rearing them is likely to be difficult.

Literature Cited

Carlson, B.A. 1975. A scarlet hawkfish from the Fiji Islands. *Tropical Fish Hobbyist,* 23(8):36-40.

Fourmanoir, P. 1971. Notes Ichthyologiques (III). *Cah. O.R.S.T.O.M., Ser. Oceanogr.,* 9(2):267-278.

Fourmanoir, P. 1973. Notes Ichthyologiques (V). *Cah. O.R.S.T.O.M., Ser. Oceanogr.,* 11(1):33-39.

Fröiland, O. 1976. Litoralfische der Malediven. V. The hawkfishes of the family Cirrhitidae (Pisces: Perciformes: Percoidei). *Senckenberg. biol.,* 57:15-23.

Lobel, P.S. 1974. Sea Spawnings—hawkfish. *Octopus,* 1(7):23.

Randall, J.E. 1963. Review of the hawkfishes (family Cirrhitidae). *Proc. U.S. Nat. Mus.,* 114(3472): 389-401.

Savitt, D. 1976. Hawk fish. *Marine Aquarist,* 7(4): 17-26.

Schultz, L.P. and collaborators. 1960. Fishes of the Marshall and Marianas Islands. *U.S. Nat. Mus. Bull.,* (202):438 pp.

Takeshita, G.Y. 1975. Long-snouted hawkfish. *Marine Aquarist,* 6(6):27-31.

Morwongs
(Cheilodactylidae)

Cheilodactylids are laterally compressed bottom-oriented fishes that are primarily found in the southern hemisphere and only sporadically on the reef. There are about 15 species in six genera, of which the best known are *Cheilodactylus* and *Goniistius*, the latter being the genus most often found on the reef. The fishes are characterized by an elongate body, prominent down-turned terminal mouth, a long dorsal fin with conspicuous spines, and large pelvic fins with several free rays. Because of the latter, the cheilodactylids are often placed close to the hawkfishes systematically, as hawkfishes also have free pectoral rays. Cheilodactylids in general are benthic carnivores (e.g., Bell, 1979) and appear to be most active at dusk.

Nothing is known about reproduction of reef-associated cheilodactylids. The biology of the temperate species, however, has been studied in some detail. Spawning by such species takes place in the summer (Han, 1964a & b; Tong & Vooren, 1972; Robertson, 1978). Females tend to be larger than males and to mature at a later age (five years, as opposed to four). Despite the size difference, there is no evidence of hermaphroditism (Han, 1964b; Tong & Vooren, 1972). Spawning takes place at night at a depth of 50 to 70 m, during which time they release positively buoyant spherical eggs 0.9 to 1.13 mm in diameter (Robertson, 1975 & 1978; Mito, 1966). Robertson (1978) described yolk-sac larvae 20 hours after hatching as elongate, with a small oval yolk sac containing an oil globule at the anteriormost point and numerous dark pigment spots. Older larvae have not been described.

Literature Cited

Bell, J.D. 1979. Observations on the diet of red morwong, *Cheilodactylus fuscus* Castelnau (Pisces: Cheilodactylidae). *Aust. J. Mar. Freshwater Res.*, 30:129-133.

Han, V.C.F. 1964a. Jackass Fish (*Nemadactylus macropterus*). *Rpt. Div. Fish. Oceanogr.*, CSIRO Aust., 1963-64:37-38.

Han, V.C.F. 1964b. Studies on the biology and fishery of the jackass fish (*Nemadactylus macropterus* Bloch and Schneider, 1801) in eastern Australia. Master's Thesis, Univ. Sydney, Australia.

Mito, S. 1966. Fish eggs and larvae. *Illustrations of the Marine Plankton of Japan*, 7:74 pp.

Robertson, D.A. 1975. A key to the planktonic eggs of some New Zealand marine teleosts. *Occas. Publ. #9, Fish. Res. Div., Ministry Agricult. Fish., N.Z.* 19 pp.

Robertson, D.A. 1978. Spawning of tarakihi (Pisces: Cheilodactylidae) in New Zealand waters. *N.Z. J. Mar. Freshwater Res.*, 12:277-286.

Tong, L.J. and C.M. Vooren. 1972. The biology of the New Zealand tarakihi *Cheilodactylus macropterus* (Bloch and Schneider). *Fish. Res. Bull. (N.Z. Ministry Agric. Fish.)*, (6):60 pp.

Groupers and Rock Cod
(Serranidae: Epinephelinae)

The epinephelines are the large (often very large) sea basses that are the most common top level predators on coral reefs. Variously referred to as groupers, coral trout, rock cod, and gag, all are characterized by a relatively elongate heavy body and large prominent jaws, along with the advanced percoid features that typify the Serranidae. Most are also cryptically colored, usually gray, brown, or black, and often with spots or a mottled pattern. Such coloration befits their roles as ambush predators, striking from cover at the smaller fishes and crustaceans that make up the bulk of their diet (Hobson, 1968). A few species are quite colorful, such as the red and blue-spotted Indo-Pacific coral trout, *Cephalopholis miniatus*, while others are polymorphic, exhibiting an often wide range of color patterns. Why such fishes should differ from the epinepheline norm is not known.

Groupers range in maximum size from only about 25 cm to well over 3 m, with a weight to over 400 kg. The "giant groupers," *Epinephelus itajara* in the western Atlantic and *E. tauvina* in the Indo-West Pacific, are truly awesome animals and are common favorites in oceanarium-style public aquaria. Despite their intimidating size, fearsome appearance, and normal epinepheline curiosity (in undisturbed areas, groupers often allow divers to approach closely and even emerge from their daytime refuges to "inspect" the strange bubble-producing apparition), even such large species are harmless (despite occasional stories about divers being swallowed whole). The larger fishes, in fact, are usually quite shy.

Because of their commercial and sporting importance, a fair amount of fisheries-type work has been done on the biology of the epinephelines. Consequently, at least the basics of their reproductive biology have been determined in some detail, though information about actual courtship and spawning is still sparse.

SEXUAL DIMORPHISM

Most, if not all, groupers are protogynous (female to male) hermaphrodites. Such hermaphroditism has been documented for a variety of American species of *Epinephelus, Mycteroperca,* and *Alphestes* by Smith (1959, 1961, 1964, 1965, 1971), McErlean & Smith (1964), Moe (1969), and Nagelkerken (1979); for two Mediterranean species of *Epinephelus* by Brustle & Brustle (1975, 1976); for the Indo-Pacific species E. *tauvina* by Tan & Tan (1974) and Chen, et al. (1977); and *Plectropomus leopardus* by Goeden (1978); and has been suggested for the American water-column foraging species *Paranthias furcifer* by Smith (1965). (Data for the last species are ambiguous and *Paranthias* may be secondarily gonochoristic.) Protogyny is also characteristic of two distantly related genera found in more temperate areas, *Centropristes* and *Paralabrax* (Lavenda, 1949, and Smith & Young, 1966, respectively).

The hermaphroditic gonads of a grouper are paired, lie below and slightly behind the swim bladder, and unite posteriorly to form a common oviduct. Outside of the spawning season those of males and females are virtually indistinguishable without microscopic examination (Thompson & Munro, 1978). During the spawning season, however, ovaries are easily recognizable due to their distension with eggs. In cross section each gonad is a hollow sac with numerous lamellae surrounding a central ovarian lumen and, posteriorly, a sperm duct. Unlike the gonads of the related serranines, the ovotestes of a grouper are not divided into distinct ovarian and testicular zones, but rather the spermatogenic tissue exists in dormant crypts on the lamellae of an active female. Upon sex reversal, the oocytes degenerate, the spermatogonia expand dramatically, and the sperm duct fills. Remnants of oocytes and the ovarian lumen are readily visible in functional males.

Given the widespread occurrence of sequential hermaphroditism in the subfamily, it is not surprising that, on the average, the sexes differ significantly in size. In general, the range of overlap between the sexes is considerable, however, and it appears that some females in at least a few species never change sex. Moe (1969), for example, found that active males of *Epinephelus morio* ranged in standard length from 42.5 to 70.1 cm, whereas females ranged from 27.5 to 72.2 cm; sex changes, on the average, occurred between 45.0 and 65.0 cm. Similar overlaps in size ranges are reported for various species by Burnett-Herkes (1975), Thompson & Munro (1978), Goeden (1978) and Nagelkerken (1979). In contrast, McErlean & Smith (1964), based on a small sample size, found that the males of *Mycteroperca microlepis* were near the upper limits of the species' size range; they suggested, however, that this was less a difference in sizes in the sexes than it was due to age. The youngest male they examined was 13 years old; the oldest female, in contrast, was only 11, and most were less than six years old. Similarly, in *Epinephelus morio* most individuals less than 17 years old are female, most over 17 are male, and those between 15 and 17 are about half male and half female.

The factors that have selected for protogynous hermaphroditism in the epinephelines are not well understood. Most workers (e.g., Burnett-Herkes, 1965; Thompson & Munro, 1978) suggested some social control of sex change similar to that documented for a distantly related anthiine by Fishelson (1970). The conditions characteristic of a group of anthiines, such as frequent social interactions, frequent spawnings, and high density populations in small, often discrete colonies, clearly do not characterize the larger and usually solitary epinephelines, and it is unlikely that similar social controls operate.

Beyond the average size difference between the sexes, there are no reports of conspicuous permanent differences between the sexes of any epinephelines. Temporary color changes associated with courtship and spawning, however, occur in at least a few species.

SPAWNING SEASON

Groupers typically spawn from early spring to early summer, the timing varying to a certain extent latitudinally. In the Caribbean most workers (e.g., Erdman, 1956; Randall, cited in Burnett-Herkes, 1975; Munro, et al., 1973; Thompson & Munro, 1978) reported the various species of *Epinephelus*, *Mycteroperca*, and *Paranthias* to be ripe from roughly December to late April and early May, with a peak in January to March (though see also Nagelkerken, 1979, who reported spawning by *E. cruentatus* at Curacao to occur later in the year, with a peak in August and September); spawning in the Bahamas also peaks in January and February (Thompson, cited by Burnett-Herkes, 1975). In more temperate regions, however, spawning occurs later in the year: March through July for *E. morio* in the Gulf of Mexico (Moe, 1969); April through August for various species at Bermuda (Smith, 1958); and May to July for *E. guttatus* at Bermuda (Burnett-Herkes, 1975). Similar patterns occur in the southern hemisphere. Various species of *Epinephelus*, *Cephalopholis*, and *Variola* spawn from early spring to summer off the east African coast (Nzioka, 1979), while at least one species, *Plectropomus leopardus*, is ripe only in early summer at Heron Island, near the temperate limit of the species' range (Goeden, 1978). Randall & Brock (1960), however, reported ripe specimens of *E. merra* late in the summer in Tahiti, indicating that any latitudinal effects are only a trend and clearly not a hard-and-fast rule for the subfamily. Burnett-Herkes (1975) compared the spawning seasons of Bermuda and Caribbean groupers in an attempt to determine those factors that stimulated spawning. He found no correlation between level of spawning activity and length of day, water temperature, or period of greatest abundance of plankton. The apparent latitudinal trend in spawning period suggests that tropical fishes are spawning at temperatures below the annual maximum, while more poleward populations are spawning at temperatures nearer the local maximum.

Several studies, including one on a temperate species (e.g., Neill, 1966), indicate a pronounced lunar influence on the spawning activity of groupers. Peaks of activity at or near the night of the new moon were suggested for *Epinephelus striatus* in the Bahamas by Smith (1972) and for a number of species at Palau by Johannes (1978, 1981). Similar peaks occurring at or near the full moon have been reported for *E. merra* (Randall & Brock, 1960, and Johannes, 1978, 1981) and for *E. striatus* off Belize (Gibson, cited in Johannes, 1978) and the Virgin Islands (Olsen, D.A. and J.A. LaPlace, 1979). Burnett-Herkes (1975), however, found that local fishermen at Bermuda gave inconsistent reports regarding full or new moon peaks in activity of *E. guttatus* and could find no peaks in reproductive activity between the last week in May and the first week in August. Again, a latitudinal effect may be occurring.

REPRODUCTIVE BEHAVIOR

Spawning by groupers characteristically takes place at localized "spawning grounds" to which fishes migrate for the relatively brief spawning periods. Several workers have noted that local fishermen are often well aware of such areas and have apparently fished them for generations, indicating that such areas are used repeatedly and consistently by the fishes. The use of such specific spawning areas has been reported for various species of *Epinephelus* by Bardach, et al. (1958), Moe (1969), Smith (1972), Burnett-Herkes (1975), Thompson & Munro (1978), and Johannes (1978, 1981), for *Plectropomus leopardus* by Johannes (1978, 1981), and for *Paranthias furcifer* by Burnett-Herkes (1975). Olsen & LaPlace (1979) described a spawning aggregation of approximately 2000 *E. striatus* at St. Thomas, Virgin Islands. The aggregation was cone-shaped and extended from the bottom (about 50 m) to within 20 m of the surface, and was located over a small reef separated by a narrow sand bank from surrounding reef and on the edge of the 100 fathom depth curve. The fishes near the bottom of the aggregation were color changing, as described by Smith (1972), whereas those in the water column were hovering quietly. No actual spawning was observed. Burnett-Herkes noted that at Bermuda all four local species spawn at approximately the same time but generally do so in different, though nearby, areas; *E. striatus*, for example, aggregates at 33 to 37 m while *E. guttatus* is found in depths of 18 to 27 m. The only species that showed a broad overlap in spawning areas were the benthic *E. fulvus* and the water-column forager *Paranthias furcifer*; interestingly, these are the only two epinephelines for which naturally occurring hybrids have been reported (Smith, 1966; Thompson & Munro, 1974). On the basis of these reports it is tempting to suggest that the use of specific spawning sites is a universal characteristic of epinephelines. However, identification of those species that use such areas is relatively easy and straightforward; species that do not migrate to spawn will be more difficult to identify, even if they are in the majority.

Thus far, all reported spawning grounds share two features: all are on the seaward slope of the reef, and all are close to deep water. In most cases the bottom is also well pitted with caves and ledges within which the groupers shelter during the day. Johannes (1978, 1981) also reported that such areas are usually near the mouths of channels through the reef of Palau, suggesting that the fishes select areas that facilitate off-shore transport of eggs. If so, selection of such areas has not yet been noted by other workers.

Like many pelagic-spawning reef fishes, spawning by epinephelines occurs at dusk (e.g., Ukawa, et al., 1966; Burnett-Herkes, 1975; pers. obs.), though some courtship occasionally occurs during the day (e.g., Goeden, 1978). There are only a few first-hand descriptions of spawning by groupers. The earliest report is by Ukawa, et al. (1966), who observed the Japanese species *Epinephelus akaara* spawning in outdoor ponds. Spawning occurs in pairs and is initiated by the male approaching the female and pushing his operculum against hers. The pair then spiral up into the water column, dash to the surface, and jump two or three times while releasing gametes at the surface (such jumping has not yet been reported for any other epinephelines and may be a consequence of the relatively shallow water the *E. akaara* were held in). The entire process took only one to two minutes.

Generally similar paired spawning occurs in *E. andersoni* according to aquarium observations made by Ballard (1970), and in a number of species at Palau according to Johannes (1981). A small species of *Epinephelus* observed at Heron Island, Great Barrier Reef, also pair-spawned, but did so in a somewhat different fashion (pers. obs). Late at dusk on the night of the full moon a pair consisting of one dark and one paler individual was observed on the outer reef slope at a depth of approximately 17 m. The darker individual (the male?) courted the paler one by means of an exaggerated swimming in place while oriented parallel and anti-parallel to her. It then rotated on its side and swam over the paler fish's back, tight against it, with a quickly vibrating, exaggerated "flutter-swimming." The moment it came down completely on the other side of the stationary fish, the pair suddenly dashed upward about a meter and a half into the water column, with the darker fish going up and then around the paler one before both headed back to the bottom. A gamete cloud was visible at the peak of the rush. The pair only spawned once and no other pairs were present.

Generally similar types of courtship may be widespread in the Indo-Pacific species. During courtship the male of *Epinephelus fasciatus* pales and swims in an exaggerated fashion alongside the female while tilted at an angle of 90° so that his dorsal fin is toward her (A. Gronell, pers. comm.). Similarly, *Plectropomus leopardus* males turn pale gray except for strikingly conspicuous black spots at the outer tips of the caudal fin and black edging on the dorsal and anal fins

(Goeden, 1978; pers. obs.). Again, the male rolls 90° so that either his dorsal fin or ventral surface is oriented toward the female while he swims slowly and in an exaggerated fashion alongside her. At dusk such males patrol a broad area along the edge of the outer reef slope at One Tree Island, Great Barrier Reef, assuming the courting colors and performing apparent courtship displays whenever one approaches a normally colored female, many of which were conspicuously distended with eggs. The overall behavior of the fish suggested a system based on temporary spawning territories established and patrolled by the males, between which females moved freely. To date, spawning itself has not been observed, though Johannes (1981) suggests that it involves a paired dash up into the water column.

In contrast, the common western Atlantic species *Epinephelus striatus* apparently does not form pairs while spawning. In a brief description of spawning, Manday & Fernandez (1966) noted that for several days prior to spawning the males "butted" their heads against the sides of the females. Subsequently, Smith (1972) observed an apparent spawning aggregation consisting of approximately 30,000 to 100,000 *E. striatus* milling about on the edge of a bank near Bimini, Bahamas. Roughly a third of the fish were in a conspicuous "bicolor" pattern: dark dorsally, pale ventrally, with a conspicuous white line running up the forehead (normal color pattern is pale brown with irregular vertical bands and a dark eye-line). No spawning was observed. Cousteau, however, in a popular television program entitled "The fish that swallowed Jonah," filmed such an aggregation off the coast of Belize, confirming the bicolor pattern and apparently witnessing spawning. Prior to spawning, color patterns vary widely; some fish are almost black, others white, and still others are in the bicolor pattern. As spawning approaches color changes become more frequent. During actual spawning the males are in the bicolor pattern and circle above the bottom, while the females, massively swollen with eggs and lying close to the bottom, are white. Such females rise in waves off the bottom, and as each ascends she apparently sheds her eggs into the water column while the males circle about shedding milt. Cousteau found only occasional indications of pairing and saw no circling or spiraling like that described for other species. Cousteau also reported spawning a pair in captivity in which the male, in the bicolor pattern, "mouthed" the side of a female prior to spawning, a description similar to that of Manday & Fernandez.

Female fecundity varies with size and apparently among individuals within a size-class (Burnett-Herkes, 1975; Goeden, 1978; Thompson & Munro, 1978; Nagelkerken, 1979). Several workers (e.g., Smith, 1961; McErlean, 1963; Moe, 1969; Burnett-Herkes, 1975) reported individual fecundities in excess of 1,000,000 eggs, with typical spawnings on the order of several hundred thousand eggs per female probably the norm. Moe (1969) and Nagelkerken (1979) suggested that each female spawns only once annually, but other workers all suggested multiple spawnings.

EGGS AND LARVAE

Thus far only the eggs of the genus *Epinephelus* have been described; those of other genera are presumably similar. All are spherical, pelagic, and non-adhesive, with a granular yolk and a single oil droplet. The egg is colorless. Reported egg diameters range from 0.50 to 0.65 mm (for *E. fulvus*, Thompson & Munro, 1978) to 1.024 mm (for *E. striatus*, Manday & Fernandez, 1966). Most reported diameters, however, are in the range of 0.70 to 0.90 mm (e.g., Lo Bianco, 1933; Bardach, 1958; Ukawa, et al., 1966; Barnabe, 1974; Hussain, et al., 1975; Chen, et al., 1977). Development of the egg to hatching is rapid, taking approximately a day at temperatures of 25.1 to 27.0°C (e.g., Manday & Fernandez, 1966; Ukawa, et al., 1966; Hussain, et al., 1975; Chen, et al., 1977).

Newly hatched larvae, ranging in length from 1.45 to 1.7 mm, are elongate, with a large yolk sac, no fins, unpigmented eyes, and a single oil droplet in the yolk. The oil droplet of *E. tauvina* is near the center of the yolk, while that of *E. akaara* is near its posterior margin. Feeding begins around the fourth or fifth day after hatching. Older epepheline larvae are characterized by three extremely prominent spines: one in the dorsal fin (the second spine—the third may also be elongate, but it is never more than a quarter the length of the second spine) and one in each pelvic fin. The operculum is also spiny and, like the fin spines, has prominent serrations (Ukawa, et al., 1966; Mito, et al., 1967; Presley, 1970; Aboussovan, 1972; Chen, et al., 1977; Hussian & Higuchi, 1980; Kendall, 1979; Kendall & Fahay, 1979). Such larvae are elongate, slender, and "kite"-shaped, with the body deepest at about the level of the dorsal spine. Pigmentation usually consists of a few spots on the head, a prominent spot on the caudal peduncle, and several internal spots near the intestines. Kendall & Fahay reported that the larva of the distantly related *Gonioplectrus hispanus* is similar but also has many anthiine fea-

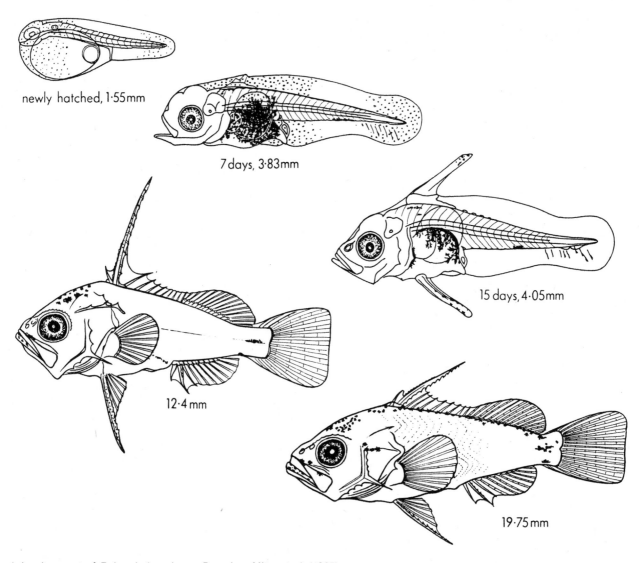

Larval development of *Epinephelus akaara*. Based on Mito, et al. (1967).

tures, such as a deep and robust body and second and third dorsal spines about equal in length.

The duration of the planktonic larval stage is roughly 30 to 40 days, near the end of which the elongate spines become progressively shorter and the larva assumes the appearance of a semitransparent juvenile. Newly settled individuals range in size from 20 to 35 mm, depending upon the species. Not much is known about the biology of such small fishes since they are generally extremely secretive. They are occasionally encountered by divers under rocks or among coral and seem to be most common in shallow water. Juvenile *Epinephelus* are cryptically colored (usually mottled brown); juvenile *Mycteroperca* are often more strik-

ingly colored, usually black with one or more white or yellow marks. Sexual maturity is normally reached in four to five years (Moe, 1969; Nagelkerken, 1979).

REPRODUCTION IN CAPTIVITY

Because of their value to commercial and sport fishing, there have been several attempts made to spawn and rear epinephelines in captivity. Successful spawnings have been made through use of very large tanks (e.g., Manday & Fernandez, 1966; Ballard, 1970; Hussain, et al., 1975) or outdoor ponds (Ukawa, et al., 1966), fish collected just at the point where they would have spawned in the field, or the use of hormones and stripping of eggs and milt (e.g., Chen, et al., 1977). The possibility of spawning even the smaller species in home aquaria is remote.

Epinepheline larvae have been reared by Mito, et al. (1967), Chen, et al (1977), and Hussain & Higuchi (1980), all using very large aquaria, and by Chua & Teng (1980), using a floating cage. Such larvae were initially fed the rotifer *Brachionus plicatilis* and then either brine shrimp nauplii or the marine cladoceran *Diaphosoma* sp. Mortality is heaviest between days five and seven, shortly after feeding begins.

Literature Cited

Aboussovan, A. 1972. Oeufs et larves de Teleosteens de l'Ouest Africain. XI. Larves serraniformes. *Bull. de Inst. Fr. Afr. Noire, Ser. A.*, 4:485-502.

Ballard, J. 1970. Aquarium briefs—mating in captivity. *Bull. South Afr. Assoc. Mar. Biol. Res.*, (8):31-33.

Bardach, J.E., C.L. Smith, and D.W. Menzel. 1958. *Bermuda Fisheries Research Program Final Report*. Bermuda Trade Development Board, Hamilton, Bermuda. 59 pp.

Barnabe, G. 1974. La reproduction du merou *Epinephelus gigas*: Observations préliminaires de terrain. *Aquacult.*, 4:363-367.

Brustle, J. and S. Brustle. 1975. Ovarian and testicular intersexuality in two protogynous Mediterranean groupers, *Epinephelus aeneus* and *Epinephelus guaza*. Pp. 222-227. *In: Intersexuality in the Animal Kingdom*, (R. Reinboth, Ed.). Springer Verlag, Berlin.

Brustle, J. and S. Brustle. 1976. Contribution à l'étude de la reproduction de deux espéces de Merous (*Epinephelus aeneus* et *Ep. guaza*) des cotes de Tunisie. *Rapp. Comm. Int. Mer. Medit.*, 23:49-50.

Burnett-Herkes, J. 1975. Contributions to the biology of the red hind, *Epinephelus guttatus*, a commercially important serranid fish from the tropical western Atlantic. Ph. D. Diss., Univ. of Miami, Coral Gables, Florida. 154 pp.

Chen, F.Y., M. Chow, T.M. Chao, and R. Lim. 1977. Artificial spawning and larval rearing of the grouper, *Epinephelus tauvina* (Forskal) in Singapore. *Singapore J. Pri. Ind.*, 5:1-21.

Chua, T.E. and S.K. Teng. 1980. Economic production of estuary grouper, *Epinephelus salmoides* Maxwell, reared in floating net cages. *Aquacult.*, 20:187-228.

Erdman, D.S. 1956. Recent fish records from Puerto Rico. *Bull. Mar. Sci.*, 6:315-340.

Fishelson, L. 1970. Protogynous sex reversal in the fish *Anthias squamipinnis* (Teleostei, Anthiidae) regulated by the presence or absence of a male fish. *Nature, Lond.*, 227:90-91.

Goeden, G.B. 1978. A monograph of the coral trout, *Plectropomus leopardus* (Lacépède). *Queensland Fish. Serv. Res. Bull.*, (1):1-42.

Hobson, E.S. 1968. Predatory behavior of some shore fishes in the Gulf of California. *Bur. Sport Fish. Wildl., Res. Rpt.* #73, 92 pp.

Hussain, N., M. Saif, and M. Ukawa. 1975. On the culture of *Epinephelus tauvina* (Forskal). *Kuwait Inst. Sci. Res., State of Kuwait.* 14 pp.

Hussain, N. and M. Higuchi. 1980. Larval rearing and development of the brown spotted grouper, *Epinephelus tauvina* (Forskal). *Aquacult.*, 19:339-350.

Johannes, R.E. 1978. Reproductive strategies of coastal marine fishes in the tropics. *Env. Biol. Fish*, 3:65-84.

Johannes, R.E. 1981. *Words of the Lagoon*. Univ. Calif. Press. Los Angeles, Calif. 320 pp.

Kendall, A.W., Jr. 1979. Morphological comparisons of North American sea bass larvae. *Tech. Rpt. N.M.F.S. Circ.* (428):50 pp.

Kendall, A.W., Jr. and M.P. Fahay. 1979. Larva of the serranid fish *Gonioplectrus hispanus* with comments on its relationships. *Bull. Mar. Sci.*, 29:117-121.

Lavenda, N. 1949. Sexual differences and normal protogynous hermaphroditism in the Atlantic sea bass, *Centropristes striatus*. *Copeia*, 1949:185-194.

Lo Bianco, S. 1933. Uova, larve e stadi giovanili de Teleostei. *Publ. Staz. Zool. Napoli, Monogr.*, (38) Part 2:384 pp.

Manday, D. Giutart, and M. Juarez Fernandez. 1966. Desarrollo embrionario y primeros estadios larvales de la Cherna Criolla, *Epinephelus striatus* (Bloch) (Perciformes: Serranidae). *Acad. Cien. Cuba, Estud. Inst. Oceanol.*, 1:35-45.

McErlean, A.J. 1963. A study of the age and growth of the gag, *Mycteroperca microlepis* Goode and Bean (Pisces: Serranidae) on the west coast of Florida. *Fla. Bd. Conserv., Mar. Lab. Tech. Ser.*, (41):1-29.

McErlean, A.J. and C.L. Smith. 1964. The age of sexual succession in the protogynous hermaphrodite, *Mycteroperca microlepis*. *Trans. Amer. Fish. Soc.*, 93:301-302.

Mito, S., M. Ukawa, and M. Higuchi. 1967. On the larval and young stages of a serranid fish, *Epinephelus akaara* (Temminck et Schlegel). *Bull. Naikai Region. Fish. Res. Lab.*, 25:337-347.

Moe, M.A., Jr. 1969. Biology of the red grouper,

Epinephelus morio (Valenciennes) from the eastern Gulf of Mexico. *Fla. Dept. Nat. Resources Mar. Res. Lab., Prof. Pap.,* (10):95 pp.

Munro, J.L., V.C. Gaut, R. Thompson, and P.H. Reeson. 1973. The spawning seasons of Caribbean reef fishes. *J. Fish Biol.,* 5:69-84.

Nagelkerken, P. 1979. Biology of the graysby, *Epinephelus cruentatus,* of the coral reef of Curacao. *Stud. Fauna Curacao,* (60):1-118.

Neill, S.R. St.J. 1966. Observations on the behaviour of the grouper species *Epinephelus guaza* and *E. alexandrinus* (Serranidae). Pp. 101-106. *In: Underwater Association Report,* (J.N. Lythgoe and J.D. Woods, Eds.). T.G.W. Industrial and Research Promotions Ltd., London.

Nzioka, R.M. 1979. Observations on the spawning seasons of East African reef fishes. *J. Fish Biol.,* 14:329-342.

Olsen, D.A. and J.A. LaPlace. 1979. A study of a Virgin Island grouper fishery based on a breeding aggregation. *Proc. Gulf Caribb. Fish. Inst.,* 31:130-144.

Presley, R.F. 1970. Larval snowy grouper, *Epinephelus niveatus* (Valenciennes, 1828), from the Florida straits. *Fla. Dept. Nat. Resources, Leaflet Ser.,* IV (18):1-6.

Randall, J.E. and V.E. Brock. 1960. Observations on the ecology of epinepheline and lutjanid fishes of the Society Islands, with emphasis on food habits. *Trans. Amer. Fish. Soc.,* 89:9-16.

Smith, C.L. 1958. The groupers of Bermuda. *In:* Bardach, J.L., C.L. Smith, and D.W. Menzel, *Bermuda Fish. Res. Prog. Final Rpt.,* Bermuda Trade Development Board, Hamilton, Bermuda. 59 pp.

Smith, C.L. 1959. Hermaphroditism in some serranid fishes from Bermuda. *Pap. Mich. Acad. Sci. Arts Letters,* 44:111-118.

Smith, C.L. 1961. Synopsis of biological data on groupers (*Epinephelus* and allied genera) of the western North Atlantic. *F.A.O. Rome, Fish. Biol. Synopsis,* 23:1-61.

Smith, C.L. 1964. Hermaphroditism in Bahama groupers. *Nat. Hist.,* 73:42-47.

Smith, C.L. 1965. The patterns of sexuality and the patterns of classification of serranid fishes. *Amer. Mus. Novitat.,* (2207):1-20.

Smith, C.L. 1966. *Menophorus* Poey, a serranid genus based on two hybrids of *Cephalopholis fulva* and *Paranthias furcifer,* with comments on the systematic placement of *Paranthias. Amer. Mus. Novitat.,* (2276):1-20.

Smith, C.L. 1967. Contribution to a theory of hermaphroditism. *J. Theor. Biol.,* 17:76-90.

Smith, C.L. 1971. A revision of the American groupers: *Epinephelus* and allied genera. *Bull. Amer. Mus. Nat. Hist.,* 146:1-241.

Smith, C.L. 1972. A spawning aggregation of Nassau grouper, *Epinephelus striatus* (Bloch). *Trans. Amer. Fish. Soc.,* 1972:257-261.

Smith, C.L. and P.H. Young. 1966. Gonad structure and the reproductive cycle of the kelp bass, *Paralabrax clathratus* (Girard), with comments on the relationships of the serranid genus *Paralabrax. Calif. Fish and Game,* 52:283-292.

Tan, S.M. and K.S. Tan. 1974. Biology of tropical grouper *Epinephelus tauvina* (Forskal) I. A preliminary study on hermaphroditism in *E. tauvina. Singapore J. Pri. Ind.,* 2:123-133.

Thompson, R. and J.L. Munro. 1978. Aspects of the biology and ecology of Caribbean reef fishes: Serranidae (hinds and groupers). *J. Fish Biol.,* 12:115-146.

Ukawa, M., M. Higuchi, and S. Mito. 1966. Spawning habits and early life history of a serranid fish, *Epinephelus akaara* (Temminck et Schlegel). *Japan. J. Ichthyol.,* 13:156-161.

Sea Basses
(Serranidae: Serraninae)

The serranines are small and active sea basses that are difficult to specifically characterize morphologically. Along with the general serranid features, such as a complete lateral line, small and somewhat inconspicuous scales, and a prominent spiny section to the dorsal fin, the serranines in general tend to be rather heavy-bodied and torpedo-shaped with a prominent set of terminal jaws. Most (for example the various species in *Hypoplectrus*, *Hypoplectrodes*, and most *Serranus*) are generalized small predators that feed mainly on small crustaceans and occasionally on fishes. As such, they tend to be cryptically colored, usually some mottled pattern of grays, black, white, and browns. A few, however, have become specialized as planktivores, feeding usually only a foot or two off the bottom and, as a consequence, have become somewhat more laterally compressed and have evolved a smaller mouth and a conspicuously forked caudal fin. Examples of this group include *Serranus tortugarum* and *Schultzea beta*, both found in the tropical western Atlantic (see discussion by Robins & Starck, 1961).

The serranines are distributed on and near reefs circumtropically but have also invaded a few temperate regions, such as the Mediterranean Sea and the southern coast of Australia. They seem to reach their peak in relative abundance, however, in the tropical western Atlantic, where they are the dominant small predators. One American genus, *Diplectrum*, has even successfully invaded the broad sand flats around reefs and has converged in behavior and gross morphology with the parapercids, an ecologically similar Indo-West Pacific family.

Serranine reproduction has been studied since at least the mid-1800's, with a number of European workers looking at various Mediterranean species of *Serranus* (e.g., Dufosse, 1856; Raffaele, 1888). As a result, serranines were among the first fishes known to be normally hermaphroditic. Work on the reef-dwelling members of the subfamily is far more recent, and

descriptions of spawning in the field have been made for only a few species within the last few years. Most such work has dealt with the western Atlantic species.

SEXUAL DIMORPHISM

As in other serranid subfamilies, all serranines that have been examined have proven to be hermaphrodites (*Serranus*—Dufosse, 1856; Longley & Hildebrand, 1941; Clark, 1959; Smith, 1959; Robins & Starck, 1961; Reinboth, 1962; Salenkova, 1963; Smith, 1965; Hastings & Bortone, 1980; *Hypoplectrus*—Smith, 1959, 1965; Fischer, 1980a; *Diplectrum*—Longley & Hildebrand, 1941; Bortone, 1977; Touart & Bortone, 1980; *Bullisichthys*—Smith & Erdman, 1973). Unlike other serranids, however, the serranines are simultaneous, rather than sequential, hermaphrodites; that is, each mature individual produces both eggs and sperm. The Y-shaped gonads consist of two hollow tubes that fuse posteriorly. The bulk of the tubes are ovarian, with numerous ovarian lamellae projecting into the central cavity. Spermatogonia are confined to a narrow loop of tissue that circles around the rear edge of each lobe of the gonad (this separation of ovarian and testicular tissue is referred to as "territorial hermaphroditism"). Sperm and eggs are shed through different openings, which prevents their mixing and subsequent self-fertilization. Of the three openings visible just ahead of the anal fin of a serranine, the anteriormost is the ovarian pore, the second is the opening of the sperm duct, and the third is the anus. Smith (1965; Smith & Atz, 1969) suggested that this type of gonad is a primitive one in the serranids and that sequential hermaphroditism has developed from it.

The factors that have lead to the evolution of simultaneous hermaphroditism in the serranines are not well understood. Such bisexuality is normally considered stable only in low density populations, where it is adaptive for each individual to be able to mate with any other individual it meets (Ghiselin, 1969). Such an argument clearly doesn't apply to many serranines, which are often quite common on the reef. Fischer (1980a, 1981) has examined the field biology and spawning system of the western Atlantic species *Hypoplectrus nigricans* in an attempt to determine those factors that stabilize its hermaphroditism, that is, with so many mates possible, why aren't the hermaphrodites outcompeted by and replaced by normal males and females? Based on his field work, Fischer suggested that each hermaphrodite protects its interest both as a male and as a female by engaging in a complex spawning behavior. First, each individual engages in a long courtship sequence before spawning as a female (which makes it difficult for a "male" to spawn with it and then quickly dash off to spawn with another female); second, as a female each individual spawns several times nightly and releases only a few eggs each time, so that if an individual cheats it doesn't gain much from it; and finally, the members of a spawning pair alternate roles, with first one shedding eggs while the other sheds sperm, and then vice versa. This last ensures that each individual will get at least some of its eggs fertilized (thus protecting its investment as a female) without risking too much if its partner deserts it. Through such "egg-trading," each hermaphrodite can concentrate most of its reproductive effort into egg production and only a small amount into sperm production without running the risk of loosing out with either. As a result, such hermaphrodites have a higher fecundity than a comparable non-hermaphrodite and will maintain themselves in the population even with high population densities. This complex schema has yet to be tested on other serranines, which presumably have some similar mechanism to maintain themselves as hermaphrodites, and only accounts for the stability of the hermaphroditism; that is, it does not explain why the fish evolved simultaneous hermaphroditism in the first place.

As simultaneous hermaphrodites there is no permanent sexual dimorphism in the serranines (mainly because there are no permanent sexes). Non-sexual variations in color pattern have been reported in a few species and may be mimetic (Thresher, 1978; Fischer, 1980b). Clark (1959, 1965) reported that during courtship *Serranus subligarius*, a small species common in the Gulf of Mexico, changes color depending on the sex role. Fish acting as males develop a vertically banded color pattern, while those acting as females initially blanch and then develop a barring that is the opposite of the "males." No such temporary dichromatism has yet been reported for other serranines.

SPAWNING SEASONS

Little information is available regarding serranine spawning seasons; what is available suggests year-round spawning in tropical areas and summer spawning in more temperate areas. Summer spawning, for example, is characteristic of the Mediterranean species of *Serranus* (e.g., Dufosse, 1856; Raffaele, 1888) and also of *Serranus subligarius* off the coast of Florida, a warm temperate area (Clark, 1959; Hastings & Bortone, 1980). Similarly, the related *S. tigrinus* forms conspicuous pairs (occasionally triads) only in the summer off Florida, but can be found in such pairs year-round deeper in the tropics. Year-round spawning also appears to be the case in the genus *Hypoplectrus* (Fischer, 1980a; pers. obs.).

There are no definitive data regarding possible lunar cycles of activity for members of the subfamily. *Hypoplectrus*, at least, spawns throughout the month with no conspicuous peaks in activity. Pressley (1981) indicated frequency of spawning by pairs of *Serranus tigrinus* was highest near the full moon, but insufficient data were presented to test for a strong lunar cycle.

REPRODUCTIVE BEHAVIOR

Spawning by serranines has been observed since at least the late 1800's, with early descriptions of aquarium spawnings of Mediterranean species by Dufosse (1856) and Raffaele (1888). Until recently the only report of spawning by a more tropical species was that of Clark (1959, 1965) on the small Gulf of Mexico species *Serranus subligarius*. *S. subligarius* normally pair-spawns at dusk, as do all other serranines that have been observed. Courtship, which was observed in both the field and aquaria, is initiated by an individual that is clearly distended with eggs. Such an individual approaches a prospective mate, S-curves before it with its fins fully erect, and quivers slightly. If the chosen mate is interested, it follows the displayer in slow jerky motions about the area. The mate playing the "male" role frequently touches its jaws to the dorsal region of the "female" or nudges its distended abdomen. At this point the "female" blanches and the "male" develops a conspicuous barred pattern. S-curving and jerk-swimming accelerate in frequency, with each member of the pair keeping its bright white abdomen clearly in sight of the other. Spawning is initiated by the "male" nudging the "female's" back, after which the pair rush up off the bottom. At the peak of the ascent the pair, female first, sharply turn back to the bottom while releasing gametes. After spawning the pair resume courtship, with the second fish now taking the "female" role. Successive spawnings continue for several hours, with pair members alternating the male and female roles.

Similar spawning behavior occurs in other species of *Serranus*. In *S. tigrinus* pairs are commonly seen throughout the day, rather than forming only at dusk (see Pressley, 1981), and the spawning ascent is fast and straight up off the bottom. Pressley (1981) reported occasional "group" spawning by triads of individuals in this species. In *S. fasciatus*, an eastern Pacific species, spawning was observed (at midday) only twice in the same pair and involved one large fish and one smaller one (pers. obs.). In both spawning rushes the smaller individual was under the larger one and nestled between its pelvic fins. The substantial size difference between the pair members is unusual and may indicate a sexual size dimorphism and possibly sequential hermaphroditism. Finally, spawning by *S. tortugarum* differs in a few respects from that of its congeners (E. Fischer, pers. comm.). Unlike the previous species discussed, the chalk bass is a planktivore that is typically found in small groups hovering over a prominent coral head or rock outcrop. At dusk the fish form conspicuous pairs, with each pair hovering a foot or so off the bottom and a similar distance from other pairs. Spawning is extremely fast and involves no evident courtship. Each pair as it spawns is frequently joined at the peak of the spawning rush by one to three "streakers," "males" from neighboring pairs that dash into the point of gamete release, even to the point of breaking apart the spawning pair. Such streakers are presumably shedding sperm on the chance that they might fertilize some eggs at their neighbors' expense. Spawning by this species, then, can be viewed as a competitive situation, one in which each pair endeavors to spawn without attracting the attention of other fish (hence the near total absence of courtship, which would signal the pair's intent to spawn) while at the same time endeavoring to take part in spawning by other pairs. In an apparent "counter-streaking" strategy, pairs occasionally begin spawning rushes but abort them part way. Such incomplete rushes nevertheless bring numerous streakers, who rush through the point at which the pair was apparently aimed. Although it may be only incidental, the net effect is that streakers waste gametes and so are "punished" for streaking. The total

system is a fascinating one to watch. Detailed description of the social behavior of this species is in preparation (E. Fischer, pers. comm.).

To date, courtship and spawning have been observed in only one other serranine genus, *Hypoplectrus*, the hamlets. Such activity by these western Atlantic fishes involves conspicuous and vigorous displaying and, for a pelagic spawner, a slow spawning ascent (Barlow, 1975; Thresher, 1976; Fischer, 1980a). During the day each hamlet maintains a vigorously defended territory; at dusk, however, they form pairs and begin courtship. Again, courtship is initiated by the "female," though sex roles switch back and forth as courtship and successive spawnings continue. Courtship is basically similar to that of *S. subligarius*, consisting of vigorous, jerky chasing back and forth over the reef by the two fish and a characteristic quivering movement in which the displayer forms its body into an "S" and poses head up or head down in front of its mate. After courting for as long as an hour, the pair eventually move close together, and then leap up off the bottom into a spawning embrace. The "female" bends herself in a shallow "S", and the "male" wraps his body crosswise around her such that his caudal fin is tight against her nape and his head is against her abdomen. The fish hold this position for 10 to 20 seconds while slowly drifting toward the bottom and, usually, rolling to one side. The pair tremble slightly (apparently signalling gamete release) and then return to the bottom. Courtship then begins again, and a few minutes later they spawn again and do so repeatedly until dark.

Courtship and spawning by the hamlets have aroused particular interest because of the confused taxonomy of the genus. There are 12 distinctly different and widely recognized "species" of *Hypoplectrus*, but the differences between them are almost entirely based on color pattern. As a result, there is some question as to whether or not the variously colored animals are valid species (see Thresher, 1978; Fischer, 1980a; Graves & Rosenblatt, 1980). One test of such validity is to determine whether broadly sympatric color forms interbreed in the field. In fact, most spawning occurs between fish with the same color pattern; crossing does occur, however, accounting for about 5% of the observed spawnings. Such crosses are not randomly distributed but tend to concentrate between specific pairs of similarly colored "species," such as *Hypoplectrus chlorurus* and *H. nigricans*, and *H. puella* and *H. unicolor*. Based on the pattern of these crosses, Fischer proposed a scheme of genetic relationships between the various species. Their validity as species will not be completely tested until the relative viability of offspring from same-color spawnings and cross-color spawnings are compared.

Finally, there are a few reports (e.g., Dufosse, 1854; Fleishmann & Kann, 1937; Clark, 1959) of self-fertilization in the subfamily. Clark, for example, reported that *S. subligarius* in isolation S-curve, quiver, and release gametes that result in viable offspring. All such reports, however, are from laboratory observations and have not been confirmed in the field. The universal occurrence of separate openings for eggs and sperm in the serranines suggests selection has been against self-fertilization, which in turn suggests that the observed cases are only laboratory artifacts. Fischer (1979) recently confirmed this observation in *Hypoplectrus* kept in captivity; very ripe individuals held in isolation but allowed to see a potential mate will release gametes, but only eggs or sperm, never both. In subsequent spawnings, however, the other gamete may be released, resulting in artificial fertilization.

EGGS AND LARVAE

There is no information available on the eggs of tropical serranines, but according to Raffaele (1888) and Lo Bianco (1933), the eggs of the Mediterranean species of *Serranus* are pelagic, spherical, and non-adhesive. Each contains a single oil droplet roughly 0.175 mm in diameter. The eggs range in diameter from 0.78 to 0.90 mm. Clark (1959) provided no details but stated that the eggs of the subtropical *S. subligarius* are similar and hatch in 18 to 22 hours at 28 to 31°C.

Newly hatched larvae have a large yolk sac that contains a single oil droplet at its anterior end; as a consequence, such prolarvae probably float head up. According to Lo Bianco (1933), 5 mm larvae of the Mediterranean species *S. scriba* and *S. cabrilla* have pigmented eyes, functional jaws, and elongate pelvic fin spines. The dorsal fin spine elongates by the time they reach a length of 6.5 mm and only begins to shorten (relative to other spines in the dorsal fin) at a length of 16 mm. Kendall (1977, 1979), however, reported that the larvae of the American species of *Serranus* lack such elongate spines and show few of the specializations characteristic of larvae of other serranid lineages. Development of the American larvae is relatively direct and the body proportions of the larvae are approximately those of the adult. Larvae of most genera are characterized by blunt points, rather than spines, on the preoperculum and pigment spots along the ventral midline. Other pigment spots on the fins and body differ in intensity and placement between species. Kendall (1979) also noted that larval *Hypoplectrus* differ from those of other serranines (shorter pelvic fin rays, head and mouth fleshier), and suggested their serranine affinities may bear closer scrutiny.

Serranines settle to the bottom at approximately 15 to 20 mm and in most species are colored much like the adults. With growth there is usually an intensification of colors and a slight increase in the complexity of the pattern. Juveniles of *Hypoplectrus*, however, develop their adult color patterns only gradually. All small hamlets appear to have the same color pattern: pale gray to brown with a tripartite spot on the caudal peduncle (two white spots, one above the other, with a black spot paired with the upper white one). Apparent intermediates between this pattern and the adult patterns characteristic of *H. puella*, *H. guttavarius*, *H. gemma*, and *H. nigricans* have been observed.

Small subtropical serranines apparently live from two (*Serraniculus pumilio* – Hastings, 1973) to six years (*Diplectrum formosum* – Bortone, 1971). No comparable information is available for strictly tropical species. Sexual maturity is apparently reached in one year.

REPRODUCTION IN CAPTIVITY

Regular spawning in aquaria has been reported for several Mediterranean species of *Serranus* by Dufosse (1854) and Raffaele (1888) and for the western Atlantic *S. subligarius* by Clark (1959, 1965). Few details on tank conditions are available, but the general impression is that little special treatment was required to stimulate spawning beyond a private tank for the spawners, clean and warm (at least 25°C) water, and heavy feedings. Photographs in Clark (1965) suggest that the breeding *S. subligarius* were kept in small (about 75-liter) all-glass aquaria aerated with an air stone and containing only a sand bottom and a few broken shells for cover. Given this success with such spartan conditions, aquarists should be able to spawn most species of *Serranus* with relative ease.

Spawning by *Hypoplectrus* and some of the other larger genera, such as *Diplectrum*, is likely to be more difficult and may require larger volume aquaria. Fischer (1979) found that various species of *Hypoplectrus* spawn readily in small aquaria for one to two nights after capture but thereafter spawn only infrequently or not at all.

Rearing of serranine larvae has not yet been reported. Their relatively direct development and their similarity to epinepheline larvae, some of which have been reared, suggest one might be successful with a diet of rotifers (*Brachionis plicatilis*) and brine shrimp.

Literature Cited

Barlow, G.W. 1965. On the sociobiology of some hermaphroditic serranids, the hamlets, in Puerto Rico. *Mar. Biol.*, 33:295-300.

Bortone, S.A. 1971. Studies on the biology of the sand perch, *Diplectrum formosum* (Perciformes: Serranidae). *Fla. Dept. Nat. Resour., Mar. Res. Lab., Tech. Ser.* (65):1-27.

Bortone, S.A. 1977. Gonad morphology of the hermaphroditic fish *Diplectrum pacificum* (Serranidae). *Copeia*, 1977:448-483.

Clark, E. 1959. Functional hermaphroditism and self-fertilization in a serranid fish. *Science*, 129:215-216.

Clark, E. 1965. Mating of groupers. *Natur. Hist.*, 74:22-25.

Dufosse, 1856. L'hermaphrodisme chez certains vertébrés. *Ann. Sci. Nat.*, (b) (4)5:295-332.

Fischer, E.A. 1979. Mating system and simultaneous hermaphroditism in *Hypoplectrus nigricans* (Serranidae), with a discussion of the systematic status of the species in the genus *Hypoplectrus*. Ph. D. Diss., Univ. Calif., Berkeley.

Fischer, E.A. 1980a. The relationship between mating system and simultaneous hermaphroditism in the coral reef fish, *Hypoplectrus nigricans* (Serranidae). *Anim. Behav.*, 28:620-633.

Fischer, E.A. 1980b. Speciation in the hamlets (*Hypoplectrus*: Serranidae)—a continuing enigma. *Copeia*, 1980:649-659.

Fischer, E.A. 1981. Sexual allocation in a simultaneously hermaphroditic coral reef fish. *Amer. Natur.*, 117:64-82.

Fleischmann, W. and S. Kann. 1937. Wirkung von Hypophysen-hormonen auf den Forbwechsel einiger Adriafischr. *Zeit. Vergl. Physiol.*, 25:251-255.

Ghiselin, M.T. 1969. The evolution of hermaphroditism among animals. *Q. Rev. Biol.*, 44:189-208.

Graves, J. and R. Rosenblatt. 1980. Genetic relationships of the color morphs of the serranid fish *Hypoplectrus unicolor*. *Evolution*, 34:240-245.

Hastings, P.A. and S.A. Bortone. 1980. Observations on the life history of the belted sandfish, *Serranus subligarius* (Serranidae). *Env. Biol. Fish.*, 5:365-374.

Hastings, R.W. 1973. Biology of the pygmy sea bass, *Serraniculus pumilio* (Pisces:Serranidae). *Fish. Bull.* (*U.S.*), 71:235-242.

Kendall, A.W., Jr. 1977. Relationships among American serranid fishes based on the morphology of their larvae. Ph. D. Diss., Univ. Calif. San Diego.

Kendall, A.W., Jr. 1979. Morphological comparisons of North American sea bass larvae (Pisces:Serranidae). *Tech. Rpt. N.M.F.S. Circ.* (428):50 pp.

Lo Bianco, S. 1933. Uova, larve e stadi giovanili de Teleostei. *Publ. Staz. Zool. Napoli, Monogr.*, (38), Part 2:384 pp.

Longley, W.H. and S.F. Hildebrand. 1941. Systematic catalogue of the fishes of Tortugas, Florida. *Papers Tortugas Lab., Carnegie Inst. Wash.*, 34:331 pp.

Pressley, P.H. 1981. Pair formation and joint territoriality in a simultaneous hermaphrodite: the coral reef fish *Serranus tigrinus*. *Z. Tierpsychol.*, 56:33-46.

Raffaele, F. 1888. Le uova gallegianti e le larve dei Teleostei del Golfo di Napoli. *Mitt. Zool. Sta. Neapel.*, 8:1-84.

Robins, C.R. and W.A. Starck, II. 1961. Materials for a revision of *Serranus* and related fish genera. *Proc. Acad. Nat. Sci. Phila.*, 113:259-314.

Salenkova, L.P. 1963. On the self-fertilization and development of self-fertilized eggs of *Serranus scriba* (L.). *Vop. Ikhtiol.*, 3:275-287.

Smith, C.L. 1959. Hermaphroditism in some serranid fishes from Bermuda. *Pap. Mich. Acad. Sci. Arts Letters*, 44:111-118.

Smith, C.L. 1965. The patterns of sexuality and the classification of serranid fishes. *Amer. Mus. Novitat.*, (2207):1-20.

Smith, C.L. and E.H. Atz. 1969. The sexual mechanism of the reef bass *Pseudogramma bermudensis* and its implications in the classification of the Pseudogrammidae (Pisces: Perciformes). *Zeit. Morph. Tiere*, 65:315-326.

Smith, C.L. and D.S. Erdman. 1973. Reproductive anatomy and color pattern of *Bullisichthys caribbaeus* (Pisces: Serranidae). *Copeia*, 1973:149-151.

Thresher, R.E. 1976. Hamlets. *Mar. Aquar.*, 7:21-28.

Thresher, R.E. 1978. Polymorphism, mimicry and the evolution of the hamlets (*Hyploplectrus*, Serranidae). *Bull. Mar. Sci.*, 28:345-353.

Touart, L.W. and S.A. Bortone. 1980. The accessory reproductive structure in the simultaneous hermaphrodite *Diplectrum bivittatum*. *J. Fish Biol.*, 16:397-404.

Sea Perches
(Serranidae: Anthiinae)

The anthiines or sea perches are the most popular of the sea basses for aquaria, largely because of their relatively small sizes (few exceed 14 cm in total length), bright colors, and "delicate" feeding habits (anthiines for the most part are planktivores and consequently are far less likely than other serranids to eat other aquarium inhabitants). As planktivores, anthiines share a number of features with other water-column foragers, including a slender, laterally compressed body, a forked or lunate caudal fin, long dorsal and anal fins, and a small terminal mouth. The family also includes a few benthic representatives, principally in the genus *Ellerkeldia*, that differ substantially in overall appearance from the more typical anthiines and more closely resemble the western Atlantic serranines in the genera *Serranus* and *Hypoplectrus*. Planktivorous anthiines, including the genera *Anthias* and *Mirolabrichthys*, are commonly observed in small to sometimes very large aggregates stationed permanently over a large coral outcrop or along the edge of a drop-off. They are, in general, more closely associated with the bottom than are many other planktivores, and this site attachment is a key feature in the structuring of the social systems of those species thus far studied.

Sea perches are found circumtropically but are a conspicuous element of the shallow reef fauna only in the Indo-Pacific. In other areas, such as the tropical western Atlantic, they are found only on the very deep reef, well below normal diving ranges (e.g., Colin, 1976). In some areas, such as southern Australia and the Mediterranean Sea, anthiines have also successfully invaded temperate areas.

Until recently, relatively little was known about reproduction of anthiines beyond descriptions of their usually conspicuous sexual dimorphism and early descriptions of the structure of their hermaphroditic gonads. A recent surge in interest in hermaphroditism has stimulated increased work on the group, but information obtained is still spottily distributed (the mechanics of sex change have been looked into in some detail, for example, but little is known about larval development) and is concentrated on only a few species.

SEXUAL DIMORPHISM

Like the related epinephelines, all anthiines thus far examined have proven to be protogynous (female to male) hermaphrodites, with all males derived from females (i.e., they are monandric). The gonads are elongate and bilobed, fusing posteriorly, and lie near the posterior end of the abdominal cavity, between the intestines and the swim bladder (Suzuki, et al., 1978; Shapiro, 1981). Ovaries are large, round, and hollow, containing a large lumen; testes are small, flattened, and largely solid (collapsed remnants of the lumen are discernible in cross sections, as are the degenerated remnants of oocytes). In the ovaries of *Anthias anthias* and *Sacura margaritacea*, testicular tissue is located in a narrow band that skirts the posterior edge of the active ovarian tissue, fusing posteriorly into a sperm duct (Reinboth, 1963, 1964; Okada, 1965a & b; Suzuki, et al., 1974). Reinboth found some evidence that even active females produce some sperm cells sporadically, but such individuals may have been in the initial stages of sex change. During this change the testicular tissue proliferates, beginning first at the periphery of the gonad and gradually moving toward the center, replacing the degenerating ovarian tissue. A similarly structured gonad also occurs in the soapfishes (Grammistidae), which led Smith (1965) to suggest a common ancestry for the two groups. A very different type of gonad, however, is present in *A. squamipinnis* (Gunderman, 1972; Fishelson, 1975; Shapiro, 1981), and in *A. nobilis* (as *Franzia squamipinnis*—*A. nobilis* is a Japanese species that is similar in appearance to the widely distributed *A. squamipinnis* and until recently has been synonymized with it) (Suzuki, et al., 1978). In both species testicular tissue does not proliferate from conspicuous bands in the gonad but rather develops from small crypts along its periphery and near the genital pore. Shapiro (1981) also reported that there is no evidence of a permanent sperm duct in *A. squamipinnis* and hypothesized instead that sperm in this species are transported posteriorly through temporary passages in the testes that form by disruption of connective tissue membranes that had previously divided the testes into multiple, enclosed sperm chambers. These passages are then suggested to close off after emptying due to the expansion of other developing spermatogonia and by the reestablishment of sperm chamber walls. This very different gonad structure has not yet been examined in light of the hypothesized affinities of anthiines and grammistids.

The hormonal mechanisms underlying sex change in anthiines have been examined in *A. squamipinnis* by Fishelson (1975). By adding testosterone in different dosages to aquaria containing females, Fishelson was able to stimulate the development of varying degrees of male coloration and behavior and the onset of spermatogenesis in the gonads. Returning the fish to normal sea water resulted in a reversal to female characteristics in those fishes treated with only small dosages of testosterone but had no effect on those given higher dosages, all of which completed sex reversal. In another experiment, the addition of estradiol (a female gonadal hormone) delayed the development of male characteristics in sex-changing fish.

More detailed work has been done on the environmental, and especially the social, factors that result in sex change. Six general findings that relate to sex change in anthiines have been reported: 1) with rare exceptions (e.g., Yogo, in prep.), small individuals in a population are female, whereas males occur only in the larger size ranges (Popper & Fishelson, 1973; Fishelson, 1975; Gunderman, 1972; Suzuki, et al., 1974, 1978; Shapiro, 1981); 2) sex reversal occurs most commonly outside of the spawning season (Fishelson, 1975; Suzuki, et al., 1974, 1978; Yogo, in prep.); 3) in single-male social groups, removal of the male results in the sex change of the largest dominant female (Fishelson, 1970; Suzuki, et al., 1978; Shapiro, 1979, 1981, in press); 4) in at least some cases, males arise spontaneously in artificially created all-female groups (Suzuki, et al., 1978); 5) in large groups with many males and many females, removal of one male leads to the sex change of one female, whereas removal of several males leads to the sex reversal of

about the same number of females (Shapiro & Lubbock, 1980; Shapiro, in press); and 6) the sex ratio is usually biased toward females, the amount of bias differing between species and varying with group size and time of year (Shapiro, 1977; Yogo, in prep.).

The generally larger sizes of males than females suggest that sex change in anthiines may be determined, at least in part, by size and/or age, such that the seasonal occurrence of such sex changes during non-spawning periods might be the result of annual growth cycles (Yogo, in prep.). Considerable evidence, however, indicates that male removal stimulates an immediate sex change by a female, regardless of her age or size. Fishelson (1970) first suggested that sex reversal by female anthiines is constrained by aggressive dominance of the male, or males, present and that removal of a male, by eliminating the source of such inhibition, results in the dominant female changing sex and inhibiting sex change by the females she "outranks." This hypothesis has been widely accepted, perhaps because of its apparent simplicity and its similarity to comparable mechanisms in other reef fishes, such as labrids, but it is likely to be far too simplistic to account for the often complex social systems of anthiines (e.g., Shapiro, 1977), nor has it been adequately tested (Reinboth, 1980). Shapiro (1979) proposed as an alternative the "priming" hypothesis, which argues that the proximate stimulus for sex change by a female is not male removal, per se, but rather is the change in the pattern of social interactions that the dominant female participates in. As a female ascends in dominance, this patterning changes, such that she is preconditioned, or primed, for the final change. Shapiro (1979) presented data supporting this priming hypothesis; however, see also Reinboth (1980) for a discussion of the question of social control of sex change in general and comments about the utility of the priming hypothesis in particular. Shapiro also argued that a careful description of the structure of social groups is needed to explain precisely how sex reversal operates in large groups. Shapiro (1977), for example, noted that many large groups are, in fact, subdivided into two or three subgroups, based on the spacing of discrete aggregates and the rate of movement of individuals between them. Such structuring may have a bearing on functional sex ratios and the likelihood that a female in any given subgroup will change sex when a male is removed. Shapiro's point is certainly valid that workers generally assume an overly simplified social unit in these fishes; his specific

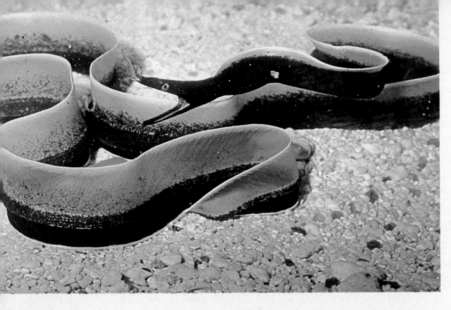

Taxonomic difficulties arose in *Rhinomuraena quaesita* because of its color changes. The black form is the juvenile. Photo by Dr. Fujio Yasuda.

The black juveniles transform into blue males as seen here. The blue form was described as *R. ambonensis*, a name that is now placed in synonymy of *R. quaesita*. Photo by Dr. Fujio Yasuda.

Eventually the blue male turns into a yellow female. The eel is thus a protandrous hermaphrodite. Photo by Dr. Shih-chien Shen.

Courtship activities in *Gymnothorax kikado* included this pose wherein the pair raised their heads off the bottom, facing each other with jaws agape. Photo by Noriaki Yamamoto.

Settlement to the bottom by larval lizardfishes occurs at about 30 to 35 mm. The arrangement and size of the black spots are very useful characters in determining the identity of the larvae. Photo by Charles Arneson.

data, however, are weakened by the absence of information on functional sex ratios and individual movements during spawning periods (Shapiro's data are based on daytime, rather than evening, observations), when conditions could be quite different. Yogo (in prep.), for example, in a detailed study of spawning behavior and patterns of male and female success in *A. nobilis*, reports movement between apparently discrete groups during evening courtship periods. Indeed, without male coercion to force spawnings, sexual selection would seem to favor such movements by females if they resulted in spawnings with "higher quality" males. Although the environmental determinants of sexuality in anthiines have already been studied in some detail, there remains considerable room for additional work on the subject.

As a consequence of protogyny, male anthiines are usually larger than females, with the exact size of sex change varying both between species and between populations within species. In *Sacura margaritacea*, females range in length from 64.4 to 124 mm fork length, males from 117.5 to 146.0 mm, and transitionals from 111.1 to 127 mm (Suzuki, et al., 1974). In *Anthias nobilis*, sex change occurs between 83.0 and 113.0 mm fork length (Suzuki, et al., 1978). Finally, in *Anthias squamipinnis* size at sex change varies conspicuously among populations. Fishelson (1975) reported it at 65 to 86 mm for a Red Sea population, Popper & Fishelson (1973), working in the same area, reported it to be at least 100 mm, and Shapiro (1981) indicated sex change occurs at 53 to 62 mm and 40 to 45 mm, respectively, for two different populations at Aldabra Atoll. The factors underlying these differences are not known, though it is clear that in at least some cases they are not genetically based. Rather, they are more likely functions of differential growth rates, recruitment rates, and rates of adult (and especially male) mortality (Shapiro, 1981).

Other than size, the sexes in some, if not most, species also differ conspicuously in color patterns and often finnage (Smith, 1961; Heemstra, 1973). Males, as a general rule, are more conspicuously and complexly colored than are females. Female *A. squamipinnis*, for example, are golden yellow to orange with a red line running down and to the rear from each eye; the male is darker, often purple, with a darker head than body, prominent blue-lined pelvic fins, a dark area near the posterior edge of the dorsal fin, and large and conspicuous spots on each pectoral fin (for a detailed description of color changes in this species, see Shapiro, 1981). In addition, the male has an elongated third dorsal spine that is lacking in most females. Development of male external features is affected strongly by an individual's social environment according to Shapiro (1981) and may not be a precise indicator of sex. Suzuki, et al. (1978) suggested that elongation of the third dorsal spine in *Anthias nobilis* begins shortly after gonadal change and is an accurate indicator of an individual's sex long before color change begins. Shapiro (pers. comm.) has not found such a correlation in *A. squamipinnis*.

SPAWNING SEASONS

There are only a few reports concerning anthiine spawning seasons, and these often differ, even within a species, suggesting considerable flexibility in this regard. In the Red Sea, *Anthias squamipinnis* apparently courts year-round but actually spawns only during the cooler months of the year—November through June, with a peak in activity in February—at water temperatures of approximately 20°C (summer temperatures reach 26°C) (Popper & Fishelson, 1973; Fishelson, 1975). In contrast, the same species spawns in the summer at One Tree Island, Great Barrier Reef, when water temperatures are near an annual peak (25 to 26°C). In a parallel fashion, *Anthias nobilis* spawns during the warmest part of the year off southern Japan (Suzuki, et al., 1978), but does so primarily in early summer, well before seasonally high water temperatures, slightly farther north (Yogo, in prep.). Late summer spawning (from August to November, with a peak in September) is also the case for another Japanese species, *Sacura margaritacea* (Suzuki, et al., 1974).

Anthiines appear to spawn nightly throughout the spawning season with no indication of any lunar-associated cycle of activity (e.g., Yogo, in prep.). There is as yet no information on the frequency with which each female spawns (successful males often spawn several times nightly), but Fishelson (1975) suggested that the daily pattern of oocyte development in *A. squamipinnis* indicates that each female spawns at least several times each season.

REPRODUCTIVE BEHAVIOR

Courtship has thus far been described for only three anthiine species: *Sacura margaritacea*, by Suzuki, et al. (1974); *Anthias squamipinnis*, by Popper & Fishelson (1973) and Fishelson (1975); and *Anthias nobilis*, by Suzuki, et al. (1978) (as *Franzia squamipinnis*) and Yogo (in prep.). Spawning has been described only for the latter two species.

Like most other planktivorous anthiines, *Sacura margaritacea* forage in large schools in the water column during non-reproductive periods. Females roam well off the bottom, while males remain closer to cover. As the spawning season approaches, small groups of males and females split off from the main school and aggregate near the bottom. Males are apparently only weakly territorial and are frequently observed performing "display dances" toward various females. Spawning has not been described but probably occurs at dusk and involves male-female pairs.

Somewhat more detail is available concerning social behavior and reproduction in *Anthias squamipinnis*. As in the previous species, the fish forage in the water column during non-spawning periods, with females roaming farther away from the bottom than males. In contrast to the previous species, however, at least in some areas there are two classes of males: non-territorial and territorial. The former are usually located near the base of the school and appear to be prevented from establishing territories by the less numerous but more dominant territory holders. The latter are loosely dispersed throughout the main school of females, usually remaining, however, near the upper parts of coral heads and the like, on which they defend areas ranging in size from 0.5 to 3 square meters. This distinction between non-territorial and territorial males may well vary geographically and is likely to be a function of the size of the social unit.

As described by Popper & Fishelson (1973), spawning by *A. squamipinnis* occurs at dusk and is preceded by "zig-zag" swimming by territorial males. Each male slowly ascends into the water column over his territory while making a series of increasingly more vigorous dashes back and forth. At a height of about 2 m the male dashes back to the bottom, in a deep U-shaped dip, coming back to his starting height at its end. The "U-swim" is performed with the pelvic fins opening and closing intermittently, the rear halves of the dorsal, anal, and caudal fins fully spread, and the pectoral fins widespread, fully exposing the conspicuous dark spots each bears. Such zig-zag swimming goes on for several minutes until a female approaches. The male and female then come close together for a few seconds while swimming ahead, and gametes are released. During this pairing the male's jaws remain agape. Spawning by several pairs usually takes place nearly simultaneously, with bouts of intense activity occurring every few minutes until dark. Females apparently spawn only once nightly, whereas successful males spawn many times. Nonter-

ritorial males probably spawn only sporadically.

Spawning by *A. squamipinnis* at One Tree Island, Great Barrier Reef, is similar to that described above for the Red Sea fishes, but differs in the actual spawning act. U-swimming is apparently a more prominent activity during dusk-spawning periods, with males regularly U-swimming toward approaching females. During spawning, the male dives down onto a female that has moved below and usually to one side of him. At the trough of the U the pair join, partially twisting their bodies together while the male gapes his jaws. Contact lasts only a second or two, during which gametes are shed and after which the female swims off and the male, completing his U, returns to a higher level in the water column.

Finally, spawning by the similar-appearing *Anthias nobilis* resembles that of *A. squamipinnis*. The social organization, spawning behavior, and patterns of male and female success in this species have been examined in the field by Yogo (in prep.). As in the previous species, successful males maintain permanent territories near the top of prominent rock and coral outcrops, with females and lower ranking males below. U-swimming is common. Spawning takes place near the bottom of the dip, with the male and female curling about one another momentarily. Unlike *A. squamipinnis*, males of *A. nobilis* develop a temporary spawning color consisting of pale vertical bands overlying their normal deep gold body color. Male-male interactions are frequent and vigorous, with dominant males chasing both other dominant males and subordinates that ascend too far off the bottom. Spawning by these subordinate males occurs primarily during "epidemic" spawning bouts in which several females ascend to spawn simultaneously and the dominant males are occupied.

There is thus far no available information on spawning by bottom-oriented anthiines such as *Ellerkeldia*. Pair spawning at dusk, however, is likely to be the case.

EGGS AND LARVAE

Eggs and early stage larvae of anthiines have been described for only two species: *Sacura margaritacea* (Suzuki, et al., 1974) and *Anthias nobilis* (as *Franzia squamipinnis*) (Suzuki, et al., 1978). In both cases material examined was obtained by stripping gametes from ripe individuals. In both species eggs are transparent, nonadhesive, buoyant spheres each containing a single oil droplet. Those of *S. margaritacea* are 0.78 to 0.80 mm in diameter and those of *A. nobilis* 0.65 to

Divers discovered a group of small (male?) *Gymnothorax brunneus* biting a larger one (female?) just behind the head. Photo by Roger Steene.

The entire group dashed up toward the surface, leaving behind a conspicuous cloud of floating eggs. Photo by Roger Steene.

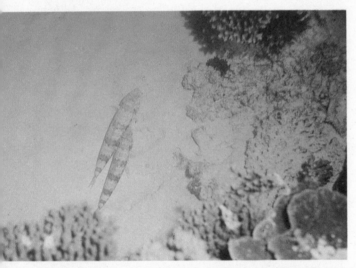

Apparent courtship in a lizardfish, *Synodus* sp. The presumed male is lying beside and behind the female. Photo by Ann Gronell.

Pairing and temporary sexual dichromatism in a species of *Apogon.* The male is lighter in color than the females. Photo by Dr. R. E. Thresher.

Pairing at dusk by *Apogon gracilis.* Photo by Dr. R. E. Thresher.

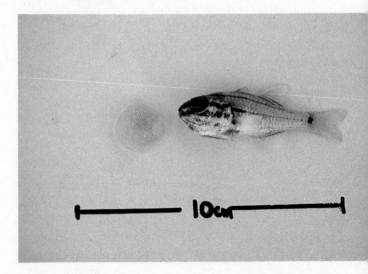

An egg mass removed from the mouth of *Apogon doderleini* (?). Photo by William Douglas.

Courtship and spawning in cardinalfishes are always paired activities. Shown here is *Apogon notatus*. Photo by Roger Steene.

Courtship generally includes circling behavior, with the female on the outside prodding the male. Photo by Roger Steene.

Apogon notatus courtship and spawning are unusual in that they take place off the bottom. Note the lighter female below. Photo by Dr. J. E. Randall.

A species of *Apogon* carrying a mouthful of eggs. Note the distended gular area. Photo by Roger Steene.

Courtship in *Apogon doderleini*. The female is on the outside pushing the male in a circle. Photo by Dr. R. E. Thresher.

Post-spawning pair of *A. doderleini*. The male is above the female and is carrying the eggs. Photo by Dr. R. E. Thresher.

0.67 mm in diameter. Hatching in both species occurred in 15 to 16 hours at water temperatures that ranged from 24 to 28°C.

Newly hatched larvae in both species (and apparently also of *Pseudanthias elongatus*, according to Suzuki, et al., 1978) are long and slender with a large yolk sac that extends anteriorly beyond the head of the prolarva. At the anteriormost tip of the yolk sac sits a single oil droplet; because of its position, the larvae float head upward. The newly hatched larva of *S. margaritacea* is 1.46 to 1.52 mm in total length, whereas those of *A. nobilis* are 1.22 to 1.28 mm long. In both species the yolk sac is absorbed, the eyes are pigmented, and the jaws are apparently functional three days after hatching.

Older anthiine larvae are routinely collected in plankton tows and are basically similar to those of the closely related epinephelines. Both, for example, have long spines in the dorsal and pelvic fins (Lo Bianco, 1933; Fourmanoir, 1976; Kendall, 1977, 1979; Kendall & Fahay, 1979). In anthiines, however, the third spine is the longest in the dorsal fin, rather than the second as in the epinephelines. Anthiine larvae are also more robust and deep-bodied than epinepheline larvae, have a much more elongate spine on the interopercle, and have pigment spots on various parts of the head and body (Kendall, 1979; Kendall & Fahay, 1979). Duration of the planktonic larval stage is not known. Newly settled juveniles, however, are roughly 10 to 15 mm long, colored like the females, and remain close to cover. Growth rate is not known; sexual maturity (as a female) probably occurs in a year or so.

REPRODUCTION IN CAPTIVITY

Rather suprisingly, there are few published reports of aquarium spawning by anthiines, although it probably occurs regularly in large public aquaria. Spawning these fishes has been reported only once, by Suzuki, et al. (1978), who spawned *Anthias nobilis* in a 3500-liter (approximately 1000-gallon) aquarium. Few details are provided. Spawning in smaller aquaria seems possible, considering the small size of the species at sexual maturity (as little as 17 mm for females and 43 mm for males), their relatively simple spawning behavior, and site-attachment. Although often delicate, the fishes also adapt well to captivity, feed readily, and are not overwhelmingly aggressive. The fact that most, if not all, species change sex also facilitates obtaining a spawnable pair (though you will probably be best off purchasing fish that are conspicuously of each sex).

Spawning *A. squamipinnis*, the most widely available member of the subfamily, should require only establishment of a moderate-sized group (two or three females and one male should be adequate) in a tank of 200 liters (44 gallons) or more that is high enough so the fish can spawn off the bottom. They should have privacy and heavy feedings with live and prepared foods. Spawning should occur at dusk and might best be stimulated by a gradual decrease in light levels rather than the abrupt onset of darkness.

It should be noted, however, that Popper & Fishelson (1973) found frequent spawnings in large schools (consisting of "hundreds" of individuals) but did not see it in smaller groups. Whether this indicates that some minimum number of individuals is required before spawning will occur or simply that it is less frequent and more easily overlooked in such small groups is not known.

Anthiines have not yet been reared in captivity. They should do well, however, given conditions similar to those used to rear epinepheline larvae.

Literature Cited

Colin, P.L. 1976. Observations of deep-reef fishes in the Tongue-of-the-Ocean, Bahamas. *Bull. Mar. Sci.*, 26:603-605.

Fishelson, L. 1970. Protogynous sex reversal in the fish *Anthias squamipinnis* (Teleostei, Anthiidae) regulated by the presence or absence of a male fish. *Nature, Lond.*, 227:90-91.

Fishelson, L. 1975. Ecology and physiology of sex reversal in *Anthias squamipinnis* (Peters), (Teleostei: Anthiidae). Pp. 284-294. *In: Intersexuality in the Animal Kingdom*, (R. Reinboth, Ed.). Springer Verlag, Berlin.

Fourmanoir, P. 1976. Formes post-larvaires et juveniles de poissons cotiers pris au chalut pelagique dans le sud-ouest Pacifique. *Cahiers du Pacifique*, (19):47-88.

Gundermann, N. 1972. The reproductive cycle and sex inversion of *Anthias squamipinnis* (Peters). Master's Thesis, Tel Aviv Univ., Israel.

Heemstra, P.C. 1973. *Anthias conspicuus* sp. nova (Perciformes: Serranidae) from the Indian Ocean, with comments on related species. *Copeia*, 1973:200-210.

Kendall, A.W., Jr. 1977. Relationships among American serranid fishes based on the morphology of their larvae. Ph. D. Diss., Univ. Calif., San Diego.

Kendall, A.W., Jr. 1979. Morphological comparisons of North American sea bass larvae (Pisces: Serranidae). *Tech. Rpt. N.M.F.S. Circ.* (428):50 pp.

Kendall, A.W., Jr. and M.P. Fahay. 1979. Larva of the serranid fish *Gonioplectrus hispanus* with comments on its relationships. *Bull. Mar. Sci.*, 29:117-121.

Lo Bianco, S. 1933. Uova, larve e stadi giovanili de Teleostei. *Publ. Staz. Zool. Napoli, Monogr.*, (38) Part 2:384 pp.

Okada, Y.K. 1965a. Sex reversal in the serranid fish *Sacura margaritacea*. I. Sex characters and changes in gonads during reversal. *Proc. Japan. Acad.*, 41:727-740.

Okada, Y.K. 1965b. Sex reversal in the serranid fish *Sacura margaritacea*. II. Seasonal variations in gonads in relation to sex reversal. *Proc. Japan. Acad.*, 41:741-745.

Popper, D. and L. Fishelson. 1973. Ecology and behavior of *Anthias squamipinnis* (Peters, 1855) (Anthiidae, Teleostei) in the coral habitat of Eilat (Red Sea). *J. Exp. Zool.*, 184:409-424.

Reinboth, R. 1963. Naturlicher Geschlechtswechsel bei *Sacura margaritacea* (Hilgendorf) (Serranidae). *Ann. Zool. Japan.*, 36:173.

Reinboth, R. 1964. Inversion du sexe chez *Anthias anthias* (Serranidae). *Vie et Millieu, Suppl.*, 17:499-503.

Reinboth, R. 1980. Can sex inversion be environmentally induced? *Biol. Reprod.*, 22:49-59.

Shapiro, D.Y. 1977. The structure and growth of social groups of the hermaphroditic fish *Anthias squamipinnis* (Peters). *Proc. Third Intern. Coral Reef Symp.*, 1:571-578.

Shapiro, D.Y. 1979. Social behavior, group structure, and the control of sex reversal in hermaphroditic fish. Pp. 43-103. *In: Advances in the Study of Behavior*, Vol. 10, (J.S. Rosenblatt, R.A. Hinde, C. Beer, M.-C. Busnell, Eds.). Academic Press, N.Y.

Shapiro, D.Y. 1980. Serial female sex reversals following the simultaneous removal of males from social groups of a coral reef fish. *Science*, 209:1136-1137.

Shapiro, D.Y. 1981a. Size, maturation and the social control of sex reversal in the coral reef fish, *Anthias squamipinnis* (Peters). *J. Zool. Lond.*, 193:105-128.

Shapiro, D.Y. 1981b. The sequence of coloration changes during sex reversal in the tropical marine fish *Anthias squamipinnis* (Peters). *Bull. Mar. Sci.*, 31:383-398.

Shapiro, D.Y. and R. Lubbock. 1980. Group sex ratio and sex reversal. *J. Theoret. Biol.*, 82:411-426.

Smith, C.L. 1965. The patterns of sexuality and the classification of serranoid fishes. *Amer. Mus. Novitat.*, (2207):1-20.

Smith, J.L.B. 1961. Fishes of the family Anthiidae. *Ichthyol. Bull., Rhodes Univ.*, 21:359-369.

Suzuki, K., K. Kobayashi, S. Hioki, and T. Sakamoto. 1974. Ecological studies of the anthiine fish *Sacura margaritacea* in Suruga Bay, Japan. *Japan. J. Ichthyol.*, 21:21-33.

Suzuki, K., K. Kobayashi, S. Hioki, and T. Sakamoto. 1978. Ecological studies of the anthiine fish, *Franzia squamipinnis*, in Suruga Bay, Japan. *Japan. J. Ichthyol.*, 25:124-140.

Yogo, Y. (In Prep.). Reproductive behavior of the anthiine fish *Anthias nobilis* Franz, at Miyake-jima, Japan. Ph. D. Diss., Kyushu Univ., Japan.

Spawning sequence of *Apogon cyanosoma.* All photos by Dr. R. E. Thresher.

The female nuzzles the male in the general region of his operculum.

Male and female in pre-spawning position—side-by-side with vents together.

At this point the egg ball is being extruded by the female.

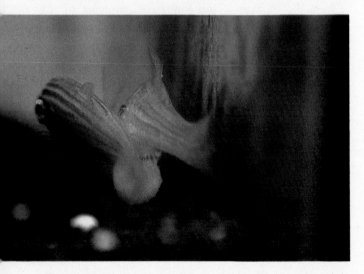

The egg ball is almost completely extruded. Note how large it is.

The male turns and picks up the egg ball in his mouth.

The male just after engulfing the egg ball. The female hovers above looking somewhat depleted.

The male may eject the eggs to rearrange them. Hopefully he will pick them up again before other fishes make a meal of them.

Spawning is completed, and the male, with the eggs properly arranged, settles down for the incubation period.

The same species, *Apogon cyanosoma*, holding eggs in the wild. Photo by Roger Steene.

Apparent courtship in *Cheilodipterus quinquelineata*. The female is below and angled toward the male, pushing him up and around. Photo by Dr. R. E. Thresher.

Cheilodipterus macrodon male that was collected while holding a batch of eggs (in front of fish). Photo by William Douglas.

Soapfishes
(Grammistidae)

The grammistids or soapfishes and their allies are a poorly known group of generally secretive fishes thought to be a recent offshoot of the main line of serranid evolution. The status of the family, if such it is, has been widely discussed, and at times its members have been placed in as many as four different families. Gosline (1960) proposed erection of the family Grammistidae, in which he included *Rypticus, Grammistes, Pogonoperca, Grammistops,* and the "pseudogrammid" genera *Pseudogramma, Suttonia, Rhegma,* and *Aporops* on the basis of anatomical features that were common among them and that set them off from the serranids. At the same time Gosline noted that their point of closest contact with the serranids was in *Diploprion bifasciatum,* a common Indo-West Pacific species which is occasionally placed in its own family (Diploprionidae) but which is also considered by some workers to be a grammistid (e.g., Randall, et al., 1971). Smith (1965) and Smith & Atz (1969) concluded, on the basis of gonad morphology, that *Pseudogramma* and *Rypticus* represented different divergences from the main serranid line and consequently erected Pseudogrammidae for the former. Most recently, Kendall (1977, 1979) looked at larval development in a number of serranoid genera and suggested that first, the nominal serranid genus *Liopropoma* was most closely allied to the grammistins, and second that the whole group of fishes was best treated as simply a diverse family Serranidae.

While refusing (probably wisely) to take a position on the whole thing (although the pseudogrammids appear to spawn quite differently from the other genera), the family Grammistidae will be defined here in the broad sense, including *Diploprion, Lipropoma, Rypticus* and its close relatives, as well as *Pseudogramma* and its relatives. As so defined, the family contains about ten genera and about 40 species. All are secretive fishes (usually seen by day deep in crevices or caves) that become active at dusk. The species in *Pseudogramma* and *Liopropoma* are particularly retiring and are often not seen unless driven out with ichthyocides. All are predators, the larger species eating fishes and crustaceans, the smaller ones feeding on a variety of invertebrates. Most species are relatively drably colored. Only the species in *Liopropoma* and *Grammistes* are colorful, and both contain popular aquarium fishes.

SEXUAL DIMORPHISM

There are no reports in the literature of sexual dimorphism in grammistids, although, for reasons to be made clear below, males probably on the average attain a larger size than females in most species. Such certainly appears to be the case in *Diploprion bifasciatum* based on observations of spawning pairs. In the Indo-West Pacific species *Pseudogramma polyacanthus* size differences between the sexes are minimal, based on specimens collected from rubble mounds in the One Tree Island lagoon. Such individuals exhibited two clearly different color patterns, however: some fish were very dark brown with only a faint indication of barring while others were lighter brown, more clearly barred, and had a bright orange wash across the lower half of the head (below the eyes) and along the ventral surface. Examination of gonads indicated that of nine dark fish, seven were mature males and two apparently immature, whereas of thirteen orange fish, seven were conspicuously female (one was full of very ripe eggs), two were males, and four lacked obvious gonads. The dark males ranged in size from 42.9 to 46.5 mm SL, the orange males from 43.6 to 46 mm, and the orange females from 43.7 to 46.7 mm; i.e., size differences between the sexes were slight and overlap in size ranges nearly complete. This broad overlap in size ranges is especially interesting in light of Smith & Atz's evidence of sequential hermaphroditism in the genus, as discussed below.

To date gonad structure has been examined in three grammistid genera: *Rypticus* by Smith (1965), *Pseudogramma* by Smith & Atz (1966), and *Liopropoma* by Smith (1971). As indicated above, the first two genera were clearly hermaphroditic, while the third appeared to be a secondary gonochorist, that is, not hermaphroditic but clearly derived from hermaphroditic ancestors. In all three cases both testes and ovaries are paired hollow organs. The structure of the ovaries is typical of percoids, i.e., circular in cross section with a central lumen into which project egg-bearing lamellae. The structure of the testes, however, varies among the three genera, which, as Smith (1965) and Smith & Atz (1969) emphasized, suggests that they are not closely related. In *Rypticus* the testicular tissue is limited to a pair of narrow bands at the base of the lamellae. After sex change, sperm pass down crypts in this tissue to enter a posterior sperm duct. Smith suggested that this gonad condition is intermediate between that of the serranines and the epinephelines. In contrast, the testicular tissue in *Pseudogramma* is found in a narrow lobe that projects from the dorsal wall of the common oviduct. In both genera mature testes routinely contain degenerate oocytes. The occurrence of sequential hermaphroditism in *Pseudogramma* may account for the overlap in color patterns of the two sexes of *P. polyacanthus*; that is, the "orange" males may be fish that have recently changed sex but have not yet changed color pattern. On the other hand, the lack of a significant size difference between the sexes is puzzling.

The species in *Liopropoma* apparently differ from other grammistids, and indeed most serranids, in having separate sexes. In their general morphology the testes are essentially ovarian (i.e., paired hollow sacs), indicating that hermaphroditism is, at least, an ancestral condition in the genus. Testicular tissue is located dorsally in the gonad and empties into sinuses that, in turn, merge to form a common sperm duct. At the base of the genital papilla this duct enlarges to form a "sperm reservoir," a feature thus far not found on any other percoid species. Given this morphology, it is difficult to account for one particularly puzzling experience with these fishes. A *L. rubre*, obtained on a collecting trip in the Bahamas and kept isolated after collecting in a transparent plastic bag, was found to have released eggs after a week in isolation (pers. obs.). The spherical transparent eggs were roughly 1 mm in diameter and were clearly developing—several were eyed. The production of viable eggs by a solitary individual suggests either that the fish are indeed hermaphroditic or that the eggs are capable of parthenogenic development.

SPAWNING SEASON

Nothing definitive is known about the spawning seasons of any grammistids. Very small juvenile *Liopropoma* can be collected year-round in the Bahamas, suggesting that some reproduction occurs year-round. *Diploprion bifasciatum* off One Tree Island spawn during the summer, at least, but data for other times of the year have not been collected. There is no information on possible lunar cycles of activity for any species in the family.

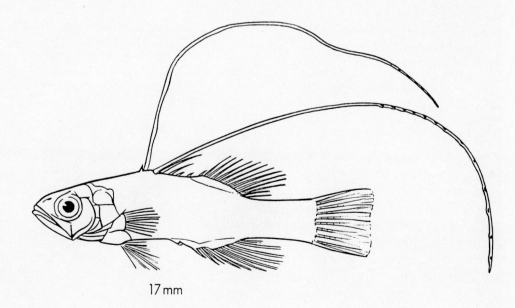

Postlarval basslet (*Liopropoma* sp.) Based on Fourmanoir (1971).

17 mm

Male *Sphaeramia orbicularis* incubating eggs (note the extended gular area). The incubation period of this species is said to be about eight days. Photo by Dr. G. R. Allen.

Postlarval western Atlantic squirrelfish just prior to metamorphosis. Photo by Charles Arneson.

Small juvenile *Pristigenys alta,* apparently recently settled. Settlement occurs at about 65 mm. Photo by Aaron Norman.

Juvenile *Plotosus lineatus* are almost always seen in small schools or "balls" of numerous individuals on the reef. Photo by Dr. G. R. Allen.

REPRODUCTIVE BEHAVIOR

For most species only scattered notes are available concerning the social and reproductive behavior of grammistids. Most appear to be normally solitary and apparently territorial. In aquaria, *Pseudogramma polyacanthus* sort themselves into apparently heterosexual pairs (one dark individual and one light one)(pers. obs.), suggesting they may normally be paired in the field. The sex ratio of mature fish collected on a series of rubble mounds at One Tree Island was close to unity (nine males to seven females), which seems to support this idea.

Spawning has thus far been observed in only two species. Most information is available concerning the relatively large and bold Indo-West Pacific species *Diploprion bifasciatum* off One Tree Island, Great Barrier Reef. During the day individuals are relatively retiring, only occasionally moving away from the shelter of caves and crevices; at dusk, however, they emerge and begin to forage. On several dives made along the outer, windward edge of the reef at dusk, large *Diploprion* were observed along the edge of the outer reef. At the approach of another individual, the patroller, apparently a male, would swim toward the intruder and begin to "dance" in front of it, using a quick, exaggerated swimming motion. If a ripe female, the new fish ascended slightly and the presumed male moved to a position about a foot directly above it, angled slightly head-up. There was no consistent color difference between the sexes, though the male was often pale relative to the female. Spawning began with the pair still arranged one above the other and swimming upward into the water col-

umn, initially at an angle of about 30° from horizontal. As they ascended the pair moved increasingly faster, drew closer together until the smaller fish was just below and to one side and slightly behind the larger one, and the angle of ascent became increasingly steeper. Ultimately, the fish were swimming at full speed directly up into the water column. At a height that varied from 7 to 14 m, the pair abruptly turned and dashed back to the bottom, leaving behind a large, conspicuous white cloud of gametes. After spawning, the female moved off while the male continued to patrol his area. One male was observed to spawn three times in a single evening.

Based on these observations, the reproductive system of *Diploprion* appears to be based on male control of temporary spawning territories located along the outer reef edge, within which each courts passing females. On three occasions males were seen to dash up into the water column alone, which may represent some type of courtship display.

Diploprion apparently produces a pelagic egg. That this is true for smaller grammistids such as *Pseudogramma* is doubtful. The ripe ova of *P. polyacanthus* are large and bright red. Most pelagic eggs are both small and transparent, which apparently minimizes the likelihood that they will be eaten. In contrast, demersal eggs are often colorful, suggesting that *P. polyacanthus* may produce demersal eggs.

Spawning has also been observed on one occasion in the western Atlantic grammistid *Liopropoma rubre*. According to I. Clavijo (pers. comm.), a pair of similar-sized individuals were seen to dash rapidly up off the bottom at dusk on a reef off Puerto Rico, shedding a cloud of gametes at the peak of the ascent. Pair-

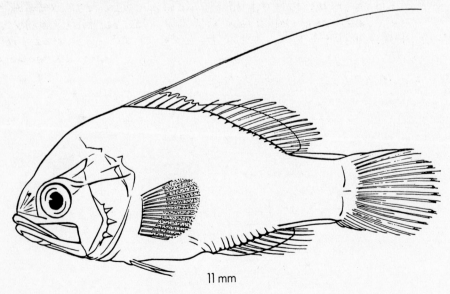

11 mm

Postlarval grammistid, *Grammistes sexlineatus*. Based on Fourmanoir (1976).

ing by these normally solitary fish at dusk has also been observed for *Liopropoma mowbrayi*, suggesting it, too, spawns at that time (Thresher, 1980).

EGGS AND LARVAE

Diploprion and *Liopropoma* appear to have pelagic eggs, whereas *Pseudogramma* either has unusually brightly colored pelagic eggs or spawns demersally. Otherwise nothing is known about grammistid eggs or their development. Similarly, small larvae of such fishes have not yet been identified. Older larvae, however, have been described for *Grammistes* (Fourmanoir, 1971a, 1976), *Diploprion* (Hubbs & Chu, 1934), *Rypticus* (Aboussovan, 1972; Kendall, 1979), *Aporops* (Fourmanoir, 1976), *Pseudogramma* (Kendall, 1979), and *Liopropoma* (Fourmanoir, 1971b; Kendall, 1979); such larvae are reviewed by Kendall (1977, 1979) and compared with other serranid larvae. All appear to be characterized by a long, tubular body, one or two extremely elongate spines near the leading edge of the dorsal fin, large pectoral fins that develop early, and a complete lack of pigmentation on the body. The extreme development of the dorsal fin spines led to these larval fishes being placed erroneously in their own genus, *Flagelloserranus* (Kotthaus, 1970). Postlarval grammistids are in general similar in appearance to their adults. Those of *Grammistes, Diploprion, Rypticus, Pseudogramma* and *Aporops* have the first dorsal spine elongated; postlarval *Liopropoma* have the first two dorsal spines elongated. Duration of the planktonic stages is not known. Juveniles in general resemble the adults in morphology and color pattern and are, if anything, even more secretive than the adults. Juvenile *Rypticus* often have a white line up their foreheads, a feature they share with a number of other small predatory fishes.

REPRODUCTION IN CAPTIVITY

There are no reports of spawning by grammistids in aquaria, although at least the smaller species seem to be likely candidates for such spawnings. The smaller species of *Liopropoma* (e.g., *L. rubre, L. mowbrayi*) are small at maturity (less than 10 cm) and adapt well to standard aquarium conditions. *Pseudogramma* is also likely to be spawnable, but given its extremely retiring nature it is likely to spawn only when given complete privacy, numerous shelter holes, and copious feedings of live food.

Grammistid larvae have not yet been reared in captivity. No attempts to do so have been reported.

Literature Cited

Aboussovan, A. 1972. Oeufs et larvae de Teleosteens de l'Ouest Africain. XI. Larves serraniformes. *Bull. de IFAN*, 34, Ser. A. (2):485-502.

Fourmanoir, P. 1971a. Notes Ichthyologiques (4). *Cah. O.R.S.T.O.M., Ser. Oceanogr.*, 9:491-500.

Fourmanoir, P. 1971b. Notes Ichthyologiques (3). *Cah. O.R.S.T.O.M., Ser. Oceanogr.*, 9:267-278.

Fourmanoir, P. 1976. Formes post-larvaires et juveniles de poissons cotiers pris au chalut pelagique dans le sud-ouest Pacifique. *Cah. du Pacifique*, (19):47-88.

Gosline, W.A. 1960. A new Hawaiian percoid fish, *Suttonia lineata*, with a discussion of its relationships and a definition of the family Grammistidae. *Pac. Sci.*, 14:28-38.

Hubbs, C.L. and Y.T. Chu. 1934. Asiatic fishes (*Diploprion* and *Laeops*) having a greatly elongated dorsal ray in very large post-larvae. *Occas. Pap. Mus. Zool., Univ. Mich.*, (299):1-7.

Kendall, A.W., Jr. 1977. Relationships among American serranid fishes based on the morphology of their larvae. Ph. D. Diss., Univ. Calif. San Diego.

Kendall, A.W., Jr. 1979. Morphological comparisons of North American sea bass larvae (Pisces:Serranidae). *Tech. Rpt. N.M.F.S. Circ.* (428):50 pp.

Kotthaus, A. 1970. *Flagelloserranus*, a new genus of serranid fishes with the description of two new species (Pisces, Percomorphi). *Dana Rpt.*, 78:31 pp.

Randall, J.E., K. Aida, T. Hibiya, N. Mitsuura, H. Kamiya, and Y. Hashimoto. 1971. Grammistin, the skin toxin of soapfishes, and its significance in the classification of the Grammistidae. *Publ. Seto Mar. Biol. Lab.*, 19:157-190.

Smith, C.L. 1965. The patterns of sexuality and the classification of serranid fishes. *Amer. Mus. Novitat.*, (2207):20 pp.

Smith, C.L. 1971. Secondary gonochorism in the serranid genus *Liopropoma. Copeia*, 1971:316-319.

Smith, C.L. and E.H. Atz. 1969. The sexual mechanism of the reef bass *Pseudogramma bermudensis* and its implications in the classification of the Pseudogrammidae (Pisces: Perciformes). *Zeit. Morph. Tiere*, 65:315-326.

Thresher, R.E. 1980. *Reef Fish*. Palmetto Publ., St. Petersburg, Fla. 172 pp.

About eight to 12 hours prior to spawning, the female *Antennarius zebrinus* begins to fill up with eggs. This proceeds at a rapid rate so that shortly before spawning she is so distended it is hard for her to maintain her position on the bottom of the tank. She becomes buoyant (tail up as shown) and is followed closely around the tank by the male. Photo by U. Erich Friese.

The male continues to nudge the female in the abdomen, and they move quickly to the surface, where spawning occurs. Photo by U. Erich Friese.

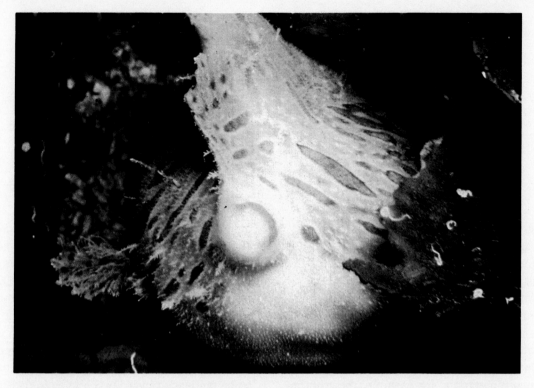

Egg-laying occurs in an almost explosive fashion, in some instances in as little as 0.4 seconds. There is some indication that the male, following close behind her, actually aids in pulling the large egg mass from her as he sheds his milt. Photo by U. Erich Friese.

The egg mass after extrusion. As can be seen by comparison with the hand holding it, the egg mass is quite large. The female often appears dazed or exhausted after she returns to the bottom once the eggs are released. Photo by U. Erich Friese.

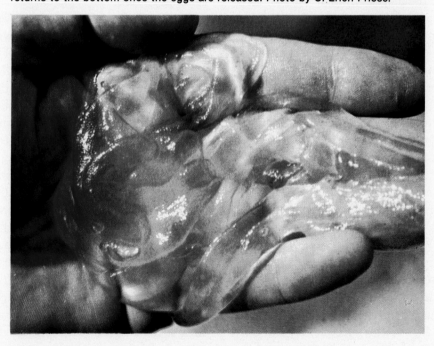

Basslets
(Pseudochromoids)

Basslets are small, often colorful, reef fishes characterized by an elongate, robust body, a long and undivided dorsal fin, pelvic fins inserted ahead of the pectoral fins, and an interrupted or incomplete lateral line. Popular and well-known examples of the group include the royal gramma (*Gramma loreto*), the black-cap basslet (*G. melacara*), the marine betta (*Calloplesiops altivelis*), and the various species of *Pseudochromis*. Currently the systematics of the group are poorly known (one family, for example, the acanthoclinids, has even been considered blennies, e.g., Breder & Rosen, 1966; Jillett, 1968a & b) and are clearly in need of revision. The last comprehensive review of the group recognized seven families (Pseudochromidae, Pseudogrammidae, Grammidae, Plesiopidae, Pseudoplesiopidae, Anisochromidae, and Acanthoclinidae) though also suggesting that some may not be valid and others may not belong in the group (Böhlke, 1960). Smith & Atz (1969) subsequently suggested that the pseudogrammids are more closely related to the Serranidae than to the basslets, per se, a move that has generally been well received. More recently, Springer, et al. (1977) merged Anisochromidae, Pseudoplesiopidae, and Pseudochromidae into a single larger family Pseudochromidae (the oldest name for the three available at that level) since they considered all three nominal families to be closely related and clearly of common origin. They were unable to determine relationships with other pseudochromoids, however, but did point out that other nominal families considered in the past to belong to this group (e.g., Opistognathidae and Owstoniidae) should be included in any revision of the overall status.

The level at which clearly related fishes are separated into different families and not simply considered genera within a family is largely a matter of judgment by systematists who have worked with the fishes (which, to a large extent, accounts for the frequent changes in familial status that occur and the alternative classifications that have been erected). In theory it makes little difference where one draws the line; in practice, however, it simplifies things greatly for the non-specialist to have a single stable name which identifies a clearly distinct and obviously closely related group of fishes. The common morphology of the various "families" of basslets and their common modes of reproduction (production of demersal eggs and a tendency toward oral incubation) clearly set the basslets as a whole off from the serranids (to which they are thought related and all of which produce pelagic eggs) and suggest that they be treated as a single family within which several distinct lines of development are visible. The evidence of widespread oral incubation in the group is especially compelling, since such parental care is rare in reef fishes (it occurs otherwise only in the apogonids) and it is unlikely that it has arisen independently in each of the nominal families.

Consequently, the reproductive biology of the various pseudochromoid families will be considered as a single unit, with one exception. The jawfishes, family Opistognathidae, will be treated separately only because there is enough information on that group alone to justify a separate section; in terms of their reproduction they share many features with the other pseudochromoid "families" and clearly belong in that group.

The pseudochromoids (other than jawfishes) consist of approximately 20 genera and 70 species. Most are found in the Indo-West Pacific, where they are abundant on coral reefs. A few, such as *Acanthoclinus quadridactylus*, and the genera *Trachinops* and *Paraplesiops*, have also invaded more temperate regions. Most are also rather retiring, typically seen hovering in caves or dashing from cover to cover seeking the small invertebrates that make up the bulk of their diets. Few exceed a total length of 20 cm; most, in fact, are only 5 to 10 cm long at sexual maturity.

SEXUAL DIMORPHISM

The degree and type of sexual dimorphism exhibited by most species of basslets are not known, but the few that have been examined suggest that the various species vary widely in the degree of dimorphism present. On one extreme, at least one species, *Acanthoclinus quadridactylus*, is monomorphic – the sexes do not differ in color nor are there significant size differences between them (Jillett, 1968a). On the other extreme, some species of *Pseudochromis* (e.g., *P. fridmani, P. magnificus, P. longipinnis, P. fuscus*) and both species of *Anisochromis* are conspicuously dichromatic (Schultz, 1967; Lubbock, 1975, 1977; pers. obs.; and J.L.B. Smith, 1955; M.M. Smith, 1977; Springer, et al., 1977, respectively). In all such cases the male is the more colorful of the two sexes. At least one species of *Pseudochromis* (*P. dilectus*) is not sexually dichromatic (Lubbock, 1976), however, which complicates making generalizations even about that genus. To date, dichromatic species appear to be a minority in the genus, but more detailed examination of the various nominal species may well indicate that some are only the different sexes of far fewer species. Sexual dichromatism has also been suggested for a western Atlantic grammid (*Lipogramma regia*) by Robins & Colin (1979); some individuals have a white bar on the body, whereas on others the bar is blue.

Lubbock (1975) reported some subtle differences between the sexes of a few species of *Pseudochromis*. In *P. fridmani*, and possibly also *P. sankeyi*, the caudal fin of the male is relatively longer than that of the female. In all species the sexes can theoretically be distinguished by the shape of the genital papilla – that of the male is more pointed than that of the female. As in other groups, however, this difference is clear only when the papilla is erect, and it requires a practiced eye to use the character.

The only form of sexual dimorphism that is likely to be widespread, though not universal, in the pseudochromoids is based on size. Males are known to attain larger average sizes than females in *Pseudochromis* (Lubbock, 1975), *Anisochromis* (J.L.B. Smith, 1955; Springer, et al., 1977), *Gramma* (Corsten-Hulsmans & Corsten, 1974), and *Assessor* (pers. obs.). Such size differences are likely to be demonstrated in species in other genera but, as noted, are not universal. The sexes of *Acanthoclinus quadridactylus*, a subtidal species found in New Zealand, overlap broadly in size ranges and are nearly identical in mean lengths (Jillett, 1968a). Similarly, the sexes of several species of *Pseudochromis* on the Great Barrier Reef appear to overlap broadly in size (pers. obs.)

The occurrence of size dimorphism in at least some species and the presumed close systematic affinities of the group with the hermaphroditic serranoids suggest that the pseudochromoids may be sequential hermaphrodites. Thus far gonads have been examined histologically for only two species: *Gramma loreto* (Corsten-Hulsmans & Corsten, 1974) and *Anisochromis straussi* (Springer, et al., 1977). Both found evidence of hermaphroditism, in the form of degenerate oocytes in the testes of mature males, testicular regions in the ovarian wall (in *A. straussi* these were located in the single germinal ridge along the dorsal side of each gonad), and individuals that appeared to be transitional between sexes. Despite this evidence, it is premature to conclude that hermaphroditism is characteristic of pseudochromoids; the broad overlap in size ranges of the sexes of *Acanthoclinus quadridactylus*, for example, suggest it to be gonochoristic.

SPAWNING SEASON

Little is known about the spawning seasons of tropical pseudochromoids. *Gramma loreto*, a common western Atlantic species, recruits to the reefs as tiny juveniles throughout the year off Curacao, suggesting some spawning occurs throughout the year, though there are peaks in recruit abundance in late spring and late summer (Luckhurst & Luckhurst, 1977). Allen & Kuiter (1976) suggested that *Assessor macneilli*, a western Pacific plesiopid, begins to spawn ". . . in mid-November, at the onset of warm summer weather . . .", but observations on the occurrence of mouth-brooding males at One Tree Island, Great Barrier Reef, strongly suggest spawning typically occurs in late winter and early spring (pers. obs.). Winter spawning is also characteristic of the temperate species *Acanthoclinus quadridactylus* at New Zealand (Jillett, 1968b).

The only information relevant to possible lunar cycles of reproductive activity is from Lubbock (1975), who noted that a pair of *Pseudochromis flavivertex* in captivity spawned every two weeks for a year, suggesting a semilunar cycle.

REPRODUCTIVE BEHAVIOR

Considering their abundance on the reef and, for at least some species, their conspicuousness, remarkably

Some carangid juveniles are commonly seen seeking protection under a variety of floating objects, including under the bells of jellyfishes. Here *Chloroscombrus chrysurus* is associated with the jellyfish *Aurelia aurita*. Photo by Charles Arneson.

Adult carangids are generally large, silver-colored fishes often found in small to very large schools. Spawning presumably occurs between school members. Photo by Wade Doak.

The juvenile *Equetus acuminatus* has elongate dorsal, caudal, and pelvic fin rays. As it grows these become proportionately shorter in comparison to the body. Photo by Dr. David L. Ballantine.

The adult *Equetus acuminatus* has comparatively short fins and has developed some extra horizontal black lines. Spawning was observed for this species in an aquarium and involved three individuals between 12 and 15 cm long. Photo by A. van den Nieuwenhuizen.

little is known about spawning of pseudochromoids or even about their general social behavior. Lubbock (1975) suggested that various species of *Pseudochromis* may be found in pairs, though such pairings may not be conspicuous. The planktivorous species, such as *Gramma loreto* and *Assessor macneilli*, are usually found in small to large aggregations (up to a hundred individuals) and are presumably promiscuous spawners. Occasional aggressive displays observed in *G. loreto* suggest that it is either territorial or, more likely, organized into some size-based dominance hierarchy. Most other, more solitary, species appear to be territorial.

Courtship and spawning have thus far been observed only in four species, all in the genus *Pseudochromis*. Lubbock (1975) reported spawning *P. flavivertex* and *P. olivaceus* in captivity. Shortly after the introduction of the fish into the aquarium, the male excavated a burrow under a piece of *Porites* (such burrow construction is common in pseudochromoids in captivity) which subsequently became the focus of his territory. The occurrence of spawning was signalled by a decline in the male's general aggressiveness toward the female and by the repeated performance of a "leading" display. The male swam toward the female, stopping about 10 cm from her, and then turned and swam back to his burrow with an exaggerated, undulating motion. Initially, the female did not visibly respond, but finally she followed the male to his burrow and entered it. Once inside she turned upside-down and the male moved next to her, without turning over, and pressed his genital region next to hers. After several such pressings extrusion of the eggs began and took from two to three hours to complete. As the eggs were extruded they adhered to the female's genital papilla until a spherical egg ball 20 to 30 mm in diameter had developed. When completed, the ball dropped to the floor. The male then chased off the female and began to fan the eggs with his pectoral fins.

Generally similar, though incomplete, courtship sequences have also been observed for *P. fuscus* and *P. mccullochi* on the reef (pers. obs.). In the former a courting pair consisting of a dusky yellow male and a black-bodied-white-tailed female was encountered at the edge of a shallow patch reef in early afternoon. The male was darker colored than normal and his caudal fin lighter, such that his overall pattern was more similar to the female's than normal; the female, in contrast, was normally colored but was also conspicuously distended with eggs. The male swam ahead of the female, angled slightly head-down, to a small coral rock located about half a meter off the main reef. He hovered, with his caudal fin slightly curled and his pelvic fins fully spread, over a conspicuous dark hole in the top center of the rock. As the female approached, the male periodically quivered violently while holding his position. After several such bouts of quivering, he slowly swam around the female, remaining about 30 cm from her. The female, which had remained motionless during this display, suddenly dashed into the dark crevice, quickly followed by the male. The male subsequently reappeared briefly at 19 minutes and at 27 minutes after entering the rock; shortly thereafter it began emerging sporadically to display aggressively to other approaching fishes and to chase some off. It was especially aggressive toward wrasses. Fifty-one minutes after entering the rock the female, much thinner than when she entered, emerged. After hesitating for a minute, she returned to the main reef, being harassed by the male on the way. The male quickly returned to the rock and remained there for the next several days. Although that rock was too massive to be opened for inspection, another similarly located and similar sized rock nearby was opened when it was observed that a male *P. fuscus* was spending an unusually large amount of time there. Inside this second rock was found a central cavity which contained a broad fluffy egg mass that covered the floor and sides of the cavity. The male entered the center of the mass sporadically, fanned the eggs, and prodded them with his jaws.

A similar but shorter sequence was observed in *P. mccullochi*. The normally blood-red male was brighter colored than normal, and the black line that edged his anal fin was more conspicuous than normal. The male was observed darting back and forth between a dark green female and a small dark bore-hole ahead of her. The male frequently stopped and hovered, with his tail slightly curled, his unpaired fins fully spread, and his head toward the bore-hole. The female approached the hole, at which point the male darted to her and pressed his jaws against her abdomen. Immediately, the pair darted into the bore-hole. The female reemerged in ten minutes.

Spawning by *Acanthoclinus quadridactylus* is somewhat different, according to Graham (1939). He suggested that the female chooses the spawning site and scoops out a shallow hole under a rock in which she deposits her ball of eggs. The male fertilizes the eggs and then seals the entrance to the burrow after the female departs. He apparently remains with the eggs in the closed chamber until they hatch. Nest construction, though not spawning, has also been observed in two western Atlantic species, *Gramma loreto* (Rosti, 1967; Corsten-Hulsmans & Corsten, 1974) and *Lipogramma trilineata* (Thresher, 1980). In the former the nest is a small hole in coral rock which is carefully lined with bits of algae; spawning apparently took place in the burrow. Nest construction by *L. trilineata* occurred on a flat algae-covered rock. The single fish observed pulled at and prodded the algae until it formed a shallow double-ended tube.

Following spawning, pseudochromoid eggs are either orally-brooded (e.g., *Gramma loreto*–Rosti, 1967; *Assessor macneilli*–Allen & Kuiter, 1976), exist as a small spherical mass inside a cave or burrow (e.g., several species of *Pseudochromis*–Lubbock, 1975; P. Doherty, pers. comm.; *Acanthoclinus quadridactylus*–Graham, 1939; Jillett, 1968a), form a fluffy mass inside the spawning cavity (e.g., *Pseudochromis fuscus*–pers. obs.; *Trachinops taeniatus*–R. Kuiter, pers. comm.), or are deposited in a single layer on the underside of a rock (e.g., *Plesiops semeion*–Mito, 1955). In all cases where the sex of the tending adult has been determined, it has been the male. The occurrence of spherical egg balls, which are normally associated with oral-brooding fishes, in nonbrooding fishes suggests that oral incubation is the primitive condition in the group. If so, the broad fluffy egg masses of *P. fuscus* and *Trachinops taeniatus* may represent the advanced condition. Thus far, detailed observations on the orally brooding male are available only for *Assessor macneilli*. When brooding, the gular region of the male is notably, though not overwhelmingly, distended. Such males are not conspicuous in an aggregation of hovering individuals but can be seen if specifically searched for. Brooding males act much like nonbrooding individuals but apparently do not feed. They are also more skittish than other individuals and are usually the first to dash for cover if the aggregation is threatened.

EGGS AND LARVAE

Pseudochromoid egg balls range in diameter from 7 mm, containing about 60 eggs (*Assessor macneilli*, pers. obs.) to 5-8 cm, containing 8,200 to 17,500 eggs (*Acanthoclinus quadridactylus*, Jillett, 1968a). Measurements have been made on the eggs of only a few species: *Assessor macneilli* eggs are 0.89 to 0.95 mm in diameter (pers. obs.); *Plesiops semeion* eggs are 0.9 by 0.6 mm (Mito, 1955); *Pseudochromis fuscus* eggs

are 1.25 mm (pers. obs.); and *Acanthoclinus quadridactylus* eggs are 1.43 mm (Jillett, 1968a). The eggs are slightly elongate spheroids from which emerge several "sticky" threads. Inconspicuous channels run between the eggs in an egg ball, permitting water circulation in the mass. Egg development is discussed briefly by Jillett (1968a). Incubation periods range from three to five days in *P. fuscus* at approximately 29°C to 16 days for *A. macneilli* (no temperature given). In all cases hatching occurred at night; in most, if not all, cases it occurs shortly after sunset.

The newly hatched larvae are relatively large, ranging in length from 2.5 mm (*P. fuscus*) to 4.75 mm (*A. quadridactylus*). The yolk sac is small, the eyes are pigmented, and the jaws are well formed. Feeding by tropical species begins on the first day after hatching (e.g., Lubbock, 1975), while in the temperate *A. quadridactylus* the yolk sac was not fully absorbed until six days after hatching. Older larvae have been described only for the latter species by Jillett (1968a), who maintained a few until day 11. Such larvae were 5.5 mm long, relatively robust, and heavily pigmented. Jillett estimated that the animals remain planktonic for three months, based on the interval between first observation of egg masses and the first arrival of new young. Sexual maturity for *A. quadridactylus* is reached, on the average, in three years.

REPRODUCTION IN CAPTIVITY

To date, five pseudochromoids are known to have spawned in captivity: *Pseudochromis flavivertex* and *P. olivaceus* (Lubbock, 1975), *Assessor macneilli* (Allen & Kuiter, 1976), *Gramma loreto* (Rosti, 1967), and *Trachinops taeniatus* (R. Kuiter, pers. comm.). These represent the nominal families Pseudochromidae, Plesiopidae, and Grammidae. Spawning by other species is likely to have also occurred in aquaria, but because of the relative inconspicuousness of mouthbrooding males and the tendency for other species to spawn well back in caves, such spawnings are likely to have been missed. Spawning at least the smaller species appears to be relatively easy, requiring little more than an aquarium large enough for the fish to establish individual territories (160 liters should suffice for all but the largest species) and heavy feedings. Most of the commonly available pseudochromoids are relatively hardy fishes that feed readily; bringing them into spawning condition, therefore, should not be a problem. The only specific requirement such fishes are likely to have is the need for a suitable spawning

Development in mugilids is fairly rapid, and within two weeks transformation into silvery schooling juveniles occurs. Photo by Dr. Fujio Yasuda.

Adult *Mugil cephalus* are commonly spawned by using injections of hormones and stripping gametes. Photo by Dr. Fujio Yasuda.

Little is known about barracuda spawning behavior. In general, they migrate into specific spawning areas where they accumulate in large numbers. Photo of *Sphyraena* sp. by Helmut Debelius.

An *Aulostomus chinensis* pair immediately prior to spawning, showing sexual dichromatism: male with temporary banded courtship color and yellow female heavy with eggs. Photo by Ann Gronell.

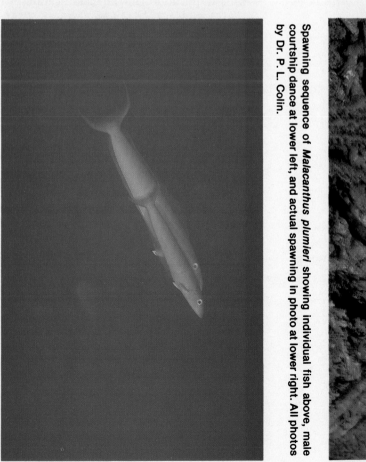

Spawning sequence of *Malacanthus plumieri* showing individual fish above, male courtship dance at lower left, and actual spawning in photo at lower right. All photos by Dr. P. L. Colin.

site; i.e., a deep hole or cave, preferably closed at one end. Since most species are quite ready to dig their own shelter holes, a decent sized overturned clam shell should do nicely.

Pseudochromoid larvae have not yet been reared in captivity despite efforts to do so by Jillett (1968a) and Lubbock (1975). Despite these failures, the large size of the larvae at hatching would seem to bode well for rearing with a standard larval diet of *Brachionis plicatilis* followed by newly hatched brine shrimp.

Literature Cited

Allen, G.R. and R.H. Kuiter. 1976. A review of the plesiopid fish genus *Assessor*, with descriptions of two new species. *Rec. West. Aust. Mus.*, 4:201-215.

Böhlke, J.E. 1960. Comments on serranoid fishes with disjunct lateral lines, with the description of a new one from the Bahamas. *Notulae Naturae*, (330):11 pp.

Breder, C.M., Jr. and D.E. Rosen. 1966. *Modes of Reproduction in Fishes*. Natural History Press, Garden City, N.Y. 941 pp.

Corsten-Hulsmans, C.J.F. and A.J.A. Corsten. 1974. *Gramma loreto*; een hermafrodiete koraalvis van Curacao: Oecologische aspecten en gevolgen van bevissing. *Rpt. Zool. Lab., Dept. Dieroecologie, K.U.Nijmegen*, (92):64 pp.

Graham, D.H. 1939. Breeding habits of the fishes of Otago Harbour and adjacent seas. *Trans. Roy. Soc. New Zealand*, 69:361-372.

Jillett, J.B. 1968a. The biology of *Acanthoclinus quadridactylus* (Bloch and Schneider) (Teleostei-Blennioidea) II. Breeding and development. *Aust. J. Mar. Freshwater Res.*, 19:9-18.

Jillett, J.B. 1968b. The biology of *Acanthoclinus quadridactylus* (Bloch and Schneider) (Teleostei-Blennioidea) I. Age, growth, and food. *Aust. J. Mar. Freshwater Res.*, 19:1-8.

Lubbock, R. 1975. Fishes of the family Pseudochromidae (Perciformes) in the northwest Indian Ocean and Red Sea. *J. Zool., London*, 176:115-157.

Lubbock, R. 1976. Fishes of the family Pseudochromidae in the central Indian Ocean. *J. Nat. Hist.*, 10:167-177.

Lubbock, R. 1977. Fishes of the family Pseudochromidae (Perciformes) in the western Indian Ocean. *Ichthyol. Bull. J.L.B. Smith Inst. Ichthyol.*, (35):1-21.

Luckhurst, B.E. and K. Luckhurst. 1977. Recruitment patterns of coral reef fishes on the fringing reef of Curacao, Netherland Antilles. *Can. J. Zool.*, 55:681-689.

Mito, S. 1955. Breeding habits of a percoid fish, *Plesiops semeion*. *Sci. Bull. Fac. Agric., Kyushu Univ.*, 15:95-99.

Robins, C.R. and P.L. Colin. 1979. Three new grammid fishes from the Caribbean Sea. *Bull. Mar. Sci.*, 29:41-52.

Rosti, P. 1967. Breeding the royal gramma. *Salt Water Aquarium*, 2:106-108.

Schultz, L.P. 1967. A review of the fish genus *Labracinus* Schlegel, family Pseudochromidae, with notes on and illustrations of some related serranoid fishes. *Ichthyologica, The Aquarium J.*, 39:19-40.

Smith, C.L. and E.H. Atz. 1969. The sexual mechanism of the reef bass *Pseudogramma bermudensis* and its implications in the classification of the Pseudogrammidae (Pisces: Perciformes). *Z. Morph. Tiere.*, 65:315-326.

Smith, J.L.B. 1955. An especially colorful new pseudochromid fish. *Ann. Mag. Nat. Hist.*, 8:145-148.

Smith, M.M. 1977. A note on *Anisochromis kenyae* Smith, 1954. *Ichthyol. Bull. J.L.B. Smith Inst. Ichthyol.*, (35):22-23.

Springer, V.G., C.L. Smith, and T.F. Fraser. 1977. *Anisochromis straussi*, new species of protogynous hermaphroditic fish, and synonomy of Anisochromidae, Pseudoplesiopidae, and Pseudochromidae. *Smithson. Contrib. Zool.*, (252):15 pp.

Thresher, R.E. 1980. *Reef Fish*. Palmetto Publ., St. Petersburg, Fla. 172 pp.

Jawfishes
(Opistognathidae)

In the past, the jawfishes have systematically floated somewhere between the serranoids and the trachinoids. They are, however, clearly allied to the pseudochromoid fishes, sharing with them such features as an interrupted lateral line, pelvic fins inserted ahead of the pectorals, a tendency to dig burrows, and oral incubation. Unlike other pseudochromoids, the jawfishes inhabit the sand and silt plains that frequently surround reefs, where they dig narrow vertical burrows and feed on smaller animals. Burrow construction is a complex process in which small rocks, pieces of coral, and other debris are retrieved from the surrounding area and used to reinforce the burrow's entrance (see Colin, 1973).

As the family is currently recognized, there are three genera (*Opistognathus*, the largest; *Lonchopisthus*; and *Stalix*) and perhaps 30 species. Except for the eastern Atlantic, the family is distributed circumtropically. Most species are small, with few exceeding a length of 50 cm and most only about half that. Most are also rather drab fishes, typically mottled brown, gray, or black, although a few, such as the yellowhead jawfish, *Opistognathus aurifrons*, are colorful and popular aquarium fishes.

SEXUAL DIMORPHISM

The degree and type of sexual dimorphism vary among jawfishes. Most, it appears, including the yellowhead jawfish *Opistognathus aurifrons*, *Lonchopisthus micrognathus*, and the majority of the drab species of *Opistognathus*, are sexually monomorphic. Sexual dichromatism, however, is exhibited by a few species. In *Opistognathus* sp., the eastern Pacific bluespotted jawfish, such color differences are temporary and are assumed by the male during courtship. His color changes from the normal blue spots on a yellow background to a vividly contrasting white anterior and black posterior (A. Kerstitch, pers. comm.). Black and white color patterns are also characteristic of the few

species that are permanently dichromatic. Males of the western Atlantic species *Opistognathus gilberti* have a dusky white head, a black body, and, except for a few black marks, stark white fins (Böhlke, 1957); females are uniformly dull yellow. The male's color pattern intensifies during courtship. Similarly, both sexes of another undescribed eastern Pacific species of *Opistognathus* are pale anteriorly and black posteriorly; males, however, have black dorsal and anal fins (Rosenblatt, pers. comm.). This predominance of black and white sexual dichromatism in jawfishes is an interesting contrast to other reef fishes such as labrids and pomacanthids, in which sexual differences typically involve complex patterns of many colors. It would be of interest to know if the jawfishes differ from other reef-associated fishes in spectral sensitivity, either permanently or during courtship.

The sexes of some, but not all, cryptic species of *Opistognathus* differ in jaw morphology. In the western Atlantic species *Opistognathus macrognathus*, the maxilla (the upper jaw bone) extends well beyond the eye of the male but only reaches to the edge of the eye in the female (Smith-Vaniz & Colin, in press). *Opistognathus melachasme* is similar but is also sexually dichromatic: the male is darker overall, has a large black spot on the dorsal fin that is lacking in the female, and has more extensive black markings around the maxilla (Anderson & Smith-Vaniz, 1976).

The structure of jawfish gonads has not been examined yet. The similar sizes of males and females of most species, however, suggest gonochorism, despite their apparent systematic affinities to the hermaphroditic serranoids.

SPAWNING SEASON

Only two reports on spawning seasons of opistognathids are available. Colin (1972) indicated that the yellowhead jawfish, *O. aurifrons*, spawns from at least spring to late autumn off Florida; it probably breeds

Dendrochirus zebra courtship and spawning at Miyake-jima, Japan. All photos in this sequence by Jack T. Moyer. Here the male circles the female (with whiter face and swollen abdomen). Females develop the white face on the day of spawning.

During the day males seek out females who show signs of spawning and roost with them all day, spawning around sunset. Here the male rests with the female (white face) at her roost.

Courtship commences. The male follows the female out to sea (the spawning site), frequently circling her (see above).

Courtship continues, and there is occasional tactile stimulation with the fins.

Occasionally two males will fight over a female (smaller fish off to the lower left awaiting the results).

The male again circling the female during the courtship sequence. The white face of the female is quite noticeable.

The upper male in this photo actually stabbed the other male (lower fish) with his dorsal spines (it took a full second to dislodge the spines). The attacked male managed to spawn with another female seconds later.

The male prepares to brush the female with his fins, again using tactile stimulation.

Another male-against-male confrontation. The female is out of sight below the coral ledge. Note the flared opercula.

At last an actual spawning ascent. The eggs are evident in the swollen abdomen of the female (smaller individual).

During spawning, the male swings over the female and down, the female following immediately. It happened extremely quickly, and the action could not be determined until frozen by the camera's flash.

A pair of long-nosed hawkfish, *Oxycirrhites typus,* on the reef. Photo by Roger Steene.

111

year-round deeper in the tropics. Similarly, *Loncho-pisthus micrognathus* appears to spawn year-round off Puerto Rico (Colin & Arneson, 1978).

Colin (1972) reported that *O. macrognathus* spawn regularly at about two-week intervals in captivity, suggesting a semilunar spawning cycle. Otherwise, no information is available regarding such cycles in these fishes.

REPRODUCTIVE BEHAVIOR

Most, if not all, jawfishes are territorial, vigorously defending the immediate vicinity of the burrow from both conspecific and some interspecific intruders (Colin, 1971a). Beyond this, the social systems of jawfishes are poorly known. A few, such as *O. aurifrons* and *L. micrognathus*, are colonial; that is, large numbers can usually be found in close proximity to one another. In the latter species two individuals may even share the same burrow (heterosexual pairs?) (Colin & Arneson, 1978). Most species are presumably promiscuous, though insufficient work has been done to rule out significant mate fidelity. *O.*

aurifrons, however, maintains long-term pairs in which a male and female not only mate with one another regularly but also warn one another about approaching predators and help dig each other out if the burrow entrance is damaged (Colin, 1971b, 1972).

Courtship and spawning have been described in detail only for *O. aurifrons*, although some more limited data are available for other species. For *O. aurifrons*, *O. gilberti*, and the eastern Pacific blue-spotted jawfish, all small noncryptic species, courtship takes place in the water column above the burrow entrance. Because of the brilliant male colors, courtship by the last two species is particularly striking. Males of the blue-spotted jawfish court by ascending 0.3 to 1.0 m off the bottom and hovering motionless with fins fully spread and in full black and white courtship colors. After hovering for three to five seconds the male suddenly dashes back to his burrow, "flashing" white against black as he does so. This flashing sequence is repeated every four to five minutes for hours, ending only when the female follows the male back to his burrow for spawning (A. Kerstitch, pers. comm.). Courtship by *O. gilberti* is

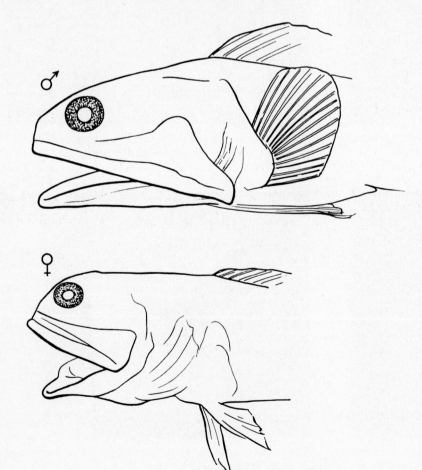

Sexual dimorphism in jaw morphology in the jawfish *Opistognathus melachasme*. Based on Anderson & Smith-Vaniz (1976).

Courtship in *Opistognathus aurifrons* involves fully spread vertical fins, and the mouth is often wide open. The branchiostegal rays are spread, causing the black markings under the jaw to be prominently displayed. Photo by H. Hansen, Aquarium Berlin.

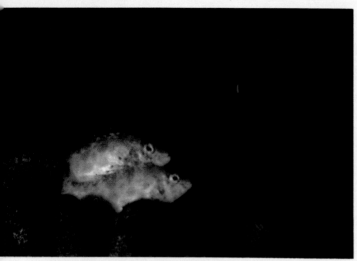

A pair of *Cirrhitichthys oxycephalus* in a pre-ascent position on a coral head. Photo by Dr. R. E. Thresher.

The same pair in their actual spawning ascent above the coral head. Photo by Dr. R. E. Thresher.

Part of a spawning aggregation of Nassau groupers (*Epinephelus striatus*) at Little Cayman Island. Photo by Dr. P. L. Colin.

Several studies have indicated a pronounced lunar influence on spawning in the Nassau grouper, with peaks of activity around the night of the new moon. Photo by Dr. P. L. Colin.

Spawning aggregations of approximately 30,000 to 100,000 individuals of *E. striatus* were seen on the edge of a bank at Bimini, Bahamas. Photo by Dr. P. L. Colin.

Epinephelus fasciatus courtship. The male in his courtship colors approaches the rear of the female. Photo by Ann Gronell.

114

A juvenile *Epinephelus flavocoeruleus* of about 35 mm from Sydney, Australia, apparently not long out of the plankton. At this age small groupers are generally very secretive. Photo by Rudie Kuiter.

An adult of apparently the same species. The spotted pattern is evident in both adult and juvenile although there are other color changes. Photo by Scott Johnson.

Serranines generally settle out at a smaller size than epinephelines, i.e., 15-20 mm as compared to 20-35 mm. Shown here is a very young *Serranus incisus.* Photo by Dr. P. L. Colin.

A male *Opistognathus aurifrons* hovering over his burrow while holding a mouthful of eggs. Photo by H. Hansen, Aquarium Berlin.

It will occasionally agitate the eggs or actually spit them out as shown, possibly to ventilate them. Photo by H. Hansen, Aquarium Berlin.

similar. The male emerges from his burrow with his white fins folded against his black body. The fins are then spread and closed quickly, flashing white for a few seconds, and then the male dashes back to his burrow (P.L. Colin, pers. comm.).

Courtship and spawning by the yellowhead jawfish is described by Leong (1967) and Colin (1972). With the species being a planktivore that normally hovers a meter or so off the bottom, courtship not surprisingly takes place in the water column. Approaching the female, a male orients broadside to her and assumes a horizontal position. His head and tail are bent upward (Colin referred to the display as an "arch"), all fins but the pelvics are fully spread, the jaws are open, and the branchiostegal rays are spread. In fishes from the Bahamas (indicated by a dusky head) thin black lines under the jaws and under the branchiostegals are exposed during the display. After an arch, if the female is ready she follows the male either to his burrow or occasionally to a third "neutral" burrow. The male enters first, the female follows, and spawning presumably takes place in the terminal cavity of the burrow. Leong suggested that spawning by the yellowhead is signalled by the female entering, briefly,

the male's burrow more often than usual. Colin found that arching occurred throughout the day but was most common at dawn and dusk, suggesting that most spawning occurs at these times.

Following spawning, the eggs are tended and guarded by the male (at least in those cases where the sex of the tending adult has been determined). Most, if not all, jawfishes orally brood their eggs, with such eggs usually protruding slightly from the jaws and easily visible (e.g., Böhlke & Chaplin, 1957; Leong, 1967). In at least two species, however, the eggs are apparently set down in the burrow occasionally (*Opistognathus aurifrons*—Leong, 1967; *Lonchopisthus micrognathus*—Colin & Arneson, 1978), a situation reminiscent of the males tending demersal egg balls in the nominal families Pseudochromidae and Acanthoclinidae. In at least some species the egg masses are occasionally agitated back and forth in the mouth, apparently to ventilate them (Colin & Arneson, 1978).

EGGS AND LARVAE

Jawfish eggs are slightly elongated spheroids approximately 0.8 to 0.9 mm in diameter (Colin & Arne-

In times of threat of danger, the jawfish will retreat to the safety of his burrow. Photo by H. Hansen, Aquarium Berlin.

It is not often that spawning in marine fishes can be documented so vividly in their natural environment as seen here on this and the following pages. Involved are several species of hamlets (if they are indeed separate species): *Hypoplectrus gummigutta* (yellow fish), *H. nigricans* (brownish violet), and *H. unicolor* (spot on caudal peduncle). All spawning photos by Charles Arneson.

son, 1978; Smith-Vaniz & Colin, in press). Hatching by *O. aurifrons* occurs in seven to nine days at 25-26°C.

Larval development has thus far been described only for *O. aurifrons* (Smith-Vaniz & Colin, in press), though Vatanachi (1974) illustrated an 8.0 mm larva of an unidentified species of *Opistognathus*. Shortly after hatching, larval *O. aurifrons* are just under 4 mm long and have pigmented eyes, well developed jaws, rudiments of the pectoral and caudal fins, and only a small yolk sac. Development is rapid, and the fins are fully formed by the end of nine days, at which time the larvae are readily recognizable as small jawfish. Settlement to the bottom occurs in 15 days for *O. aurifrons* and in 18 days for *O. macrognathus* (Colin, 1972). Newly settled jawfishes are roughly 10 to 15 mm long and immediately begin burrowing. The juveniles of *O. aurifrons* and *L. micrognathus* appear to settle preferentially near adults. The mechanism that triggers such settlement patterns has not been determined but is probably based on some olfactory cue. Age or size at sexual maturity is not known for any species.

REPRODUCTION IN CAPTIVITY

The smaller jawfishes, such as *O. aurifrons* and *O. macrognathus*, spawn readily in captivity and require little more than a fairly large aquarium to themselves, a deep gravel bed in which they can dig adequate burrows, and heavy feedings. Because of its small size, adaptability to aquarium conditions, and mild disposition, the yellowhead jawfish, *O. aurifrons*, has proven especially attractive to prospective breeders. The best way to facilitate spawning in this species is to establish a large colony in a relatively large aquarium (400 + liters) and let the fish pair naturally. Under such conditions the fish usually spawn regularly.

The only reports of successful rearing of jawfish larvae in captivity are those of Colin (1972) for *O. macrognathus* and Smith-Vaniz & Colin (in press) for *O. aurifrons*. In both cases the larvae were reared in small bare aquaria and fed size-sorted wild-caught plankton. Although not reported in the literature due to its potential commercial value, commercial breeders have been rearing *O. aurifrons* for some time, the basic approach apparently involving use of semi-closed systems and feedings with the rotifer *Brachionis*

plicatilis, size-sorted plankton, and (later) *Artemia*. The large size of the larvae at hatching and the speed of their development into benthic juveniles suggest such fishes will be relatively easy even for amateur aquarists to rear.

Literature Cited

Anderson, W.D., Jr. and W.F. Smith-Vaniz. 1976. Sexual dimorphism in the jawfish *Opistognathus melachasme*. *Copeia*, 1976:202-204.

Böhlke, J.E. 1967. A new sexually dimorphic jawfish (*Opistognathus*, Opistognathidae) from the Bahamas. *Notulae Naturae*, (407):1-12.

Böhlke, J.E. and C.C.G. Chaplin. 1957. Oral incubation in Bahamian jawfishes *Opisthognathus whitehursti* and *O. maxillosus*. *Science, N.Y.*, 125:353.

Böhlke, J.E. and C.C.G. Chaplin. 1968. *Fishes of the Bahamas and adjacent tropical waters*. Livingston Press, Wynnewood, Pennsylvania. 772 pp.

Colin, P.L. 1971a. Interspecific relationships of the yellowhead jawfish, *Opistognathus aurifrons* (Pisces, Opistognathidae). *Copeia*, 1971:469-473.

Colin, P.L. 1971b. The other reef. *Sea Frontiers*, 17: 160-170.

Colin, P.L. 1972. Daily activity patterns and effects of environmental conditions on the behavior of the yellowhead jawfish, *Opistognathus aurifrons*, with notes on its ecology. *Zoologica, N.Y.*, 57:137-169.

Colin, P.L. 1973. Burrowing behavior of the yellowhead jawfish, *Opistognathus aurifrons*. *Copeia*, 1973:84-89.

Colin, P.L. and D.W. Arneson. 1978. Aspects of the natural history of the swordtail jawfish, *Lonchopisthus micrognathus* (Poey) (Pisces: Opistognathidae), in southwestern Puerto Rico. *J. Natur. Hist.*, 12:689-697.

Leong, D. 1967. Breeding and territorial behaviour in *Opisthognathus aurifrons* (Opisthognathidae). *Naturwissenschaften*, 54:97.

Smith-Vaniz, W. and P.L. Colin. (In Press). Systematics and biology of western Atlantic jawfish (Pisces: Opistognathidae). *Proc. Acad. Natur. Sci. Phila.*

Vatanachai, S. 1974. The identification of fish eggs and larvae obtained from the survey cruises in the South China Sea. *Proc. Indo-Pacific Fish. Council*, 15:111-130.

Snappers
(Lutjanidae)

Snappers are ubiquitous, usually mid-sized predators that are common, often in very large numbers, not only on the reef but also in most other shallow-water tropical habitats. By day they mill about in loose aggregations of up to several hundred fish, usually close to cover, and at night they fan out from these shelter sites to search out the benthic crustaceans and small fishes that make up the bulk of their diets. Starck & Schroeder (1970) suggested that young gray snappers, *Lutjanus griseus*, roam a mile or more from their shelter sites each night.

Lutjanids are found circumtropically and are a speciose group with about two dozen genera and about ten times that many species. All are perch-like and characterized by a single dorsal fin with well developed spines, a laterally compressed body, well developed jaws that usually bear conspicuous canine teeth, and a head that is triangular in profile. In benthic species the lower jaw usually projects somewhat beyond the upper one, giving the fish a somewhat pugnacious look. Many snappers are also quite colorful, especially as juveniles, but as a whole the group tends to grays and browns and is of interest more to fishermen than to aquarists.

SEXUAL DIMORPHISM

As a rule there are no conspicuous external differences between the sexes in lutjanids. Sexual dimorphism has thus far been reported only for two species of *Pristipomoides,* a genus of deep-water fishes (Kami, 1973). In both *P. filamentosus* and *P. auricilla* from Guam, large males have more yellow in their unpaired fins than do small males and females. The function of such color differences is not known, and at least two other species of *Pristipomoides* at Guam are sexually monomorphic.

Starck and Schroeder (1970) suggested that females of the western Atlantic species *Lutjanus griseus* reach a slightly larger maximum size than do males. More detailed work, however, indicates that in most, if not all, species, males average larger than females with a broad overlap in size ranges (e.g., Almeida, 1965; Thompson & Munro, 1974).

Starck and Schroeder also noted a difference in habitat preference by the two sexes of *L. griseus:* except for spawning periods, where the sexes obviously come together, males dominated the population on the reef, while most fishes inshore were female. Camber (1955) reported that the proportion of females in a population of *Lutjanus campechanus* (as *L. aya*) decreases with age but offered no explanation for the trend.

The sexes can be readily identified based on the shape of the gonads. In both sexes the gonads are paired and found at the rear of the abdominal cavity just below the swim bladder. Ovaries are round and tubular; testes are flattened, ribbon-like, and lack a lumen. Early work with the group suggested that lutjanids were hermaphroditic, but more recent studies indicate that most, if not all, are normally gonochoristic (see discussion in Atz, 1964).

SPAWNING SEASON

Spawning for most tropical snappers seems to occur over a large part of the year and may take place year-round for many species. Spawning peaks, however, generally coincide with periods of warm water temperature, though not necessarily the warmest part of the year. In the western Atlantic, for example, spawning reaches a peak in the summer near the northern limits of the family's range (e.g., *Lutjanus griseus* and *L. synagris* off Florida—Starck & Schroeder, 1970; Reshetnikov & Claro, 1976; *L. campechanus* (as *L. aya*) off Texas—Mosley, 1966) but peaks in the spring or is bimodal with peaks in the spring and fall in the tropics (Munro, et al., 1973; Thompson & Munro, 1974). Long spawning periods with spring peaks of activity are also reported for various snappers in other

Each hamlet maintains an individual territory during the day. At dusk, however, pairs form and individuals chase one another across the reef, first one and then the other actively courting.

In spawning, the pair "leap" off the bottom, the "female" flexing its body into an "S" shape while the "male" wraps its body around "hers." After 10-20 seconds, the embrace breaks apart and the pair rush back to the bottom to begin courtship again.

Spawning clasp in the butter hamlet, *Hypoplectrus unicolor*. Photo by R. E. Thresher.

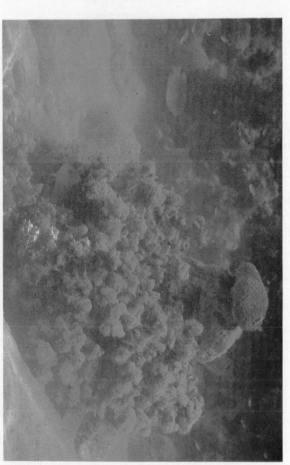

A male *Pseudochromis fuscus* peering out of a cave in which spawning took place. Photo by Dr. R. E. Thresher.

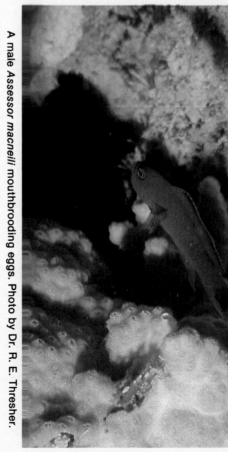

A male *Assessor macneili* mouthbrooding eggs. Photo by Dr. R. E. Thresher.

A male *P. fuscus* following a female after spawning. Photo by Dr. R. E. Thresher.

tropical areas by Rangarajan (1971), Nzioka (1979), and Johannes (1981).

The limited data available also suggest a pronounced lunar component to the timing of lutjanid reproduction. Peak spawning on or close to the night of the full moon has been suggested for *Lutjanus synagris* off Cuba, *L. griseus* off Florida, *L. vaigiensis* off Tahiti, and several species of *Lutjanus* and a *Symphorus* off Palau (Reshetnikov & Claro, 1976; Starck & Schroeder, 1970; Randall & Brock, 1960; and Johannes, 1978, respectively). *Aprion virescens* off Palau, however, appears to spawn just after the new, rather than the full, moon (Johannes, 1981). Similar lunar rhythms are probably universal in the family.

REPRODUCTIVE BEHAVIOR

A key feature of reproduction by inshore-dwelling lutjanids is an extensive spawning migration to select areas along the outer reef in the week or so prior to the full moon. Again, data are limited but consistent. Such spawning migrations have been reported for *Lutjanus griseus* (Croker, 1960; Starck & Schroeder, 1970) and for several species off Palau (Johannes, 1978, 1981). Similarly, Moe (1963) reported two areas of concentrated spawning activity of *L. campechanus* (as *L. aya*) off the northern coast of Florida, each at the edge of deep water between 20 and 40 m. Not surprisingly, deeper water species which are normally found on the outer reef apparently move little or not at all to spawn (e.g., Wicklund, 1969; Nzioka, 1979). All observations made thus far indicate spawning occurs late in the evening or early at night.

Spawning by lutjanids has been observed three times, twice in the field and once in captivity. All three involve group spawning at dusk. Wicklund (1969) observed groups of *Lutjanus synagris* in 15 m of water off Florida become restless and increase their rate of milling about half an hour before sunset. Groups of five to ten fish split off from the main aggregate, with each group focused on and pursuing a lighter colored individual (presumably a female). There was occasional brief pairing, and several of the males frequently "nuzzled" the anal region of the female. As milling activity became more vigorous the groups began to ascend slowly into the water column. About 2.2 m up each group suddenly drew together, became very active, and then burst apart, leaving behind a milky cloud (presumably gametes). Spawning activity ended just before dark as the fish dispersed to forage. Virtually identical spawning behavior has been recently reported for *Lutjanus kasmira* in

a large aquarium (Suzuki & Hioki, 1979). There was brief pairing, nuzzling of the female's anal region, and a brief ascent to group spawn slightly off the bottom. The behavior differed only in that the fish spiraled off the bottom rather than ascending directly, and it appeared that several females were involved in each spawning group rather than one. Similar milling and formation of aggregates at dusk have also been reported for *L. griseus* off Florida (Starck & Schroeder, 1970) and for *L. amabilis* off One Tree Island, Great Barrier Reef (pers. obs.), again near the night of the full moon, but no conspicuous spawning was evident. Finally, *Pterocaesio diagramma* spawn in a similar, if not identical, manner off One Tree Island. At dusk, one commonly sees the fish in pairs, with one individual vigorously chasing another. In early summer, shortly before the full moon, P. Doherty (pers. comm.) observed a large school of *P. diagramma* to rush together a few feet off the bottom, draw into a compact unit, and then dash apart leaving behind a white cloud. Such activity was seen only once, on the windward side of the reef in approximately 18 m of water.

EGGS AND LARVAE

Although Starck & Schroeder (1970) suggested otherwise, based on observation of infertile eggs of *Lutjanus griseus*, lutjanids produce a pelagic egg (Suzuki & Hioki, 1979). The transparent and spherical eggs of *Lutjanus kasmira* are 0.78 to 0.85 mm in diameter, with each containing a single oil droplet. Hatching occurs in 18 hours at 22.3 to 25°C. Egg development is detailed by Suzuki & Hioki (1979).

Larval development has been described in *Lutjanus kasmira* (Suzuki & Hioki, 1979), in *L. griseus* (Richards & Saksena, 1980), and, based on specimens obtained in plankton tows, in *Rhomboplites aurorubens* (La Roche, 1977) and *L. campechanus* (Collins, et al., 1980). Newly hatched *L. kasmira* are 1.83 mm long and have a large, elliptical yolk sac that has an oil droplet at its anteriormost point. By the third day after hatching the larvae have reached a length of slightly over 3 mm, have completely absorbed the yolk sac, and have begun feeding. Feeding begins at about 3 mm in *L. griseus* as well. By the time they reach a length of 4 mm larval snappers are elongate with a large head, prominent eyes, spines in the dorsal fin, smaller spines on the opercula, and a continuous fin fold around the rear part of the body. By 4.7 mm pelvic and anal spines develop and scattered pigment spots begin to appear. Later-stage larvae are characterized by a large head, a laterally compressed

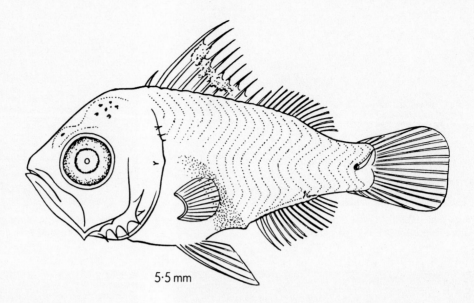

Postlarval snapper (*Lutjanus* sp.). Based on Vatanachai (1974).

5·5 mm

body, conspicuous preopercular spines, and spines in the dorsal and anal fins, including a particularly long and smooth second dorsal fin spine (Starck & Schroeder, 1970; Fourmanoir, 1973; Heemstra, 1974; Vatanachai, 1974; Musiy & Sergiyenko, 1977). The duration of the planktonic larval stage of *L. griseus* reared in captivity is about four weeks. Metamorphosis to the juvenile stage for most species occurs at a length of approximately 12 to 20 mm.

This length is comparable to that of juveniles of *Lutjanus griseus* when they first appear in grass flats, still showing some postlarval characteristics such as long spines in the fins and only slight pigmentation (Starck & Schroeder, 1970). Initially, such juveniles remain in the grass beds, foraging as individuals during the day. At a length of approximately 80 mm they begin to aggregate around debris and channel edges and forage at night. Sexual maturity occurs in three years when they reach a length of 17.5 to 18.0 cm. Sexual maturity in other western Atlantic species occurs at lengths ranging from about 25 to 50 cm SL, with females usually maturing at a smaller size than males (Thompson & Munro, 1974). Almeida (1965) suggested that females of *L. campechanus* (as *L. aya*) grow faster than the males and mature at an earlier age.

REPRODUCTION IN CAPTIVITY

Snappers have only been spawned once in captivity. This was accomplished by Suzuki & Hioki (1979), who spawned *Lutjanus kasmira* in a large observation tank. This concrete and glass aquarium had a volume of 19.7 cubic feet (about 650 liters), was filtered by means of an external sand filter, and contained a varie-

ty of relatively large predatory fishes along with the 20 snappers. Water temperature averaged 23.3°C, pH 7.7, and salinity 33.35 parts per thousand. Adults ranged in size from 24.5 to 33.6 cm. Spawning occurred at dusk on five days in late summer and, as described above, was always a group affair. Aside from routine heavy feedings, presumably with fish, the snappers apparently received no special conditioning treatments, which suggests that similar spawnings of snappers may be occurring unobserved or unrecognized in other large public aquaria. Spawning of snappers in tanks much smaller than these is unlikely given the large size of the adults at maturity.

There are two reports of rearing eggs and larvae of snappers: Suzuki & Hioki (1979) with *L. kasmira*, and Richards & Saksena (1980) with *L. griseus*. The spawned eggs of *L. kasmira* were netted out of the observation tank and transferred to smaller tanks (approximately 50 and 415 liters) for rearing. The authors used a closed system with clear water (no algal culture) and weak aeration. Hatching occurred in 18 hours, and development to three-day-old larvae appeared normal. At three days of age the yolk sacs were fully absorbed and the larvae began to feed on the oyster larvae (*Crassostrea gigas*) offered to them. The oldest lasted only seven days, however, suggesting that the offered food was not adequate. Richards & Saksena (1980) reared *L. griseus* to metamorphosis by feeding them size-sorted zooplankton (initially 0.35 to 0.73 mm, then 0.73 to 1.10 mm) collected in plankton tows. These larvae were reared in a 75-liter aquarium at temperatures of 26 to 28°C, a salinity of 32 to 36 parts per thousand, and in constant illumination.

125

A large male *Plectropomus leopardus* exhibiting courtship coloration. Photo by Dr. R. E. Thresher.

A male *Anthias squamipinnis* swimming among a group of females. Photo by Dr. R. E. Thresher.

During courtship, the male swims above a female, displaying with its dorsal fin folded and its pectoral fins prominently spread. Spawning is initiated by the male dashing rapidly downward toward the female. Photo sequence by Dr. R. E. Thresher.

At the end of the downward rush, the male moves close to the female. Spawning takes place with the male beside and slightly twisted around the female. Note the male's open jaws. After spawning, the male swims back up to his courting position.

Mirolabrichthys dispar, one of the planktivorous anthiines, is commonly seen in small to large aggregations over the reef. Here is an apparent male with what appear to be four females. Photo by Allan Power.

Literature Cited

Almeida, N. 1965. Estudos preliminaires sobre a primeira maturacao sexual, epoca de desova e sex-ratio do Pargo (*Lutianus aya*) no Nordeste. *Bol. Est. Pesca, Recife*, 5:7-17.

Atz, J.W. 1964. Intersexuality in Fishes. Pp. 145-232. *In: Intersexuality in Vertebrates, including Man.* (C.N. Armstrong & A.J. Marshall, Eds.). Academic Press, N.Y.

Camber, C.I. 1955. A survey of the red snapper fishery of the Gulf of Mexico, with special reference to the Campeche Banks. *Tech. Ser. Fla. St. Bd. Conserv.*, (10):1-64.

Collins, L.A., J.H. Finucane, and L.E. Barger. 1980. Description of larval and juvenile red snapper, *Lutjanus campechanus. Fish. Bull. (U.S.)*, 77:965-974.

Croker, R.A. 1960. A contribution to the life history of the gray (mangrove) snapper, *Lutjanus griseus* (Linnaeus). Master's Thesis, Univ. Miami, Florida.

Fourmanoir, P. 1973. Notes ichthyologiques (V.). *Cahiers O.R.S.T.O.M., Ser. Oceanogr.*, 9:33-39.

Heemstra, P.C. 1974. On the identity of certain Eastern Pacific and Caribbean post-larval fishes (Perciformes) described by Henry Fowler. *Proc. Acad. Nat. Sci. Phila.*, 126:21-26.

Johannes, R.E. 1978. Reproductive strategies of coastal marine fishes in the tropics. *Env. Biol. Fishes*, 3:65-84.

Johannes, R.E. 1981. *Words of the Lagoon.* Univ. Calif. Press., Los Angeles, Calif. 320 pp.

Kami, H.T. 1973. The *Pristipomoides* (Pisces: Lutjanidae) of Guam with notes on their biology. *Micronesica*, 9:97-118.

La Roche, W.A. 1977. Description of larval and early juvenile vermilion snapper, *Rhomboplites aurorubens. Fish. Bull. (U.S.)*, 75:547-554.

Moe, M.A., Jr. 1963. A survey of offshore fishing in Florida. *Prof. Papers Ser., Mar. Lab. Fla.*, (4):1-117.

Mosley, F.N. 1966. Biology of the red snapper, *Lutjanus aya* Bloch, off the northwestern Gulf of Mexico. *Publ. Inst. Mar. Sci., Univ. Texas*, 11:90-101.

Munro, J.L., V.C. Gaut, R. Thompson, and P.H. Reeson. 1973. The spawning seasons of Caribbean reef fishes. *J. Fish Biol.*, 5:69-84.

Musiy, Y.I. and V.A. Sergiyenko. 1977. Fingerlings of the genus *Lutjanus* (Lutjanidae, Perciformes) from the Gulf of Aden. *J. Ichthyol.*, 17:151-154.

Nzioka, R.M. 1979. Observations on the spawning seasons of East African reef fishes. *J. Fish Biol.*, 14:329-342.

Randall, J.E. and V.E. Brock. 1960. Observations on the ecology of epinepheline and lutjanid fishes in the Society Islands, with emphasis on food habits. *Trans. Amer. Fish. Soc.*, 89:9-16.

Rangarajan, K. 1971. Maturity and spawning of the snapper, *Lutjanus kasmira* (Forskal) from the Andaman Sea. *Indian J. Fish.*, 18:114-125.

Reshetnikov, Y.S. & R.M. Claro. 1976. Cycles of biological processes in tropical fishes with special reference to *Lutjanus synagris. J. Ichthyol.*, 16:711-723.

Richards, W.J. and V.P. Saksena. 1980. Description of larvae and early juveniles of laboratory-reared gray snapper, *Lutjanus griseus* (Linnaeus) (Pisces, Lutjanidae). *Bull. Mar. Sci.*, 30:515-521.

Starck, W.A., II and R.E. Schroeder. 1970. Investigations of the gray snapper, *Lutjanus griseus. Stud. Trop. Oceanogr.*, (10):224 pp.

Suzuki, K. and S. Hioki. 1979. Spawning behavior, eggs, and larvae of the lutjanid fish, *Lutjanus kasmira*, in an aquarium. *Japan. J. Ichthyol.*, 26:161-166.

Thompson, R. and J.L. Munro. 1974. The biology, ecology and bionomics of Caribbean reef fishes: Lutjanidae (Snappers). *Res. Rpt. Zool. Dept. Univ. West Indies*, (3):69 pp.

Vatanachai, S. 1974. The identification of fish eggs and larvae obtained from the survey cruises in the South China Sea. *Proc. Indo-Pacific Fish. Council*, 15:111-130.

Wicklund, R. 1969. Observations on spawning of the lane snapper. *Underwater Natur.*, 6:40.

Lutjanoid Fishes (Lethrinidae, Sparidae, Nemipteridae, Emmelichthyidae, Inermiidae, Scolopsidae, Pentapodidae, Gaterinidae)

Closely related to the true snappers, family Lutjanidae, are a host of snapper-like fishes that in the past have been treated in a bewildering variety of nominal families. Like the snappers, such fishes tend to be laterally compressed, have strongly sloping foreheads and large jaws, and have a continuous dorsal fin with a prominent spiny portion. The emmelichthyids (Emmelichthyidae and Inermiidae) appear to be a specialized subgroup of these fishes adapted to foraging in the water column and so are more streamlined than the other species and lack many of these otherwise unifying characteristics. The grunts of the nominal family Haemulidae (Pomadasyidae) also belong in this general group of fishes and in fact are often included in the same family as some of them; they will be treated separately here only because of the amount of work that has been done on them. The remaining lutjanoid fishes, as the group is defined here, consist of around 50 genera and perhaps 200 species in eight nominal families. Most species are in the family Sparidae. These fishes are distributed in shallow temperate and tropical seas worldwide and are often commercially important. Only a few species (mainly juvenile gaterinids) are commonly kept in aquaria.

SEXUAL DIMORPHISM

Most, if not all, lutjanoid fishes appear to lack conspicuous external differences between the sexes. Temporary color differences involving male-specific color patterns during courtship and spawning occur in at least two species, however, both of which will be described below. It is apparently characteristic of such fishes that males tend to be larger on average than females, usually with a considerable overlap in size ranges. Such differences have been reported for the Gaterinidae (Druzhinin & Filatova, 1979), the Lethrinidae (Toor, 1965a; Lebeau & Cueff, 1975; Aldonov & Druzhinin, 1978; Young & Martin, in press), and the Scolopsidae and Nemipteridae (P. Young, in prep.).

Such consistent size differences between the sexes could reflect either differential growth rates of males and females or sequential hermaphroditism. In fact, hermaphroditism has been widely demonstrated in the Sparidae, a lutjanoid family which probably has the most diverse pattern of sexuality of any extant teleost family. Sparids range from gonochorists to both protandrous and protogynous species (the extensive literature is reviewed by Atz, 1964, and Reinboth, 1970). Reinboth (in Atz, 1965) noted that, in general, the pelagic spawning sparids are protandrous and the demersal spawning species protogynous. Sparid hermaphroditism has also been examined hormonally by various workers, including Reinboth (1962), Colombo, et al. (1972), and Eckstein, et al. (1978).

Far less information is available on sexuality of other lutjanoid fishes. According to Smith (1975), at least three species of emmelichthyids have proven to be protogynous. Protogyny also seems likely for several species of Scolopsidae and Nemipteridae examined by P. Young (in prep.). The most detailed work to date is on seven species of *Lethrinus* (Lethrinidae) examined histologically by Young & Martin (in press). The testes of all species examined showed typical "secondary male" characteristics, including a central lumen and the presence of degenerate oocytes. In four species, individuals with apparent intersex gonads were collected.

SPAWNING SEASON

Temperate lutjanoid fishes (mainly sparids) vary with respect to spawning season, though most spawn in spring and summer. Information for tropical species is sparse but in general suggests long spawning seasons. Spring spawning has been documented for sparids (Matsuura, 1972; Munro, et al., 1973) and

Most of the jawfishes are sexually monomorphic, but sexual dichromatism is exhibited in a few species such as *Opistognathus gilberti* on this page and *O.* sp. on the opposite page. The male *O. gilberti* is at the left and the female *O. gilberti* is below. Both photos by Dr. P. L. Colin at Puerto Rico.

To the right is the normal male threat display of *O*. sp., the blue-spotted jawfish. Below, he exhibits his courtship color as he hovers motionless up to a meter above his burrow. After a few seconds he will dash back to his burrow "flashing" white against black. Both photos by Alex Kerstitch at Sonora, Mexico.

nemipterids (Kwan-ming, 1954); long spawning seasons, running from spring through to at least early fall, have been suggested for lethrinids (Johannes, 1981; with spring and fall peaks – Toor, 1965a; Nzioka, 1979), gaterinids (Nzioka, 1979), and pentapodids (Nzioka, 1979). Johannes (1981) suggests that both *Lethrinus* and *Monotaxis* at Palau have peaks in spawning activity at the time of the new moon.

REPRODUCTIVE BEHAVIOR

Little is known about spawning in most lutjanoid families. With the exception of a few sparids (discussed below), all apparently produce pelagic eggs presumably shed in the water column some distance off the bottom. Johannes (1981) provides the only available information on the spawning behavior of lethrinids (spawning occurs after dark, in most species in large spawning aggregations along the inner or outer edge of the fringing reef following migration), gaterinids (two species reported to form spawning aggregations, with at least one spawning on the outer reef slope), and pentapodids (*Monotaxis grandoculis* forming spawning aggregations near the bottom of the reef slope).

Spawning by sparids is somewhat better known, with more or less detailed information available for three species. According to Cassie (1956a), the southern temperate species *Chrysophrys auratus* spawns near the surface in deep water. When spawning, a pair of individuals "drift lethargically" along near the surface while discharging gametes. Offshore spawning is also suggested for *Lagodon rhomboides*, an inshore tropical species (Caldwell, 1957). Inshore spawning, however, is implied by the observations of courtship and spawning of the Japanese species *Gymnocranius griseus* (Suzuki & Hioki, 1978). Spawning by these species was observed in a large public aquarium, the action taking place at dusk in late spring. Courtship begins with the male developing a distinctive pattern of horizontal wavy silver lines on his side. He approaches a school of individuals on the bottom and apparently solicits spawning by blocking a female's way and by "tapping" her abdomen with his snout. If ready, the pair ascend slowly toward the surface, with the male below and slightly behind the female. Usually there are several false starts, but then the pair ascend to within 1 to 2 meters of the surface (mean aquarium depth about 6 m) and move into a side-by-side position, at which time the gametes are shed. After spawning both fish dash for the bottom.

The family Sparidae is one of the few families that has members that spawn pelagic eggs, as described above, and also members that produce demersal eggs. Demersal spawning, involving male preparation of a nest and subsequent defense of eggs, is described for *Spondyliosoma cantharus*, a northern temperate species (Raffaele, 1898; Wilson, 1958), and for *S. emarginatum*, a southern temperate species (van Bruggen, 1965; Penrith, 1972). Spawning is preceded by nest-digging by the male in the form of vigorous swimming in place over a flat, sandy spot on the bottom until a shallow, roughly circular depression is formed. In *S. emarginatum* several males often prepare nests close to one another. Spawning apparently takes place late in the evening or during the night. Details are lacking, but in *S. cantharus* the male becomes very dark while courting and develops a conspicuous white bar just ahead of his anal fin. He dashes at apparent females in the school passing by, and if one is ready, she follows him back to the nest site and inspects it. The male swims actively about her and repeatedly nuzzles the region of her anal fin. The female apparently deposits the adhesive eggs on the bottom, producing an irregular whitish patch "about a foot in diameter." On several occasions a male was observed with two different-aged egg patches. After spawning, the male fans and guards the eggs until hatching, a matter of about nine days at 13°C for *S. cantharus*.

Toor (1965) estimated that female *Lethrinus lentjan* produce between 12,000 and 77,000 eggs per spawning season. Sexual maturity occurs in about three years and at about 28.0 cm for males and 30.0 cm for females (Toor, 1964b).

EGGS AND LARVAE

The eggs of the demersally spawning *Spondyliosoma cantharus* are about a millimeter in diameter and flattened on the bottom. Each contains a single yellow oil globule. Hatching takes place in nine days at 13° C, producing a prolarva that has a prominent yolk sac and an already pigmented eye (Wilson, 1958).

The eggs of pelagic-spawning lutjanoids range in diameter from 0.65 mm to 0.9 mm (Aoyama & Satogaki, 1955; Cassie, 1956b; Mito, 1956; Lumare & Villani, 1970; Ciechomski & Weiss, 1973; Suzuki & Hioki, 1978). Hatching takes between 28 and 45 hours, depending upon water temperature and egg size. Newly hatched larvae have been described for *Gymnocranius griseus* (Suzuki & Hioki, 1978), *Lethrinus nematacanthus* (Mito, 1956), *Archosargus rhomboidalis* (Houde & Potthoff, 1976), and several temperate species (e.g., Cassie, 1956b). Such larvae have unpigmented eyes and a moderate-sized, somewhat elongate yolk sac. At the lower front of the yolk sac is an oil globule that partially protrudes out of the sac. The yolk is fully absorbed in about 25 hours at a larval length of about 2.4 mm. Development of older larvae of the western Atlantic sparid *Archosargus probatocephalus* is described in great detail by Mook (1977). Settlement to the bottom by this species occurs at a length of about 25 mm. Juveniles of most species have color patterns similar to the adults; juveniles of various species of *Gaterin*, however, have complex patterns of spots and stripes very different from the adults (e.g., Smith, 1962).

SPAWNING IN CAPTIVITY

Species of only two genera, *Spondyliosoma* (Wilson, 1958; van Bruggen, 1965) and *Gymnocranius* (Suzuki & Hioki, 1978), have thus far spawned naturally in captivity. In all three cases spawning took place in large oceanarium-style community aquaria. Spawning in smaller aquaria for any species in the group is unlikely. Hormone-induced spawning has been conducted on the northern temperate species *Sparus aurata* (e.g., Lumare & Villani, 1971; Zohar & Billard, 1978). Several sparids have been reared in captivity, usually using feedings of a mixture of cultured and wild-caught foods (e.g., Kasahara, et al., 1960; Houde, 1975; Houde & Potthoff, 1976).

Literature Cited

Aldonov, V.K. and A.D. Druzhinin. 1978. Some data on scavengers (family Lethrinidae) from the Gulf of Aden region. *J. Ichthyol.*, 18:527-535.

Aoyama, T. and N. Satogaki. 1955. On the development of the eggs of *Nemipterus virgatus* (Houttuyn). *Japan. J. Ichthyol.*, 4:130-132.

Atz, J.W. 1964. Intersexuality in fishes. Pp. 145-232. *In: Intersexuality in Vertebrates, including Man.* (C.N. Armstrong and A.J. Marshall, eds.). Academic Press, New York.

Atz, J.W. 1965. Hermaphroditic fish. *Science*, 150:789-797.

Caldwell, D.K. 1957. The biology and systematics of the pinfish, *Lagodon rhomboides* (Linnaeus). *Bull. Florida State Mus., Gainesville, Biol. Sci.*, 2:77-173.

Cassie, R.M. 1956a. Spawning of the snapper, *Chrysophrys auratus* Forster, in the Hauraki Gulf. *Trans. Roy. Soc. New Zealand, Wellington*, 84:309-328.

Cassie, R.M. 1956b. Early development of the snapper, *Chrysophrys auratus* Forster. *Trans. Roy. Soc. New Zealand, Wellington*, 83:705-713.

Ciechomski, J.D. de and G. Weiss. 1973. Desove y desarrollo embrionarie de besugo, *Pagrus pagrus* (Linne) en el Mar Argentino. Sparidae. Pisces. *Physis (B. Aires)*, 32:481-487.

Colombo, L., E. Del Conte, and P. Clemenze. 1972. Steroid biosynthesis in vitro by the gonads of *Sparus auratus* L. (Teleostei) at different stages during natural sex reversal. *Gen. Comp. Endocrinol.*, 19:26-36.

Druzhinin, A.D. and N.A. Filatova. 1979. Some data on *Plectorhinchus pictus* of the family Pomadasyidae. *Vopr. Ikhtiol. (J. Ichthyol.)*, 19:154-155.

Eckstein, B., M. Abraham, and Y. Zohar. 1978. Production of steroid hormones by male and female gonads of *Sparus aurata* (Teleostei, Sparidae). *Comp. Biochem. Physiol.*, 60B:93-97.

Houde, E. 1975. Effects of stocking density and food density on survival, growth and yield of laboratory-reared larvae of sea bream *Archosargus rhomboidalis* (L.) (Sparidae). *J. Fish. Biol.*, 7:115-127.

Houde, E. and T. Potthoff. 1976. Egg and larval development of the sea bream *Archosargus rhomboidalis* (Linnaeus): Pisces, Sparidae. *Bull. Mar. Sci.*, 26:506-529.

Johannes, R.E. 1981. *Words of the Lagoon.* Univ. Calif. Press, Los Angeles. 320 pp.

Kasahara, S., R. Hirano, and Y. Oshima. 1960. A study of the growth and rearing methods of the fry of Black Porgy, *Mylio macrocephalus* (Basilewsky). *Bull. Japan. Soc. Sci. Fish.*, 26:239.

Kwan-ming, L. 1954. An account of the golden thread group fishery in Hong Kong, and a preliminary note on the biology of *Nemipterus virgatus* (Houttuyn). *Hong Kong Univ. Fish. J.*, 1:1-18.

Lebeau, A. and J.C. Cueff. 1975. Biologie et peche due capitaine *Lethrinus enigmaticus* (Smith) 1959 du banc de Saya de Mahla. *Rev. Trav. Inst. Peches Marit.*, 39:415-442.

Lumare, F. and P. Villani. 1970. Contributo alla conoscenze delle uova e dei primi stadi larvali di *Sparus aurata* (L.). *Pubbl. Stn. Zool. Napoli.*, 38:364-369.

Matsuura, S. 1972. Fecundity and maturation process of ovarian eggs of sea bream, *Pagrus major* (Temminck et Schlegel). *Sci. Bull. Fac. Agric., Kyushu Univ.*, 26:203-215.

Mito, S. 1956. On the egg development and hatched larvae of *Lethrinus nematacanthus* Bleeker. *Sci. Bull. Fac. Agric., Kyushu Univ.*, 15:497-500.

Mook, D. 1977. Larval and osteological development of the sheepshead, *Archosargus probatocephalus*

Juvenile grunts typically differ from the adults, usually being horizontally striped. Shown here is a juvenile *Haemulon flavolineatum.* Photo by Dr. Herbert R. Axelrod.

The adults of *H. flavolineatum* are quite different from juveniles, having an obliquely yellow-striped pattern. Photo by Dr. W. A. Starck II.

Male *Lutjanus griseus* dominated the reef population, while females were more common inshore. Photo by Dr. Herbert R. Axelrod.

Chrysophrys auratus, a southern temperate water species, spawns near the surface in deep water. A pair will "drift lethargically" as the gametes are discharged. Photo by Wade Doak.

Courtship and spawning of *Gymnocranius griseus* have been observed in large public aquaria in Japan. The action took place at dusk late in the spring. Photo by Dr. Fujio Yasuda.

(Pisces: Sparidae). *Copeia*, 1977:126-133.

Munro, J.L., V.C. Gaut, R. Thompson, and P.H. Reeson. 1973. The spawning seasons of Caribbean reef fishes. *J. Fish Biol.*, 5:69-84.

Nzioka, R.M. 1979. Observations on the spawning season of East African reef fishes. *J. Fish Biol.*, 14:329-342.

Penrith, M.J. 1972. The behaviour of reef-dwelling sparid fishes. *Zoologica Africana*, 7:43-48.

Raffaele, F. 1898. Osservazioni sulle uova di fondo dei pesci ossei del Golfo di Napoli e mari adiacenti. *Boll. Notiz. agr. Anno XX.*, 8:325-335.

Reinboth, R. 1962. Morphologische und funktionelle Zweigeschlechtlichkeit bei marinen Teleostiern (Serranidae, Sparidae, Centracanthidae, Labridae). *Zool. Jb. Abe. Allg. Zool. Physiol.*, 69:405-480.

Reinboth, R. 1970. Intersexuality in fishes. *In:* Hormones and the environment. *Mem. Soc. Endocrinol.*, 18:516-543.

Smith, J.L.B. 1962. Fishes of the family Gaterinidae of the western Indian Ocean and the Red Sea. *Ichthyol. Bull., Dept. Ichthyol. Rhodes Univ.*, (25):468-502.

Smith, C.L. 1975. The evolution of hermaphroditism in fishes. Pp. 295-310. *In: Intersexuality in the Animal Kingdom.* (R. Reinboth, Ed.). Springer Verlag, Berlin, Heidelberg, New York.

Suzuki, K. and S. Hioki. 1978. Spawning behavior, eggs, and larvae of the sea bream, *Gymnocranius griseus*, in an aquarium. *Japan. J. Ichthyol.*, 24:271-277.

Toor, H.S. 1964a. Biology and fishery of the pig-face bream, *Lethrinus lentjan* Lacepede. II. Maturation and spawning. *Ind. J. Fish.*, 11:581-596.

Toor, H.S. 1964b. Biology and fishery of the pig-face bream, *Lethrinus lentjan* Lacepede, from Indian waters. III. Age and growth. *Ind. J. Fish.*, 11:597-620.

Toor, H.S. 1965. Biology and fishery of the pig-face bream, *Lethrinus lentjan*, from Indian waters. 4. Fecundity. *Res. Bull. Punjab Univ. Sci. N.S.*, 16:165-178.

van Bruggen, A.C. 1965. Records and observations in the Port Elizabeth Oceanarium in 1960. *Zool. Gart. Lpz.*, 31:184-202.

Wilson, D.P. 1958. Notes from the Plymouth Aquarium. III. *J. Mar. Biol. Ass. U.K.*, 37:299-307.

Young, P.C. and R.B. Martin. (In Press). Evidence for protogynous hermaphroditism in some lethrinid fishes, and its implication for stock fecundity in a trawl fishery. *J. Fish Biol.*

Zohar, Y. and Billard, R. 1979. New data on the possibilities of controlling reproduction in teleost fish by hormonal treatment. *Coll. Aquac. Thon, Sete, CNEXO Edt.*

Grunts
(Haemulidae)

Grunts, so called because of a low frequency "grunting" noise many emit when handled, are ubiquitous predators commonly found on or around reefs. Like the similar appearing snappers (Lutjanidae), they are primarily nocturnal foragers and spend the day hovering, often in groups of several hundred, around rock and coral outcrops, under docks, or around virtually any other cover available. At night they spread out along well traveled routes to forage over sand and grass flats for the crustaceans and fishes upon which they feed.

Like snappers, grunts are relatively advanced fishes with distinct spiny and soft portions to the dorsal fin and a laterally compressed body. They differ from the snappers in lacking prominent canine teeth and in having a somewhat differently shaped head. Where snappers have a distinctly triangular head with straight upper and lower margins, the upper margin of a grunt's head tends to be slightly convex, the snout is blunter, and the upper jaw usually protrudes slightly beyond the lower. The entire impression is of an animal with a rounder and less aggressive appearance than a snapper. Grunts range in shape from heavy-bodied and blunt-headed benthic predators to fusiform water column foragers. They range in size at maturity from approximately 10 cm to over 50 cm. The family is distributed circumtropically and contains about 20 genera and 175 species.

SEXUAL DIMORPHISM

There are no reports of sexual differences in shape, color, or behavior for any of the grunts, and most, if not all, appear to be monomorphic. Detailed work on the group has yet to be conducted, however.

SPAWNING SEASON

Limited data on spawning seasons are available for only a few western Atlantic species. Three independent reports (Hildebrand & Schroeder, 1928; Munro, et al., 1973; and Saksena & Richards, 1975) all suggest peak spawning of the various species in early to late spring. Some spawning probably occurs year-round, however, in the deep tropics. There is no information on possible lunar cycles, though given the general biology of the fishes a peak in spawning activity near the full moon seems likely.

REPRODUCTIVE BEHAVIOR

Spawning has not yet been observed for any species of grunt and there is relatively little information available about reproduction in the group. Hildebrand & Cable (1930) suggested that spawning by the temperate species *Orthopristis chrysoptera* occurs in estuaries off North Carolina and probably takes place at night. For reef-dwelling species, spawning probably parallels that of other roving benthic predators such as snappers and groupers; that is, there is probably a general migration to select spawning sites on the outer reef and spawning is probably a group activity that occurs at dusk. In support of this hypothesis, Moe (1966) reported large concentrations of ripe *Haemulon plumieri* off the coast of Florida in May.

EGGS AND LARVAE

Grunts produce spherical pelagic eggs that contain a single oil droplet. Egg diameter for *Orthopristis chrysoptera* is approximately 0.75 mm (Hildebrand and Cable, 1930); that of *Haemulon plumieri* is about 0.90 to 0.97 mm (Saksena and Richards, 1975); and that of *Parapristipoma humile* is 0.82 to 0.95 mm (Podosinnikov, 1977). Respective incubation times reported are 36 to 72 hours at 15 to 37° C, 20 to 24 hours at 24° C, and 22 to 23 hours at 20 to 25° C.

Newly hatched *P. humile* are 1.7 to 1.9 mm long and have a moderate sized yolk with an oil droplet at its posteror end. Newly hatched *H. plumieri* are similar but larger (2.7 to 2.8 mm) and have a large ellipsoidal yolk sac that contains a single oil droplet at its

A territory-holding male *Parupeneus trifasciatus* displaying over a gravid female. Photo by Dr. R. E. Thresher.

The pair begins their spawning ascent. Note that the male is much larger than the female. Photo by Dr. R. E. Thresher.

The male showing territorial display colors. These are lost during courtship and spawning. Photo by Dr. R. E. Thresher.

The pair near the peak of their ascent. This sequence was photographed at Enewetak. Photo by Dr. R. E. Thresher.

Chromis albomaculata spawning. Note the temporary dichromatism of the male (silvery operculum). Photo by Jack T. Moyer.

Chromis albomaculata spawning. This pair was photographed by Jack T. Moyer at Miyake-jima, Japan.

Chromis cyanea with visible urogenital papilla. Photo by Dr. R. E. Thresher.

Chromis caerulea male on the Great Barrier Reef showing courtship coloration. Note that this color is different from that in presumably the same species photographed in other areas. Photo by Roger Steene.

Chromis caerulea pair spawning (male is above female). Photo by Roger Steene.

Chromis atrilobata male in spawning colors. The male is signal jumping to the female in the background. Photo by Dr. R. E. Thresher.

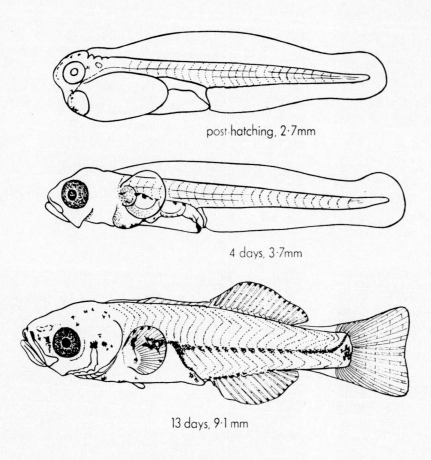

post-hatching, 2·7mm

4 days, 3·7mm

13 days, 9·1 mm

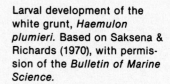

Larval development of the white grunt, *Haemulon plumieri*. Based on Saksena & Richards (1970), with permission of the *Bulletin of Marine Science*.

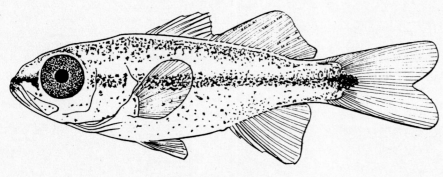

40 days, 13·6mm

anteriormost point. Eyes become pigmented within 24 hours, and the yolk is absorbed and the larva begins to feed within 48 hours. Growth is rapid. Despite their similarities as adults, the larvae of snappers and grunts differ markedly. Larval snappers are large-headed and have long spines in the dorsal and pelvic fins. Larval grunts have a slender, somewhat pointed, head and lack prominent spines in the fins; they are also laterally compressed, elongate, and have prominent pigmented eyes, an upturned terminal mouth, and a lower jaw that protrudes slightly beyond the upper. Duration of the planktonic stage for *H.*

flavolineatum is approximately 15 days, and the larvae settle to the bottom as benthic postlarvae at a length of approximately 6 mm (McFarland, 1980; Brothers & McFarland, 1981.). Juvenile grunts are commonly found in small groups on grass flats, near mangroves, and in other inshore areas. The juveniles typically differ in color pattern from the adults, are usually horizontally striped, and are frequently difficult to identify to the species level. Keys to juveniles of tropical western Atlantic species are provided by Courtenay (1961) and Böhlke & Chaplin (1968); information on juveniles of eastern Pacific species is given by Hong (1977).

Cummings, et al. (1966) reported that sexual maturity of *Haemulon album* occurs at approximately 37.5 cm.

REPRODUCTION IN CAPTIVITY

There are no reports of spawning by grunts in aquaria, and such observations seem unlikely in anything but large public aquaria. Saksena & Richards (1975) reared larval *H. plumieri* in a semi-closed system by feeding them with rotifers (*Brachionis plicatilis*) and wild plankton.

Literature Cited

Böhlke, J.E. and C.C.G. Chaplin. 1968. *Fishes of the Bahamas and adjacent tropical waters*. Livingston Publ. Co., Wynnewood, Pa. 771 pp.

Brothers, E.B. and W.N. McFarland. 1981. Correlations between otolith microstructure, growth, and life history transitions in newly recruited french grunts [*Haemulon flavolineatum* (Desmarest), Haemulidae]. *Rapp. P.-v. Re'un. Cons. int. Explor. Mer*, 178:369-374.

Courtenay, W.R., Jr. 1961. Western Atlantic fishes of the genus *Haemulon* (Pomadasyidae): systematic status and juvenile pigmentation. *Bull. Mar. Sci.*, 11(1):66-149.

Cummings, W.C., B.D. Brahy, and J.Y. Spires. 1966. Sound production, schooling and feeding habits of the margate, *Haemulon album* Cuvier, off North Bimini, Bahamas. *Bull. Mar. Sci.*, 16(3):626-640.

Hildebrand, S.F. and L.F. Cable. 1930. Development and life history of fourteen teleostean fishes at Beaufort, North Carolina. *Bull. U.S. Bur. Fish.*, 46:383-488.

Hildebrand, S.F. and W.C. Schroeder. 1928. Fishes of Chesapeake Bay. *Bull. U.S. Bur. Fish.*, 43:366 pp.

Hong, S.L. 1977. Review of eastern Pacific *Haemulon* with notes on juvenile pigmentation. *Copeia*, 1977:493-501.

McFarland, W.N. 1980. Observations on recruitment in haemulid fishes. *Proc. Gulf Caribb. Fish. Inst.* 32:132-138.

Moe, M.A., Jr. 1966. Tagging fishes in Florida offshore waters. *Tech. Ser. Fla. St. Bd. Conserv.*, 49:1-40.

Munro, J.L., V.C. Gaut, R. Thompson, and P.H. Reeson. 1973. The spawning seasons of Caribbean reef fishes. *J. Fish Biol.*, 5:69-84.

Podosinnikov, A.Y. 1977. Early ontogeny of the 'striped grunt', *Parapristipoma humile* (Pomadasyidae, Pisces). *J. Ichthyol.*, 17:683-684.

Saksena, V.P. and W.J. Richards. 1975. Description of eggs and larvae of laboratory-reared white grunt, *Haemulon plumieri* (Lacépède) (Pisces, Pomadasyidae). *Bull. Mar. Sci.*, 25(4):523-536.

An aggregation of *Chromis caerulea* that has selected turtle grass (*Syringodium* sp.) as a spawning site. The apparent male (dark dorsal fin) is displaying to some females above him in the water column (above), and apparently succeeding (below). Photos by Bruce Carlson at Fiji.

Actual spawning may be accomplished on the leaves of the turtle grass or on rocks, etc. in the area. Photo by Bruce Carlson.

An apparent spawning aggregation of *Chromis caerulea* at Enewetak. Note the dark upper portion of the pectoral fin and the yellow anal fin of the supposed males. Spawning will occur on the coral. Photo by Scott Johnson.

Goatfishes
(Mullidae)

Goatfishes are elongate heavy-bodied fishes that are characterized by two clearly separate dorsal fins, a forked caudal fin, and a pair of barbels under the lower jaw. The barbels, which fold back into special grooves when not in use, are used to probe sand and mud bottoms for the small soft-bodied invertebrates and crustaceans that make up the goatfishes' diets. Most often the fishes are seen singly or in small groups swimming close over a sand bottom, probing into it with wriggling barbels, and pouncing sporadically upon some uncovered tidbit. Goatfishes seem to feed almost constantly in the field and doubtless consume vast numbers of invertebrates.

The goatfishes are a numerically small family (about 50 to 60 species in six genera according to Lachner, 1960) but have successfully radiated into all shallow tropical seas and several more temperate regions (such as the North Atlantic). Until recently aspects of reproduction were known only for these temperate species, with scattered information on the Mediterranean and Black Sea species of *Mullus* available since at least the late 1800's. More detailed information on the tropical species has been collected only within the last 20 years. Most of this information concerns the tropical western Atlantic species.

SEXUAL DIMORPHISM

Permanent sexual dimorphism in goatfishes is poorly known, with detailed information thus far available only for a pair of tropical western Atlantic species. In *Mulloidichthys martinicus,* females average a centimeter or so larger than the males, with a broad overlap in size between the two sexes; mature females range in size from 17 to 25 cm Fork Length (FL), whereas males range from 16 to 23.5 cm (Munro, 1974, 1976). At lengths less than 21 cm FL males are also slightly deeper bodied and heavier than females of the same length. In contrast, the male is the larger sex in *Pseudupeneus maculatus.* Males range from 16.5 to

26.5 cm FL, with a peak between 19 and 23 cm, whereas females range from 16.5 to 24.5 cm, with a peak around 19 to 20 cm (Munro, 1974, 1976). Males also have a higher arched back than females and a more angular facial profile, while females tend to be more fusiform (Caldwell, 1962). The sexes also differ in dentition: large males have two rows of teeth in the upper jaw, while females and small males have only a single row (Rosenblatt & Hoese, 1968). The male also seems to be the larger sex in two western Pacific species, *Parupeneus multifasciatus* (Lobel, 1978) and *P. trifasciatus* (pers. obs.), based on observations of spawning pairs.

Why there should be within-family differences in the identity of the larger sex in goatfishes is not known. Munro (1974) suggested that it relates to differences in growth rates and age at sexual maturity, but this seems only to push the question back to another level. Given that goatfishes exhibit both pair and group spawning, it is interesting to speculate that large female size is associated with species that predominantly group spawn, and large male size occurs in species that predominantly pair spawn, due to differences in the patterns of sexual selection in the two modes of spawning. There is, to date, no information relative to the sexuality of goatfishes, but the broad overlap in size ranges of the sexes suggests that the fishes are gonochoristic.

SPAWNING SEASON

Only limited information is available regarding mullid spawning seasons. In the tropical western Atlantic both shallow-water species, *Mulloidichthys martinicus* and *Pseudupeneus maculatus,* have a long spawning season with peaks of activity in the spring (roughly February to May) and in the fall (roughly September and October) (Erdman, 1956; Cervigon, 1966; Munro, et al., 1973; Munro, 1974). *Upeneus tragula* has a similar season off the coast of India, with spawning peaks in July and September (Thomas, 1969). Long spawning seasons are also characteristic of three species of *Parupeneus* off the east coast of Africa, but they do not appear to show a bimodal peak in activity (Nzioka, 1979). In more temperate regions goatfishes spawn in the summer (e.g., Essipov, 1934).

There is as yet no detailed information available regarding possible relations of lunar cycles to mullid spawning activity, though given their general reproductive behavior (offshore migrations combined with production of pelagic eggs), such cycles might be expected. Lobel (1978) reported observing spawning by *Parupeneus multifasciatus* off Hawaii two days before the full moon but provided no information as to whether spawning may have occurred at other times as well. According to the native fishermen at Palau, two local species, *Mulloidichthys flavolineatus* and *Upeneus arge*, spawn only within a few days of the new moon (Johannes, 1981).

REPRODUCTIVE BEHAVIOR

Spawning by goatfishes in general seems to occur at select points along the outer reef edge to which the fishes migrate from their normal inshore feeding areas (e.g., Essipov, 1934; Colin & Clavijo, 1978; Johannes, 1978). Spawning is either as pairs or in small groups. Only one species *(Pseudupeneus maculatus)* has been observed to do both, but other species might reasonably be expected to show such a range of behavior. In either case spawning in most species occurs at dusk (Colin & Clavijo, 1978; Lobel, 1978). The only known exception is a single report of a pair spawning at peak high tide, in midday, by *Parupeneus cyclostomus* (A. Gronell & B. Goldman, pers. comm.). The pair was seen spawning at the mouth of a break in the reef line off Lizard Island, Great Barrier Reef, an area subject to strong currents. *Parupeneus trifasciatus* at Enewatak Atoll, in an area of equally strong currents, spawns at dusk.

Spawning has thus far been observed in only four species: *Pseudupeneus maculatus* (Randall & Randall, 1963; Colin & Clavijo, 1978; Colin & Gronell, pers. comm.); *Parupeneus cyclostomus* off Lizard Island (A. Gronell & B. Goldman, pers. comm.); the Hawaiian *Parupeneus multifasciatus* (Lobel, 1978); and *Parupeneus trifasciatus* at Enewetak Atoll (pers. obs.). Few details are available on the spawning of *P. cyclostomus*. A pair of gold individuals was observed to rapidly ascend 3-4 m off the substratum (approximately 6 m deep), before turning sharply back to the reef and leaving behind a visible gamete cloud. The pair

was seen spawning only once; there was no obvious difference between the fish in size or color (they were observed from a distance, however); and no information was gathered on social organization. In the three other species pair spawning is basically similar and much like that reported for labrids, scarids, and other reef fishes. Prior to spawning the male patrols back and forth 1 to 2 m above the reef, either over an aggregation of presumed females or along a reef edge area, apparently awaiting the arrival of females. In *P. maculatus* such males bear a distinctive spotted pattern, while females maintain a "plaid" pattern; in *P. multifasciatus* the male is darker than the females; and in *P. trifasciatus* patrolling males maintain a bright red flush (this red color is lost during courtship and spawning however). Spawning is initiated by a female that either approaches a patrolling male or ascends off the bottom to join him. In *P. maculatus* the male first circles around the female at a distance of about 3 m while swimming in an exaggerated fashion. In all three species the male then swims to the ascending female, moves by her side in a head-to-head orientation, and the pair dash upward to release gametes. In *P. maculatus* and *P. trifasciatus* the ascent is a short one, is angled about 50° upward, and ends with a quick dash to the bottom. In *P. multifasciatus* the pair ascends almost to the surface and gamete release takes place while the pair is swimming side-by-side just beneath the surface.

Group spawning by *Pseudupeneus maculatus* is described by Colin & Clavijo (1978), who observed an aggregation of 300 to 400 individuals at a depth of 21 m off the coast of the U.S. Virgin Islands. The area was also used as a spawning site by *Sparisoma rubripinne* and, possibly, *Priacanthus cruentatus*. Of the fish observed, only a portion were actively engaged in courtship and spawning; the rest remained quietly on the reef. The spawning individuals were, in general, lightly colored with indistinct spots on their sides and a conspicuous darkening to the edges of the caudal fins. Each spawning ascent was initiated by a pair of individuals approaching one another and then swimming horizontally about a meter off the bottom in a head-to-head, parallel orientation. Typically several other individuals join the pair, all with their heads together, and the group then dash en masse toward the surface. After a short ascent only about two meters high, visible gamete release occurred and the individuals involved dispersed and rejoined the mass of milling fish below. Several groups of fishes often spawned simultaneously, after which there was usual-

Chromis caerulea male (?) with genital papilla very obvious. Note lack of black dorsal but presence of dusky upper part of pectoral fin. Photo by Bruce Carlson.

It was noted by some workers that females of *Chromis* spawn only briefly before returning to the water column. Here a pair of *Chromis caerulea* prepare to spawn. Photo by Bruce Carlson.

For actual spawning a temporary territory is established by normally non-territorial pomacentrids such as species of *Chromis.* This pair of *C. caerulea* is continuing the spawning sequence shown on these two pages. Photo by Bruce Carlson.

Actual egg deposition and fertilization in this pair of *C. caerulea.* The female will eventually move off while the male guards the nest of eggs as he tries to solicit spawning with another female. Photo by Bruce Carlson.

ly a lull in activity.

Group spawning is also referred to by Randall & Randall (1963) but apparently involved only a single pair of *P. maculatus* that ascended from within a group of milling individuals (Colin & Clavijo, 1978).

The only estimate of fecundity in goatfishes is provided by Thomas (1969). Females ranging in size from 12.9 to 20.0 cm total length ranged in fecundity from 19,000 to 92,800 eggs; there was no relationship between fecundity and female weight and only a weak relationship between fecundity and female length.

EGGS AND LARVAE

Goatfish eggs are typical of pelagic eggs: they are spherical, transparent, non-adhesive, and contain a single oil droplet. Only a few measurements of egg diameter are available: the egg of *Upeneus bensasi* from off Japan ranges in diameter from 0.63 to 0.90 mm (Mito, 1966); that of *Mullus surmuletus*, a temperate species, ranges from 0.85 to 0.93 mm in diameter (Raffaele, 1888; Montalenti, 1937). The latter hatches in three days (no temperature given) to produce a slender prolarva that carries a moderate-sized, elongate yolk containing an oil droplet at its anterior-most point. The newly hatched larvae of *Upeneus bensasi* are similar and 2.12 to 2.30 mm long.

Different stages in the development of the planktonic larvae, postlarvae, and prejuveniles of various tropical goatfishes are described by Caldwell (1962), Hunter (1967), Aboussouvan (1972), and Vatanachi (1974); those of a temperate species are described by Montalenti (1937). Goatfishes, in general, appear to have a long larval development. The yolk is absorbed and the jaws have developed by the time they reach 4 to 5 mm in length, and fin rays and spines are present by a length of 6 to 7 mm. Larger individuals (over 18 mm) are elongate, slender fishes that are unmistakably goatfishes. These prejuveniles, however, usually lack barbels (they are occasionally present on very large individuals) and are silvery with a dark blue sheen. Settlement to the bottom occurs at a length of roughly 40 to 60 mm, after which the juveniles gradually lose their silvery color and assume adult colors. By a size of 90 to 100 mm they have developed full adult proportions and appearance. According to Munro (1974, 1976), at least the two shallow-water western Atlantic species reach sexual maturity at a length of approximately 18 cm and an age of a year or so. Munro also suggested that few live more than three years. Thomas (1969) similarly suggested that the Indian Ocean species *Upeneus tragula* reaches sexual maturity at an age of approximately one year and a length of about 12 cm.

REPRODUCTION IN CAPTIVITY

Goatfishes generally do not do well in aquaria. Juveniles especially need to be fed large amounts of food often and frequently starve to death despite all efforts by the aquarist to the contrary. For this reason and because of their large size at maturity it is unlikely that goatfishes will be spawned in captivity (short of artificial fertilization using stripped eggs and milt). Similarly, the apparent long duration of the larval stages suggests difficulty in rearing the young to metamorphosis.

Literature Cited

Aboussouan, A. 1972. Oeufs et larves de Teleosteens de l'Ouest Africain. X. Larves d' Ophidioidei (*Oligopus, Ophidion* et *Carapus*) et de Percoidei (*Pseudupeneus*). *Bull. de I.F.A.N., 34, Ser. A,* (1):169-178.

Caldwell, M.C. 1962. Development and distribution of larval and juvenile fishes of the family Mullidae of the western North Atlantic. *Fish. Bull. (U.S.),* 62:403-457.

Cervigon, M.F. 1966. *Los Peces Marinos de Venezuela, Vols. I & II.* Fundacion la Salle de Ciencias Naturales, Caracas, Venezuela.

Colin, P.L., and I.E. Clavijo. 1978. Mass spawning by the spotted goatfish, *Pseudupeneus maculatus* (Bloch) (Pisces: Mullidae). *Bull. Mar. Sci.,* 28:780-782.

Erdman, D.S. 1956. Recent fish records from Puerto Rico. *Bull. Mar. Sci.,* 6:315-340.

Essipov, B. 1934. Le rouget (*M. barbatus*) dans le region de Ketch. *Zool. Zhur., Moscow,* 13:97-116.

Hunter, J.R. 1967. Color changes of pelagic prejuvenile goatfish, *Pseudupeneus grandisquamis,* after confinement in a shipboard aquarium. *Copeia,* 1967:850-852.

Johannes, R.E. 1978. Reproductive strategies of coastal marine fishes in the tropics. *Env. Biol. Fishes,* 3:65-84.

Johannes, R.E. 1981. *Words of the Lagoon.* Univ. Calif. Press, Los Angeles, Calif. 320 pp.

Lachner, E. 1960. Family Mullidae: Goatfishes. Pp. 1-46. *In:* Schultz, L.P., et al., 1960. Fishes of the Marshall and Marianas Islands, Vol. 2. *Bull. U.S. Nat. Mus.,* (202):438 pp.

Lobel, P.S. 1978. Diel, lunar, and seasonal periodicity in the reproductive behavior of a pomacanthid fish,

Centropyge potteri, and some other reef fishes in Hawaii. *Pac. Sci.* 32:193-207.

Mito, S. 1966. Fish eggs and larvae. *Illustrations of the Marine Plankton of Japan,* 7:74 pp.

Montalenti, G. 1937. Mullidae. *In:* Lo Bianco, S. Uova, larve e stadi giovanili de Teleostei. *Publ. Staz. Zool. Napoli, Monogr.* (38) Part 3:391-398.

Munro, J.L. 1974. The biology, ecology, and bionomics of Caribbean reef fishes: Mullidae (Goatfishes). *Res. Rpt. Zool. Dept., Univ. West Indies,* (3):44 pp.

Munro, J.L. 1976. Aspects of the biology and ecology of Caribbean fishes: Mullidae (goatfishes). *J. Fish Biol.,* 9:79-97.

Munro, J.L., V.C. Gaut, R. Thompson, and P.H. Reeson. 1973. The spawning seasons of Caribbean reef fishes. *J. Fish Biol.,* 5: 69-84.

Nzioka, R.M. 1979. Observations on the spawning seasons of East African Reef fishes. *J. Fish Biol.,* 14:329-342.

Raffaele, F. 1888. Le uovagallegianti e le larve dei Teleostei nel Golfo di Napoli. *Mitt. Zool. Sta. Neapel.,* 8:1-84.

Randall, J.E. and H.A. Randall. 1963. The spawning and early development of the Atlantic Parrot Fish, *Sparisoma rubripinne,* with notes on other scarid and labrid fishes. *Zoologica,* N.Y., 48:49-60.

Rosenblatt, R.H. and D.F. Hoese. 1968. Sexual dimorphism in the dentition of *Pseudupeneus,* and its bearing on the generic classification of the Mullidae. *Copeia.* 1968:175-176.

Thomas, P.A. 1969. Goatfishes (family Mullidae) of the Indian Seas. *Mem. Mar. Biol. Assoc. India,* III: 174 pp.

Vatanachi, S. 1974. The identification of fish eggs and larvae obtained from the survey cruises in the South China Sea. *Proc. Indo-Pac. Fish. Council,* 15, Sect. 3:111-130.

Chromis atripectoralis male in courtship coloration. Photo by Dr. R. E. Thresher.

Male C. atripectoralis (blotchy color) leading female to nest. Photo by Roger Steene.

C atripectoralis female in inverted position laying eggs as the male remains close by. Photo by Roger Steene.

The female continues to deposit eggs around the selected piece of coral rubble. Photo by Roger Steene.

As the female nears the completion of the bout of egg-laying, the male prepares to fertilize them. Photo by Roger Steene.

As the female moves off, the male fertilizes the eggs. His yellow genital papilla is quite evident in this photo by Roger Steene.

A large spawning aggregation of *Chromis atripectoralis* (note black spot in axil of pectoral fin). Spawning is occurring both on the coral and in the algal mat near its base (note pair spawning in the mat). Photo by Walter Deas at Heron Island.

Damselfishes
(Pomacentridae)

The damselfishes are one of the most successful and best studied groups of reef-dwelling fishes. Many are also popular as aquarium fishes due to their bright colors, small size, and remarkable hardiness. As a whole, the family is characterized by a combination of features, including a single nostril on either side of the snout (most reef-associated fishes have two nasal openings on each side), an interrupted lateral line, scales that usually extend onto the fins, and slight lateral compression. Most are also small at maturity; the largest species, the eastern Pacific *Microspathodon dorsalis,* reaches a maximum length of only 30 cm (Thomson, et al., 1979).

This chapter deals with the reef-dwelling damselfishes other than anemonefishes. The latter, a group of closely related species that differ significantly and consistently in reproductive biology from other damselfishes, will be treated separately in the next chapter.

There are about 225 species in the family Pomacentridae in approximately 25 genera. Damselfishes, like many other reef fishes, reach a peak in abundance and diversity on the Indo-Pacific reefs, where Allen (1975) recorded 162 species in 21 genera. An additional 42 species in 7 genera are found in the Americas, with smaller groups of endemic species at Hawaii and in the eastern Atlantic. Although primarily a tropical family, at least two genera, *Parma* and *Hypsypops,* are distributed mainly in temperate seas, along with representatives of several other genera, including *Chromis* and *Abudefduf.*

The reproductive biology of the pomacentrids is, in general, well studied. Even 20 years ago Reese (1964) could state that more was known about reproduction in damselfishes than in any other family of reef fishes, although his review covered only nine species in five genera. A few years later Breder & Rosen (1966) discussed the reproduction of 12 species in five genera. Since then, so much work has been done on the behavior of the pomacentrids that, even excluding the anemonefishes (subfamily Amphiprioninae), information on reproduction is available for 36 species in 11 genera. Most of these species are closely associated with coral reefs. Information on reproduction of temperate species is provided by Abel (1961), Turner & Ebert (1962), Limbaugh (1964), Clark (1971), Russell (1971), Mapstone & Wood (1975), and Nakazono, et al. (1979).

SEXUAL DIMORPHISM

Damselfishes can be readily sexed by the appearance of the gonads. Testes are long paired organs that are located dorsally and posteriorly in the abdominal cavity, each ending in a narrow tube, the gonoduct, that in turn fuse to form a common urogenital pore. In at least the sergeant majors, *Abudefduf* spp., the testes are white, triangular in cross section, and asymmetrical; that is, the left testis is larger than the right one and lies slightly below it in the abdominal cavity (Helfrich, 1958; Cummings, 1968). Ovaries are usually similar in appearance to testes but are circular in cross section and range in color from pale translucent gray (when the eggs are immature) to salmon, pink, and deep red (when the eggs are mature).

Determination of the sexes of most species of damselfishes is difficult based on external characteristics and may even be impossible outside of brief spawning periods. Cummings (1968) compared a variety of morphological features on males and females of the western Atlantic sergeant major, *Abudefduf saxatilis,* and was able to find only two features that differed significantly, if not consistently, between them: females had slightly longer pelvic fins than males relative to body length, and the upper caudal lobe of the female was longer than the lower lobe by a greater percentage in the female than in the male. Neither feature is particularly useful in sexing a free-

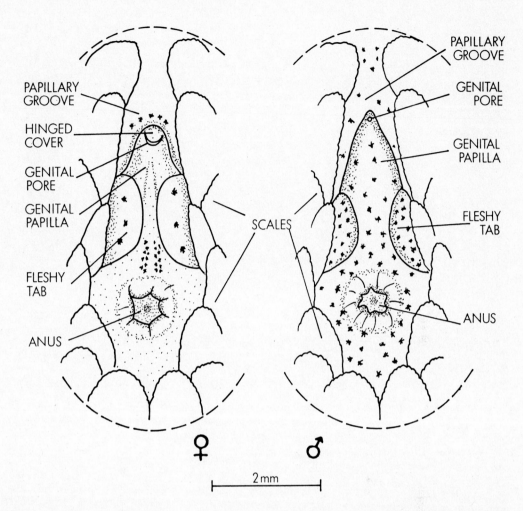

PAPILLARY GROOVE

HINGED COVER

GENITAL PORE

GENITAL PAPILLA

FLESHY TAB

ANUS

PAPILLARY GROOVE

GENITAL PORE

GENITAL PAPILLA

FLESHY TAB

ANUS

SCALES

♀   ♂

├── 2 mm ──┤

Fine structure of the male and female urogenital papillae of the damselfish *Abudefduf saxatilis*. Based on Cummings (1968).

swimming individual. Honda & Imai (1973) also reported a morphological difference between the sexes of the Japanese species *Pomacentrus nagasakiensis:* males have a more convex head profile than females. There have also been three reports of permanent color differences between damselfish sexes, only one of which has been confirmed in the field. The male of *Dascyllus melanurus* reputedly has a white leading edge on each pelvic fin, a mark lacking in the female (Kattman & Kattman, 1974). Similarly, Etherington (1972) suggested that the dorsal fin of male *Pomacentrus* sp. (probably *P. coelestis*) is edged in black, while the female's fin is edged in yellow. J. Moyer (pers. comm.) however, reports this to be an unreliable character, suggesting instead that males in breeding condition have blue edges on the caudal fin. The only definite sexual dichromatism, however, is also the most conspicuous. Males of the Indo-West Pacific species *Glyphidodontops cyaneus*, in some areas, are more brightly colored than the females and have a vibrant yellow-orange caudal fin lacking in the females;

females, in turn, often have a yellow wash along the underside that is lacking in the male (Allen, 1975). Thresher & Moyer (in press) examine in detail the socio-sexual biology of this species and relate its extreme dichromatism to strong mate selection in the species.

In contrast to the general lack of permanent sexual dichromatism, most damselfishes are dichromatic during courtship and spawning. The temporary development of male-specific "spawning colors" occurs in at least *Pomacentrus, Eupomacentrus*[*], *Glyphidodontops, Paraglyphidodon, Dischistodus, Microspathodon, Abudefduf, Chromis,* and *Dascyllus* (e.g., Albrecht, 1969; Keenleyside, 1972; Myrberg, 1972), though the conspicuousness of the color change varies widely. In *Abudefduf,* males of many species darken during courtship, developing a deep metallic blue coloration that largely obliterates their normal vertical barring. In most genera, however, the temporary color change

[*]*Stegastes* of some authors; some authors also consider *Eupomacentrus* a synonym of *Pomacentrus.*

153

Glyphidodontops hemicyaneus is small at maturity and has spawned in small (less than 150 liters) aquaria. This pair has selected a flowerpot laid on its side as their spawning site and are depositing the eggs on the more protected "roof." The lighting from the camera flash is of course only temporary. Photos by Nelson Hertwig.

A male *Amblyglyphidodon leucogaster* picking at his nest site. Photo by Roger Steene.

The courtship colors of a male *A. leucogaster.* Photo by Roger Steene.

A male *A. leucogaster* (blotchy pattern) courting a female. Photo by Roger Steene.

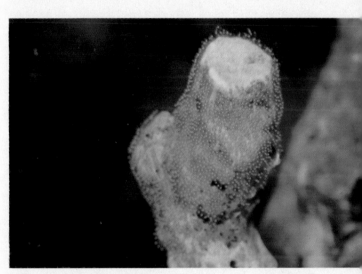

Close-up of *Amblyglyphidodon leucogaster* eggs on the tip of a dead coral branch. Photo by Roger Steene.

A pair of *Amblyglyphidodon* sp. spawning on a very small diameter coral "twig." Photo by Dr. J. E. Randall.

Approximately 4-day-old eggs of *Glyphidodontops leucopomus.* They are 1.23 mm long. Photo by Dr. R. E. Thresher.

155

typically involves a paling of all or part of the body, coupled with the development of one or more horizontal or vertical bars. Although information is still sparse, there appear to be consistent geographic and perhaps systematic differences in the pattern of such spawning colors. In the genus *Eupomacentrus*, for example, western Atlantic species all develop some form of "mask" across the eyes or covering the head and dorsal part of the body; many eastern Pacific species, however, retain normal body colors during courtship and instead develop a vividly contrasting iris around each pupil. Similarly, at One Tree Island, Great Barrier Reef, all species of *Pomacentrus* thus far examined develop a nearly identical courtship pattern involving a general darkening of the body and the development of a series of large pale spots across the top of the head and back. *Pomacentrus nagasakiensis* and several other species from off Japan apparently develop a similar color pattern (Moyer, 1975, pers. comm.), as does at least one species in another genus *(Glyphidodontops biocellatus)* at One Tree Island (Keenleyside, 1972; Thresher & Moyer, in press). This general pattern of pale dorsal spots on a dark body is also characteristic of courtship colors of the deep-water species of western Atlantic *Chromis,* suggesting convergence on a particularly effective color pattern.

Though such male-specific courtship colors will probably be documented for most pomacentrids, it is not universal. According to Kattman & Kattman (1974) and Kattman (1977), the female develops a distinctive courting color in *Dascyllus melanurus.* Thomson, et al. (1979) reported a similar female color change in *Eupomacentrus rectifraenum* (personal observations in the field, however, indicate that the male of this species normally develops such spawning colors). In *Dascyllus trimaculatus* both sexes lighten during courtship and spawning (Fricke, 1973). Finally, neither sex develops any notably distinctive coloration in *Acanthochromis polyacanthus* (Robertson, 1973), *Chromis multilineata* (Albrecht, 1969), *Glyphidodontops rollandi* (Thresher & Moyer, in press), and apparently *Eupomacentrus beebei* (as *Pomacentrus leucoris*) (Breder & Coates, 1933).

Aside from noting such temporary color patterns, or actually seeing the fish spawning, there is only one positive means of externally sexing many pomacentrids. Unfortunately this technique, examination of the urogenital papilla, is a difficult one to use and requires considerable experience before reliable sex determination is possible. In part this is because the papilla is quite small (roughly 4 to 5 mm long in a 12 cm *Abudefduf saxatilis* (Cummings, 1968), for example) and is readily visible only during spawning periods. At other times the papilla, which is located between the anus and the first anal spine, lies against the body in a recessed groove surrounded by scales; it can, however, be teased away from the body with a probe or needle and, in a ripe fish, will erect if slight pressure is applied to the abdomen. The male papilla is long, narrow, pointed, and ends with a tiny urogenital pore. The female papilla is similar but tends to be shorter, stouter, more rounded at the tip, and has a much larger urogenital pore that is set back just short of the tip of the papilla.

The only other form of sexual dimorphism in damselfishes is also a relative one. As a general rule mature males tend to be larger than mature females in, at least, *Abudefduf, Chromis, Dascyllus, Eupomacentrus,* and *Pomacentrus* (the opposite is true in anemonefishes). This difference in size at maturity, along with a sex ratio that favors females, led Schwarz (1980) to suggest that several species of *Dascyllus* are protogynous (female to male) hermaphrodites. Fricke & Holzberg (1974) had previously suggested that small individuals of *Dascyllus aruanus* were hermaphroditic, arguing that such sexual ambivalence enabled immigrating juveniles to enter the social system as females if the position of dominant male was already filled. They stopped short, however, of suggesting that the dominant male was the result of the sex inversion of a previous female, apparently because they found no evidence for such a change in the male's gonads. Further histological and experimental studies, similar to those conducted on labrids, are needed before a conclusion of widespread hermaphroditism in the genus is warranted. To date there is no conclusive evidence of hermaphroditism in any other genus of damselfish (outside of the anemonefishes).

SPAWNING SEASON

In general, damselfishes spawn sporadically all year-round with a pronounced peak in activity in early summer, a pattern that has been documented for various species of *Abudefduf* (Helfrich, 1958, 1959; Cummings, 1968; Fishelson, 1970), *Acanthochromis* (Thresher, in prep.), *Eupomacentrus* (Myrberg, 1972; Schmale, 1981), *Pomacentrus* (Doherty, 1981), *Microspathodon* (MacDonald, 1973), and *Dascyllus* (Stevenson, 1963; Fricke, 1973). Cummings (1968) reported a significant positive correlation between water temperature and number of egg clutches in a colony of captive *Abudefduf saxatilis*. As a variation on this general pattern, many species in the subtropics spawn only during the warmer months of year (e.g., Allen, 1975; Moyer, 1975; Sale, 1970; Gronell, 1978; Nakazono, et al., 1979). A few species, however, differ from the norm for reasons that have not yet been investigated. *Plectroglyphidodon johnstonianus* at Hawaii has semiannual peaks in reproductive activity in early spring and fall (MacDonald, 1976). *Chromis nitida* at the southern end of the Great Barrier Reef spawns in the winter and is the only damselfish locally that does so.

Lunar periodicity of reproductive activities appears to be a common, though not universal, characteristic of reef-dwelling pomacentrids. The type and magnitude of the rhythm vary widely, however. Clear unimodel lunar cycles have been documented for *Chromis multilineata* by Albrecht (1969) and for *Pomacentrus nagasakiensis* by Moyer (1975); Fishelson's (1970) data for *Abudefduf saxatilis* in the Red Sea also suggest a lunar cycle, though a somewhat looser one. Semilunar cycles, with peaks of activity near the new and full moons, also appear to be widespread, having been reported for *Chromis notata* by Nakazono, et al. (1979), for two species of *Pomacentrus* by Doherty (1981), for *Eupomacentrus partitus* by Schmale (1981), for *Microspathodon chrysurus* by Pressley (1980), and for *Abudefduf abdominalis* by Helfrich (1958). May (1967) subsequently looked at *A. abdominalis* in different areas and found that the peaks in each area were often out of phase with those in nearby areas, suggesting that they are not cued by a single pervasive factor. Widespread synchronous spawning has also been reported for *Abudefduf saxatilis* in the western Atlantic by Cummings (1968) and for *Glyphidodontops biocellatus* (as *Abudefduf zonatus*) by Keenleyside (1972); neither indicated whether such synchrony was in phase with lunar cycles (though for the latter, see Thresher & Moyer, in press). At the other end of the spectrum, neither *Plectroglyphidodon johnstonianus* nor *Dascyllus trimaculatus* demonstrate a lunar cycle in reproductive activity (MacDonald, 1976; Fricke, 1973, respectively). *D. trimaculatus* is of particular interest in that the degree of reproductive synchrony exhibited by a group of these fish apparently varies with group size. In a group of 80 individuals, all spawning (there were several pairs present) regularly occurred on the same days, at four-day intervals in the summer and at ten-day intervals in the winter. In a colony of only 12 fish, however, spawning was largely asynchronous and occurred at irregular intervals. Fricke

suggested that synchrony results from reciprocal stimulation of members of a colony; in small colonies not enough fish are present to generate a "critical mass" of mutual stimulation. The idea is an interesting one and should clearly be followed up by more detailed work.

REPRODUCTIVE BEHAVIOR

In the few studies that have looked at it, spawning by damselfishes characteristically occurs in the early morning and rarely continues past midday (e.g., Albrecht, 1969; Fricke, 1973; MacDonald, 1973; Nakazono, et al., 1979; Thresher & Moyer, in press). Spawning late in the day, however, has been reported for western Atlantic *A. saxatilis* by Cummings (1968), for *Glyphidodontops biocellatus* by Keenleyside (1972), for *Pomacentrus nagasakiensis* by J. Moyer (pers. comm.) and for *Eupomacentrus flavilatus* by A. Gronell (pers. comm.). Both Cummings and Albrecht suggested that *A. saxatilis* begins spawning at night, but nocturnal observations could not confirm this and it does not seem likely. The closely related *A. troschelii*, on the other hand, begins to spawn at first light and spawning is well under way by sunrise (pers. obs.).

So far as is known, all species of damselfish produce demersal eggs that are tended and guarded until hatching. Prespawning activity, therefore, varies depending on whether the fish are permanently territorial, and thus already have a defended spawning site, or are normally not territorial (most of these are planktivores that forage well off the bottom) and must establish a temporary territory before courtship and spawning begin. Examples of the former include most species of *Eupomacentrus*, *Microspathodon*, *Parma*, *Plectroglyphidodon*, and *Pomacentrus*. Examples of the latter include most species of *Abudefduf*, *Neopomacentrus*, and *Chromis*. Species of *Dascyllus* are somewhat intermediate, permanently site-attached but also water column foragers.

The temporarily territorial species are the best studied in terms of reproductive behavior, in large part, one suspects, because their spawning activities readily attract attention as a dramatic and conspicuous change from their normal activities. Detailed work, however, has thus far been concentrated on two groups: the sergeant major complex, i.e., *Abudefduf saxatilis* and its sibling species (Helfrich, 1958, 1959; Walker, 1967; Cummings, 1968; Albrecht, 1969; Fishelson, 1970; and Thresher, in prep.), and various species of *Chromis* (Myrberg, et al. 1967; Albrecht, 1969; Sverdloff, 1970a & b; Sale, 1971; and De Boer,

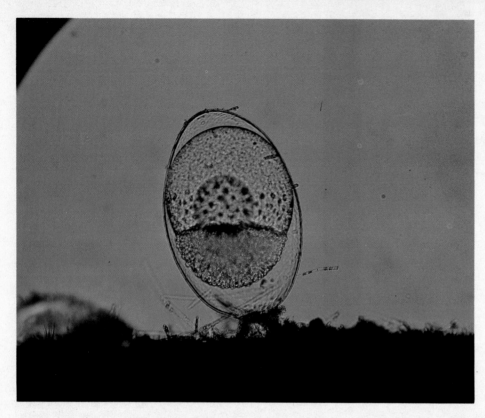

Close-up of a typical developing egg of a pomacentrid (in this example it is an egg of *Chromis atripectoralis*). Note the cluster of attaching threads. Photo by Bruce Carlson.

A group of small juvenile *Amblyglyphidodon aureus*. Newly settled juveniles of some species are more commonly seen in such groups than as solitary individuals. Photo by Wade Doak, Solomon Islands.

Eupomacentrus flavilatus eggs about a day old that were deposited on the roof of a cave. Photo by Dr. R. E. Thresher.

Mixed school of juvenile pomacentrids, including *Acanthochromis polyacanthus* (largest), *Chromis caerulea,* and *C. nitida.* Photo by Ann Gronell.

Pomacentrus nagasakiensis male guarding eggs spawned in a piece of pipe. Photo by Jack T. Moyer.

Newly settled juvenile *Paraglyphidodon melas.* Photo by Dr. R. E. Thresher.

Spawning by *Abudefduf troscheli.* The new eggs being laid are brighter in color than the day-old eggs next to them. Photo by Dr. R. E. Thresher.

Spawning by *Abudefduf troscheli.* The male in this photo and the photo to the left is the darker individual. Photo by Dr. R. E. Thresher.

159

Spawning on sand by *Chromis albomaculata*. Photo by Jack T. Moyer.

1980; and to a lesser extent Longley & Hildebrand, 1941; and Thresher, 1975, 1980a), including several temperate species (Abel, 1961; Turner & Ebert, 1962; Limbaugh, 1964; Russell, 1971; Mapstone & Wood, 1975; and Nakazono, et al., 1979).Detailed information is also provided on the similar acting *Pomacentrus nagasakiensis* by Moyer (1975).

Spawning behavior is similar for all such species studied, involving selection of the spawning site by a male, preparation of the site, courtship, egglaying (usually involving several females sequentially), and then male care and guarding of the eggs until hatching. Spawning sites vary from species to species, ranging from an algal turf in *Chromis atripectoralis* (as *C. caeruleus*) (Sale, 1971) to flat surfaces on or under rocks and ledges (e.g., most species of *Abudefduf* and many *Chromis*). Spawning is usually either solitary, i.e., only one territorial male is present, as in *Chromis cyanea* (Albrecht, 1969; De Boer, 1978), *C. caeruleus* (Swerdloff, 1970a & b), *C. atripectoralis* (Sale, 1971), *Pomacentrus nagasakiensis* (Moyer, 1975), and *Acanthochromis polyacanthus* (Robertson, 1973), or communal, as in the species of the sergeant major complex (e.g., Fishelson, 1970) and *Chromis atrilobata*. Most species seem to be characteristically one or the other, but Myrberg, et al. (1967) and Albrecht (1969) reported that both are characteristic of the western Atlantic species *Chromis multilineata;* single territorial

males can be observed throughout the year, while mass spawning, involving up to 1,000 courting and nest-guarding males, regularly occurs in areas of high population density.

Detailed descriptions of nest-site selection and prespawning activities are provided for *Pomacentrus nagasakiensis*, a solitary spawner (Moyer, 1975), and for *Abudefduf saxatilis*, a communal spawner (Fishelson, 1970). For the former, the spawning season begins with the males descending to the bottom, either to sites they occupied the previous year or, especially for younger males, to new sites. Initially there is no territoriality and individuals mill in tight aggregations while exhibiting a distinctive blue-headed "appeasement" color. Such "clustering" is common to many territorial damselfishes and may serve to facilitate individual recognition of neighbors and to stabilize the social system (see Thresher, 1979b). Following selection of a nest-site (usually the exposed underside of a boulder), the frequency of clustering decreases and each male spends hours each day preparing the spawning site, not only picking off and removing algae and encrusting invertebrates, but also digging out sand, shells, and small stones to enlarge the spawning cavity. The male also becomes increasingly more territorial. Soon he begins courting passing females.

Nest-site selection by the communal spawning

Abudefduf saxatilis is very different. During nonreproductive periods, the fish forage in small to large aggregates in the water column and approach the bottom only sporadically to feed and, at dusk, to seek shelter for the night. As the reproductive period approaches the loose aggregates merge into a single, more cohesive, school and engage in "exploratory swimming." Such exporation tends to occur at two time periods, mid-morning and early dusk. The school, consisting of 150 to 200 individuals and perhaps three to five fish wide and 25 to 35 m long, is lead by the mature males, each already showing traces of the steel-blue spawning colors and a partially erect urogenital papilla. The more numerous females, clearly distended with eggs, follow close behind, with subadults, juveniles, and a few scattered females bringing up the rear. The mature males display to one another sporadically (laterally displaying with fully erected fins) as the school moves across the outer reef. The males as a group are attracted to bare and eroded spots with many caves and holes. They approach such areas closely, sometimes followed by a few females, and inspect any small caves and crevices. If the area is found lacking it is quickly abandoned and the males accelerate to rejoin the mass of females, etc., that have moved past them. If it is acceptable, however, the disappearing school is "ignored" and the males begin the establishment of spawning territories. Each gradually stops swimming widely about and begins to concentrate his activity on a particular spot, usually about 6 to 8 m². His steel-blue color darkens and he begins to defend his area with steadily increasing vigor. Within 12 to 15 minutes after arrival the colony of ripe males is essentially established, with territorial aggression gradually changing from physical contact to brief threats and chases and eventually to acceptance and disregard of familiar neighbors. Throughout the spawning period new males establish territories each day, usually near one another and at the edge of the area occupied by "old" males, and the center of most intense spawning activity may gradually shift from one part of the colony to another. The spawning surface, in the center of the territory, is equally likely to be the top or underside of a boulder and is carefully cleared of everything but short algal stubble. One and a half to two hours after colonization, courtship begins.

Permanently territorial species, of course, bypass this prespawning activity, though males of many species become more aggressive and expand their territories during reproductive periods (e.g., Myrberg &

Thresher, 1974; Moran & Sale, 1977; Gronell, 1978), and most appear to spawn on the same surface repeatedly (Myrberg, 1972; Gronell, 1978; Schmale, 1981). Spawning sites vary from empty gastropod and pelecypod shells to the sides and undersides of rocks, inside caves (usually on the roof), and cleaned coral branches.

Courtship by damselfishes involves a variety of motor, sonic, and color patterns. These were analyzed in greatest detail for the western Atlantic species *Eupomacentrus partitus* by Myrberg (1972) and *Chromis cyanea* by De Boer (1980), and for the Indo-Pacific species *Dascyllus trimaculatus* by Fricke (1973) and *D. marginatus* by Holzberg (1973). The primary motor patterns used in courtship by damselfishes in general are "leading," "signal-jumping," and "dipping," the precise limits of which are often not clearly defined. As a result, there has been some ambiguity in the literature. The simplest and most direct form of courtship is leading (also referred to as "enticing" by Moyer, 1975), in which a nest-tending male swims directly toward a female, turns abruptly in front of her, and returns directly to the spawning site, often while swimming slowly in an exaggerated sinusoidal motion. As commonly used (e.g., Keenleyside, 1972; Myrberg, 1972; MacDonald, 1973; Moyer, 1975), leading involves movement in a horizontal plane and occurs close to the bottom. Signal-jumping (after Abel, 1961, "signalsprunge") is a virtually identical movement whose only distinction is that it is directed at females in the water column (e.g., Fishelson, 1970; Sale, 1971). If signal-jumping is to have any validity as anything other than a purely descriptive term (i.e., denoting vertical rather than horizontal movement), the term should be limited to those movements made in the absence of females; in such cases signal-jumping may have some communicatory value beyond its obvious role in leading. Fricke (1973), for example, suggested that such "unoriented" signal-jumps in *Dascyllus trimaculatus* may facilitate reproductive synchrony in a colony. In other species it may serve as a signal between males denoting locations of territories.

The final widespread form of courtship in damselfishes is "dipping" as described by Emery (1968) and Myrberg (1972). Unlike signal-jumping, which is a rapid movement both up and down, a dip begins as a gradual ascent off the bottom which is followed by an abrupt turn downward and a dash toward the bottom. Again, the exact line between dipping and signal-jumping is a blurred one based on the

Male *Abudefduf sordidus* caring for eggs that were deposited on the piling. Eggs that die or fungus are usually removed by the attending parent. Photo by Dr. Herbert R. Axelrod in the Maldive Islands.

Eupomacentrus nigricans male in courtship colors. Photo by Roger Steene.

Eupomacentrus nigricans male in normal color pattern. Photo by Roger Steene.

Eupomacentrus leucorus male in spawning colors (note the blue eye ring). Photo by Dr. R. E. Thresher.

Eupomacentrus rectifraenum male in courting colors (the normal color is all black). Photo by Dr. R. E. Thresher.

Spawning of *Dascyllus trimaculatus*. The female is laying the eggs. Photo by Roger Steene.

Spawning of *Dascyllus trimaculatus*. The male is fertilizing the eggs. Photo by Roger Steene.

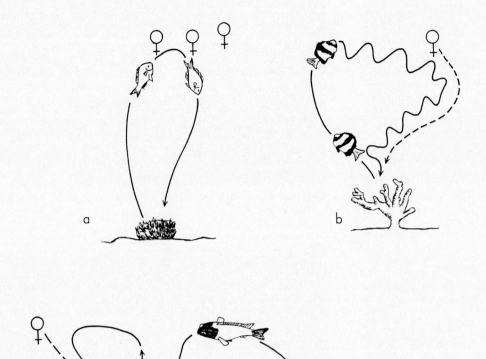

Signal jumping by *Chromis* (a) and *Dascyllus* (b). Both based on Sale (1971), with permission of author. Dipping by *Eupomacentrus partitus* (c). Based on Myrberg (1972a).

ascent time of the male; indeed, dipping, as described, is likely to be only the endpoint of a continuous range of motor patterns whose other end is the lightning fast signal-jumps of some species of *Chromis*. Like signal-jumps, dips occur sporadically in the absence of females, but increase in frequency and speed when one approaches. Dipping also occurs occasionally during leading, the male dipping while approaching the female and then leading her back to the nest-site while swimming in an exaggerated motion.

There appear to be consistent systematic differences among genera of damselfishes with respect to the motor patterns the males characteristically use to court females. Not surprisingly, signal-jumping is characteristic mainly of the water column foragers such as *Abudefduf* and *Chromis* and can be expected to occur in other genera such as *Neopomacentrus* and *Lepidozygus*. In that respect it would be interesting to examine the courtship movements of some of the benthic species in these typically water column foraging genera, such as *Abudefduf taurus* and its near relatives.

Among the typically benthic genera, dipping per se is confined largely to *Dascyllus* and *Eupomacentrus*, though at least two species of *Pomacentrus* dip—*P. nagasakiensis* (according to Moyer, 1975, though he referred to it as signal-jumping) and *P. pavo*—and at least one species of *Eupomacentrus*, *E. flaviatus*, rarely, if ever, dips (A. Gronell, pers. comm.). Males of most other benthic genera characteristically lead females to the nest without first dipping or signal-jumping, or they combine leading with some form of stationary quivering. Examples include most species of *Pomacentrus* and most, if not all, species of *Glyphidodontops*, *Paraglyphidodon*, and *Dischistodus* (pers. obs.).

During courtship males of most species assume a distinctive color pattern (see section on *Sexual Dimorphism*) which may have a communicatory value in its own right, as well as increasing the conspicuousness of the motor patterns. Sound production during courtship has also been documented for fishes in several genera by Garnaud (1957a), Emery (1968),

Spanier (1970), and Moyer (1975). The most detailed work on sonic activity associated with courtship, however, has been done on various western Atlantic species of *Eupomacentrus* by Myrberg (1972a & b, 1978), Myrberg & Spires, 1972), Ha (1973), and Spanier (1979). *Eupomacentrus partitus,* a small ubiquitous planktivore similar in many respects to various species of *Dascyllus,* produces three different sounds during various stages of courtship. A "chirp" is a 3-pulse staccato with a total duration of approximately 0.2 seconds and a peak frequency range of 600 to 1200 hertz. Chirps are produced by males during a dip, though the two are not invariably correlated (fish do not always chirp while dipping, but they never chirp at any other time). A "long chirp" is a longer duration (up to 0.6 seconds) 4- to 6-pulse staccato, similar in spectral composition to a chirp, that occurs during "flutter," a motor pattern characteristic of a male apparently attempting to entice a departing female back to the spawning site. Finally, a "grunt" is a lower frequency sound of variable pulse number and duration that is produced by the male as he and the female circle one another in the vicinity of the nest. Subsequent study has indicated that similar sounds are characteristic for all western Atlantic species of *Eupomacentrus* thus far examined. Playback of recorded sounds by means of an underwater speaker has subsequently demonstrated that such damselfishes discriminate between sounds associated with courtship and those associated with other behaviors (such as aggression), and further, that courtship sounds alone will stimulate dipping by a ripe male even in the absence of a female. By using artificial and experimentally modified sounds in the playbacks, Ha demonstrated that different aspects of the courtship sound (chirp) convey different information: the number of pulses (ranging from one to four) conveys the signal producer's major behavioral state; chirp rate (number of chirps per minute) indicates relative arousal level of the signal producer; and pulse interval probably provides species specificity. The last was examined in greater detail by Spanier, who played back similarly modified sounds from both conspecifics and congeners. Each species responded best to sounds produced by conspecifics, indicating a degree of species specificity. Such discrimination was best when the chirps played contained the species-typical number of pulses but occurred even if the same number of pulses was used for each of the four test species used. Finally, changing the pulse interval without changing other characteristics of the sound

resulted in a misidentification of the chirp, confirming that these intervals indeed carried the information upon which species discrimination is based.

Following this complex suite of courtship movements, colors, and sounds, spawning itself is relatively uncomplicated. The male first approaches the spawning site closely and "skims" it; that is, he moves smoothly across it with his venter (and possibly urogenital papilla) in contact with the spawning site. The female approaches, the male withdraws slightly, and she begins the first of many spawning passes. Eggs are deposited in long, even rows, one on each pass over the nest, with the rows gradually overlapping and crossing to such a degree that a solid, uniform, and single-layered mass of eggs results. The eggs are adhesive at one end and become firmly attached to the substrate, but not to one another (Helfrich, 1958). While the female is spawning the male circles about her, clearly and vigorously defending her and the nest-site from both egg predators and other males (who attempt to sneak in and fertilize the eggs before the guarding male can do so—see Albrecht, 1969; Fishelson, 1970). Periodically he approaches the female and, either beside or just behind her, fertilizes the eggs.

Spawning-bout duration varies widely. Several workers (e.g., Myrberg, et al., 1967; Sale, 1971) noted that females of *Chromis* spawn only briefly before returning to the water column; whether or not such females returned to spawn with a given male repeatedly, however, was not determined. At the other extreme, spawnings involving a pair of permanently territorial species may last several hours (A. Gronell, pers. comm.), suggesting that the female deposits all her eggs in the nest of one male (see also Brinley, 1939; Doherty, 1981). In the only detailed analysis of clutch sizes to date, Thresher (in prep.) found evidence that females of *Abudefduf troschelii* spawn several times during a reproductive period, each time with a different male. Each time she spawns she deposits slightly more than half of the eggs she is carrying, i.e., about 55% of the number she starts with in the first spawning, about 30% in the second, and about 15% in the third. This pattern appears to result from competition between the sexes for avoidance of parental care, along the lines suggested by Maynard-Smith (1977). By leaving only as many eggs as she can afford to lose with each male, a female always retains the option of deserting the eggs. A male, therefore, can never desert without condemning his spawn.

Estimates of female fecundity for damselfishes are

Male *Dascyllus aruanus* guarding eggs spawned in an aquarium. Photo by Dr. R. E. Thresher.

Amphiprion clarkii biting the substrate, apparently in preparation for spawning. Photo by Jack T. Moyer.

Amphiprion clarkii spawning on rock immediately adjacent to an anemone. Photo by Jack T. Moyer.

Amphiprion clarkii continuing to spawn. Note the sexual dichromatism in the caudal fin. Photo by Jack T. Moyer.

Amphiprion polymnus standing guard over a batch of eggs. Photo by Roger Steene.

The same individual, apparently a male, mouthing the eggs. Photo by Roger Steene.

Sexual dichromatism in *Amphiprion clarkii,* with the male sporting yellow-orange upper and lower edges to the caudal fin. This same tendency has also been noted in the Yaeyama Islands, but other populations may differ in the type of color differences. Photo by K. H. Choo at Taiwan.

A pair of *Amphiprion clarkii* with a batch of eggs on a flat surface adjacent to their anemone. Photo by A. van den Nieuwenhuizen.

not common. Those available range from only 100 to 150 eggs per clutch for *Acanthochromis polyacanthus* (Robertson, 1973) to over 40,000 eggs per clutch for large females of *Abudefduf abdominalis* (Helfrich, 1958) and *Dascyllus albisella* (Stevenson, 1963). Both Helfrich and Stevenson demonstrated that clutch size is positively correlated with female body size in their respective species. Other estimates of clutch size include 400 to 500 eggs for *Eupomacentrus beebei* (as *Pomacentrus leucoris*) in captivity (Breder & Coates, 1933); 1,000 to 5,000 eggs for *Pomacentrus coelestis* (Matsuoka, 1962); 12,000 to 32,000 eggs for *Chromis notata* (Nakazono, et al. 1979); 20,000 to 25,000 eggs for *Dascyllus trimaculatus* (Garnaud, 1975); an average of 20,000 for *Abudefduf saxatilis* (Cummings, 1968); and 27,000 for the slightly larger *Abudefduf troschelii* (Thresher, in prep.). Honda & Imai (1973) estimated total annual fecundity of *Pomacentrus nagasakiensis* at 211,000 eggs.

Each nest, however, frequently contains many more eggs than this as males frequently spawn with many females at the same nest site. Sequential polygamy is especially conspicuous in temporarily territorial species; Sale (1971), for example, reported male *Chromis atripectoralis* spawning with 20 to 30 females at the same time. Estimates of the total number of eggs in such a nest range as high as 200,000 (Fishelson, 1970), and broods of 40,000 to 100,000 are probably common. In permanently territorial species fewer females tend to be present in a small area and the opportunities for polygamy are consequently reduced; broods on the order of only a few thousand are more common for such species.

After spawning the eggs are tended and guarded by one or (rarely) both parents. The latter has thus far been reported only for *Eupomacentrus beebei* (Breder & Coates, 1933), based on a single observation in captivity, and for *Acanthochromis polyacanthus* (Robertson, 1973; Allen, 1975). The latter species is unique among damselfishes in terms of its reproductive biology and will be discussed in detail below. In all other species nest care is performed by the male alone (e.g., Myrberg, 1972; Sale, 1971; Keenleyside, 1970). Such care includes keeping it clear of sand and detritus, mouthing the eggs, and apparently removing (eating?) those that are infertile or fungused, fanning the eggs, and defending the nest from egg predators. Fanning has been examined in some detail by Helfrich (1958) and Albrecht (1969), both of whom looked at a species of sergeant major (*Abudefduf abdominalis* and *A. saxatilis*, respectively). Helfrich reported a steady increase in the frequency and intensity of fanning that correlated with the age of the eggs, an observation very reminiscent of similar observations made by Tinbergen (1953) in a classic study of stickleback behavior. Both suggested that this increased fanning is a response to increased oxygen demands by the developing eggs. Albrecht, however, could not confirm this correlation and, in general, discounted any effect of oxygen consumption on fanning rate by the male. He did note, however, that one day after spawning the fanning rate showed a conspicuous depth distribution (peaking at 4 to 5 m) and also increased significantly at night. The former he could not explain; the latter he suggested might relate to predation on nests at night by echinoderms.

Vigorous territorial defense by nest-tending males has been reported, at least in passing, by virtually every paper on reproduction in damselfishes. Detailed accounts of such defense are provided by Cummings (1968), Low (1971), Myrberg & Thresher (1974), Moyer (1975), and Thresher (1976).

As noted, parental care by *Acanthochromis polyacanthus* differs markedly from that reported for other damselfishes. In all other species care and defense by the male end at hatching, with the larvae ascending into the water column to begin a planktonic stage. In *A. polyacanthus*, care and defense persist since the young are demersal and remain with the parents (*A. polyacanthus* is thus far the only reef fish verified to bypass the planktonic larval stage). Prior to hatching, parental activity is limited to defense of the developing eggs; Robertson (1973) reported that neither fanning nor cleaning the nest is observed. After hatching, the large young remain with the parents for three to six weeks. Newly hatched fry, about 5 mm long, remain in a tight school close to the spawning cave and are vigorously guarded against a variety of fishes (Thresher, in press). The significance of this defense is readily demonstrated by removing the parents—the young are quickly devoured (Allen, 1975). Robertson also noted fry "glancing" off the bodies of the adults as well as a few passing fishes and suggested that they might be feeding on the mucous coat of the adults. Such has previously been demonstrated in cichlids (Ward & Barlow, 1967), and mucophagy has also been suggested for juveniles of other reef fishes (Thresher, 1979b). As the fry grow, the school spreads out and defense by the parents lessens in intensity. At 15 to 20 mm TL the young begin to develop adult coloration, and by 25 to 30 mm they leave the parents to form loose aggregations with young of other broods.

EGGS AND LARVAE

Damselfish eggs are smoothly elliptical and are attached to the substrate by means of a cluster of fine threads at one end of the egg. Garnaud (1957a) reported that female *Dascyllus trimaculatus* deposit an adhesive substance on the substrate to which the eggs subsequently stick. This observation has yet to be confirmed, however; rather, it is common for females to make several "false" spawning passes before actually depositing eggs, and Garnaud may have mistaken this behavior for deposition of such an adhesive base. Helfrich (1958) also looked at the mechanism that attaches damselfish eggs to the nest and found, instead of a special cement, an "adhesive cap" on one pole of the egg of *Abudefduf abdominalis*. This cap adheres to the first point of contact with the substrate (but not other eggs) and gradually everts to firmly anchor the egg as it rocks back and forth due to water movements.

Egg dimensions are available for a number of genera (see list) and range from 0.49 mm to 2.3 mm along the longest dimension for species with planktonic larvae and 4.5 mm for *Acanthochromis polyacanthus*, which lacks a pelagic larval stage. There is a general trend for eggs to be larger for subtropical and temperate species than for tropical species. The eggs of two species of temperate *Chromis*, for example, are 0.9 mm and 1.14 to 1.32 mm long, respectively, larger than any tropical species yet measured (Turner & Ebert, 1962; Russell, 1971); the 2.3 mm long eggs of the subtropical *Pomacentrus nagasakiensis* are larger than those of other members of the genus, which range from 1.2 to 1.75 mm (Honda & Imai, 1973). The eggs of *Hypsypops rubicunda*, a temperate species, are larger than those of any tropical species with pelagic larvae, averaging 2.0 mm long and 1.0 mm wide (Limbaugh, 1964).

Egg development is described in varying detail and completeness for *Chromis* (Fujita, 1957; Myrberg, et al., 1967; Donato, 1967); *Eupomacentrus* (Breder & Coates, 1933; Brinley, 1939); *Pomacentrus* (Honda & Imai, 1973; Doherty, 1980); and *Abudefduf* (Shaw, 1955; Helfrich, 1958). At species-normal temperatures incubation times differ consistently among genera, with incubation taking longer the larger the egg (Thresher, in prep). *Chromis* require two to three days to hatch, *Dascyllus* three days, *Eupomacentrus* and *Pomacentrus* three and a half to six days (usually five or six), and *Abudefduf* five to seven days. Incubation time, of course, decreases with increasing water temperature: *A. saxatilis* hatches in five to six days at 28 to 29°C and in six and a half days at 24°C. Limited data suggest that hatching in most, if not all, species occurs at dusk (e.g., Fricke, 1973; Doherty, 1980) or at night (e.g., Albrecht, 1969; Moyer, 1975).

Newly hatched planktonic larvae range in total length from 1.1 mm (*Chromis dispilis*, Russell, 1971) to 3.06 mm (*Chromis punctipinna*, Turner & Ebert, 1962). By comparison, the newly hatched non-planktonic juveniles of *Acanthochromis polyacanthus* are huge, approximately 5 mm long, and heavy-bodied. With this exception, newly hatched damselfishes are long and slender with a blunt head (usually pigmented on top), well developed jaws, pigmented eyes, a continuous vertical finfold, and a small yolk sac (e.g., Miller, et al., 1979). The yolk is fully absorbed in about three days, at which time feeding begins. A detailed description of larval development is thus far available for only two species. In *Abudefduf abdominalis* development is essentially continuous with no abrupt metamorphosis (Helfrich, 1958). Dorsal fin rays are evident by day five, and by day eight several dorsal spines are present. Between days five and eight, the larva's body deepens and the fin regions become clearly defined. The pelvic fins grow especially large and soon develop conspicuous black and yellow pigments. By day 13 the larvae average 6 mm total length and have developed the general shape and proportions of the adults, except for the enlarged pelvic fins. At about 20 days, at a length of 12 mm, the larvae begin to develop the vertically barred color pattern of the adult, migrate into shallow water, and begin to settle (settling can begin as early as day seven). Development of *Pomacentrus nagasakiensis* is similar according to Honda & Imai (1973). *P. nagasakiensis* larvae reared in captivity settled to the bottom 15 to 20 days after hatching, at a length of approximately 8 mm. Several studies (e.g., Emery, 1968; Nolan, 1975; Williams, 1980) suggest that settlement occurs mainly at dusk and at night. Size at settlement ranges from 7 to 15 mm.

A strikingly different pattern of larval development has been suggested for *Chromis cyanea*, a western Atlantic species. De Boer (1978) suggested that newly hatched larvae head directly into the reef after hatching and remain there feeding on plankton; that is, they bypass the planktonic stage. This seems unlikely given the small size of the larvae at hatching, the high risk they run of predation, and the absence of parental care in the species, but it should clearly be examined in greater detail.

Though little is known about the larval ecology of

The spawning pair of *Amphiprion clarkii* keep a close watch on their eggs. They will stay close to the eggs, fanning them and keeping them clean. This care lasts until the eggs hatch, which, in the case of *A. clarkii,* may vary from six and a half days (at 27°C) to two weeks (at 21°C). Photos by Jan Carlen.

The following sequence shows spawning in *Amphiprion frenatus.* In this first photo the pair are preparing the nest site for spawning by cleaning it.

The egg-laying commences as the female makes a pass over the site, depositing the adhesive eggs as she goes.

With spawning completed the parents stand guard until the eggs hatch. As in other species of anemonefishes, the eggs are laid very close to the anemone itself. All photos courtesy Midori Shobo, *Fish Magazine,* Japan.

List of Egg Characteristics for Tropical Damselfish Genera

Genus	Longest Dimension	Color	Reference
Chromis	0.6 to 0.78 mm	transparent	Longley & Hildebrand (1941); Fujita (1957); Myrberg, et al. (1967); Swerdloff (1970a); Sale (1971)
Dascyllus	0.49 to 0.85 mm	transparent to straw-colored	Garnaud (1957a); Stevenson (1963)
Acanthochromis	4.5 mm	pink	Thresher (in prep.)
Abudefduf (*saxatilis* complex)	1.1 to 1.3 mm	red to violet or brown	Shaw (1955); Cummings (1968); Albrecht (1969); Helfrich (1958)
Abudefduf (other)	0.8 mm	red or green	Cummings (1968); Albrecht (1969)
Eupomacentrus	0.8 to 0.99 mm	white, pink, or orange	Breder & Coates (1933); Brinley (1939)
Pomacentrus	1.2 to 2.3 mm	white or pink	Honda & Imai (1973); Doherty (1981); pers. obs.
Plectroglyphidodon	0.85 to 0.90 mm	pink	pers. obs.
Glyphidodontops	1.05 to 1.40 mm	pink	pers. obs.
Amblyglyphidodon	1.5 to 1.6 mm	pink	pers. obs.
Paraglyphidodon	1.6 to 1.65 mm	pink	pers. obs.
Microspathodon	slightly over 1 mm	gray-green	Cummings (1968); MacDonald (1973)

damselfishes, considerable work and discussion have centered on the stimuli that trigger their settling on the reef and that, in part, determine the distribution of species on the reef. Two general hypotheses have been put forward. The "order" hypothesis proposes that habitat partitioning, largely as the result of differential settlement by larvae of different species, is the major factor underlying continued coexistence of closely related species of damselfishes on the reef (see reviews by Smith, 1978; Dale, 1978; Robertson & Lassig, 1980). The "lottery" hypothesis, in contrast, proposes only limited differences in settlement patterns and, consequently, broad overlaps in preferred areas. Coexistence is seen as the result of the unpredictable availability of living space, limited rates of recruitment, and the consequent inability of any single species to completely exclude and out-compete others (see reviews by Sale, 1977, 1978). Distribution of adult pomacentrids in relatively discrete species-specific habitats has been documented by Fricke (1975), Clarke (1977), Itzkowitz (1977), and Robertson & Lassig (1980); broad overlap in habitat usage has been documented mainly by Sale and his students (Sale, 1974, 1975; Sale, et al., 1980; Dybdahl, 1975, 1978; Doherty, 1980; Williams, 1980).

Whether large-scale differences in the distribution of different species of damselfishes on the reef reflect differential settlement patterns or differential survival of randomly settling larvae (or some combination of the two) is still an open question awaiting definitive experimental work. Nolan (1975) and Williams (1980) provided evidence that settling postlarvae are not using the presence of conspecific adults to trigger settlement. Williams did indicate, however, that for several Great Barrier Reef species some coral outcrops have consistently higher rates of settlement than others. Whether such a "preference" is the result of active selection by the larvae or results from the interaction of current patterns and larval distribution in the water column is still not clear.

Following settlement, a variety of processes seem to be occurring in the newly settled juveniles. In at least some species recruits seem to aggregate; juveniles of many species of *Chromis*, *Neopomacentrus*, and even of some benthic species such as *Pomacentrus coelestis* are commonly found in small widely scattered groups rather more often than they are seen as solitary individuals. This suggests that after settlement such

A postlarval damselfish, 14.2 mm total length, collected near a submerged light at night off Papua New Guinea. Photo by Dr. R. E. Thresher.

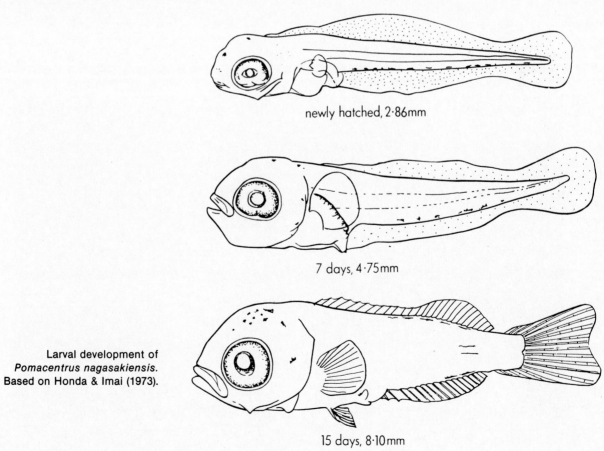

newly hatched, 2·86mm

7 days, 4·75mm

Larval development of
Pomacentrus nagasakiensis.
Based on Honda & Imai (1973).

15 days, 8·10mm

Amphiprion chrysopterus picking at a spawn of eggs at Enewetak, Marshall Islands, the area where Dr. G. R. Allen made many of his observations while preparing his doctoral dissertation. Photo by Dr. G. R. Allen.

The eggs of this *Amphiprion chrysopterus* are protected by both the parents and the stinging anemone in close approximation. The parents also benefit from the protection of the anemone. Photo by Dr. G. R. Allen.

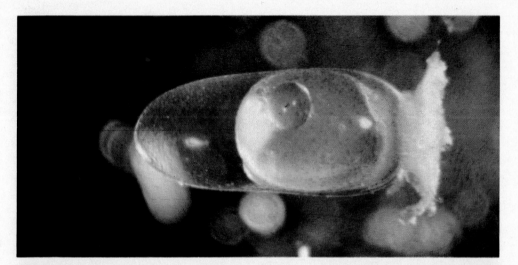

A

B

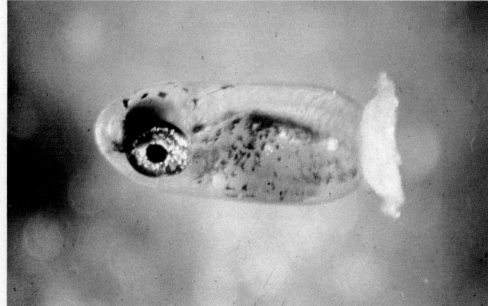

Stages in the development of the egg of *Amphiprion chrysopterus:* a) 36 hours; b) 104 hours; c) 126 hours. Photos by Dr. G. R. Allen.

C

juveniles move about seeking conspecifics and become site-attached only when in the presence of conspecifics. While the risk of movement is doubtless high, it may be outweighed by the disadvantage, in terms of reproductive success, of settling in an area that lacks suitable mates. Alternatively, in many benthic species in particular, there is considerable competition for space among juveniles, and such juveniles may be effectively excluded from some areas by the actions of adults (e.g., Sale, 1976; Godsey, et al., 1980). Juveniles of many species, especially those of the benthic territorial species, are more brightly and differently colored than the adults, often in colors that maximize conspicuousness (Thresher, 1977). This color difference, combined with the frequent observation of adults vigorously chasing juveniles, led Thresher (1978) to suggest that such juvenile-specific colors facilitate habituation of the adults to the juveniles. Alternatively, such colors may function in interjuvenile competition, perhaps by advertising the presence of one juvenile to others and warning them away from its territory, along the lines first suggested by Lorenz (1962) (in aggregating species, discussed above, one could also argue that the colors facilitate the recruits locating one another). As not all species have juvenile-specific colors and those that do vary widely in conspicuousness, there would seem to be considerable opportunity for profitable correlational and experimental studies of the role(s) of such coloration (see the final chapter).

The size at which juveniles assume adult coloration varies widely within a species (e.g., Emery, 1968; Thresher, 1979b), and the factors that trigger such a change are poorly understood (e.g., Patterson, 1975). For the western Atlantic species of *Eupomacentrus*, sexual maturity is probably not reached until an age of two years, with a probable total life span of six to eight years (Gronell, 1978; pers. obs.).

REPRODUCTION IN CAPTIVITY

Spawning in aquaria has been reported for a scattering of damselfishes: *Eupomacentrus beebei* (as *Pomacentrus leucoris*) by Breder & Coates (1933); *E. partitus* by Myrberg (1972); *Pomacentrus nagasakiensis* by Honda & Imai (1973); *Pomacentrus* sp. (probably *P. coelestis*) by Etherington (1972); *P. coelestis* by Matsuoka (1962); *Glyphidodontops hemicyaneus* by Walker & Herwig (1976) and Walker (1977); and *Chromis notata* by Nakazono, et al. (1979). The only genus frequently and even regularly spawned in captivity, however, is *Dascyllus* (Garnaud, 1957a & b; Koenig, 1958; Kattman & Kattman, 1975; Kattman, 1977; Walker, 1977).

Stimulating *Dascyllus* to spawn in an aquarium requires little more than placing two or more fish (one individual clearly larger than the others) in a 150- to 200-liter aquarium that offers adequate coral cover. Good water conditions, relatively warm temperatures (24 to 28°C), and heavy feedings with live and flake foods are usually enough to elicit spawning. Although spawning frequently occurs in large community aquaria, most consistent results are obtained by isolating the spawning group. Prespawning activities include nest preparation by the male (mainly picking at, fanning, and cleaning a spot of coral) and shallow dipping by the male toward the female whenever she approaches the area. One to two days prior to spawning the female will swell with eggs, but unless you've been watching her closely, this may not be obvious. Shortly before spawning begins, often early in the morning, the urogenital papillae of both male and female will erect. During actual spawning male and female will alternate swimming slowly over the chosen spot, each skimming closely over it while leaving gametes behind. After spawning the male drives the female off and he begins to tend the almost invisible eggs. Because of its usual occurrence in early morning and because of the small size of the eggs, it is likely that spawning by *Dascyllus* occurs far more often than it is observed; the average aquarist, in fact, may have a pair breeding regularly without his ever being aware of it. Such a pair will produce a new clutch of eggs every two to three weeks almost indefinitely.

Aquarium spawning by other species is far less common and far more fortuitous. Breder & Coates, for example, had only two specimens of *Eupomacentrus beebei* survive from a collecting trip. Without further thought they placed them in a large (about 1500-liter) aquarium and subsequently discovered that the pair had spawned. The use of a very large aquarium was also necessary to spawn *E. partitus*. *Pomacentrus coelestis* (?) and *Glyphidodontops hemicyaneus*, however, are small at maturity and have both spawned in small (less than 150 liters) aquaria. Both are probably spawnable by a moderately determined aquarist, as are a variety of other small species in the two genera. Again, the keys to successful spawning are likely to be: (1) isolation of the spawning group; (2) enough room so that territorial fish can feel comfortable; (3) heavy feeding with live foods; (4) good water conditions; and (5) a suitable spawning site (a well-cleaned overturned clam shell is ideal). Walker also noted that his G. hemicyaneus spawned shortly after he'd reduced the specific gravity in his tank from above to below 1.020, and he thought that this may have stimulated spawning.

There is only one published report of successful rearing of damselfish larvae in captivitiy. A scattering of unpublished information suggests that the sergeant majors (Abudefduf saxatilis and its near relatives) have been successfully reared several times using more or less standard techniques (feedings with rotifers and wild-caught plankton, open water systems, constant illumination). Honda & Imai (1973) reared Pomacentrus nagasakiensis through to settlement by keeping them in 500-liter aquaria and feeding them the fertile eggs of the sea urchin Heliocidaris, rotifers (Brachionis plicatilis), and Artemia nauplii. All of the juveniles died by day 30, however. In most other attempts to rear the larvae of benthic species the larvae inexplicably die at around two weeks after hatching, well after they had begun feeding. The only report of successful rearing of larvae to metamorphosis is a brief popular account by Moe & Young (1981). These authors reared large numbers of the western Atlantic species Microspathodon chrysurus. They provide no information on rearing techniques, but it is likely they used techniques similar to those above.

Literature Cited

Abel, E.F. 1961. Freiwasserstudien über das Fortpflanzungsverhalten des Monchfisches Chromis chromis Linne, einem Vertreter der Pomacentriden im Mittelmeer. Z. Tierpsychol., 18:441-449.

Albrecht, H. 1969. Behavior of four species of Atlantic damselfishes from Columbia, South America (Abudefduf saxatilis, A. taurus, Chromis multilineata, C. cyanea; Pisces: Pomacentridae). Z. Tierpsychol, 26:662-676.

Allen, G.R. 1975. *Damselfishes of the South Seas.* T.F.H. Publ., Neptune City, New Jersey. 240 pp.

Breder, C.M., Jr. and C.W. Coates. 1933. Reproduction and eggs of Pomacentrus leucoris Gilbert. Amer. Mus. Novitates, (612):1-6.

Breder, C.M., Jr. and D.E. Rosen. 1966. *Modes of Reproduction in Fishes.* Natural History Press, Garden City, N.Y. 941 pp.

Brinley, F.J. 1939. Spawning habits and development of beau-gregory (Pomacentrus leucostictus). Copeia, 1939:185-188.

Clark, T.A. 1971. Territory boundaries, courtship and social behavior in the garibaldi, Hypsypops rubicunda (Pomacentridae). Copeia, 1971:295-299.

Clarke, R.D. 1977. Habitat distribution and species diversity of chaetodontid and pomacentrid fishes near Bimini, Bahamas. Mar. Biol., 40:277-289.

Cummings, W.C. 1968. Reproductive habits of the sergeant major, Abudefduf saxatilis, (Pisces, Pomacentridae), with comparative notes on four other damselfishes in the Bahama Islands. Ph. D. Diss., Univ. Miami, Fla.

Dale, G. 1978. Money-in-the-bank: a model for coral reef fish coexistence. Env. Biol. Fishes, 3:103-108.

De Boer, B.A. 1978. Factors affecting the distribution of the damselfish Chromis cyanea (Poey), Pomacentridae, at a reef at Curacao, Netherland Antilles. Bull. Mar. Sci., 28:550-565.

Doherty, P.J. 1980. Biological and physical constraints on populations of two sympatric territorial damselfishes on the southern Great Barrier Reef. Ph. D. Diss., Univ. Sydney.

Donato, A. 1967. Yolk-formation in Chromis chromis Cuv. (Teleostei, Labridae). Boll. Zool., 34:115-116.

Etherington, W. 1972. Mating behavior of the Fiji damsel. Marine Aquarist, 3(1):32-34.

Fishelson, L. 1970. Behaviour and ecology of a population of Abudefduf saxatilis (Pomacentridae, Teleostei). Anim. Behav., 18:225-237.

Fricke, H.-W. 1973. Okologie und Sozialverhalten des Korallenbarsches Dascyllus trimaculatus (Pisces, Pomacentridae). Z. Tierpsychol., 32:225-256.

Fricke, H.-W. 1975. Sozialstruktur und okologische Spezialisierung von verwandten Fischen (Pomacentridae). Z. Tierpsychol., 39:492-520.

Fricke, H.-W. and S. Holzberg. 1974. Social units and hermaphroditism in a pomacentrid fish. Naturwissenschaften, 61:367-368.

Fujita, S. 1957. On the egg development and prelarval stage of a damselfish, Chromis notatus (Temminck et Schlegel). Japan. J. Ichthyol., 6:87-90.

Garnaud, J. 1957a. Ethologie de Dascyllus trimaculatus (Rüppell). Bull. Inst. Oceanogr. Monaco, 54(1096):1-10.

Garnaud, J. 1957b. Breeding the black damselfish. The Aquarium, Philadelphia, 26(7):211-213.

Godsey, M.S., S. Tanaka, and E.A. Aughney. 1980. An experimental analysis of the factors influencing habitat selection by juvenile Sea of Cortez damselfish, Eupomacentrus rectifraenum (Pomacentridae). Bull. Mar. Sci., 30:326.

Gronell, A.M. 1978. Home-ranging behavior of cocoa damselfish, Eupomacentrus variabilis, off the coast of Florida. Master's Thesis, Univ. Miami, Fla.

Amphiprion perideraion guarding a small clutch of eggs. Photo by Dr. G. R. Allen.

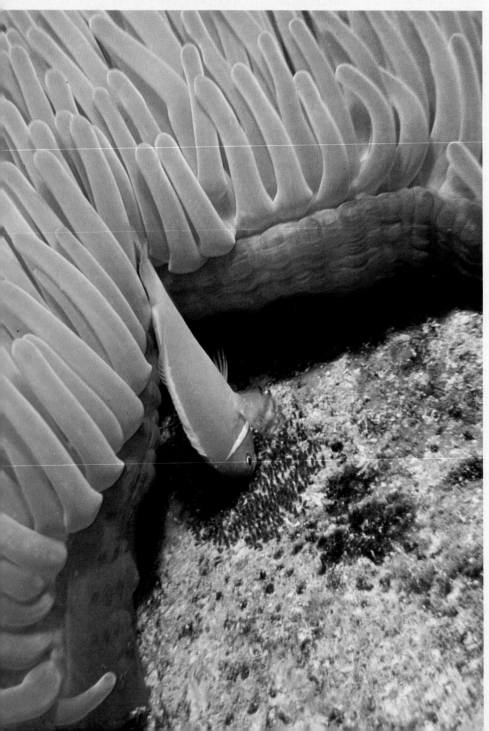

In *A. perideraion* there is some sexual dichromatism. The male has a pale orange trim along the upper and lower edges of the caudal fin as well as the edge of the soft dorsal. Photo by Dr. G. R. Allen.

Once the eggs are deposited, their care is similar to that of other anemonefishes. *A. percula* has been successfully spawned many times and is actually currently being produced on a commercial basis. Tank-raised individuals are usually a bit more costly than wild specimens. Photo by Maulenasacht-Madson.

Amphiprion percula (*A. ocellaris* of some authors) in the midst of spawning. The female is quite heavy-bodied, with the ovipositor visible against the dark background. Photo by A. van den Nieuwenhuizen.

Ha, S.J. 1973. Aspects of sound communication in the damselfish *Eupomacentrus partitus*. Ph. D. Diss., Univ. Miami, Fla.

Helfrich, P. 1958. The early life history and reproductive behavior of the maomao, *Abudefduf abdominalis* (Quoy & Gaimard). Ph. D. Diss., Univ. Hawaii, Honolulu.

Helfrich, P. 1959. Observations on the reproductive behavior of the maomao, a Hawaiian damselfish. *Proc. Hawaiian Acad. Sci.*, 34:22.

Holzberg, S. 1973. Beobachtungen zur Okologie und zum Sozialverhalten des Korallenbarsches *Dascyllus marginatus* Rüppell (Pisces; Pomacentridae). *Zeit. Tierpsychol.*, 33:492-513.

Honda, S. and S. Imai. 1973. Breeding and early development of a pomacentrid, *Pomacentrus nagasakiensis* Tanaka in the aquarium. *Mem. Fac. Fish., Kagoshima Univ.*, 20:95-106.

Itzkowitz, M. 1977. Spatial organization of the Jamaican damselfish community. *J. exp. Mar. Biol. Ecol.*, 28:217-241.

Kattman, D.L. 1977. The way to do it from one inch to parenthood. *Marine Hobbyist News*, 5(1):1,3.

Kattman, D. and D. Kattman. 1974. The courtship and spawning of *Dascyllus melanurus*. *Octopus*, 1(8):20-23.

Keenleyside, M.H.A. 1972. The behavior of *Abudefduf zonatus* (Pisces, Pomacentridae) at Heron Island, Great Barrier Reef. *Anim. Behav.*, 20:763-774.

Koenig, O. 1958. Ein Beitrag zur Fortpflanzungsbiologie von *Dascyllus trimaculatus*. *Die Aquar. und Terrar. Z.*, 11:107-111.

Limbaugh, C. 1964. Notes on the life history of two Californian pomacentrids: Garibaldis, *Hypsypops rubicunda* (Girard), and Blacksmiths, *Chromis punctipinna* (Cooper). *Pac. Sci.*, 17:41-50.

Longley, W.H. and S.F. Hildebrand. 1941. Systematic catalogue of the fishes of Tortugas, Florida. *Papers Tortugas Lab., Carnegie Inst. Wash.*, 34:331 pp.

Lorenz, K. 1962. The function of color in coral reef fishes. *Proc. Roy. Inst. Great Britain*, 39:282-296.

Low, R.M. 1971. Interspecific territoriality in a pomacentrid reef fish *Pomacentrus flavicauda* Whitley. *Ecology*, 52:648-654.

MacDonald, C.D. 1973. Reproductive behavior and social dynamics of the yellowtail damselfish, *Microspathodon chrysurus* (Perciformes: Pomacentridae). Master's Thesis, Univ. Puerto Rico, Mayaguez, Puerto Rico.

MacDonald, C.D. 1976. Nesting rhythmicity in the damselfish *Plectroglyphidodon johnstonianus* (Perciformes, Pomacentridae) in Hawaii. *Pac. Sci.*, 30:216.

Mapstone, G.M. and E.M. Wood. 1975. The ethology of *Abudefduf luridus* and *Chromis chromis* (Pisces: Pomacentridae) from the Azores. *J. Zool. London*, 175:179-199.

Matsuoka, T. 1962. On the spawning of *Pomacentrus coelestis* Jordan and Stark. *The Aquacultur.*, 10(3):1-6.

May, R.C. 1967. Larval survival in the maomao, *Abudefduf abdominalis* (Quoy and Gaimard). Master's Thesis, Univ. Hawaii. 88 pp.

Maynard-Smith, J. 1977. Parental investment: a prospective analysis. *Anim. Behav.*, 25:1-9.

Miller, J.M., W. Watson and J.M. Leis. 1979. An atlas of common nearshore marine fish larvae of the Hawaiian islands. *Sea Grant Misc. Rept.*, (INIHI-Sea Grant-MR-80-02).

Moe, M.A. and F.A. Young. 1981. Spawning the jewels of the reef. *Freshwater and Mar. Aquar.*, 4:24-25 + 82-84.

Moran, M.J. and P.F. Sale. 1977. Seasonal variation in territorial response, and other aspects of the ecology of the Australian temperate pomacentrid fish, *Parma microlepis*. *Mar. Biol.*, 39:121-128.

Moyer, J.T. 1975. Reproductive behavior of the damselfish, *Pomacentrus nagasakiensis* at Miyakejima, Japan. *Japan. J. Ichthyol.*, 22(3):151-163.

Myrberg, A.A., Jr. 1972a. Ethology of the bicolor damselfish, *Eupomacentrus partitus* (Pisces: Pomacentridae): A comparative analysis of laboratory and field behavior. *Anim. Behav. Monogr.*, 5:199-283.

Myrberg, A.A., Jr. 1972b. Using sound to influence the behavior of free-ranging marine animals. *In: Behavior of Marine Animals* (H.E. Winn & B. Olla, Eds.), 2:435-468.

Myrberg, A.A., Jr. 1978. Ocean noise and the behavior of marine animals: relationships and their implications. Pp. 169-208. *In: Effects of Noise on Wildlife*, (J.F. Fletcher & R.G. Busnel, Eds.). Academic Press, N.Y.

Myrberg, A.A., Jr. and J.Y. Spires. 1972. Sound discrimination by the bicolor damselfish, *Eupomacentrus partitus*. *J. Exp. Biol.*, 57:727-735.

Myrberg, A.A., Jr. and R.E. Thresher. 1974. Interspecific aggression and its relevance to the concept of territoriality in reef fishes. *Amer. Zool.*, 14:81-96.

Myrberg, A.A., Jr., B.D. Brahy, and A.R. Emery.

1967. Field observations on reproduction of the damselfish, *Chromis multilineata* (Pomacentridae), with additional notes on general behavior. *Copeia* 1967:819-827.

Nakazono, A., H. Takeya, and H. Tsukahara. 1979. Studies on the spawning behavior of *Chromis notata* (Temminck et Schlegel). *Sci. Bull. Fac. Agr., Kyushu Univ.*, 1/2:29-37.

Nolan, R. 1975. The ecology of patch reef fishes. Ph. D. Diss., Univ. Calif., San Diego. 230 pp.

Patterson, S. 1975. The ontogeny of territoriality in juvenile three-spot damselfish, *Eupomacentrus planifrons*. Master's Thesis, Univ. Miami, Fla.

Reese, E.S. 1964. Ethology and marine zoology. *Ann. Rev. Oceanogr. Mar. Biol.*, 2:455-488.

Robertson, D.R. 1973. Field observations on the reproduction of a pomacentrid fish, *Acanthochromis polyacanthus*. *Z. Tierpsychol.*, 32:319-324.

Robertson, D.R. and B. Lassig. 1980. Spatial distribution patterns and coexistence of a group of territorial damselfishes from the Great Barrier Reef. *Bull. Mar. Sci.*, 30:187-203.

Russel, B.C. 1971. Underwater observations on the reproductive activity of the demoiselle *Chromis dispilis* (Pisces: Pomacentridae). *Mar. Biol.*, 10:22-29.

Sale, P.F. 1970. Behavior of the humbug fish. *Aust. Nat. Hist.*, 16:362-366.

Sale, P.F. 1971. The reproductive behavior of the pomacentrid fish, *Chromis caeruleus*. *Z. Tierpsychol.*, 29:156-164.

Sale, P.F. 1974. Mechanisms of co-existence in a guild of territorial fishes at Heron Island. *Proc. Second Internat. Coral Reef Symp.*, 1:195-206.

Sale, P.F. 1975. Patterns of use of space in a guild of territorial reef fishes. *Mar. Biol.*, 29:89-97.

Sale, P.F. 1976. The effect of territorial adult pomacentrid fishes on the recruitment and survival of juveniles on patches of coral rubble. *J. exp. Mar. Biol. Eco.*, 24:297-306.

Sale, P.F. 1977. Maintenance of high diversity in coral reef fish communities. *Amer. Natur.*, 111:337-359.

Sale, P.F. 1978. Coexistence of coral reef fishes – a lottery for living space. *Env. Biol. Fishes*, 3(1):85-102.

Sale, P.F. and R. Dybdahl. 1975. Determinants of community structure for coral reef fishes in an experimental habitat. *Ecology*, 5b:1343-1355.

Sale, P.F. and R. Dybdahl. 1978. Determinants of community structure for coral reef fishes in isolated coral heads at lagoonal and reef slope sites. *Oecologia*, 34:57-74.

Sale, P.F., P.J. Doherty, and W.A. Douglas. 1980. Juvenile recruitment strategies and the coexistence of territorial pomacentrid fishes. *Bull. Mar. Sci.*, 30:147-158. :

Schmale, M. 1980. A preliminary report on sexual selection in the bicolor damselfish, *Eupomacentrus partitus. Bull. Mar. Sci.*, 30:328.

Schmale, M. 1981. Sexual selection and reproductive success in males of the bicolor damselfish, *Eupomacentrus partitus* (Pisces: Pomacentridae). *Anim. Behav.*, 29:1172-1184.

Schwarz, A.L. 1980. Almost all *Dascyllus reticulatus* are girls! *Bull. Mar. Sci.*, 30:328.

Shaw, E.S. 1955. The embryology of the sergeant major, *Abudefduf saxatilis. Copeia*, 1955:85-89.

Smith, C.L. 1978. Coral reef fish communities: a compromise view. *Env. Biol. Fishes*, 3:109-128.

Spanier, E. 1970. Analysis of sounds and associated behavior of the domino damselfish, *Dascyllus trimaculatus* (Rüppell, 1828) (Pomacentridae). Master's Thesis, Tel-Aviv Univ., Israel. 80 pp.

Spanier, E. 1979. Aspects of species recognition by sound in four species of damselfishes, genus *Eupomacentrus* (Pisces, Pomacentridae). *Zeit. Tierpsychol.*, 51:301-303.

Stevenson, R.A. 1963. Life history and behavior of *Dascyllus albisella* Gill, a pomacentrid reef fish. Ph. D. Diss., Univ. Hawaii, Honolulu, Hawaii.

Swerdloff, S.N. 1970a. Behavioral observations on Eniwetok damselfishes (Pomacentridae: *Chromis*) with special reference to the spawning of *Chromis caeruleus. Copeia*, 1970:371-374.

Swerdloff, S.N. 1970b. The comparative biology of two Hawaiian species of the damselfish genus *Chromis* (Pomacentridae). Ph. D. Diss., Univ. Hawaii, Honolulu, Hawaii.

Thomson, D.A., L.T. Findley, and A.N. Kerstitch. 1979. *Reef Fishes of the Sea of Cortez*. Wiley, N.Y. 302 pp.

Thomson, D.A. & N. McKibbin. 1976. *Gulf of California Fishwatcher's Guide*. Golden Puffer Press, Tucson, Arizona. 75 pp.

Thresher, R.E. 1975. Atlantic Chromis. *Marine Aquarist*, 6(8):23-32.

Thresher, R.E. 1976. Field analysis of the territoriality of the threespot damselfish, *Eupomacentrus planifrons* (Pomacentridae). *Copeia*, 1976:266-276.

Thresher, R.E. 1977. Eye ornamentation of Caribbean reef fishes. *Z. Tierpsychol.*, 43:152-158.

Thresher, R.E. 1978. Territoriality and aggression

Wrasses are well known for having two (or more) color patterns in the same species. Shown here is *Halichoeres maculipinna*, the more brightly colored male above and behind the female. Photo by Charles Arneson.

This pair of *Pseudolabrus luculentus* was photographed by Wade Doak at Lord Howe Island. The larger male is in the center of the photo while the female is the smaller fish in front of him.

The female begins to move upward, still being vigorously courted by the male.

Pteragogus flagillifera males aggressively displaying at a mutual territorial border. All photos of this sequence by J. T. Moyer.

The male moves beside the smaller female as the pair begin their upward spawning ascent.

Male *P. flagillifera* circling above a female, courting her, as she approaches the spawning site.

Pseudojuloides elongatus male courting an egg-swollen female. This species is haremic and vividly dichromatic. Photo by Jack T. Moyer.

Actual spawning, with a cloud of gametes visible at the peak of the ascent.

in the threespot damselfish (Pisces; Pomacentridae): an experimental study of causation. *Z. Tierpsychol.*, 46:401-434.

Thresher, R.E. 1980a. *Reef Fish*. Palmetto Publ., St. Petersburg, Fla. 172 pp.

Thresher, R.E. 1979a. Possibly mucophagy by juvenile *Holacanthus tricolor* (Pisces; Pomacanthidae). *Copeia*, 1979(1):160-162.

Thresher, R.E. 1979b. The role of individual recognition in the territorial behavior of the threespot damselfish. *Mar. Behav. Physiol.*, 6:83-94.

Thresher, R.E. 1980b. Clustering: Non-agonistic group contact in territorial reef fishes, with special reference to the Sea of Cortez Damselfish, *Eupomacentrus rectifraenum*. *Bull. Mar. Sci.*, 30:252-260.

Thresher, R.E. In Press. Habitat effects on reproductive success of the coral reef fish *Acanthochromis polyacanthus* (Pomacentridae). *Ecology*.

Thresher, R.E. and J.T. Moyer. In Press. Male success, courtship complexity and patterns of sexual selection in three congeneric species of sexually monochromatic and dichromatic damselfishes (Pisces: Pomacentridae). *Anim. Behav.*

Tinbergen, N. 1953. *Social Behavior of Animals with special Reference to Vertebrates*. Methuen & Co., London. 150 pp.

Turner, C.H. and E.E. Ebert. 1962. The nesting of *Chromis punctipinnis* (Cooper) and a description of their eggs and larvae. *Calif. Fish and Game*, 48:243-248.

Walker, C.K. 1967. Nest-guarding behavior of the male maomao, *Abudefduf abdominalis* (Quoy and Gaimard). Master's Thesis, Univ. Hawaii, Honolulu.

Walker, S. 1977. Walker successfully spawns five species. *Mar. Hobbyist News*, 5(9):1-4.

Walker, S. and N. Herwig. 1976. Salinity and spawning. *Mar. Aquar.*, 7(2):45-50.

Ward, J.A. and G.W. Barlow. 1967. The maturation and regulation of glancing off the parents by young of orange chromides (*Etroplus maculatus*: Pisces-Cichlidae). *Behaviour*, 29:1-56.

Williams, D.M. 1980. The dynamics of the pomacentrid community on small patch reefs in One Tree Island lagoon (Great Barrier Reef). *Bull. Mar. Sci.*, 30:159-170.

Anemonefishes (Pomacentridae: Amphiprioninae)

Anemonefishes, or clownfishes, are brilliantly colored and relatively small damselfishes that are without doubt the most popular and widely kept group of marine fishes. In part their bright colors, hardiness, and relative peacefulness explain their popularity; a larger part, however, must be due to the fascinatingly close relationship these fishes have developed with sea anemones. Of the other damselfishes only a few species of *Dascyllus* (e.g., Stevenson, 1963) regularly associate with anemones, but none have developed as complex and as close a relationship as the anemonefishes. The mechanisms involved in this symbiosis and the factors that may have led to its evolution have been examined in detail (see reviews by Mariscal, 1970; Allen, 1972), but such knowledge does not detract from the fascination of watching an anemonefish "bathe" in tentacles whose slightest touch would kill other small fishes.

The anemonefishes constitute the pomacentrid subfamily Amphiprioninae, involving 29 known species in two genera: one species in *Premnas (P. biaculeatus)* and 28 in *Amphiprion* itself (Allen, 1975a & b; Burgess, 1981). All are confined to the Indo-West Pacific, roughly from the Red Sea to the Central Pacific (anemonefishes are not found, for example, at Hawaii, Easter Island, or other eastern Pacific islands), with a peak of abundance and diversity in the Indo-Australian-Philippine area. Ecologically all are associated with warm shallow tropical seas (the most cold resistant members of the genus only reach southern Japan to the north and southeastern Australia to the south) and with anemones. The degree that the fishes depend on the anemone varies widely. While it is generally assumed that all would be quickly eaten by predators if removed from the protection of the anemone (though see Moyer, 1980), within this constraint some species, such as *A. percula* and *A. sandaracinos,* are relatively poor swimmers and are closely tied to their hosts, while others, such as *A. clarkii*

and *A. bicinctus,* are stronger swimmers and frequently forage well away from the host.

Anemonefishes adapt well to captivity, and reports on their spawning behavior go back to at least the 1930's (Verwey, 1930; Delsman, 1930; Sachs, 1937). As with other reef fishes, however, detailed field studies have been made only in the last few years.

SEXUAL DIMORPHISM

Anemonefishes appear to be sexually monochromatic. The only known exceptions are *Amphiprion perideraion* and some populations of *A. clarkii.* There are also scattered reports of color differences between the sexes of *A. percula* (e.g., males are more yellow than females and have a black spot inside the pectoral fins – Garnaud, 1951; males have a narrower central band than the females and it projects forward more sharply – Valenti, 1968; Squires, 1971), but Allen (1972) could not verify such differences.

In *A. perideraion* the male has a pale orange trim along the upper and lower edges of the caudal fin and along the edge of the soft dorsal fin that is lacking in the female (Allen, 1972). Sexual dichromatism in *A. clarkii* is both more conspicuous and less consistent. Allen (1972) reported *A. clarkii* to be sexually monochromatic, but Tanase & Araga (1975) and Moyer (1976) subsequently demonstrated consistent color differences between the sexes in several Japanese populations. *A. clarkii* is a remarkably variable species with four distinct and differently colored populations in Japanese waters alone. One of the four, a starkly black-and-white form from the Bonin Islands, is sexually monochromatic with a uniformly white caudal fin. Specimens from three other areas, however, show varying degrees of sexual dichromatism, all involving greater amounts of orange on the male's caudal fin than on the female's (a pattern similar to that of *A. perideraion*). Fish from the Yaeyama Islands vary widely in color, but there is a general tendency for

Spawning sequence of the bluehead wrasse, *Thalassoma bifasciatum.*   At the end of a spawning rush the individuals return to the bottom, leaving behind a cloud of gametes. Photo by Dr. W. A. Starck II.

In this photo the spawning aggregation is milling about over the reef. Photo by Dr. W. A. Starck II.

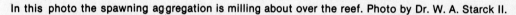

After milling about, just prior to dashing upward to spawn, the wrasses pull together into a very concentrated ball of fish. It has been suggested that these are mainly males aggregating around a single female. Photo by Dr. W. A. Starck II.

An aggregation of bluehead wrasses, *Thalassoma bifasciatum*, with the "supermale" in the center of a number of yellow females (and males?). Photo by Dr. P. L. Colin.

males to have orange upper and lower margins on a white caudal fin and for females to lack these margins. Okinawa-type males always have such upper and lower orange margins (averaging 1 to 2 mm wide); females, in contrast, always lack them and thus have uniformly white caudal fins (the same is true for Philippine specimens; pers. obs.). Finally, the sexes of the Japan-type fish differ greatly in tail color. Dominant nesting males have completely orange caudal fins; females either have uniformly white fins or show only a trace of orange along the upper and lower lobes. Small subordinate and sexually immature individuals look like Okinawa-type males. Color in Japan-type *A. clarkii*, therefore, is a sign of both sex and dominance. Moyer noted that in Okinawa and southern Japan, where sexual dichromatism is most consistent and most prominent, *A. clarkii* live in denser populations than in other areas. He suggested that such dichromatism may have evolved in these areas to facilitate sexual recognition by territorial individuals. Alternatively, such dense populations also lead to an increased frequency of polygamy in these normally monogamous fish (Moyer, 1980) that could result in increased levels of intersexual selection. The latter often leads to sexual dimorphism (Ralls, 1977).

As opposed to the few reports of color differences, most studies of anemonefish behavior comment on the size difference between mature males and females (e.g., Verwey, 1930; Moyer & Sawyers, 1973; Fricke, 1974; though see also Moyer & Steene, 1979, for a discussion of the evolution of equal-sized sexes in *A. polymnus*). Allen (1972), for example, reported that breeding females of four species at Enewetak (*A. chrysopterus, A. melanopus, A. perideraion,* and *A. tricinctus*) were usually 10 to 20 mm longer than their mates. The significance of this size difference was only recently discovered, however, when Fricke (1976) and Fricke & Fricke (1977) documented protandrous (male to female) hermaphroditism in two eastern Indian Ocean species, *A. bicinctus* and *A. akallopisos*. Similar protandry has since been documented for *A. melanopus* by Ross (1978a) and for six species in Japan (*A. clarkii, A. frenatus, A. ocellaris, A. perideraion, A. polymnus,* and *A. sandaracinos*) by Moyer & Nakazono (1978). Fricke & Fricke suggested that protandry would prove universal among anemonefishes; the evidence available seems to bear them out.

Fricke & Fricke (1977) and Moyer & Nakazono (1978) provided limited information on the morphology and histology of anemonefish gonads. Like those of other damselfishes, the gonads are paired, lie posteriorly and dorsally in the abdominal cavity, and unite posteriorly to form a common urogenital duct. Also, as in other damselfishes the urogenital papilla is sexually dimorphic: slender and pointed in the male, thicker and blunter in the female. Figures provided by Moyer & Nakazono indicate that the "ambosexual" testes are elongate with a central ovary-like lumen. The bulk of such testes in a dominant male consists of active spermatogenic tissue, with pockets of immature oocytes located along the periphery. In smaller subordinate fish neither testicular nor ovarian tissues are active, a state referred to by Fricke & Fricke as "psychophysical castration." At sex change the testicular tissue of the large male degenerates and oocytes proliferate. Ultimately, in a fully mature female there is little or no sign of spermatogenic tissue. Fricke & Fricke reported that in captivity a functional male *A. akallopisos* changed sex in less than 63 days; in a field test a functional male *A. bicinctus* laid eggs 26 days after removal of the female.

Fricke (1976) and Fricke & Fricke (1977) hypothesized that hermaphroditism is an adaptation to the specialized environment occupied by anemonefishes. Following the logic developed by Ghiselin (1969), they suggested that the often widely scattered distribution of anemones, their small size, and the apparent high risk of predation run by fish outside of the anemone result in usually no more than a single adult pair of fish occupying an anemone and great difficulty in replacing one member of such a pair should it die. If not a hermaphrodite, a juvenile that settles on such a partially vacant anemone stands only a 50:50 chance of complementing the remaining adult and reconstituting a pair. As a hermaphrodite, however, such a juvenile can fill whichever role is open, with the result that sexual flexibility by these juveniles is strongly selected for. In actual practice few anemones are occupied by only an adult pair. Rather, most contain a large dominant female, a smaller functional male that is subordinate to the female, and one or more small neuter individuals that rank below the male in the dominance hierarchy. If the female is removed the male, now dominant, changes sex and the previously ranking neuter individual replaces him as the active male. Given the opportunity, it can in turn transform into a female, with the next ambosexual individual replacing it as the male. Newly settled juveniles fill in the dominance hierarchy from below as space becomes available.

Fricke & Fricke further suggested that the direction

of sex change (male to female rather than vice versa) results from the benefits to the pair of having the larger member a female; large females produce more eggs than small ones. As a consequence, when a female dies the male maximizes his fecundity by changing sex and letting a smaller fish become the new male rather than remaining male and mating with a new female that is smaller than he is and that will produce fewer eggs. The problem, however, is not so simple. Several studies, e.g., Allen (1972) and Fricke (1974), have demonstrated that the presence of a dominant pair "stunts" juveniles and that removal of the dominants results in rapid growth by such fish. With the removal of the dominant female an ambosexual could theoretically become female, accelerate its growth rate, and quickly surpass the male in size. Whether or not a male increases his fecundity by changing sex, then, is a complex function of: (1) the difference between his size and that of the top ranking ambosexual individual; (2) the physiological cost of sex change (in theory greater for the male than an ambosexual fish); (3) the magnitude of the increase in rate of growth of the ambosexual individual (and consequently the amount of time it requires to equal the male in size); and (4) the speed with which each could begin to produce eggs (presumably longer for the male than for the ambosexual). Some of these variables might also change in magnitude with time of year: in the warmer months egg laying is more frequent and the penalty greater for a longer delay in producing eggs; conversely, growth rate may also be faster then, so that an ambosexual individual might reach the size of the male sooner. Fricke & Fricke's size advantage model logically predicts that the probability of sex change by the male will vary with the size difference between it and the largest ambosexual individual present and may also vary with time of year.

An alternative and somewhat more parsimonious explanation for protandry by anemonefishes can be based on the differences between the sexes in certainty of future reproduction. The position of a male is, at best, a tenuous one. Although a successful male exclusively fertilizes a female's eggs and has a fecundity equal to hers, he maintains his favorable position only by dominating other males or potential males. As a male he risks being displaced (see, for example, Moyer, 1980) and having his fecundity reduced to zero. As a female, however, such a drastic reduction in fecundity is nearly impossible. Since she produces the eggs that males compete to fertilize, a female is largely guaranteed continued reproductive success regardless

of future events. Sex change by a male, then, may be adaptive despite its short term cost in immediate fecundity, because in the long term it guarantees future reproduction. This alternative model predicts that a male will almost always change sex, given the option, regardless of the time of year or the size difference between it and the largest ambosexual. The limited data available seem to support this prediction. Moyer (1980) suggested a modified version of this model based on higher rates of male mortality due to nest-guarding.

SPAWNING SEASON

Verwey (1930) noted a two-month "rest period" in spawning for *Amphiprion percula* in captivity, but since then all studies of tropical anemonefishes have either suggested (e.g., Allen, 1972) or demonstrated (e.g., Fricke, 1974; Ross, 1978a) year-round spawning. The only well documented case of seasonal spawning by an anemonefish is that of *A. clarkii* off the warm temperate coast of southern Japan. Cold winter temperatures (13° to 16°C) apparently inhibit spawning from November to early April, with individuals during this period moving from the nest sites to shelter holes (Bell, 1976; Moyer & Bell, 1976). Despite this gap in reproduction the net annual fecundity of *A. clarkii* in Japan is higher than that of anemonefishes in other areas; Bell suggested that such intense summer spawning may be a local adaptation to severe winter temperatures. Presumably *A. clarkii* in more tropical areas spawn year-round and, on the average, are as fecund as other comparably sized species.

Several studies indicate a lunar influence on timing of spawning by anemonefishes. Allen (1972), Fricke (1974), and Moyer (1976) suggested that spawning, or at least the onset of spawning, most often occurs near the full moon. Allen, for example, found that of 34 spawnings observed at Enewetak, 26 occurred within six days of the full moon. Ross (1978a), on the other hand, found a semi-lunar spawning rhythm in *A. melanopus* in Guam, with peaks in activity at the first and third quarters rather than a single peak at the full moon.

Allen (1972) suggested four reasons why anemonefishes concentrate their spawning near the time of the full moon: (1) increased visibility at night facilitates nest-care by the tending male; (2) increased surface illumination combined with a positive phototaxis by the larvae keep them closer to the surface; (3) water currents are generally strongest near the full moon, facilitating larval dispersal; and (4) many invertebrates

Thalassoma lucasanum aggregating above an outcrop before spawning. Photo by Dr. R. E. Thresher.

T. lucasanum spawning. The group in the center is ascending, while the group to the side is descending. Note the visible gamete cloud. Photo by Dr. R. E. Thresher.

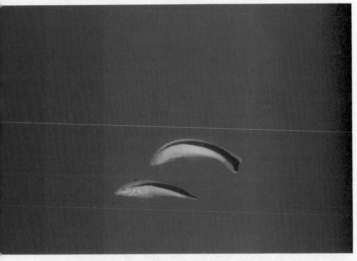

A series of photos of *Labroides dimidiatus* spawning. Here the male pursues a female.

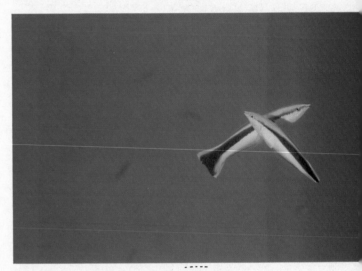

The spawning pair spirals toward the surface.

The *L. dimidiatus* pair beginning their upward movement.

The *L. dimidiatus* pair near the peak of their ascent, about to spawn. All photos in this sequence by Dr. R. E. Thresher.

Pseudojuloides cerasinus during pre-spawning courtship. Note the egg-swollen female below the colorful male. Photo by Jack T. Moyer.

A spawning pair of *Lachnolaimus maximus,* the male being the larger individual at the back. Photo by Charles Arneson.

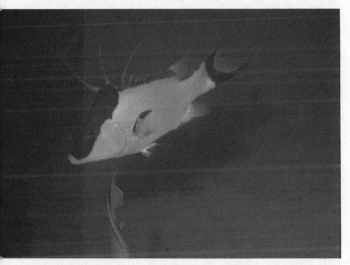

A courting male hogfish, *Lachnolaimus maximus,* with his very contrasty color pattern. Photo by Dr. P. L. Colin.

Lachnolaimus maximus pair making their spawning ascent. Photo by Dr. P. L. Colin.

A pair of sexually dichromatic *Bodianus diplotaenia* spawning. The male is above the female during the ascent. Photo by Dr. R. E. Thresher.

A pair of *Bodianus rufus* spawning. Photographed at a depth of about 60 feet by Charles Arneson.

191

spawn near the time of the full moon, so that food supplies for larvae might be greatest then. None of these hypotheses are unambiguously supported by available data. Ross (1978a) supported the third, reporting that the eggs of *A. melanopus* hatch at both the time of month and time of day (shortly after sunset) that tides are highest and tidal currents are strongest. His data, however, do not clearly support this conclusion since peak hatching does not coincide with peak early evening water levels and only one peak (new moon, following first quarter peak in spawning) occurs when tidal currents are at a maximum; the other (full moon, following third quarter peak in spawning) coincides with minimum early evening tidal currents. Fricke (1974) agreed with several of Allen's suggestions but also argued that hatching of *A. bicinctus* in the Red Sea occurs at the time of minimal tidal currents and serves to minimize dispersion rather than maximize it. No single hypothesis appears to account for the variety of lunar and semi-lunar spawning rhythms observed, and additional work with the animals in the field is needed.

REPRODUCTIVE BEHAVIOR

Reports of spawning of anemonefishes in aquaria are common; reports of observed spawning in the field are less common and most are recent, e.g., Fishelson, 1965; Allen, 1972; Fricke, 1974; Moyer & Bell, 1976; Ross, 1978a. The consistency of the behavior observed, however, suggests relatively uniform reproductive behavior within the subfamily.

The most detailed studies of anemonefish behavior in general, and reproduction in particular, are those of Fricke (1973a & b, 1974, 1975) and Fricke & Fricke (1977) on *Amphiprion bicinctus* and Moyer and his co-workers (Bell, 1976; Moyer, 1976; Moyer & Bell, 1976; Moyer & Nakazono, 1978; Moyer, 1980) on *A. clarkii.* Substantial but less complete studies include those by Allen (1972) on four species of *Amphiprion* at Enewetak, Ross (1978a & b) on *A. melanopus,* and Fricke (1979) on *A. akallopisos.*

The basic unit of anemonefish social systems is a monogamously mated adult pair cohabiting an anemone with a variable number of subordinate ambosexual individuals and juveniles. Multi-male, multi-female systems occur only occasionally (e.g., Fricke, 1974; Moyer, 1980) and are always associated with abnormally high densities of fish due to either extreme abundance or large size of the anemone hosts. In most cases pairs are stable for at least a year and may remain stable for the lifetime of the fish. Fricke (1973b,

1974) demonstrated by means of choice tests conducted on the reef that *A. bicinctus* recognize mates as individuals even when both fish are outside their normal areas and have been separated for as long as ten days. Forced to accept a new mate, a fish will learn the individual characteristics of this mate within 24 hours. Recognition is apparently based on individual differences in color pattern.

Because it is both permanently territorial and permanently paired, courtship and spawning by an anemonefish differ significantly from those of other damselfishes. Signal-jumping, for example, has been reported only once (for *A. bicinctus* by Fishelson, 1965) and has not since been verified. Permanent pairing has, in general, led to a simplification of courtship to the point that in long-mated pairs both within-pair agonism and elaborate prolonged courtship are largely eliminated (e.g., Moyer & Bell, 1976).

Spawning, which can occur throughout the day (though see Ross, 1978a), takes place at the base of the anemone, within the area swept by its tentacles. Because of this, the chosen spawning site is relatively clean compared to those used by other damselfishes, and picking at the spawning site by the pair is mainly a ritual rather than actual cleaning (Fricke, 1974). Because they are found with reef-dwelling anemones, most anemonefishes nest on rock. *A. polymnus,* however, lives in an anemone that is found on sand flats and hence does not have a spawning substrate immediately available. Moyer & Steene (1979) reported that pairs of these fish regularly drag small objects such as aluminum cans and echinoderm tests to the anemone and use these as a depository for their eggs.

The sequences of behavior that lead to spawning have been analyzed in detail by Fricke (1974) for *Amphiprion bicinctus;* similar, though less well documented, behavior is characteristic of all species thus far studied. Spawning is usually initiated by the male, who several days before spawning begins to pick at and defend the nest site. The frequency and duration of such substrate picking increase as the day of spawning approaches. On that day the male bites repeatedly at the tentacles of the anemone, which causes them to withdraw and expose the nest site completely. The male then rushes toward the female and "leads" her to the nest site by swimming back and forth between it and her until she follows. At the nest site the pair engage in prolonged bouts of substrate biting and body trembling, the tempo of which increases steadily. Fricke suggested that such behavior is a form of mutual stimulation in which each fish rouses the other to a high pitch for spawning. Spawning itself is begun by the female, who stops substrate biting and makes a false spawning pass over the nest site, dragging her venter and urogenital papilla over it without actually laying eggs. After a few such passes egg-laying begins with the female leaving short lines of relatively large orange eggs. At first she moves only slightly while laying, but gradually she moves in ever-larger circles as the size of the nest and the number of eggs laid increase. The male also makes frequent passes over the eggs, presumably fertilizing them. Both male and female take periodic breaks from spawning to bite at the substrate or at the tentacles of the anemone, the latter causing the tentacles to remain withdrawn. At first these pauses are quite short, but as spawning continues the female takes more and more frequent breaks and each time takes longer to return to the nest and resume spawning. Eventually, one and a half to two hours after she began, the female leaves for good while the male finishes his last pass over the eggs and begins brood care. Estimates of clutch size for tropical species range from 200 to 400 eggs for *A. melanopus* (Ross, 1978b) to 600 to 1600 for *A. bicinctus* (Fishelson, 1965). Clutch size of *A. clarkii* off Japan ranges from 1,000 to 2,500 eggs (Bell, 1976).

Care of the eggs is provided mainly by the male, though the female may help occasionally. The primary activities are mouthing the eggs, picking out and eating those that are infertile or fungused, and fanning the eggs. As in some other damselfishes, the frequency and duration of fanning increase as the eggs age, which might relate to increased oxygen requirements by the eggs. In contrast to other damselfishes, however, anemonefishes do not become more aggressive when tending eggs. Indeed, Allen (1972) noted that during the middle part of the incubation period the nest was frequently left unattended by either parent for three to five minutes at a time.

EGGS AND LARVAE

Anemonefish eggs are bright orange when first laid and are larger than those of other damselfishes. Reported sizes range from 2.2 mm long and 0.9 mm wide (for *A. percula*, Delsman, 1930) to 3.3 mm long and 1.2 mm wide (for *A. bicinctus*, Gohar, 1948). Like those of other damselfishes, the eggs are elliptical and attached to the substrate by means of a number of fine threads at the ventral pole. Development is discussed by Gohar (1948), Hackinger (1967), Terver (1971), and Allen (1972), the last showing several stages in the development of *A. chrysopterus* in color. The length of

the incubation period averages about seven to eight days but is clearly dependent on water temperature (Fricke, 1974; Bell, 1976). The eggs of *A. clarkii*, for example, hatch in only six and a half days at 27°C but can take up to 13.5 days at 21°C.

During development the color of the eggs changes gradually from orange to gray or brown and finally, just before hatching, the eggs develop a silver sheen due to the developing eyes. Hatching in most, if not all, cases occurs one half to two hours after sunset (e.g., Hackinger, 1967; Neugebauer, 1969; Fricke, 1974; Moyer & Bell, 1976) and is facilitated by vigorous fanning by the male. Hatching at dusk is widely considered an antipredator strategy, since by that time the numerous diurnal planktivores have retired for the night and most nocturnal planktivores can be avoided by ascending a few meters off the bottom (see Fricke, 1974, for a discussion of predation by nocturnal planktivores). Ross (1978a) suggested that early night hatching by *A. melanopus* at Guam also facilitates larval dispersal since this is the period of highest tides and strongest tidal currents.

Newly hatched larvae are both larger (4 to 5 mm TL) and better developed than those of other pomacentrids (except for the nonpelagic larvae of *Acanthochromis polyacanthus*), with well developed jaws, eyes, and pigmentation and only a small yolk sac. Each is transparent with a silvery eye and abdomen and scattered melanophores on the head (Allen, 1972).

After hatching the larvae are positively phototactic and swim toward the surface to begin a planktonic stage. The duration of this stage in the field is not known and is still a source of debate. Early workers (e.g., Verwey, 1930; Garnaud, 1951; Hackinger, 1967) assumed fairly lengthy and, for reef fishes, normal planktonic periods based on observations in aquaria. Several lines of evidence suggest, however, that under natural conditions the planktonic stage might be quite short, perhaps as little as one night (Fricke, 1974). First, the smallest larvae observed on the reef are only 2 to 3 mm longer than newly hatched larvae, usually between 6.5 and 8 mm (Fricke, 1974; Allen, 1972), a length that could be reached in ten days given even the growth rates reported for larvae in aquaria by Allen. Second, the distribution and variability of the animals suggest limited larval dispersal and a short planktonic stage (Allen, 1972). Moyer (1976), for example, found consistent and striking differences in color, size, and some meristic counts between populations of *A. clarkii* in four geographically close areas off southern Japan; these differences suggest limited gene exchange be-

A pair of *Clepticus parrae* courting, the male above. Photo by Charles Arneson.

Pair of *Bolbometopon bicolor* making their spawning ascent. Photo by Roger Steene.

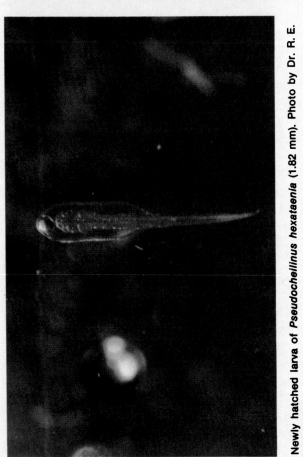

Newly hatched larva of *Pseudochellinus hexataenia* (1.82 mm). Photo by Dr. R. E. Thresher.

Male *Bolbometopon bicolor* (above) courting a female. Note the position of the tail and the open mouth of the male. Photo by Roger Steene.

Feeding aggregation of male and female *Scarus iserti.* Although the color patterns are very similar, the fish in the center of the photo appears to be a developing supermale. Photo by Dr. P. L. Colin.

A fully developed pattern of the supermale in *Scarus iserti.* This pattern is quite different from that of the individuals above. Photo by Dr. Walter A. Starck II.

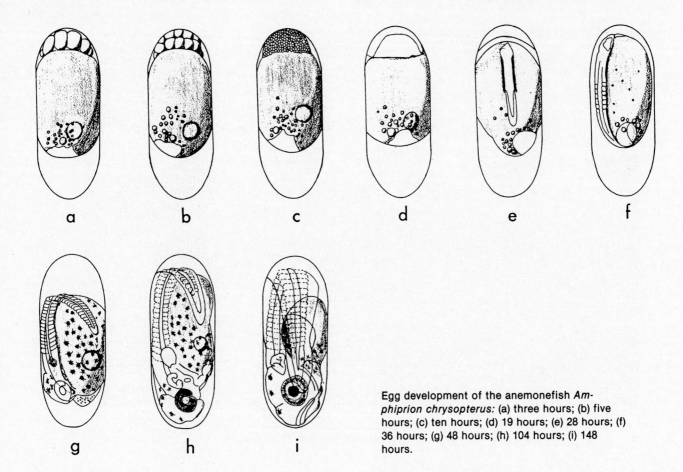

Egg development of the anemonefish *Amphiprion chrysopterus:* (a) three hours; (b) five hours; (c) ten hours; (d) 19 hours; (e) 28 hours; (f) 36 hours; (g) 48 hours; (h) 104 hours; (i) 148 hours.

tween even these close areas. Third, the larvae are extremely well developed for planktonic existence, an "atypicalness" emphasized by Fricke (1974); indeed, the larvae are in many respects more similar to those of *Acanthochromis polyacanthus,* which lacks a planktonic stage altogether, than they are to normal pomacentrid larvae. Fourth, Allen (1972) reported observing larval *A. chrysopterus* near or on the bottom of aquaria at night and feeding off the sides and bottom of the aquarium during the day, both of which suggest that the larvae remain in close association with the bottom in the field. Finally, in the laboratory juveniles often settle out within a week of hatching (e.g., Tanase & Araga, 1975).

Whatever the duration of the planktonic larval stage, initial location of an anemone by a settling juvenile is apparently based on olfactory cues. Fricke (1974) demonstrated through use of a "Y" maze that 10 to 20 mm juveniles significantly more often entered the arm with "anemone-water" than they did that with normal sea water. Prior to settlement, however, postlarval *A. clarkii* are indifferent to anemones (Tanase & Araga, 1975). Larger juveniles (60 to 80 mm) can discriminate between a preferred

and a nonpreferred species of anemone on the basis of visual cues.

Juveniles enter at the bottom of the dominance hierarchy of fish inhabiting an anemone and grow slowly so long as dominated by a pair of adults. The mechanism of such "stunting" is not known. Sexual maturity is not reached until the fish are not only large enough to develop mature gonads but can replace one member of the dominant pair. Sexual maturation of *A. chrysopterus* and *A. tricinctus* normally occurs at 60 to 70 mm SL for males and 70 to 80 mm for sex-changed females (Allen, 1972).

REPRODUCTION IN CAPTIVITY

Anemonefishes have proved to be one of the easiest groups to spawn in aquaria, probably as a result of three factors: (1) they are normally attached to only a small part of the reef and so are not unduly stressed by the confines of an aquarium; (2) they are sequential hermaphrodites, which largely eliminates the problem of obtaining a spawnable pair; and (3) they produce demersal eggs. As a consequence, there are not only a number of technical articles dealing with anemonefish reproduction in captivity (e.g., Verwey, 1930; Gohar,

1948; Garnaud, 1951; Terver, 1971; Allen, 1972; Tanase & Araga, 1975), but also numerous popular articles in aquarist-oriented magazines that discuss techniques of breeding and raising anemonefishes (e.g., Mitsch, 1941; Hackinger, 1967; Neugebauer, 1969; Squires, 1971; Schreiner, 1972; Moe, 1973; Sieswerda, 1977).

Spawning anemonefishes in aquaria requires little more than good tank conditions, healthy fish fed a varied diet (with at least some live food for best results), a reasonably large tank (minimum about 100 liters) containing suitable shelter, and patience. Garnaud (1951) spawned *A. percula* without an anemone present by providing instead small clay flower pots lying on one side within which egglaying occurred.

Nonetheless, most successful breeders provide an anemone on the premise that it results in a more natural and less stressful environment for the fish and facilitates spawning. Anemonefishes will spawn in community aquaria, but for most consistent success breeders typically isolate the pair. Finally, specific water conditions are probably not critical, though Sieswerda (1977) suggested that best results are obtained when the salinity is slightly low (specific gravity about 1.020).

Two methods can be used to obtain a breedable pair. In the first, a number of small fish are placed in a large aquarium that contains many anemones and shelter holes. Under such conditions pairing will occur naturally as the fish grow. The alternative is to ob-

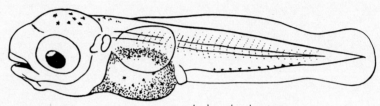

newly hatched, 4·1mm

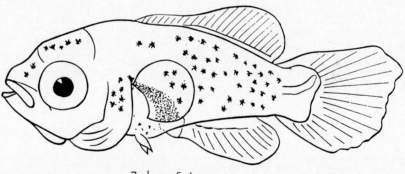

7 days, 5·4mm

Larval development of the anemonefish *Amphiprion chrysopterus*. Based on Allen (1972).

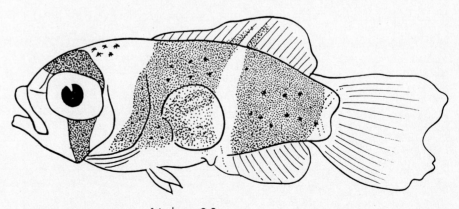

16 days, 9·2mm

A

B

C

Color changes accompanying protogynous sex reversal in *Genicanthus semifasciatus:* a) functional female (134 mm total length); b) transitional form (same individual as a; c) functional male (same individual as a and b). All photos by Katsumi Suzuki.

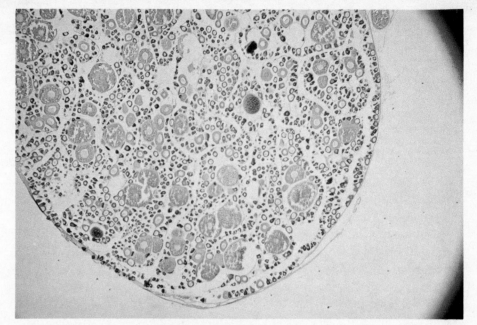

A

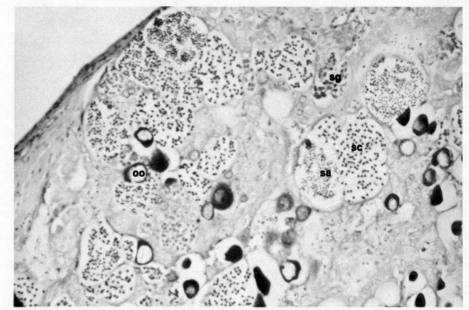

B

C

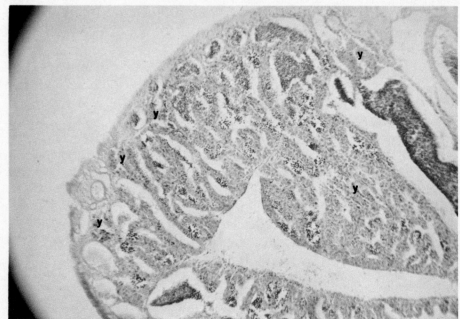

a) Functional ovary of *Genicanthus semifasciatus;* b) transitional gonad (sg = spermatogonia; sc = spermatocytes; sa = spermatids; oo = oocytes); c) functional testis (y = yellow granules). All photos by Katsumi Suzuki.

tain two very different sized fish and forceably pair them by placing them together in a small aquarium containing only one anemone. Although some aggression is likely at first, the sequential hermaphroditism characteristic of these fishes virtually guarantees a breeding pair. In a few species (*A. perideraion*, some *A. clarkii*, and possibly *A. percula*) producing such pairs is facilitated by sexual dichromatism.

As noted, anemonefish eggs hatch a few hours after sunset. In aquaria time of hatching can to a certain extent be controlled by adjusting the light cycle. Hatching can be delayed up to 12 hours by leaving a bright light on over the eggs; abruptly turning the light off routinely results in hatching within an hour (Sieswerda, 1977).

Considerable progress has also been made on the techniques of rearing larvae, so that now all anemonefishes commercially available are tank-reared and several generations removed from the ocean. Three rearing techniques have been used successfully: a partially open system which employs a slow but steady inflow of clean water and is preferred by most commercial breeders; a "green" closed system in which planktonic algae are cultured to act as a living seawater filter; and a "clean" closed system in which clear filtered sea water alone fills the tank. The latter two methods require the use of larger rearing tanks than the first for consistently good results and are used mainly by hobbyists who must worry about limited supplies of clean sea water. In all three systems, bare all-glass tanks covered with black cloth or painted black on the outside and given constant illumination are standard.

The most critical factor in rearing larval anemonefishes is the first food offered. Despite their unusually large size at hatching, larval anemonefishes are still far too small to feed immediately upon newly hatched brine shrimp. Hackinger (1967) and Allen (1972) reported limited success in rearing larvae by feeding them prepared foods finely ground and emulsified in sea water. Hackinger's mixture contained a variety of animal and plant products, vitamins, and antibiotics; Allen simply used pulverized commercially available flake food. Neugebauer (1969) successfully used cultured *Euplotes* (a marine ciliate) and a variety of wild-caught plankton to feed newly hatched larvae. Recently, however, a readily cultured marine rotifer, *Brachionis plicatilis*, has become the standard first food for anemonefish larvae. Commercial breeders often combine it with size-sorted (initially about 35 micron) wild-caught plankton. After *Brachionis* the larvae readily take *Artemia* and, eventually, flake food.

Literature Cited

Allen, G.R. 1972. *Anemonefishes*. T.F.H. Publ., Neptune City, New Jersey. 288 pp.

Allen, G.R. 1975a. *Anemonefishes*. 2nd Edition. T.F.H. Publ., Neptune City, New Jersey. 352 pp.

Allen, G.R. 1975b. *Damselfishes of the South Seas*. T.F.H. Publ., Neptune City, New Jersey. 240 pp.

Bell, L.J. 1976. Notes on the nesting success and fecundity of the anemonefish *Amphiprion clarkii* at Miyake-jima, Japan. *Japan. J. Ichthyol.*, 22:207-211.

Burgess, W.E. 1981. *Pomacentrus alleni* and *Amphiprion thiellei*, two new species of pomacentrids (Pisces: Pomacentridae) from the Indo-Pacific. *Tropical Fish Hobbyist Magazine*, 30(3):68-73.

Delsman, H.C. 1930. Fish eggs and larvae from the Java Sea. 16. *Amphiprion percula*. *Treubia*, 12:367-370.

Fishelson, L. 1965. Observations and experiments on the Red Sea anemones and their symbiotic fish *Amphiprion bicinctus*. *Bull. Sea Fish., Res. Stat. Haifa*, 39:1-14.

Fricke, H.-W. 1973a. Behaviour as part of ecological adaptation. *Helgo. Wiss. meeresunters.*, 24:120-144.

Fricke, H.-W. 1973b. Individual partner recognition in fish: field studies on *Amphiprion bicinctus*. *Naturwissenschaften*, 60:204-205.

Fricke, H.-W. 1974. Oko-ethologie des monogamen Anemonenfisches *Amphiprion bicinctus* (Freiwasseruntersuchung aus dem Roten Meer). *Z. Tierpsychol.*, 36:429-512.

Fricke, H.-W. 1975. Selectives Feinderkennen bei dem Anemonenfisch *Amphiprion bicinctus* (Rüppell). *J. exp. Mar. Biol. Ecol.*, 19:1-7.

Fricke, H.-W. 1976. *Bericht aus dem Riff*. Piper Verlag, Munich, Germany. 254 pp.

Fricke, H.-W. 1979. Mating system, resource defence and sex change in the anemonefish *Amphiprion akallopisos*. *Z. Tierpsychol.*, 50:313-326.

Fricke, H.-W and S. Fricke. 1977. Monogamy and sex change by aggressive dominance in coral reef fish. *Nature*, 266(5605):830-832.

Garnaud, J. 1951. Nouvelles donnees sur l'Ethologie d'un Pomacentridae: *Amphiprion percula* Lacépède. *Bull. Inst. Oceanogr. Monaco*, (998):1-12.

Ghiselin, M.T. 1969. The evolution of hermaphroditism among animals. *Quart. Rev. Biol.*, 44:189-208.

Hackinger, A. 1967. Anemonenfisch—im Aquarium gezüchtet. Die Aufzucht von *Amphiprion bicinctus* im Aquarium. *Aquarien Magazin, Stuttgart*,

(April):137-141.

Mariscal, R.N. 1970. The nature of the symbiosis between Indo-Pacific anemone fishes and sea anemones. *Mar. Biol.,* 6:58-65.

Mitsch, H.J. 1941. Breeding of marine clown fishes (*Amphiprion percula*). *The Aquarium, Philadelphia,* 10(3):48-50.

Moe, M.A. 1973. Breeding the clownfish, *Amphiprion ocellaris. Salt Water Aquarium,* 9(2):3-14.

Moyer, J.T. 1976. Geographical variation and social dominance in Japanese populations of the anemonefish *Amphiprion clarkii. Japan. J. Ichthyol.,* 23:12-22.

Moyer, J.T. 1980. Studies of the influence of temperate waters on the behavior of the tropical anemonefish *Amphiprion clarkii* at Miyake-jima, Japan. *Bull. Mar. Sci.,* 30:261-272.

Moyer, J.T. and L.J. Bell. 1976. Reproductive behavior of the anemonefish *Amphiprion clarkii* at Miyake-jima, Japan. *Japan. J. Ichthyol.,* 23:23-32.

Moyer, J.T. and A. Nakazono. 1978. Protandrous hermaphroditism in six species of the anemonefish genus *Amphiprion* in Japan. *Japan. J. Ichthyol.,* 25:101-106.

Moyer, J.T. and C.E. Sawyers. 1973. Territorial behavior of the anemonefish *Amphiprion xanthurus* with notes on the life history. *Japan. J. Ichthyol.,* 20:85-93.

Moyer, J.T. and R.C. Steene. 1979. Nesting behavior of the anemonefish *Amphiprion polymnus. Japan. J. Ichthyol.,* 26:209-214.

Neugebauer, W. 1969. So zuchten wir Korallenfische. *Aquarien Magazin, Stuttgart* (Dec.):483-488.

Ralls, K. 1977. Sexual dimorphism in mammals: avian models and unanswered questions. *Amer. Natur.,* 111:917-938.

Ross, R.M. 1978a. Reproductive behavior of the anemonefish *Amphiprion melanopus* on Guam. *Copeia,* 1978:103-107.

Ross, R.M. 1978b. Territorial behavior and ecology of the anemonefish *Amphiprion melanopus* on Guam. *Z. Tierpsychol.,* 46:71-83.

Sachs, W.B. 1937. Zur Pflege von Korallenfischen. *Aquar.-Terrienk.,* 48(2):25-29.

Schreiner, W. 1972. Breeding report – clownfishes. *Marine Aquarist,* 3(3):31-33.

Sieswerda, P.L. 1977. Doing what comes naturally. *Marine Aquarist,* 8(1):18-28.

Squires, K. 1971. Breeding clown fish. *Marine Aquarist,* 2(2):30-33.

Stevenson, R.A. 1963. Behavior of the pomacentrid reef fish *Dascyllus albisella* Gill in relation to the anemone *Marcanthia cookei. Copeia,* 1963:612-614.

Tanase, H. and C. Araga. 1975. Observation of breeding and taxonomy of the anemonefish, *Amphiprion clarkii* (Bennett). *J. Japan. Assoc. Zool. Gardens Aquar.,* 17:16-21.

Terver, D. 1971. Comportement de ponte, reproduction et developpement embryonnaire d'un poisson-clown (*Amphiprion polymnus*). *Pisciv. Fr.,* 26:9-23.

Valenti, R.J. 1968. *The Salt Water Aquarium Manual.* Aquar. Stock Co., N.Y.

Verwey, J. 1930. Coral reef studies I. The symbiosis between damselfishes and sea anemones in Batavia Bay. *Treubia,* 12(3-4):305-366.

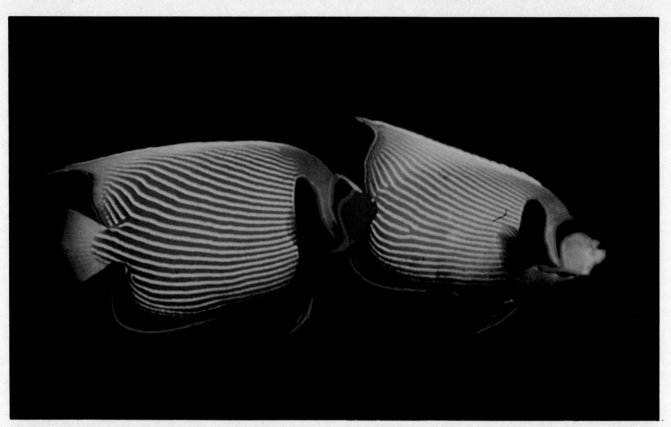

Spawning sequence of the emperor angelfish, *Pomacanthus imperator*. In the bottom photo, the larger and more brightly colored male is chasing a female across the reef, moving ahead of her to soar. In the above photo, the pair are ascending to spawn, the male behind and slightly below the female. Both photos by Dr. R. E. Thresher.

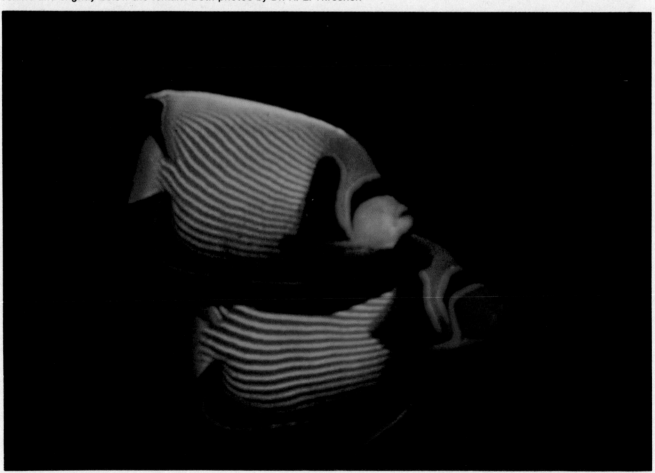

Soaring by male *Centropyge argi* (largest of three) with two females at Curacao. Photo by Jack T. Moyer.

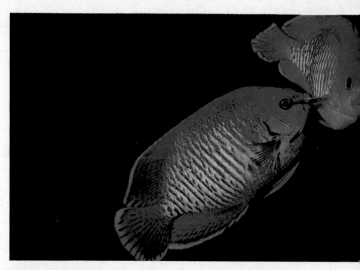

Nuzzling by a pair of *Centropyge bispinosus*. Photo by Dr. J. E. Randall.

A *Centropyge shephardi* male and *C. bispinosus* female interacting. Photo by Jack T. Moyer.

Nuzzling by *C. shephardi*. Note the pinkish abdomen of the female. Photo by Jack T. Moyer.

Sexual dichromatism in *Centropyge heraldi*. The male has the dark markings around the eye. Photo by Roger Steene.

A spawning pair of the pearlscale angelfish, *Centropyge vroliki*. Photo by Ann Gronell.

Wrasses
(Labridae)

Wrasses are among the most abundant and conspicuous fishes on the coral reef. They come in a variety of shapes, colors, and sizes, often varying widely even within a species (which has complicated their systematics considerably). They range from several dwarf species (e.g., *Pseudocheilinus* and *Doratonotus* species) that reach a maximum length of only a few centimeters to a number of giant Indo-Pacific species of *Cheilinus* that can exceed 3 m in length. Body shape varies widely from elongate and strongly laterally compressed, as typified by the razorfishes (*Hemipteronotus* spp.), to stocky and massive, as in those fishes in the genera *Cheilinus, Choerodon,* and many *Bodianus.* Most wrasses, however, are elongate and robust, tapering both anteriorly and posteriorly, a shape commonly referred to as "cigar-shaped." Representative and well-known examples of the cigar-shaped wrasses include the genera *Thalassoma, Halichoeres,* and *Labroides.* With such diversity there are few external features that specifically identify the fishes of this family. Only two, in fact, are readily apparent: all wrasses swim with a characteristic "pectoral fins only" propulsion, using the caudal fin only when an extra burst of speed is needed, and most have prominent canine teeth, often angled outward, that give them a buck-toothed appearance.

Wrasses are closely related to the parrotfishes (some ichthyologists even prefer to treat them as being in the same family) and share numerous features with them including the pectoral wing-beat style of swimming and a prominent terminal mouth. Unlike the parrotfishes, however, which are strictly tropical in their distribution, wrasses have radiated into both shallow tropical and temperate areas. The systematics of the family are still rather confused, with around 200 nominal genera (of which perhaps a third will ultimately be considered valid)(B. Russell, pers. comm.) and several times that number of species. Because of the common occurrence of several radically different color patterns in the same species, many wrasses considered as distinct species in the past are now being recognized as color phases of far fewer species (e.g., Randall, 1972), with the result that the total number of species in the family is likely to continue to decrease slowly.

Reproduction by wrasses has been studied since the early 1900's, with early workers concentrating on the temperate European and Mediterranean species, many of which produce demersal eggs and nests (e.g., Fiedler, 1964). Work on the often more complex socio-sexual systems of the typically pelagic-spawning tropical wrasses has progressed far more slowly, and most such work has been done in the last decade. As a consequence, many of the hypotheses offered concerning labrid sexuality and social behavior are still being hotly contested.

SEXUAL DIMORPHISM

Every species of tropical labrid thus far examined has proved to be a protogynous hermaphrodite; i.e., most if not all males are secondarily derived from functional females. At least two species of temperate wrasse (*Oxyjulis californica,* Diener, 1976; and *Crenilabrus melops,* Dipper & Pullin, 1979), however, are apparently not hermaphroditic, and other non-hermaphroditic species may yet be discovered.

Although wrasses, like other vertebrates, have the normal two sexes (male and female), the patterns of sexuality in the family are so diverse that additional, more precise, terms are necessary to deal with the animals. The key is the development of the male. In most labrid species there are two types of male: "primary" males, which are born male, and "secondary" males, which are born and often function as females and later transform into functional males. Species vary as to the mix of primary and secondary males. Those with both types of males (most species thus far studied) are referred to as "biandric" by Reinboth (1967) and as "diandric" by subsequent workers (e.g., Reinboth, 1970; Smith, 1975; Warner & Robertson, 1978); those with secondary males only are "monandric." Examples of monandric species include *Labroides dimidiatus* (Robertson, 1972a, 1973), *Halichoeres garnoti* (Warner & Robertson, 1978), and *Hemipteronotus novacula* (Reinboth, 1975). Warner & Robertson further distinguished between secondary males that change sex before functioning as females ("prematurational" secondary males) and those that function as females before changing ("postmaturational" secondary males). Finally, females that cannot develop into males (i.e., those that remain female for life because they lack the genetic capability to transform into a male, as opposed to those that do not change sex because they never achieve a position where sex change is adaptive) are referred to by Warner & Robertson as "primary" females; they present no evidence, however, that such females exist.

The difference between primary and secondary males is clearly reflected in gonad structure (Reinboth, 1962a, 1970). The testes of primary males are similar to those of non-hermaphroditic fishes: elongate, white, and solid, with a small tubular sperm duct extending posteriorly to the common urogenital opening. Such paired testes are located on the upper sides of the abdominal cavity between the viscera and the coelomic wall. In contrast, testes of secondary males are hollow, clearly reflecting their origin as ovaries. In a functional female such ovaries are short, thick, and often yellowish, consisting of a large central space, the lumen, surrounded by a ring of lobe-like projections known as the ovarian lamellae within which the eggs develop. Eggs burst free from the lamellae when ripe, enter the lumen and, transported posteriorly, are expelled through the urogenital opening.

In *Labroides* precursors of spermatogenic tissue are located in the periphery of the ovary. Upon appropriate stimulation these expand and develop, beginning at the periphery of the gonad and progressing toward its center, while the ovarian tissue regresses. The resulting secondary testes resemble the ovaries but, of course, contain spermatogenic rather than ovarian tissue. Other wrasses differ from this. Instead of being hollow the testes are solid and surrounded by a thin membrane derived from the ovarian wall, within which the sperm duct develops as a ramified system (Reinboth, 1970; Meyer, 1977; Nakazono, 1979).

The hormonal control of sex change in wrasses has been investigated by Stoll (1955), Okada (1964), Rein-

both (1957, 1962b, 1975), Roede (1975), and Nakazono (1979) with results that are, as yet, somewhat ambiguous. Reinboth (1957), for example, found that injecting females of the Mediterranean species *Coris julis* with testosterone had no significant effect on them; later (1962b), however, he reported that such females assumed the male color pattern and showed a range of responses in their gonad structure. Several months after injection some females had more or less completely changed sex, others were still female but showed signs of extensive destruction of tissues in the ovaries, and still others had essentially normal ovaries despite the male colors. Similar, somewhat ambiguous, results were reported for other wrasses by Okada (1964), Reinboth (1975), and Roede (1975), with some species changing color without showing comparable changes in gonad structure and others changing gonad structure without a corresponding color change. Finally, Roede (1972) reported that injection of estrogens and progesterones (female hormones) caused an inverse sex change in males of *Thalassoma bifasciatum;* Reinboth (1975), however, suggested that she misinterpreted her results and that such a change was unlikely.

Recent work has focused less on the physiology of labrid sex change and more on the environmental, specifically the social, factors that select for such a change. Robertson (1972, 1973) was the first to document the fact that sex change in a wrasse can be socially controlled (for a review of such phenomena, see Reinboth, 1980). In the cleaner wrasse *Labroides dimidiatus* maleness is correlated with social dominance; subordinate individuals are all females. The social system is basically a harem in which a large male dominates three to six females and excludes other males from access to them. The entire group is organized into a size-dependent dominance hierarchy, with the male at the top and the largest female just below him dominating other females and cohabiting the male's territory. If one of the females is removed the others jockey for position in the hierarchy and those below the removed female all generally advance one position in rank. The new opening at the bottom of the hierarcy is filled by a juvenile. If the male is removed the largest female attempts to assume his position as dominant, sometimes unsuccessfully if neighboring males can take control of the vacant harem. If the dominant female is large enough and aggressive enough to resist such takeover attempts, within one and a half to two hours after the male's removal she begins to perform a male-specific ag-

A pair of *Centropyge tibicen* descending from the spawning rise and leaving a cloud of gametes behind. Photo by Jack T. Moyer.

Centropyge tibicen nuzzling. Note the sexual dichromatism shown in these two photos. Photo by Jack T. Moyer.

Centropyge interruptus soaring. Photo by Jack T. Moyer at Miyake-jima, Japan.

Centropyge interruptus male circling the female (the male is above and behind the female). Photo by Jack T. Moyer.

C. interruptus pair engaged in mutual soaring. Photo by Jack T. Moyer.

Spawning pair of *C. interruptus* soaring; the male in this case is behind the female. Photo by Jack T. Moyer.

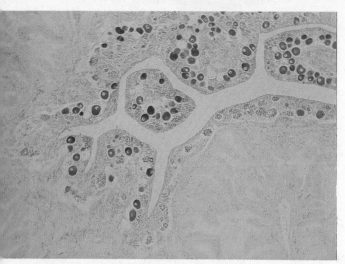

Centropyge interruptus, transitional gonad two weeks after removal of the male. Photo by A. Nakazono.

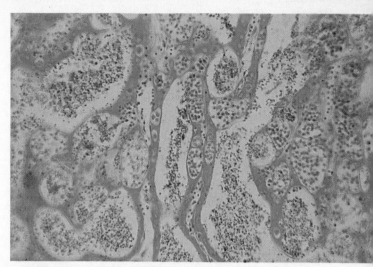

C. interruptus, mature testis (after sex change). Photo by A. Nakazono.

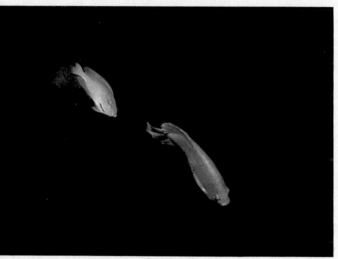

A pair of *C. interruptus* spawning. Note the gamete cloud behind the female. Photo by Jack T. Moyer.

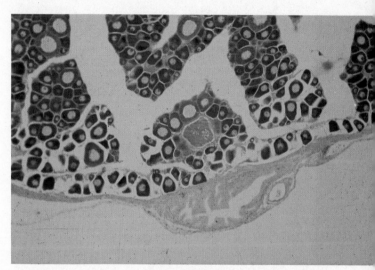

C. interruptus ovary prior to sex change. Photo by A. Nakazono.

C. interruptus nuzzling, the male pushing at the abdomen of the smaller female. Photo by Jack T. Moyer.

Another instance of nuzzling in the same species (note that the caudal fin of the male is different). Photo by Jack T. Moyer.

gressive display toward the other females. Within a few more hours she assumes completely the old male's behavior, visiting the females, aggressively dominating them, and vigorously defending the borders of the territory. Within two to four days the changeover is complete and the "new" male is actively courting and spawning with the subordinate females. Histological examination of the gonads of "old" and "new" males and females reveals that most females have ovaries containing completely enclosed pockets of spermatogenic tissue; about half of these contain active sperm. Once the female achieves a position of dominance these pockets expand. Sperm can be released in as little as 14 days after the start of sex reversal. So any "spawning" of new males between the completion of the changeover and this 14 day period will result in unfertilized eggs.

At the other extreme, maleness can also be a relatively simple function of body size or some size-related factor and may be only loosely correlated with dominance. The monandric Caribbean species *Halichoeres garnoti*, for example, is socially organized into broadly overlapping home ranges with few limited aggressive interactions, no territoriality, and no conspicuous dominance relationships (Thresher, 1979). Males (which are larger than females) and females act similarly, with the males having home ranges that overlap those of the females. Lacking specific social control of sex reversal, some less specific stimulus for sex change is operating, with a result that the proportion of males increases steadily with increasing body size class (Roede, 1972; Warner & Robertson, 1978).

Most wrasses fall somewhere between these two extremes in terms of the preciseness of social control of sex reversal. Other harem-forming species are likely to be similar to *Labroides dimidiatus;* examples include *Cirrhilabrus temminckii* (Moyer & Shepard, 1975), *C. jordani* (Takeshita, 1977), *Labroides bicolor* (Robertson & Hoffman, 1977) and probably other species of *Labroides, Hemipteronotus splendens* (Thresher, 1979b), and *Pseudocheilinus hexataenia* and *Macropharyngodon moyeri* (Moyer, pers. comm.). Species similar to *Halichoeres garnoti* include *H. bivittatus* and *H. poeyi* (Warner & Robertson, 1978), *H. maculipinna* (Thresher, 1979a), and possibly *Thalassoma lunare* (Robertson & Choat, 1974). In the western Atlantic species *Thalassoma bifasciatum* the precision of control may vary with the size of the population. In small populations fewer males can control favorable spawning sites and can actively inhibit spawning by subordinate males, resulting in relatively few males in the population. In larger populations control is more difficult and competition for favorable spawning sites more intense, resulting in a higher proportion of active males in the population (Warner & Robertson, 1978; Warner & Hoffman, 1980). Similar data have recently been provided for *T. lucasanum* by Warner (1982).

As a general rule, there are conspicuous color differences between dominant, or large, males and the smaller subordinate males, if any. Females usually resemble the subordinate males. Dominant males are typically brilliantly colored in complex patterns of red, yellow, green, blue, and black. These "terminal-phase" individuals consist of secondary males, some primary males (if the species is diandric), and in one report (for *H. garnoti* by Roede, 1972), a few large females. In most species sex reversal of a secondary male is complete before the color change begins. The smaller, often primary, males and females are often more cryptically, or at least less gaudily, colored than the terminal-phase fish. Referred to as "initial-phase," these usually outnumber terminal-phase individuals overwhelmingly (for examples see Choat, 1969; Roede, 1972; Warner & Robertson, 1978; Warner, 1982). For examples of initial-phase and terminal-phase color patterns of species in various genera see Strasburg & Hiatt (1957), Randall (1965, 1972, 1976, 1978), Randall & Böhlke (1965), Shepard & Randall (1976), Gomon & Randall (1978), and Heiser (1981).

The only documented examples of sexual monochromatism in tropical wrasses are for various species of the genus *Labroides* (Randall, 1958), for various hogfishes of the tribe Hypsigenyini (= Bodianini) (e.g., Gomon & Randall, 1978; Warner & Robertson, 1978; Gomon, 1979), and for the Japanese species *Halichoeres melanochir* (Moyer & Yogo, in press). Robertson, in a series of papers (1972, 1973, 1981; Robertson & Choat, 1974; Robertson & Hoffman, 1977), suggested that such monochromatism in *L. dimidiatus* and *L. bicolor,* and by implication the other members of the genus, is the result of limited selection of males by females. In these harem-forming species a single male monopolizes the reproductive activities of the females so that female choice of mate is not possible. At the same time each male mates almost exclusively with the members of his harem, so that his need to advertise his presence to extraharemic females is minimal. With the small social units (only three to six females per male), individual recognition of the male by females is possible and there has been no additional selection for male-specific colors. Thresher

(1979) has criticized this interpretation of monochromatism by pointing out, first, that harem-forming wrasses with small harems other than *Labroides* are conspicuously sexually dichromatic (*Cirrhilabrus temminckii*, for example, discussed by Moyer & Shepard, 1975), and second, that matings of *L. dimidiatus* males are not, in fact, confined to harem members, as emphasized by Robertson & Hoffman (1977). Rather, 2 to 3% of the matings involved extra-haremic females (Warner, et al., 1975), with the result that selection for male-specific color patterns should well occur. The absence of such colors in *Labroides* may instead be the result of this species being an obligate cleaner completely dependent on host preferences for a food supply. The dominant male, with the most to gain from sexual dichromatism, may also be the most sensitive to the behavior of the hosts, as even a slight decrease in food availability due to his "odd" colors on a cleaner could have a severe impact on his ability to maintain aggressive dominance. Similar arguments might also apply to various monochromatic hogfishes, many of which are also cleaners. The evolutionary ecology of labrid sexual dimorphism is discussed in greater detail in the final chapter.

Aside from color differences between terminal-phase males and initial-phase males and females, Warner & Robertson (1978) suggested that the sexes of *Thalassoma bifasciatum* also differ in more subtle ways, since they suspect that terminal-phase males in this species can distinguish between initial-phase males and females on the basis of inspection of their urogenital openings. Whether such discrimination is the result of differing appearances of the sexes (ripe females, perhaps, are more rounded when viewed from below than are males) or results from some olfactory cue, as suggested by Randall & Randall (1963) for the related scarids, is not known.

SPAWNING SEASON

In the tropics data for a variety of wrasses indicate year-round spawning, at least for each species as a whole if not also for individuals. Genera for which such data have been obtained include *Labroides* (Robertson & Hoffman, 1977), *Thalassoma* (Robertson & Hoffman, 1977; Warner & Robertson, 1978), *Halichoeres* (Roede, 1972; Warner & Robertson, 1978), and *Hemipteronotus* (Roede, 1972). Near the temperate limits of their ranges, however, some species (though not all—see Thresher, 1979a) restrict spawning activities to the warmer parts of the year. Examples include various Japanese species in the

genera *Thalassoma, Cirrhilabrus, Pseudolabrus,* and *Halichoeres* (e.g., Kuhn, 1973; Moyer & Shepard, 1975; Meyer, 1977; Nakazono, 1979), and *Labroides dimidiatus* at both the southern limits of the Great Barrier Reef, Australia (Robertson & Hoffman, 1977) and its northern limits off Japan (Kuwamura, 1981). The last species is of particular interest in that although egg production on the Great Barrier Reef occurs only in the warmer part of the year (October to May), individuals appear to go through the motions of spawning year-round. Such activity may be necessary for the maintenance of the rigid male-dominated social hierarchy characteristic of the species.

Data concerning possible lunar rhythmicity in the reproductive activities of wrasses are as yet ambiguous. Randall & Randall (1963) and Feddern (1965) noted a slight tendency for increased spawning activity by *Thalassoma bifasciatum* within a few days of the full moon and, to a lesser extent, the new moon. Moyer (1974) suggested a similar pattern for the Japanese species *Thalassoma cupido* based on limited observations; a later, more detailed, study of spawning by this species in the same area could not confirm this rhythm (Meyer, 1977). Roede (1972) plotted the per cent of active ovaries against moon phase for seven species of Caribbean labrids in three genera (*Thalassoma, Halichoeres,* and *Hemipteronotus*); the resulting distributions are ragged but seem to indicate increased activity during the ten-day period following the full moon for most species examined. On the other hand, *Cirrhilabrus temminckii* (Moyer & Shepard, 1975), *Thalassoma bifasciatum* (Reinboth, 1973; Robertson & Hoffman, 1977), *Labroides dimidiatus* (Robertson & Hoffman, 1977), and *Halichoeres melanochir* (Moyer & Yogo, in press) spawn daily throughout the lunar cycle. Whether such activity shows a lunar rhythm of intensity is not reported.

REPRODUCTIVE BEHAVIOR

Spawning by wrasses characteristically, if not invariably, occurs along the outer edge of a patch reef or, on a more extensive reef complex, along the outer slope. Migrations to specific spawning areas appear to be a common feature in the family, thus far having been reported for various species of *Thalassoma, Halichoeres, Choerodon, Bodianus,* and *Hemigymnus* by Moyer (1974), Meyer (1977), Robertson & Hoffman (1977), Warner, et al. (1975), and Johannes (1981), along with personal observations. Terminal-phase males of various species of *Thalassoma, Gomphosus,* and *Labrichthys,* at least, often move to prominent

Four photos in a spawning sequence of the rock beauty, *Holacanthus tricolor.* The spawning is almost exactly like that of species of *Centropyge,* except that in *H. tricolor* (and *H. ciliaris*) there may be spawning triplets or small groups suggesting haremic spawning. Nuzzling and the spawning ascent are evident. Photos by Charles Arneson.

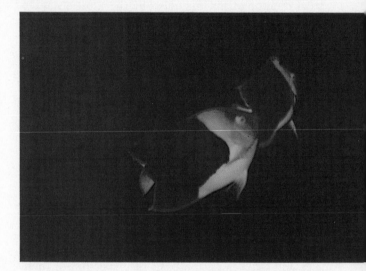

In the bottom two photos the rock beauties are exhibiting nuzzling behavior above the reef. Photo at bottom left by Dr. P. L. Colin; photo at right by Charles Arneson.

A pair of *Holacanthus passer* on the reef with the male in front of the female. Photo by Dr. R. E. Thresher.

Holacanthus passer spawning. Here a male is courting two females. Photo by Dr. R. E. Thresher.

The *H. passer* male is enticing the female to ascend toward the surface. Photo by Dr. R. E. Thresher.

A pair of *H. passer* spawning while spiraling toward the surface. Photo by Dr. R. E. Thresher.

The pair break off after spawning and return toward the bottom. Photo by Dr. R. E. Thresher.

An adult pair of *Holacanthus ciliaris.* The male is above and much larger than the female. The male was subsequently seen courting several females in his harem. Photo by Jack T. Moyer at Curacao.

coral or rock outcrops within the reef complex to spawn.

For most species spawning activity is restricted to only a brief period each day, though this period varies from species to species and may be determined by different factors for each. A strong tidal effect, with spawning occurring at any time during the day but always on an outgoing tide (usually just after peak high tide) or on a strong along-shore current, is reported for *Labroides dimidiatus* and *Thalassoma lunare* (Robertson & Choat, 1974), *T. lucasanum* (Warner, 1982; Thresher, unpubl. data), and several species of *Halichoeres* off Florida (Thresher, 1979). Tidal control is absent in many species, however, which consistently spawn at a particular time of day regardless of current pattern. Very nearly every part of the day is used for spawning by one species or another—early morning by *Pseudolabrus japonicus* (Nakazono, 1979); mid-morning by *Thalassoma cupido* (Moyer, 1974; Meyer, 1977); early afternoon by several western Atlantic species of *Halichoeres* (Warner & Robertson, 1978), *Clepticus parrae*

(Robertson & Hoffman, 1977), and *Cirrhilabrus temminckii* (Bell, in prep.); late afternoon and dusk by *Thalassoma lutescens* (pers. obs.), *Duymeria flagellifera* (Nakazono, 1979), and apparently most, if not all, species of *Bodianus* (pers. obs.). Robertson & Hoffman (1977) suggested that timing of spawning by wrasses may vary depending upon local conditions. In areas with strong and predictable tidal currents spawning would occur when currents are most favorable for transporting the fertilized eggs off the reef; that is, just after peak high tide. Where tidal currents are weak or unpredictable other factors such as time of day or light intensity may coordinate spawning activities. This hypothesis may account for apparently contradictory results for two western Atlantic species of *Halichoeres* where Warner & Robertson report spawning always in late afternoon in Panama and Thresher reports spawning at any time, but always on an outgoing tide, off Florida. Yet observations of different species of wrasses in the same area suggest that spawning for each is timed in a different manner. On Gulf of California reefs, for example, *Thalassoma*

Spawning ascent by *Halichoeres melanochir*. The male is behind and to one side of the female. The female has a tag on her to permit individual recognition in a scientific study. Note the lack of sexual dimorphism. Photo by Y. Yogo.

lucasanum spawns according to current patterns, *T. lutescens* spawns in late afternoon and at dusk, and *Halichoeres dispilis* spawns at mid-morning, although all spawn in the same areas and are subject to the same light and current regimens. Temporal segregation of spawning activities in such species may have evolved to reduce the probability of hybridization.

With one possible exception (to be discussed later), spawning of tropical wrasses is similar for all of the many species in which it has thus far been observed. Paralleling the two types of males (initial-phase and terminal-phase) are two types of spawning (referred to by Reinboth, 1973, as "dualistic" reproductive behavior)—group spawning and pair spawning.

Group, or aggregate, spawning characteristically involves initial-phase males and females; participation by terminal-phase males has thus far been reported only for *Thalassoma cupido* (Meyer, 1977), *T. lucasanum* (Hobson, cited by Meyer, 1977; Thresher, unpubl. data), and *Halichoeres bivittatus* (Warner & Robertson, 1978) and occurs only rarely in each. Spawning begins with the arrival of large numbers of initial-phase fish at prominent coral or rock outcrops, where they begin to mill about in increasingly tighter aggregations. The size of the aggregations varies from only a dozen or so fish to as many as several hundred, with several different-sized aggregates often forming within a few meters of each other. Although larger aggregations tend to remain active throughout the spawning period, fish move back and forth between nearby aggregates regularly and the smaller ones, especially, frequently disperse and are reformed. Collection of fish from the spawning aggregates reveals that males overwhelmingly predominate, outnumbering females by as much as ten to one (Roede, 1972; Robertson & Choat, 1974; Meyer, 1977), although initial-phase females typically outnumber initial-phase males in the general population. As the aggregate increases in size the fish participating begin to swim faster and faster and more erratically. *Thalassoma cupido* in such aggregates performs a characteristic "bobbing" motion while swimming, with the entire body arcing up and down (Meyer, 1977); similar behavior has not yet been reported for other species.

Spawning is preceded by a tightening of all or, in large aggregates, some of the fish into a dense ball all moving in the same direction. Each consists of a lone female at the apex of the ball closely surrounded by a variable number of males. The tight mass of fish accelerates as a unit, often in a series of hesitating movements, then dashes 0.1 to 2 m into the water col-

umn and above the remaining fish. At the peak of the spawning ascent the fish abruptly reverse direction and less cohesively return to the substrate, leaving behind a white cloud of milt and eggs that quickly disperses in the currents. In large aggregates the occurrence of one spawning ascent often triggers others, and at any one time as many as seven or eight balls may be forming and rushing off in different directions. The entire spawning sequence, from initial formation of the dense ball of fish to return to the outcrop, often takes less than a second.

Such group spawning activities have thus far been documented for several species of *Thalassoma* (numerous authors), for a few species of *Halichoeres* (Warner & Robertson, 1978; Nakazono, 1979), and for *Stethojulis interrupta* (Nakazono, 1979).

The alternate spawning system, manifested by most if not all tropical wrasses, is pair spawning, involving a terminal-phase male and an initial-phase female. Rarely it can also occur between a pair of initial-phase fish (see Meyer, 1977, and Warner & Robertson, 1978). Prior to spawning each male establishes a temporary spawning territory centered on a particularly prominent coral head or outcrop, from the vicinity of which he drives off other terminal-phase and, if possible, initial-phase males and from which he courts passing females. The permanence and degree of defense of such areas varies widely, from permanent and vigorous defense in *Labroides dimidiatus* (Robertson, 1972, 1973; Robertson & Choat, 1974), *Halichoeres maculipinna* (Thresher, 1979), and *Lachnolaimus maximus* (Colin, in press) to broadly overlapping male home ranges within which several males court and spawn with little or no aggression manifest. Examples of the latter include *Clepticus parrae* (Robertson & Hoffman, 1977) and *Halichoeres garnoti* (Thresher, 1979). Most species, however, seem to fall mid-way between these extremes and defend small territories only during brief spawning periods. Examples include a number of species of *Thalassoma* (e.g., Warner, et al., 1975; Robertson & Choat, 1974; Meyer, 1977), *Duymaeria flagellifera* (Nakazono & Tsukahara, 1974), several species of *Halichoeres* (Warner & Robertson, 1978; Nakazono, 1979) and *Pseudolabrus* (Nakazono, 1979; Jones, 1981), and apparently several species of *Bodianus* (pers. obs.). Warner & Robertson suggested that the temporary territoriality of *Thalassoma bifasciatum* is a phenomenon comparable to an avian "lek," a reproductive system in which males aggregate at a common courtship area for the sole purpose of competing for females attracted by their joint ac-

A *Centropyge* hybrid: crossing of *C. eibli* and *C. flavissimus.* Photo by Roger Steene.

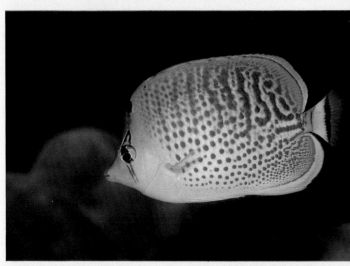

A hybrid butterflyfish: crossing of *Chaetodon punctatofasciatus* and *C. guttatissimus.* Photo by Roger Steene.

A hybrid butterflyfish: crossing of *Chaetodon lunula* and *C. auriga.* Photo by Roger Steene.

A male *Chaetodon rainfordi* courting a female by "dancing" in front of her. Photo by Dr. R. E. Thresher.

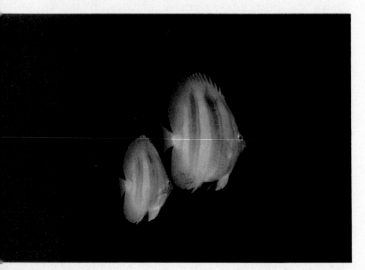

C. rainfordi pair preparing for a spawning ascent. Photo by Dr. R. E. Thresher.

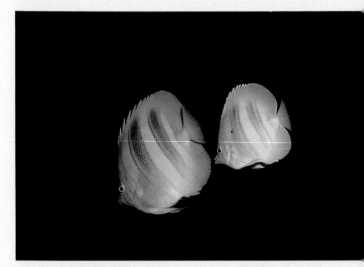

C. rainfordi male moving into spawning position. Photo by Dr. R. E. Thresher.

Newly settled juvenile of *Chaetodon auripes* still with the tholichthys plates on the head. Photo by Dr. Fujio Yasuda.

Tholichthys larva of *Chaetodon sedentarius* about to settle. This specimen was collected while night-lighting. Photo by Dr. W. E. Burgess.

Chaetodon sedentarius after attaining its full juvenile coloration. Photo by Aaron Norman.

tivities. Moyer & Yogo (in press), observing a similar spawning system in *Halichoeres melanochir* off Japan, examined the question of lekking in wrasses in great detail and concluded that although previous work has not been sufficiently detailed to demonstrate the occurrence of the phenomenon in any species of fish, *Halichoeres melanochir* does, in fact, fulfill many of the requirements usually used to define a lek-based social system.

Courtship behavior of territorial males is of two types—long distance and short distance (Robertson & Hoffman, 1977). Apparent long distance signals have thus far been described for only a few species. The most common form of such a display is "looping," a movement in which the male dashes up and down in the water column in a manner similar to actual spawning. Such looping has been described for *Thalassoma bifasciatum* (Robertson & Hoffman, 1977) and for several species of *Halichoeres* (Thresher, 1979; Moyer & Yogo, in press). Variations on the theme include a spiraling ascent off the bottom by displaying male *Cirrhilabrus temminckii* (Moyer & Shepard, 1975), a step-like up-and-down bobbing during ascent by male *Pseudolabrus japonicus* (Nakazono, 1979), and a smoothly arched ascent and descent while swimming in an exaggerated fashion in *Bodianus diplotaenia* (pers. obs.). Short distance courtship displays are more widespread. In the western Atlantic species of *Halichoeres* such a display consists of little more than a male following closely behind a female while main-

Courtship and spawning in the bluehead wrasse, *Thalassoma bifasciatum.* During courtship periods, terminal-phase males perform a long-distance courtship movement known as looping (a). At the approach of a female, the male swims vigorously back and fourth above her (b), vibrating his pectoral fins. If the female is ready, she ascends, and the pair move side by side (c) and dash up off the bottom to shed gametes (d). Afterward, the female departs (e), while the male returns to courtship behavior.

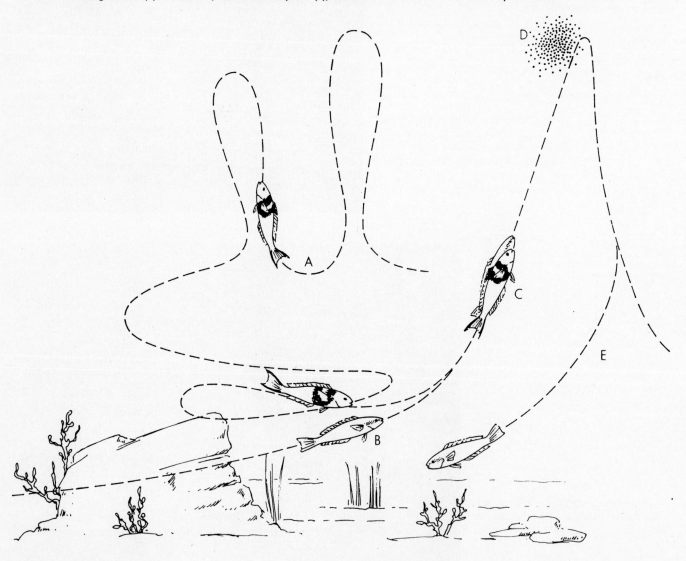

taining his colors at full intensity and with his un-paired fins fully spread. In all species of *Thalassoma* thus far studied terminal-phase males vibrate their brilliantly colored pectoral fins while swimming back and forth over a prospective mate, often combining such behavior with vibrations of the entire body or rapid beating of the caudal fin (e.g., Meyer, 1977; for a detailed discussion of this color signal, see Heiser, 1981). In *Cirrhilabrus temminckii* the males "flash" their colorful dorsal and anal fins while moving among the females, while males of *Labroides dimidiatus* "flutter-swim" to an approaching female (i.e., swimming directly at her and curving around in front of her while expanding and rapidly "fluttering" his caudal fin (Robertson & Hoffman, 1977)).

In most species it is not the courting male but rather the female who initiates spawning. In many species, e.g., *Thalassoma bifasciatum, Cirrhilabrus temminckii, Hemipteronotus splendens*, and numerous *Halichoeres*, she ascends slightly off the bottom and hovers there, usually head up, until the male moves into position beside and slightly below her. In *Labroides dimidiatus* the female "body-sigmoids" to the male, arching her body in an "S"-shape and displaying her egg-distended abdomen while at the same time paling anteriorly from black to brown (Robertson & Hoffman, 1977). In *Thalassoma cupido* females actually initiate the spawning rush, such that the male must hurry to follow behind her and be in the correct position to fertilize the eggs (Meyer, 1977).

Pair spawning consists of a rapid up-and-down dash by the male and female usually so close together as to be almost, if not actually, touching in the form of a narrow inverted "U." Gametes are released at the top of the "U" and are visible as a white cloud that rapidly disperses. The height of the spawning ascent varies from half a meter to over two meters. The positions of the male and female range from side-by-side (e.g., many *Halichoeres, Lachnolaimus maximus*) to the female below the male and nestled between his fully spread pelvic fins (e.g., *Bodianus diplotaenia*). In a few species (e.g., *Clepticus parrae* and *Hemipteronotus splendens*) the male pushes the female toward the surface before moving beside her.

Though basically simple, pair spawning is often complicated by the activities of nearby initial-phase males, which often dash in to join the pair as they begin the upward part of the ascent. Warner, et al. (1975) referred to such behavior as "streaking" and suggested that the initial-phase males are increasing their fecundity at the expense of the terminal-phase male by releas-ing milt at the peak of the ascent and thereby fertilizing some of the eggs released. Not surprisingly, terminal-phase males are particularly aggressive toward initial-phase males and vigorously chase them after interference. The latter, however, are identical in color pattern and size to the females, an apparent case of intraspecific mimicry, and quickly lose the male by mingling in among such females. In that regard, Warner & Robertson (1978) noted that there are no initial-phase males in harem-forming species and suggested that their absence results from the inability of such males to mimic females due to small harem size and individual recognition of harem members by the male. Thresher (1979a) similarly argued that primary initial-phase males are absent in some non-haremic species, such as *Halichoeres garnoti*, because such species do not spawn at a predictable site but instead spawn anywhere within their broadly overlapping home ranges. Without being able to predict when, where, or even if a streakable pair spawning will occur, initial-phase males cannot streak enough pairs to maintain themselves in the population.

The theory underlying the success of initial-phase males is treated rigorously by Warner, et al. (1975), Warner & Robertson (1978), and Warner & Hoffman (1980). Since the continued presence of such males depends upon their rate of success in streaking or group spawning, selection will favor such males if their average fecundity is higher than that of a comparable sized female and will favor elimination of such males if, on the average, the female produces more offspring. The stable point in a population will depend upon the number of terminal-phase males present, the ratio of paired to group spawning, and the success rate of the terminal-phase males in preventing interference in pair spawning by the initial-phase males. One consequence is that in large populations where terminal-phase male control of spawning is less effective than in small populations, the proportion of initial-phase males in the population is higher.

One final consequence of the different spawning strategies of initial-phase and terminal-phase males is the difference in the size of their respective gonads. Terminal-phase males frequently take part in many pair spawnings each day, but because they are so close to the female during the spawning and, aside from streaking, have exclusive access to her, each male needs to release only a relatively small amount of milt during each spawning ascent. Initial-phase males, in contrast, when participating in group spawning bouts must compete with many other males to fertilize the eggs of the

Acronurus stage of *Acanthurus triostegus*. Note that the banded pattern of both the juvenile and adult is already visible. (See photo on opposite page.) Photo by Allan Power.

Acronurus stage of a western Atlantic species of *Acanthurus*. The typical transparency of the body and the silver area of the head are larval characters. The caudal spine is already developed, and the individual is about to assume its juvenile color. Photo by Charles Arneson.

Temporary sexual dichromatism in *Acanthurus triostegus*. Photo by Dr. R. E. Thresher.

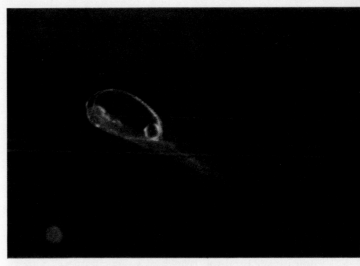

Newly hatched larva of *Zebrasoma scopas* (1.5 mm total length). Photo by Dr. R. E. Thresher.

Spawning sequence of *Ctenochaetus striatus*. Here the male is displaying over the female.

Pair of *C. striatus* at the peak of their spawning ascent. Note the third individual rushing up to join them.

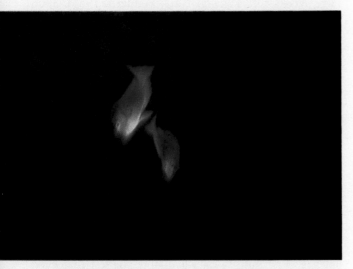

C. striatus pair descending after their spawning ascent. All photos in this sequence by Dr. R. E. Thresher.

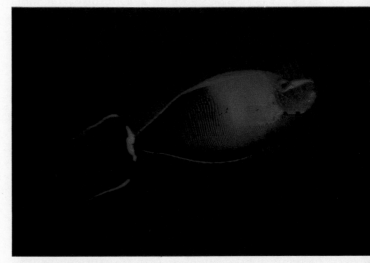

Temporary sexual dichromatism in *Naso vlamingi*. Photo by Dr. R. E. Thresher.

single female present; consequently, each male benefits by flooding the area with his sperm. Similarly, it is also beneficial to flood the area near the peak of a spawning ascent when streaking. The result, not surprisingly, is that the gonads of initial-phase males are relatively much larger than those of terminal-phase males, in some cases more than twice as large (Robertson & Choat, 1974; Warner & Robertson, 1978).

As complex as reproduction by tropical wrasses seems, there are several unifying themes: diandry, alternate spawning behaviors, the effects of male dominance, and the launch of pelagic gametes during a spawning ascent. There is, however, a single unconfirmed report of an altogether different spawning pattern in such a wrasse. De Bernardo (1975) reported that in captivity a pair of *Anampses cuvieri*, a Hawaiian representative of a widely distributed Indo-Pacific genus, produced clear adhesive eggs which were laid on a small rock and on the filter bed in a pit excavated by the male. After spawning, which was supposedly observed, the male tended and defended the eggs. Only a single batch of eggs was produced and these apparently were eaten by the male. There have been no reported observations of spawning in the field by any species of *Anampses,* although Randall (1972) reported seeing unspecified courtship in *A. chrysocephalus* and J. Moyer (pers. comm.) observed what appeared to be pre-group spawning behavior in *A. caeruleopunctatus.* The production of demersal eggs by this species seems unlikely based on their typically labrid sexual differences in color pattern and general behavior, but cannot be ruled out without direct evidence of pelagic spawning. If confirmed, *Anampses* would be the only known tropical example of demersal spawning by a labrid, and comparison of its social behavior with that of other tropical species and with the demersal spawning temperate species (e.g., Fiedler, 1964) would be of great interest.

No hybrids have been reported for tropical wrasses (which is an interesting absence considering their tendency to group spawn and to streak pair-spawners—one would expect at least occasional mistakes, e.g., Meyer, 1977), but they have been reported for two genera of temperate wrasses. In the North Atlantic several species of demersal-spawning wrasses are interfertile and naturally occurring hybrids are occasionally seen (Hagstrom & Wennerberg, 1964); off New Zealand apparent hybrids between two sympatric species of *Pseudolabrus* have been collected (Ayling, 1980).

EGGS AND LARVAE

Labrid eggs, with the exceptions noted earlier, are pelagic, spherical, 0.45 to 1.2 mm in diameter, and contain a single oil droplet (Kubo, 1939; Mito, 1962; Feddern, 1965; Watson & Leis, 1974; Kuwamura, 1981; Suzuki, et al., 1981; Colin, in press). The incubation period appears to vary with water temperature (e.g., Mito, 1962), taking slightly over 24 hours at a temperature of around 27°C. Newly hatched larvae range in size from 1.1 to 2.84 mm total length (in part depending on egg size), lack fins or pigment in the eyes, and have a moderate sized yolk sac that extends forward of the head and contains a single oil droplet (e.g., Kubo, 1939; Mito, 1962; Suzuki, et al., 1981; Colin, in press; for a temperate species, see Bolin, 1930).

Older larvae, based on collections of specimens in plankton tows, are elongate to deep-bodied, laterally compressed, and sparsely pigmented. The fins and eyes are well developed with distinctive long and low anal and dorsal fins and a square caudal fin (e.g., Vatanachi, 1974; Fourmanoir, 1976). Such larvae also have a small but prominent pointed, terminal mouth. Miller (1974) suggested that such larvae are typically found in relatively deep water.

The duration of the planktonic larval stage is not known but appears to be around a month (Colin, in prep.). The smallest juveniles seen on the reef are approximately 10 mm long, are very slender, and usually are only slightly pigmented. Most species have a characteristic juvenile coloration that differs from both the initial-phase and terminal-phase adult colors (e.g., Masuda & Tanaka, 1962; Feddern, 1963). Such juvenile-specific colors range from brilliant yellows and oranges to drab grays and browns. Their function has not been determined, though Randall & Randall (1960) suggested that the cryptic juveniles of several species of *Hemipteronotus* may be camouflaged as to resemble plant material and debris. Juveniles of a few species are cleaners (e.g., *Thalassoma bifasciatum,* many species of *Bodianus,* Losey, 1974). A few also shelter in the tentacles of sea anemones as juveniles, although they are apparently not immune to the anemones' nematocysts (Schlichter, 1970; Gendron & Mayzel, 1976).

Age or size at sexual maturity (as a female) varies, not surprisingly, with the maximum size of the species. The limited data available suggest that even very small individuals (less than 30 mm total length) of species in such genera as *Thalassoma* and

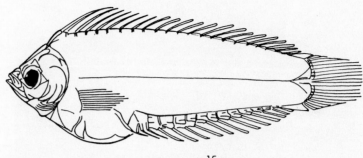

Larval wrasses. Based on
Fourmanoir (1976).

15mm

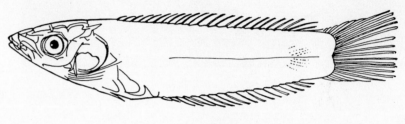

16mm

Larval wrasse, 6.8 mm total length, collected in plankton tow off Miami, Florida. Photo by Dr. R. E. Thresher.

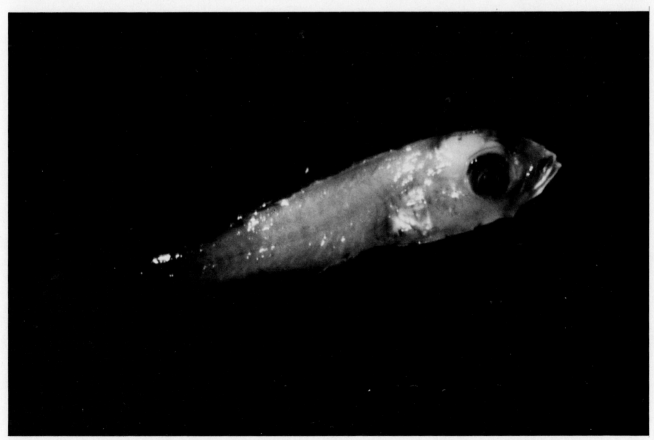

Acanthurus coeruleus goes through a change of colors from juvenile to adult. Shown here are the juvenile (yellow fish) with a subadult (blue-gray fish with yellow tail). Eventually the adult color of blue to purplish gray with grayish horizontal lines develops. Photo by Paysan.

Hybrid acanthurid. The cross is probably between *Acanthurus achilles* and *A. glaucopareius.* Photo by James H. O'Neill.

Males of the filefish *Amanses scopas* have a group of five or six long stout spines on the side of the body between the soft dorsal and anal fins as seen here. Females have a short brush-like mass of setae in the same position. Photo by Allan Power.

Halichoeres may already be sexually active (e.g., Warner & Robertson, 1978).

REPRODUCTION IN CAPTIVITY

Aside from the smallest species, spawning most wrasses in small aquaria is likely to be difficult, if only because of the relatively large size of the fishes at maturity and the demands made on space by terminal-male territoriality and the vigorous spawning ascent. There are only two reports of aquarium spawning of pelagic-spawning tropical wrasses: *Thalassoma bifasciatum* by Zumpe (1963) and *T. cupido* by Meyer (1977). Neither provides many details. Zumpe held four fish (sex or color phases not reported) in a 1.8 m³ aquarium and noted that pair-spawning was preceded by a male defending a preferred spawning site. Meyer twice spawned *T. cupido* in a 158-liter aquarium maintained at 28°C and lit by natural illumination.

As noted above, *Anampses cuvieri* was reported by De Bernardo (1975) to produce demersal eggs in captivity. The fish were approximately 9 cm long when purchased, were maintained in a 76-L aquarium at 20°C, a pH of 7.8 to 8.0, and a specific gravity of 1.019 to 1.020, and were fed regularly on frozen, fresh, and canned shrimp, a variety of frozen foods, and, at least three times weekly, live guppies. Spawning occurred seven weeks after the fish were introduced into the tank. Given the high aggressive levels of the fish, especially the male, more consistent results might be obtained through use of a larger aquarium.

Spawning in large public aquaria has been reported (Hioki, 1979) to have occurred regularly for ten species in nine genera at the Marine Science Museum of Tokai University. This museum is particularly well set up to record and observe spawnings by marine fishes, but its success in spawning labrids suggests that similar spawnings may be occurring unnoticed in other large public aquaria.

Literature Cited

Ayling, A.M. 1980. Hybridization in the genus *Pseudolabrus* (Labridae). *Copeia*, 1980:176-180.

Bolin, R.L. 1930. Embryonic development of the labrid fish *Oxyjulis californicus* Günther. *Copeia*, 1930:122-128.

Choat, J.H. 1969. Studies on the biology of labroid fishes (Labridae and Scaridae) at Heron Island, Great Barrier Reef. Part I. Taxonomy and reproductive behavior. Ph. D. Diss., Univ. of Queensland.

Colin, P.L. (In Press). Spawning and larval development of the hogfish, *Lachnolaimus maximus* (Pisces: Labridae). *Fish. Bull. (U.S.)*

De Bernardo, J.P. 1975. *Anampses cuvieri*: the firefish from Hawaii. *Trop. Fish Hobbyist*, 24(2):60-65.

Diener, D.R. 1976. Hermaphroditism in fish: a comparative study of the reproductive biology and endocrinology of the California Labridae. Ph. D. Diss., Univ. Calif., San Diego.

Dipper, F.A. and R.S.V. Pullin. 1979. Gonochorism and sex-inversion in British Labridae (Pisces). *J. Zool., London*, 187:97-111.

Feddern, H.A. 1963. Color pattern change during growth of *Bodianus pulchellus* and *Bodianus rufus* (Pisces: Labridae). *Bull. Mar. Sci.*, 13:224-241.

Feddern, H.A. 1965. The spawning, growth, and general behavior of the bluehead wrasse, *Thalassoma bifasciatum* (Pisces, Labridae). *Bull. Mar. Sci.*, 15:896-941.

Fiedler, K. 1964. Verhaltensstudien an Lippfischen der Gattung *Crenilabrus* (Labridae, Perciformes). *Z. Tierpsychol.*, 21:521-591.

Fourmanoir, P. 1976. Formes post-larvaire et juveniles de poissons cotiers pris au chalut pelagique dans le sud-ouest Pacifique. *Cahiers du Pacifique*, (19):47-88.

Gendron, R.P. and K. Mayzel. 1976. Association of *Thalassoma bifasciatum* with *Condylactis gigantea* in the Bahamas. *Copeia*, 1976:382-384.

Gomon, M.F. 1979. A revision of the labroid fish genus, *Bodianus*, with an analysis of the relationships of other members of the tribe Hypsigenyini. Ph. D. Diss., U. Miami, Miami, Fla.

Gomon, M.F. and J.E. Randall. 1978. Review of the Hawaiian fishes of the labrid tribe Bodianini. *Bull. Mar. Sci.*, 28:32-48.

Hagstrom, B.E. and C. Wennerberg. 1964. Hybridization experiments with wrasses (Labridae). *Sarsia*, 17:47-54.

Heiser, J.B. 1981. Review of the labrid genus *Thalassoma* (Pisces: Teleostei). Ph. D. Diss., Cornell Univ., Ithaca, N.Y.

Hioki, S. 1979. Spawnings by wrasses. *Fish Mag.*, April:1-7 (in Japanese).

Hobson, E.S. 1965. Diurnal-nocturnal activity of some inshore fishes in the Gulf of California. *Copeia*, 1965:291-302.

Johannes, R.E. 1978. Reproductive strategies of coastal marine fishes in the tropics. *Env. Biol. Fishes*, 3:65-84.

Johannes, R.E. 1981. *Words of the Lagoon*. Univ. Calif. Press, Los Angeles. 320 pp.

Jones, J.P. 1981. Spawning-site choice by female *Pseudolabrus celidotus* (Pisces: Labridae) and its influence on the mating system. *Behav. Ecol. Sociobiol.*, 8:129-142.

Kubo, I. 1939. Notes on the development of a teleost, *Thalassoma cupido* (Temminck and Schlegel). *Bull. Jap. Soc. Sci. Fish.*, 8:165-167.

Kuhn, R. 1973. Freiwasserbeobachtungen im Mittelmeer uber das Verhalten protogyner Labriden, unter besonderer Berucksichtigung von *Thalassoma pavo* (L.). Master's Thesis, Johannes Gutenburg-Universitat, Mainz, Germany.

Kuwamura, T. 1981. Life history and population fluctuation in the labrid fish, *Labroides dimidiatus*, near the northern limit of its range. *Publ. Seto Mar. Lab.*, 26(1/3):95-117.

Losey, G.S. 1974. Cleaning symbiosis in Puerto Rico with comparison to the tropical Pacific. *Copeia*, 1974:960-970.

Masuda, T. and K. Tanaka. 1962. Young of labroid and scaroid fishes from the central Pacific coasts of Japan. *J. Tokyo Univ. Fish.*, 48:1-98.

Meyer, K.A. 1977. Reproductive behavior and patterns of sexuality in the Japanese labrid fish *Thalassoma cupido*. *Japan. J. Ichthyol.*, 24:101-112.

Miller, J.M. 1974. Nearshore distribution of Hawaiian marine fish larvae: effects of water quality, turbidity, and currents. Pp. 217-231. *In: The Early Life History of Fish*. (J.H.S. Blaxter, Ed.). Springer Verlag, N.Y.

Mito, S. 1962. Pelagic fish eggs from Japanese waters. IV. Labrina. *Sci. Bull. Fac. Agricult. Kyushu Univ.*, 19.

Moyer, J.T. 1974. Notes on the reproductive behavior of the wrasse *Thalassoma cupido*. *Japan. J. Ichthyol.*, 21(1):34-36.

Moyer, J.T. and J.W. Shepard. 1975. Notes on the spawning behavior of the wrasse, *Cirrhilabrus temminckii*. *Japan. J. Ichthyol.*, 22(1):40-42.

Moyer, J.T. and Y. Yogo. (In Press). The lek mating system of *Halichoeres melanochir* (Pisces: Labridae) at Miyake-jima, Japan. *Z. Tierpsychol.*

Nakazono, A. 1979. Studies on the sex reversal and spawning behavior of five species of Japanese labrid fishes. *Rept. Fish. Res. Lab., Kyushu Univ.*, (4):64 pp.

Nakazono, A. and H. Tsukahara. 1974. Underwater observations on the spawning behavior of the wrasse, *Duymaeria flagellifera* (Cuvier et Valenciennes). *Rept. Fish. Res. Lab., Kyushu Univ.*, (2):1-11.

Okada, Y.K. 1964. Effects of androgen and estrogen on sex reversal in the wrasse, *Halichoeres poecilopterus*. *Proc. Japan. Acad.*, 40:541-544.

Randall, J.E. 1958. A review of the labrid fish genus *Labroides* with descriptions of two new species and notes on ecology. *Pac. Sci.*, 12:327-347.

Randall, J.E. 1965. A review of the razorfish genus *Hemipteronotus* (Labridae) of the Atlantic Ocean. *Copeia*, 1965(4):487-501.

Randall, J.E. 1972. A revision of the labrid fish genus *Anampses*. *Micronesica*, 8(1-2):151-195.

Randall, J.E. 1976. A review of the Hawaiian labrid fishes of the genus *Coris*. *Underwat. Observ.*, (26):1-10.

Randall, J.E. 1978. A revision of the Indo-Pacific labrid fish genus *Macropharygodon*, with descriptions of five new species. *Bull. Mar. Sci.*, 28(4):742-770.

Randall, J.E. and J.E. Böhlke. 1965. Review of the Atlantic labrid fishes of the genus *Halichoeres*. *Proc. Acad. Nat. Sci. Philadelphia*, 117(7):235-259.

Randall, J.E. and H.A. Randall. 1960. Examples of mimicry and protective resemblance in tropical marine fishes. *Bull. Mar. Sci.*, 10(4):444-480.

Randall, J.E. and H.A. Randall. 1963. The spawning and early development of the Atlantic parrotfish, *Sparisoma rubripinne*, with notes on other scarid and labrid fishes. *Zoologica, N.Y.*, 48:49-60.

Reinboth, R. 1957. Sur la sexualité du teleosteen *Coris julis* (L.). *Comptes Rendus Hebdomadraires de Seances de l'Academie des Sciences*, 245:1662-1665.

Reinboth, R. 1962a. Morphologische und functionelle Zweigeschlechtlichkeit bei marinen Teleostuern (Serranidae, Sparidae, Centracanthidae, Labridae). *Zool. Jahrb. Physiol.*, 69:405-480.

Reinboth, R. 1962b. The effects of testosterone on female *Coris julis* (L.), a wrasse with spontaneous sex-inversion. *Gen. Comp. Endocr.*, 2: Abstract # 39.

Reinboth, R. 1967. Biandric teleost species. *Gen. Comp. Endocr.*, 9(3):486.

Reinboth, R. 1970. Intersexuality in fishes. *Mem. Soc. Endocrinol.*, (18):515-543.

Reinboth, R. 1973. Dualistic reproductive behavior in the protogynous wrasse, *Thalassoma bifasciatum*, and some observations on its day-night changeover. *Helgo. wissenschaft. Meeresunter.*, 24:174-191.

Reinboth, R. 1975. Spontaneous and hormone-induced sex-inversion in wrasses (Labridae). *Publ. Staz. Zool. Napoli*, 39 (Suppl.):550-573.

Reinboth, R. 1980. Can sex inversion be environmentally induced? *Biol. Reprod.*, 22:49-59.

Male *Monacanthus ciliatus* have strongly pronounced ventral flaps such as the one seen in this photo.

Two individuals of *Balistoides conspicillum* show what appears to be aggressive behavior in an aquarium. Whether this is due to protection of territory or has sexual implications is not known. Photo by Earl Kennedy.

A juvenile boxfish, *Ostracion cubicus*, hiding among the branches of the coral *Acropora*. The adult is colored quite differently and the sexes show distinct sexual dichromatism. Photo by Walter Deas.

Robertson, D.R. 1972. Social control of sex reversal in a coral-reef fish. *Science, Wash.,* 177:1007-1009.

Robertson, D.R. 1973. Sex change under the waves. *New Sci., 31 May:*538-539.

Robertson, D.R. 1981. The social and mating systems of two labrid fishes, *Halichoeres maculipinna* and *H. garnoti,* off the Caribbean coast of Panama. *Mar. Biol.*

Robertson, D.R. and J.H. Choat. 1974. Protogynous hermaphroditism and social structures in labrid fish. *Proc. Second Internat. Coral Reef Symp.,* 1:217-225.

Robertson, D.R. and S.G. Hoffman. 1977. The role of female mate choice and predation in the mating systems of tropical labroid fishes. *Z. Tierpsychol.,* 45:298-320.

Roede, M.J. 1972. Color as related to size, sex, and behavior in seven Caribbean labrid species (genera *Thalassoma, Halichoeres,* and *Hemipteronotus*). *Stud. Fauna Curacao and Other Carib. Isl.,* 138:1-264.

Roede, M.J. 1975. Reversal of sex in several labroid fish species. *Publ. Staz. Zool. Napoli,* 39 (Suppl.):595-617.

Shepard, J.W. and J.E. Randall. 1976. Notes on the labrid fish *Stethojulis maculata* from Japan. *Japan. J. Ichthyol.,* 23:165-170.

Schlichter, D. 1970. *Thalassoma amblycephalus,* ein neuer Anemonenfisch-Typ. *Mar. Biol.,* 7:269-272.

Smith, C.L. 1975. The evolution of hermaphroditism in fishes. Pp. 295-310. *In: Intersexuality in the Animal Kingdom.* (R. Reinboth, Ed.). Springer Verlag, Heidelberg.

Stoll, L.M. 1955. Hormonal control of the sexually dimorphic pigmentation of *Thalassoma bifasciatum. Zoologica, N.Y.,* 40(11):125-132.

Strasburg, D.W. and R.W. Hiatt. 1957. Sexual dimorphism in the labrid fish genus *Gomphosus. Pac. Sci.,* 11(1):133-134.

Suzuki, K., S. Hioki, K. Kobayashi, and Y. Tanaka. 1981. Developing eggs and early larvae of the wrasses, *Cirrhilabrus temminckii* and *Labroides dimidiatus,* with a note on spawning behaviors. *J. Fac. Mar. Sci. Technol. Tokai Univ.,* 14:369-377.

Takeshita, G.Y. 1977. Hawaiian flame wrasse. *Marine Aquarist,* 7(8):44-48.

Thresher, R.E. 1979. Social behavior and ecology of two sympatric wrasses (Labridae: *Halichoeres* spp.) off the coast of Florida. *Mar. Biol.* 53:161-172.

Thresher, R.E. 1980. *Reef Fish.* Palmetto Publ. Co., St. Petersburg, Fla. 172 pp.

Vatanachai, S. 1974. The identification of fish eggs and larvae obtained from the survey cruises in the South China Sea. *Proc. Indo-Pac. Fish. Council,* 15:111-130.

Warner, R.R. 1982. Mating systems, sex change and sexual demography in the rainbow wrasse, *Thalassoma lucasanum. Copeia,* 1982:653-661.

Warner, R.R. and S. Hoffman. 1980. Local population size as a determinant of mating system and sexual composition in two tropical marine fishes (*Thalassoma* spp.). *Evolution,* 34:508-518.

Warner, R.R. and D.R. Robertson. 1978. Sexual patterns in the labroid fishes of the western Caribbean, I. The wrasses (Labridae). *Smithsonian Contrib. Zool.,* (254):1-27.

Warner, R.R., D.R. Robertson, and E.G. Leigh, Jr. 1975. Sex change and sexual selection. *Science, Wash.,* 190:633-638.

Watson, W., and J.M. Leis. 1974. Ichthyoplankton of Kaneohe Bay, Hawaii. *Sea Grant Tech. Rept.* (TR-75-01).

Zumpe, D. 1963. Uber das Ablaichen von *Thalassoma bifasciatum. Das Aquar.-Terrar. Z.,* 16:86-88.

Parrotfishes
(Scaridae)

Parrotfishes are large (up to 2 m) heavy-bodied browsing fishes largely confined to coral reefs and nearby areas. They are characterized by beak-like jaws (formed by fusion of the teeth), large cycloid scales, a discontinuous lateral line, and, as in the closely related wrasses, a distinctive "pectoral fins only" mode of swimming. Color patterns range from drab browns, grays, and blacks, typical mainly of juveniles and females, to complex and brilliant combinations of red, green, and blue, typical usually of the larger males.

Browsing herbivores, parrotfishes are typically seen in small to sometimes very large foraging groups moving slowly and steadily over the reef while cropping algae, coral, and sea grasses. A few species are permanently territorial, but the majority seem to be wanderers, with the size of the home range increasing with the size of the fish. Straughn (1977), for example, found that tagged eastern Pacific fish routinely moved as much as a half mile (0.8 km) from the tagging site. Because they move over such great distances, only a few of the smaller scarids have been studied in terms of social organization. Information for all but a few of the smaller and more common western Atlantic species is sketchy and mainly anecdotal.

Like the closely allied labrids, parrotfishes generally have complex socio-sexual systems based on protogynous hermaphroditism, distinctive initial-phase and terminal-phase color patterns, and dualistic reproduction. Specifics related to the scarids will be discussed in this chapter, but for a more complete discussion of these concepts the reader is directed to the labrid chapter.

SEXUAL DIMORPHISM

Protogynous sex reversal in scarids was first documented by Reinboth (1968), though several earlier workers had suggested that such reversal was likely because of parrotfishes' conspicuous sexual dichromatism and close relationship to the wrasses. In general, scarid sex changes closely parallel those of the wrasses, and the two share a number of common features. (1) Ovaries of functional females, reflecting their dual sexual nature, possess spermatogenic cells along the ovarian lamellae and periphery. Upon sex reversal these areas expand while the oogonia degenerate. (2) Males can be either primary, i.e., born male, or secondary, i.e., derived from females. (3) Primary and secondary males differ in gonad structure. The testes of primary males are relatively solid and lack any sign of ovarian origin; those of secondary males possess remnants of the ovarian lumen.

Reinboth's original paper dealt with seven species of western Atlantic and eastern Pacific parrotfishes, including representatives of the two subfamilies (Scarinae and Sparisominae). Protogyny has since been documented for an additional 16 species of *Scarus* on the Great Barrier Reef (Choat, 1969; Choat & Robertson, 1975), for *Scarus sordidus* off Japan by Yogo, et al. (1980), and for ten western Atlantic species (including three discussed by Reinboth) in the genera *Sparisoma, Scarus,* and *Cryptotomus* (Robertson & Warner, 1978). In contrast, Bebars (1978) reported finding no evidence of protogyny in 12 species of *Scarus* from the Red Sea. It seems unlikely that there would be such a strong geographic effect on what otherwise appears to be a universal trait in scarids; it would therefore be valuable for Bebars' observations to be independently confirmed.

As in the wrasses, the relative proportions of primary and secondary males in a population vary widely. On one extreme, Bebars' observations suggest that at least some Red Sea species lack secondary males altogether. On the other, several species have proved to be monandric; that is, all males are derived from females by sex change. Robertson & Warner (1978) suggested that the two parrotfish subfamilies differ in this trait. Most species of *Scarus* are diandric; indeed, only one monandric species, *Scarus niger,* has been described (Robertson & Choat, 1975). All seven

Spawning sequence of *Lactoria fornasini.* All photos by Jack T. Moyer. Shown here is the upward rise, with the male behind.

The male *L. fornasini* flashes to the female (a form of color signaling), the first stage of spawning.

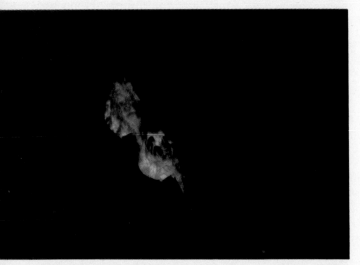

Upward rise in *L. fornasini.* The male is the larger and brighter of the two fish.

Near the peak of the upward rise, the pair start to change position.

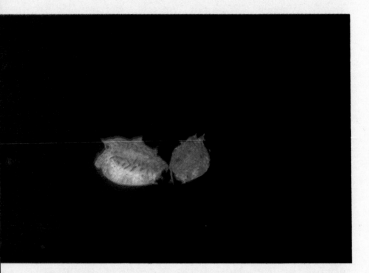

At the peak the pair turn away from each other, releasing the gametes. At this point their bodies do not touch and they cannot see each other.

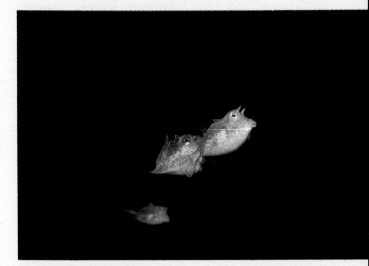

During the upward rise, in one instance another female also rose to spawn.

230

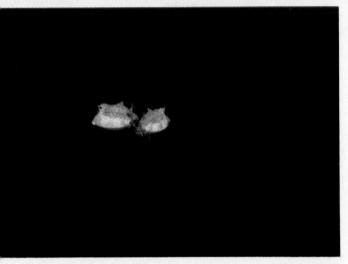

Lactoria fornasini in the actual act of spawning, with the cloud of gametes next to the spawners. Photo by Jack T. Moyer.

Lactoria diaphanus male biting the female, a common expression of male control of a harem. Such biting is common when females stray too close to neighboring territories. Photo by Jack T. Moyer.

Continued spawning of *L. fornasini,* with larger amount of gametes visible. Photo by Jack T. Moyer.

L. fornasini starting to turn away. Photo by Jack T. Moyer.

The gamete release, at least in *L. fornasini,* is said to be synchronized by a sound from the male. Photo by Jack T. Moyer.

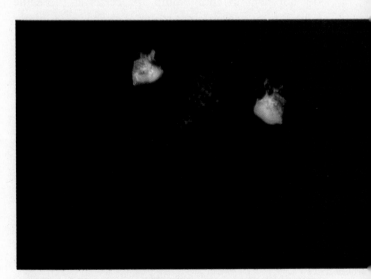

At the conclusion of spawning there is a downward rush, leaving the cloud of gametes above the reef. Photo by Jack T. Moyer.

species of sparisomine parrotfishes examined by Robertson & Warner, however, lacked primary males. They suggested that these fishes may have lost such males in the evolutionary past and that their position in the social structure has been filled by prematurational sex reversal of small females into small secondary males.

The relationship between sex and color pattern in parrotfishes is a complex one. As in the wrasses, the most widespread relationship is the presence of two characteristic adult color patterns in a species— terminal-phase males and initial-phase males and females. The former are usually gaudy combinations of green, red, blue, and yellow; the latter are usually more drab combinations of gray, brown, and black (for examples of such color patterns for various scarids see Schultz, 1958; Randall, 1963; Rosenblatt & Hobson, 1969). Such dichromatism has been documented for species in *Scarus, Sparisoma, Nicholsina, Bolbometopon,* and *Cryptotomus.*

While most parrotfishes fit the above pattern, those that differ from it are not uncommon. Monochromatic species are widespread and include *Scarus coelestinus* and *S. coeruleus* in the western Atlantic (Randall, 1968), *S. perrico* in the eastern Pacific (Rosenblatt & Hobson, 1969), and *S. niger* in the Indo-West Pacific (Choat & Robertson, 1975). In two of these species, however, the role of large male-specific colors may be served by other morphometric differences between the sexes. Large adults of both *S. coeruleus* and *S. perrico* develop more squared-off and prominent foreheads than do smaller fish. Although development of this "hump" proceeds at the same pace for both sexes, since males of both species attain larger sizes than the females the humps on these large males are best developed and most prominent. *S. niger* does not develop such a hump or any other form of sexual dimorphism other than size, which led Choat & Robertson to argue that it is harem-forming, in contrast to other scarids at Heron Island, and that the restriction of spawning to only those females in the harem results in little, if any, selection for male-specific features. Their evidence for harem-formation by *S. niger,* however, consists only of observations of occasional inter-male aggression and the apparent (not demonstrated) attachment of such males to a specific part of the reef. Such observations equally well fit a spawning system based on male defense of limited spawning sites and courtship of passing females. Their hypothesis is further weakened by the subsequent description of demonstrably haremic scarids

that are still strongly dichromatic (Robertson & Warner, 1978).

Another variation from the typical scarid pattern of dichromatism is the development of large female-specific colors in the eastern Pacific species *Scarus compressus.* Small individuals of both sexes are reddish brown with dark streaks along the scale rows and, except for red pelvics, dark fins. At approximately 30 cm SL males develop a terminal pattern of bright green and orange. At approximately 45 cm SL females become light blue (Rosenblatt & Hobson, 1969). The social significance of such, essentially, terminal-phase female coloration has not been investigated.

Finally, sexual dimorphism other than in size and coloration has been documented for a few species. The prominent forehead hump of *Scarus coeruleus* and *S. perrico* is also characteristic of *S. compressus.* The hump is always present in males and becomes proportionately larger with increasing size; in females its development is more variable (Rosenblatt & Hobson, 1969). Males of some species, e.g., *Scarus rubroviolaceus,* also develop filamentous tips on the upper and lower lobes of their caudal fins, a feature usually present but less well developed in smaller males and females.

SPAWNING SEASON

In general, spawning by parrotfishes appears to occur year-round, though it is more frequent and more vigorous in the summer (e.g., Kamiya, 1925; Randall & Randall, 1963; Reeson, 1975; Colin, 1978; Robertson & Warner, 1978; Yogo, et al., 1980; though see also Winn & Bardach, 1957). *Scarus croicensis (= S. iserti)* from off Jamaica, for example, spawns both in January and June, but spawning ascents are six times more frequent in the summer than in the winter (see below) (Colin, 1978). Robertson & Warner (1978) looked in detail at the spawning seasons of scarids off the Atlantic coast of Panama and found that although species as a whole spawned year-round, non-ripe individuals could also be collected year-round. They suggested that individuals, if not species, have discrete non-spawning periods.

No systematic data regarding lunar rhythmicity of scarid spawning activity have yet been reported. It seems noteworthy, however, that Colin's (1978) summer period, in which he saw vigorous spawning activity, centered on a full moon, while his winter period, with low spawning frequency, centered on the moon's first quarter. Although Colin interpreted his summer-winter differences as due to seasonal effects, a full moon peak and first quarter minimum of activity are consistent with lunar spawning rhythms of other reef fishes (see Johannes, 1978), and a lunar explanation for his results cannot yet be discounted. Johannes (1981) also suggested a lunar cycle to parrotfish spawning activity for several species in the South Pacific. Many species of parrotfishes, especially the smaller ones, can be routinely seen spawning at, or shortly after, high tide regardless of the time of the lunar month (pers. obs.), suggesting that lunar cycles, if they occur, are not universal in the family.

REPRODUCTIVE BEHAVIOR

Migration to specific spawning areas on the outer edge of the reef is a widespread characteristic of scarids, though not one that is universal even within some species (e.g., Buckman & Ogden, 1973; Randall, 1963; Barlow, 1975; Choat & Robertson, 1975; Robertson & Warner, 1978). Colin (1978) described the spawning migrations of the small western Atlantic species *Scarus croicensis* in some detail. During the morning large foraging groups composed of up to several hundred initial-phase fish, a few terminal-phase fish, and assorted other herbivores and small predators move about over the shallow reef and periodically descend to the bottom to feed on benthic algae. As afternoon approaches, small groups split off from the larger one and move to a deeper reef coral pinnacle where spawning occurs each day. Such small groups arrive over a "lengthy period of time," and the number of fish gradually increases until group spawning begins in mid-afternoon. After spawning, the fish return to shallow water for the night.

The use of specific spawning sites such as the deep reef coral pinnacle described above may occur daily for years (Colin, 1978). Colin observed spawning at this site regularly over a four-year period and reported that another site used by *Sparisoma rubripinne* that was observed by Randall & Randall (1963) in 1960 was still being used by similar numbers of fish 17 years later.

Scarid spawning characteristically occurs in the afternoon (e.g., Randall & Randall, 1963; Colin, 1978; Robertson & Warner, 1978). In those species that spawn throughout the day (e.g., Barlow, 1975) or at different times of the day at different reefs (e.g., *Sparisoma viride*, according to Robertson & Warner, 1978), time of spawning may be a function of local current patterns rather than time of day. Supporting this hypothesis, Choat & Robertson (1975) found that all spawning observed in Heron Island parrotfishes

occurred at or immediately after peak high tide, the optimum time for dispersal of eggs by tidal currents. Yogo, et al. (1980) reported similar high tide spawning of *Scarus sordidus* off Japan.

As in the wrasses, two forms of spawning (pair spawning and group spawning) occur in parrotfishes, the former involving typically a single terminal-phase male and a female and the latter a number of initial-phase males and females. The relative frequency of the two varies both between species and between locales within species.

Pair spawning has been reported in virtually every species studied, including members of *Cryptotomus*, *Sparisoma*, *Bolbometopon* and *Scarus*, and is probably universal in the family. Its frequency, however, may vary widely depending on local conditions (see below). The behavior involved is similar in all species studied and has been described by Randall (1963), Randall & Randall (1963), Barlow (1975), Buckman & Ogden (1973), Winn & Bardach (1960), Rosenblatt & Hobson (1969), Bruce (1978), and Yogo, et al. (1980). Terminal-phase males characteristically arrive at the spawning site before the females and establish vigorously defended temporary spawning territories. In some species male colors intensify during these bouts and remain intense throughout the spawning period. Shortly after territorial boundaries have more or less stabilized, initial-phase males and females begin arriving in small groups and roam about freely within and between the territories, feeding sporadically. When a group enters his territory a male approaches them while "bob-swimming," i.e., swimming in a smooth up-and-down fashion with the dorsal and anal fins folded and the caudal fin turned up. As he approaches the group the male carefully inspects the fish in it, chases some (presumably the males—Randall & Randall suggested that *Sparisoma viride* and perhaps other scarids can discriminate between sexes based on some olfactory cue), and begins to court others. Courtship involves either circling the female with fins fully spread or bob-swimming vigorously around her. If she is ready she ascends slightly off the bottom, the male moves next to her with his head approximately at the level of her pectoral fin (according to Rosenblatt & Hobson, 1969), and together the pair veer sharply upward in a short fast spawning ascent. One to 3 meters up the fish abruptly reverse direction and return to the bottom, leaving behind a visible whitish cloud of gametes. Winn & Bardach (1960) reported a variation on this ascent in *Sparisoma aurofrenatum* in which the pair

A pair of honeycomb cowfish, *Acanthostracion polygonius,* spawning. Photo by Charles Arneson.

A pair of *Diodon holacanthus* at dusk, the female in front. Photo by Dr. R. E. Thresher.

Two male *Canthigaster bennetti* displaying toward each other. Photo by Roger Steene.

A pair of *Canthigaster punctatissimus* with obvious sexual dichromatism. The male is on the left. Photo by Dr. R. E. Thresher.

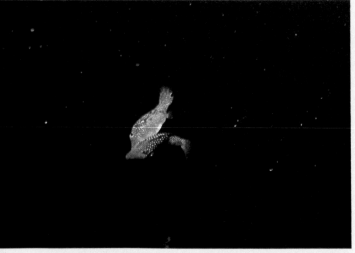

Apparent spawning of a pair of *C. punctatissimus.* The female is hanging onto the male while dashing up into the water. Photo by Dr. R. E. Thresher.

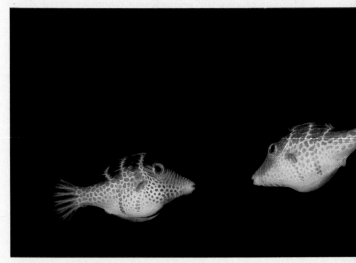

Two male *Canthigaster valentini* fighting. Photo by Dr. J. E. Randall.

Laboratory reared *Diodon holacanthus* 180 days after hatching. At this point it is about 80 mm in total length. Photo by Katsumi Suzuki.

A larval *Chilomycterus* is a very strange creature, as can be seen by this postlarva. It was once treated as a separate genus, *Lyosphaera*, a name that is still used to refer to this larval stage. Photo by Charles Arneson.

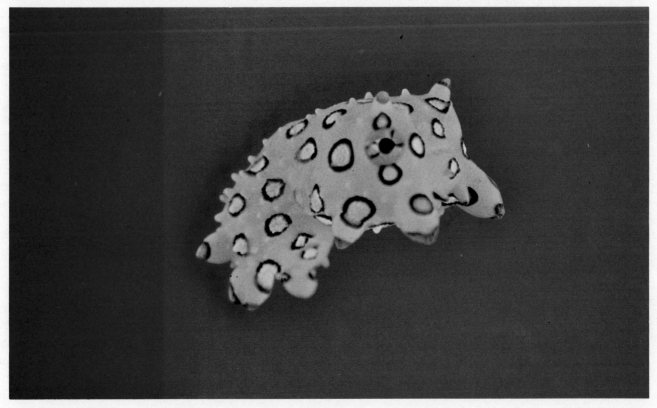

"rotate around and close to each other" on the way up; this has not subsequently been confirmed.

Though characteristically involving an initial-phase female and a terminal-phase male, pair spawning between two initial-phase fish has been reported for several species, all thus far in the genus *Sparisoma* (Barlow, 1975; Robertson & Warner, 1978). Barlow noted that such occurred only during group spawning activities and that the general behavior of the pair was more similar to that of group spawning fish than to pair spawning fish, suggesting that such apparent pair spawnings may actually represent a minimum group spawning rather than the other spawning system. "Streaking" by initial-phase males of terminal-phase pair spawning is also common in some species (Robertson & Warner, 1978).

Group spawning, in which three to approximately 20 initial-phase fish participate en masse in a spawning, has been described for eight species of *Scarus* and two of *Sparisoma* (Randall & Randall, 1963; Barlow, 1975; Choat & Robertson, 1975; Bruce, 1978; Colin, 1978; Robertson & Warner, 1978; Yogo, et al., 1980). Again, the characteristic behavior is virtually identical in all cases where it has been described. Spawning begins shortly after groups of initial-phase males and females arrive at the spawning site and start milling around in large aggregations. Terminal-phase males present occasionally participate (e.g., Choat & Robertson, 1975; Robertson & Warner, 1978) but more often act to disrupt activities, vigorously chasing initial-phase males present and attempting to break up aggregates as they form. Eventually, however, such aggregates reach some minimum size and spawning begins. Small groups break away from the larger mass of fish and begin to swim away from it in increasingly tight formations and with increasing speed. The composition of such groups is not known for sure, but Randall & Randall suggested they are composed of a single female at the lead surrounded by a variable number of males. Still accelerating and tightening into a compact school, the group begins to make abrupt lateral turns and then finally veers upward several meters into the water column. At its peak the fish suddenly reverse direction and dash back to the bottom, leaving behind a cloud of gametes. As in wrasses, ascents are often "epidemic," i.e., after one goes others frequently follow in quick succession. Colin (1978), using frame-by-frame analysis of slow motion films, calculated that the spawning ascent of *Scarus croicensis* takes as little as 0.45 seconds, with the upward movement averaging around 40 km/hour.

The relative mix of pair and group spawning in a population varies widely. *Sparisoma rubripinne*, for example, primarily group spawns in the Virgin Islands (Randall & Randall, 1963), but only pair spawns off Panama (Robertson & Warner, 1978). The lability of scarid socio-sexual systems is well exemplified by studies in various locales of the small and very common western Atlantic species *Scarus croicensis*, to date far and away the most intensively studied species of parrotfish.

In Panama *S. croicensis* has a complex social system based on three classes of individuals: territorials, stationaries, and foragers (Buckman & Ogden, 1973; Ogden & Buckman, 1973). The first group consists of fish that have established permanent territories on the shallow reef, each territory containing a single large dominant female, several subordinate females, and a single male, usually but not always in terminal-phase. Both males and females participate in territorial defense, and spawning (always paired) takes place within the territory. "Stationaries" are initial- and terminal-phase fish that remain in a single area day after day but do not defend that area. Warner & Downs (1977) found that most of these are sexually immature or inactive, i.e., not ripe, fish. Finally, foraging groups of up to 500 individuals each move in large aggregations about consistent parts of the reef. Robertson, et al. (1976) demonstrated that these groups are basically a feeding strategy. By descending en masse to feed, the invaders totally overwhelm territorial defenses by conspecifics and secure a substantial food supply before moving on. Such foraging groups consist mainly of females, about half of which are ripe at any given time; smaller foraging groups in deeper water consist almost entirely of terminal-phase males. Warner & Downs (1977) and Robertson & Warner (1978) interpreted the entire social system of *S. croicensis* as basically haremic. The territorial males are harem-masters within whose territories the females are organized into a size-based dominance hierarchy. Presumably all spawn with the harem-master, who also courts and spawns with females from the foraging groups. The latter, along with initial-phase males, migrate through the territories daily on their way to deeper water, where they participate in pair spawning and, to a limited extent, group spawning. The deepwater foraging groups are interpreted as males concentrating their resources on growth rather than immediate reproduction in order to maximize their chances of eventually controlling a territory and, consequently, a harem. Different strategies are also used by primary and secondary males. Although both develop terminal-phase colors at the same size, primary males reach this size at a younger age and apparently have exchanged reproduction while small (as a female) and long life for faster achievement of large size and control of a territory. The net lifetime fecundity of the two types of males is equal; the two simply use different strategies to reach the same end.

The socio-sexual system described above is complex, internally consistent, and completely different from that described for *S. croicensis* off Puerto Rico by Barlow (1975). Rather than being permanently territorial, the fish on the Puerto Rican reefs establish only temporary spawning territories in a deep-water spawning area to which both terminal-phase and initial-phase fish migrate daily. The social system more than anything else resembles an avian "lek" society in which males compete for spawning territories at a site to which the females come for the sole purpose of spawning. In the Puerto Rican parrotfish almost all spawning is in pairs. In essence, the fish off Puerto Rico seem to have emphasized the "foraging group" part of the Panamanian social system to the exclusion of the territorials, the stationaries, and the small deep-water foraging groups of non-territorial "celibate" males. Randall & Randall (1963) described basically the same social system for *S. croicensis* off the Virgin Islands.

Finally, Colin (1978) described a third spawning system for the fish off Jamaica. As in the system above, the foraging group aspects of the system seem to be emphasized, except now all spawning is group only. Colin reported seeing only one territorial terminal-phase male, but also reported that he was never observed spawning.

These three different spawning systems by the same species in three different areas suggest a basic flexibility of the scarid socio-sexual system, with different

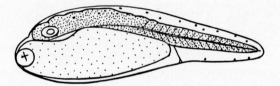

Newly hatched larva

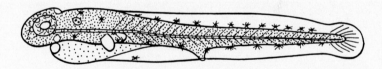

Approx. 1 day old

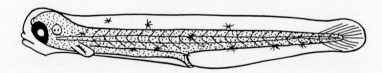

Approx. 2 days old

Early stages in the larval development of *Sparisoma rubripinne*. Based on Randall & Randall (1963).

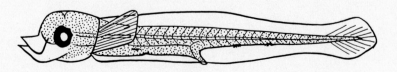

Approx. 6 days old

237

A male neon goby, *Gobiosoma oceanops,* digging a nest in an aquarium. He is using a piece of PVC pipe as part of his burrow. Photo by Dr. P. L. Colin.

An egg-laden female following the male into his nest. Photo by Dr. P. L. Colin.

A pair of *Gobiosoma xanthiprora* males on a piece of PVC pipe. The dusky individual is guarding eggs, having assumed the characteristic brooding coloration. Photo by Dr. P. L. Colin.

"Mated" pair of *Gobiosoma evelynae* in an aquarium. The male (in the pipe) is guarding eggs. Photo by Dr. P. L. Colin.

Male *Gobiosoma genie* guarding eggs in a PVC pipe. Photo by Dr. P. L. Colin.

Laboratory-reared hybrid male goby. The cross was between a *Gobiosoma oceanops* female and a *G. evelynae* male. Photo by Dr. P. L. Colin.

aspects of the basic underlying system being manifest under different combinations of population density, availability of and competition for spawning sites and other resources, and possibly even such simple factors as season and water temperature. Robertson & Warner pointed out that one explanation for the continued success of so many different types of sexuality in scarids (primary males, secondary males, initial-phase versus terminal-phase males, and so on) is that each type may be successful in a different habitat and that generalizations made about a social system based on examinations of only a few habitats may be premature. The inherent flexibility of the scarid reproductive system not only requires caution in interpretation of the system, but it may also account for the extreme success of the group on coral reefs.

EGGS AND LARVAE

Little recent work has been done on the development of either scarid eggs or larvae. According to Winn & Bardach (1960) the pelagic eggs of western Atlantic species differ between the two subfamilies. Scarine eggs (*Scarus croicensis, S. guacamaia,* and *S. vetula*) are spindle-shaped, 2.4 to 3.1 mm long, and transparent with a yellow or white oil droplet. Similar eggs are reported for two Hawaiian species of scarids by Watson & Leis (1974). Sparisomine eggs (*Sparisoma aurofrenatum, S. chrysopterum, S. viride,* and, according to Randall & Randall, 1963, *S. rubripinne*) are spherical, between 0.6 and 1.1 mm in diameter, and have a single yellow or orange oil droplet.

Parrotfish eggs hatch in approximately 25 hours at around 26°C (Kamiya, 1925; Cipria, 1937; Randall & Randall, 1963). Newly hatched larvae of *S. rubripinne* are 1.7 mm long, lack functional eyes or a mouth, are largely unpigmented, and have a prominent yolk sac that has a single oil droplet at its anteriormost point. They are positively buoyant and are weak swimmers. By day three the yolk and oil droplet disappear, the mouth and anus open, and the larvae begin to feed. The duration of the planktonic larval stage is not known.

Juvenile parrotfishes are typically drab and are often horizontally striped (e.g., Masuda & Tanaka, 1962). Many are similar in appearance, so good keys and a specimen in hand are often necessary to identify them to species level (see, for example, Böhlke & Chaplin, 1968). Such juveniles are usually found in small foraging groups and are one of the prime food sources of predatory reef fishes.

REPRODUCTION IN CAPTIVITY

The only report of anything resembling spawning of parrotfishes in captivity is by Breder & Rosen (1966), who reported observing pairs of *Sparisoma radians,* a small western Atlantic species, assuming a vent-to-vent position at the bottom of an aquarium and making sudden lunges toward the surface. No details about aquarium conditions or conditions of the fish were given, and it appears that none of the lunges resulted in release of gametes.

Due to their large size at maturity (even the smallest parrotfishes are typically 6 to 10 cm when mature as a female and larger than that as a mature male) and the space requirements for successful spawning, it is unlikely that parrotfishes will be spawned in anything less than large public aquaria.

There are no reports of successful rearing of parrotfish larvae in captivity.

Literature Cited

Barlow, G.W. 1975. On the sociobiology of four Puerto Rican parrotfishes (Scaridae). *Mar. Biol.,* 33:281-293.

Bebars, M.J. 1978. *Scarus ghardaquensis,* N.sp., a new parrotfish (Pisces, Scaridae) from the Red Sea, with a note on sexual dichromatism in the family. *Cybium,* Ser. 3:76-81.

Böhlke, J.E. and C.C.G. Chaplin. 1968. *Fishes of the Bahamas and adjacent tropical waters.* Livingston Publ., Wynnewood, Pa. 771 pp.

Breder, C.M., Jr. and D.E. Rosen. 1966. *Modes of Reproduction in Fishes.* Natural History Press, Garden City, N.Y. 941 pp.

Bruce, R.W. 1978. Intraspecific organization in parrot fish. *Physiology and Behavior of Marine Organisms, Proc. 12th Europ. Symp., Mar. Biol.:*229-236.

Buckman, N.S. and J.C. Ogden. 1973. Territorial behavior of the striped parrotfish, *Scarus croicensis* Bloch (Scaridae). *Ecology,* 54:1377-1382.

Choat, J.H. 1969. Studies on the biology of labroid fishes (Labridae and Scaridae) at Heron Island, Great Barrier Reef. Part I. Taxonomy and reproductive behavior. Ph. D. Diss., Univ. Queensland. 294 pp.

Choat, J.H. and D.R. Robertson. 1975. Protogynous hermaphroditism in fishes of the family Scaridae. Pp. 263-283. *In: Intersexuality in the Animal Kingdom.* (R. Reinboth, Ed.). Springer Verlag, Heidelberg, Germany.

Cipria, G. 1939. Uova, sviluppo embrionale e postembrionale di *Scarus cretensis* O.V. ottenute per fecondazione artificiale. *Mem. Roy. Com. Talassogr. Ital.,* (241):5 pp.

Colin, P.L. 1978. Daily and summer-winter variation in mass spawning of the striped parrotfish, *Scarus croicensis. Fish. Bull. (U.S.),* 76:117-124.

Johannes, R.E. 1978. Reproductive strategies of coastal marine fishes in the tropics. *Env. Biol. Fishes,* 3:65-84

Johannes, R.E. 1981. *Words of the Lagoon.* Univ. Calif. Press, Los Angeles, Calif. 320 pp.

Kamiya, T. 1925. The pelagic eggs and larvae of fishes in the Tateyanua Bay (Pref. Chiba). *Third Rept. J. Imp. Fish. Inst., Tokyo,* 21:27-29.

Masuda, T. and K. Tanaka. 1962. Young of labroid and scaroid fishes from the central Pacific coasts of Japan. *J. Tokyo Univ. Fish.,* 48:1-98.

Ogden, J.C. and N.S. Buckman. 1973. Movements, foraging groups, and diurnal migration of the striped parrotfish, *Scarus croicensis* Bloch (Scaridae). *Ecology,* 54:589-596.

Randall, J.E. 1963. Notes on the systematics of parrotfishes (Scaridae), with emphasis on sexual dichromatism. *Copeia,* 1963:225-237.

Randall, J.E. 1968. *Caribbean Reef Fishes.* T.F.H. Publ., Neptune City, New Jersey. 318 pp.

Randall, J.E. and H.A. Randall. 1963. The spawning and early development of the Atlantic parrotfish, *Sparisoma rubripinne,* with notes on other scarid and labrid fishes. *Zoologica, N.Y.,* 48:49-60.

Reeson, P.H. 1975. The biology, ecology, and bionomics of Caribbean Reef Fishes: Scaridae (Parrotfishes). *Res. Rpt. Zool. Dept., Univ. West Indies,* (3):49 pp.

Reinboth, R. 1968. Protogynie bei Papageifischen (Scaridae). *Z. Naturforsch., ser. B,* 23:852-855.

Robertson, D.R. and R.R. Warner. 1978. Sexual patterns of the labroid fishes of the western Caribbean, II. The parrotfishes (Scaridae). *Smithsonian Contrib. Zool.,* (255):1-26.

Robertson, D.R., H.P.A. Sweatmen, E.A. Fletcher, and M.G. Clelland. 1976. Schooling as a mechanism for circumventing the territoriality of competitors. *Ecology,* 57:1208-1220.

Rosenblatt, R.H. and E.S. Hobson. 1969. Parrotfishes (Scaridae) of the eastern Pacific, with a generic rearrangement of the Scarinae. *Copeia,* 1969:434-453.

Schultz, L.P. 1958. Review of the parrotfishes, family Scaridae. *Bull U.S. Nat. Mus.,* (214):1-143.

Straughn, S. 1978. Community structure among reef fish in the Gulf of California: the use of reef space and interspecific foraging associates. Ph. D. Diss., Univ. California, Davis.

Warner, R.R. and I.S. Downs. 1977. Comparative life histories: growth vs. reproduction in normal males and sex-changing hermaphrodites of the striped parrotfish, *Scarus croicensis. Proc. Third Internat. Coral Reef Symp.,* 1:275-281.

Watson, W. and J.M. Leis. 1974. Ichthyoplankton of Kaneohe Bay, Hawaii. *Sea Grant Tech. Rpt.* TR-75-01.

Winn, H.E. & J.E. Bardach. 1960. Some aspects of the comparative biology of parrotfishes at Bermuda. *Zoologica, N.Y.,* 45:29-34.

Yogo, Y., A. Nakazono, and H. Tsukahara. 1980. Ecological studies on the spawning of the parrotfish, *Scarus sordidus* Forsskal. *Sci. Bull. Fac. Agric., Kyushu Univ.,* 34:105-114.

Tenacigobius yongei in typical position on a sea whip. Photo by Roger Steene.

An adult pair of *Valencienna* sp. The female is the heavier (lower) fish. Photo by Dr. R. E. Thresher.

Tenacigobius yongei eggs on a cleared section of the sea whip. Photo by Roger Steene.

Male goby tending eggs laid on a partially uncovered rock. Photo by Ann Gronell.

A pair of *Tenacigobius* sp. with eggs on a section of sea whip. Photo by Roger Steene.

Gobiosoma oceanops tending eggs on the side glass of a pet-shop aquarium. Photo by Dr. R. E. Thresher.

A male *Emblemaria pandionis* displaying to a female with his raised dorsal fin. Photo by Dr. R. E. Thresher.

Ecsenius mandibularis males in courting colors competing for a female. Photo by Dr. R. E. Thresher.

Banded courtship color pattern of male *Ophioblennius steindachneri*. Photo by Dr. R. E. Thresher.

Male *Chaenopsis alepidota* in courtship display. Photo by Dr. R. E. Thresher.

A nest in the sand prepared by the triggerfish *Sufflamen verres*. Photo by Dr. R. E. Thresher.

A female *Sufflamen verres* tending the nest. Photo by Dr. R. E. Thresher.

Angelfishes
(Pomacanthidae)

Angelfishes are among the most colorful and widely recognized of all reef fishes. Few other fishes can rival the color and majesty of a pair of angelfish swimming over a reef, and it is little wonder that many have common names like queen, king, and emperor angelfish. In general appearance, angelfishes look much like butterflyfishes, to which in the past they have been considered closely related; more recently they have proven to be more distinct (see Freihofer, 1963; Burgess, 1974). Fishes in both families are strongly laterally compressed, are round in profile, and have continuous dorsal fins. To some extent they even swim and act alike. Angelfishes and butterflyfishes can readily be told apart, however, on two points: angelfishes are usually more brightly colored than butterflyfishes, and angelfishes always have a strongly developed preopercular spine that is always lacking in butterflyfishes.

Considering the circumtropical distribution of the family, the pomacanthids are a relatively small group with only seven widely recognized genera: *Centropyge, Chaetodontoplus, Euxiphipops, Genicanthus, Holacanthus, Pomacanthus,* and *Pygoplites* (Fraser-Brunner, 1933). Smith (1955) raised several subgenera to the generic level (e.g., *Pomacanthops* and *Apolemichthys*), and several workers have suggested that some genera should be merged (e.g., Tominga & Yasuda, 1973; Randall, 1975), but none of these recommendations have been widely accepted. Paralleling the relatively few genera, there are also relatively few species. *Centropyge,* the pigmy angelfishes, is by far the most speciose genus, with 29 recognized species (Randall & Klausewitz, 1976; Randall & Yasuda, 1979); at the other extreme, *Pygoplites* is monotypic, containing only the striking *P. diacanthus.* The remaining five genera consist of less than a dozen species each.

The pomacanthids are one of the best examples of the recent surge in interest in and data on reproduction of coral reef fishes. Prior to 1970 virtually nothing was known about the group despite their conspicuousness and prominence on the reef. Breder & Rosen (1966), for example, could cite only a single pair of references on angelfishes (Straughan, 1959a & b), both of which dealt with a vague description of the spawning of *Pomacanthus arcuatus* in captivity. Within the last decade, however, considerable and often detailed information has become available on the spawning of five of the seven genera. As yet, relatively little is known about *Euxiphipops* and *Pygoplites.*

The vigor of research on reproduction by pomacanthids is amply testified to by the appearance of several major papers on the group since the final version of this chapter was prepared. These papers could not be effectively integrated into the chapter at such a late date, but should be consulted by those interested in the family. They include Bauer & Bauer's (1981) study of field and aquarium spawning by several species of *Centropyge,* Neudecker & Lobel's (1982) report on spawning and social behavior of, principally, *Holacanthus tricolor* off St. Croix, Virgin Islands, and Thresher's (1982) study of reproduction by the emperor angelfish, *Pomacanthus imperator,* and several species of *Centropyge* at Enewetak Atoll. The last paper also reviews the occurence of sexual dichromatism and lunar spawning cycles in the family. Moreover, several important papers are currently in preparation. These include J. Aldenhoven's detailed study of the social behavior and reproductive biology of *Centropyge bicolor* at Lizard Island, Great Barrier Reef (Ph. D. Dissertation, MacQuarie University), A. Gronell & P. Colin's report on spawning and social behavior of *Pygoplites diacanthus* at Enewetak Atoll, and J.T. Moyer's field observations on spawning by *Genicanthus* in the Philippines.

SEXUAL DIMORPHISM

The amount and kind of sexual dimorphism exhibited in pomacanthids varies widely, from *Euxiphipops* and *Pomacanthus,* which so far as is known are monomorphic, to *Genicanthus,* all species of which exhibit pronounced sexual differences in size, shape, and coloration. External sexual dimorphism has thus far been reported for species in four genera (*Centropyge, Genicanthus, Holacanthus,* and *Chaetodontoplus*) but is universal only in *Genicanthus.*

The most detailed work to date on sexual differences has been done on various species of *Centropyge.* At least some, and possibly all, species in the genus are protogynous (female to male) hermaphrodites. Moyer & Nakazono (1978) demonstrated that in the western Pacific species *Centropyge interruptus* all males are derived from females and so are larger than females. Of 40 specimens examined, males ranged in size from 11.2 to 15.0 cm, whereas the largest female was only 13.3 cm. Pronounced size differences between the sexes also seem to occur in other species of *Centropyge* (in all reports of spawning the male is invariably the largest individual present), which suggests that similar protogyny is widespread in the genus (e.g., Randall & Yasuda, 1979).

Sex change in *Centropyge interruptus* is a socially controlled phenomenon similar to that of many wrasses. The angelfish social system is based on male dominance and defense of a harem of one to four females. Removal of the male occasionally results in a takeover of the harem by a neighboring male, especially if the harem is a small one, but most often results in the largest dominant female changing sex. Within seven days an ex-female can be acting much like a male, and within 20 days she can be fully functional as

a male. Histological examination of the Y-shaped gonads indicates that the ovaries are much like those of other teleosts; that is, they consist of a central lumen surrounded by involuted egg-producing membranes. They differ, however, in having an exterior membrane that is conspicuously thickened in several areas. These thickened areas are hollow and are surrounded by inactive spermatogonia. Upon sex change these spermatogonia proliferate, the thickened areas expand, and the ovaries collapse and degenerate.

Along with size differences, color and morphometric differences between the sizes have also been reported for a few species of *Centropyge.* Males of *C. interruptus* and *C. ferrugatus* have horizontal black and blue stripes on the soft portions of their dorsal and anal fins and heavy blue lines on their opercula; females lack both marks (Moyer & Nakazono, 1978). Males of *C. shepardi* also have short blue lines on the rear edges of their dorsal and anal fins, though they lack the opercular markings (Randall & Yasuda, 1979). Males of *C. tibicen* are dark blue in contrast to the black females (J. Moyer, pers. comm.). Other forms of conspicuous within-species color variation occur in a number of species and may represent undiscovered sexual dimorphism. Examples include the western Atlantic *C. argi,* which varies widely in the extent of blue and yellow on the head and in the completeness of the blue eye ring; *C. heraldi,* in which some individuals have prominent dark marks around the eyes and others do not (Masuda, et al., 1975, indicated only the males have the dark marks); and *C.*

Adult pair of *Euxiphipops sexstriatus* at Lizard Island. The male (larger fish) is in front of the female (smaller fish). This photo shows the sexual size differences typical of the species. Photo by Jack T. Moyer.

The courtship display of a male *Chaenopsis ocellata.* Note the bright orange spot at the anterior end of the enlarged dorsal fin. Photo by Charles Arneson.

Two males of the eastern Pacific *Emblemaria hypacanthus.* The male on the left is displaying to a drably colored female nearby (not visible in the photo). Photo by Alex Kerstitch at Guaymas, Mexico.

Exallias brevis guarding eggs. It will not abandon them even though a diver's finger is poked into the nest. Photo by Bruce Carlson at Hanauma Bay, Hawaii.

Variations in the pattern of the blenny *Blennius zvonimiri:* a) female; b) male; c & d) male collected at the time of reproduction under an overhang covered with sponges and encrusting algae. Photos by P. Louisy in an aquarium, courtesy Dr. D. Terver, Nancy Aquarium, France.

A

B

C

D

flavicauda, in which some individuals have dark-edged caudal fins and others uniformly white fins. Among the various monochromatic species some appear to develop temporary sexual dichromatism during courtship and spawning. Bauer & Klaij (1974), for example, reported that females of *C. bispinosus* lose their dark bars during spawning and pale to a uniform red; males, on the other hand, retain their normal colors. A similar lightening by the female occurs in *C. vroliki*, as well as lightening by both males and females in *C. tibicen*. At least a few species, however, are neither permanently nor temporarily dichromatic. Examples include *C. bicolor* and *C. potteri*. Why there should be so much variation in patterns of sex-specific coloration in the genus is not known.

Aside from color differences, there are also two reports of morphometric differences between the sexes in *Centropyge*. In *C. joculator*, an Indian Ocean endemic similar in color to the widely distributed *C. bicolor*, males have longer and more pointed dorsal and anal fins than females (Smith-Vaniz & Randall, 1974). In two Hawaiian species, *C. potteri* and *C. fisheri*, Lobel (1975) suggested that males have fewer but larger spines than females above the single prominent preopercular spine characteristic of the family. This difference may be size-related rather than sexually specific.

Genicanthus are also protogynous hermaphrodites according to Shen & Liu (1976), Debelius (1978), Suzuki, et al. (1978), and Bruce (1980), though they differ in gonad structure from *Centropyge* (suggesting an independent origin of hermaphroditism). Ovaries are paired and united ventrally, whereas testes are ball-shaped with a long projection extending anteriorly. Upon sex inversion spermatogenic tissue develops throughout the degenerating ovaries rather than only in previously dormant areas along the periphery. Not surprisingly, considering the hermaphroditism, *Genicanthus* males are larger than females. Male *G. semifasciatus*, for example, average 102.5 mm; females average only 76.2 mm (Suzuki, et al., 1978).

Aside from size differences, the nine species of *Genicanthus* also exhibit marked sexual differences in color pattern and morphology (reviewed by Randall, 1975). Details vary from species to species, but in general males are usually horizontally or vertically striped, are brightly colored, and have pronounced lunate caudal fins with "streamers" trailing off the upper and lower lobes. Females generally lack stripes, are more drably colored, have dark upper and lower margins on the caudal fin, and either lack or have less

pronounced caudal fin streamers.

Reports of sexual dimorphism in two additional genera, *Holacanthus* and *Chaetodontoplus*, have been made on a more scattered basis, and it appears that most members of both genera lack such dimorphism. In *Holacanthus* the Y-shaped gonads lie along the rear wall of the abdominal cavity and join ventrally to form a pore located just behind the anus (Feddern, 1968). Ovaries are short, thick, and cylindrical; testes are long and ribbon-like. Permanent external sexual dimorphism has thus far been reported for only one species of *Holacanthus*, *H. passer* in the eastern Pacific. In this strikingly colored species males are both larger (up to 50% longer and 300% heavier) than females (Strand, 1978) and differently colored (males have snow white pelvic fins; females have pale yellow pelvic fins). Similar size differences between the sexes are likely in the western Atlantic species of *Holacanthus* (which are commonly found in pairs consisting of one large and one small individual) and may turn out to be widespread in the genus. A consistent size difference also suggests sequential hermaphroditism, though such has not yet been looked for in the genus. Otherwise, the only other report of possible sexual dimorphism is by Debelius (1977) for the eastern Atlantic *Holacanthus africanus*. Large individuals (males?) typically have a broad white band on each size of the body that is lacking in smaller individuals (females?).

In *Chaetodontoplus* only two species have thus far been reported as sexually dimorphic. Males of *C. melanosoma* have a black caudal fin edged with yellow; females have all-yellow caudal fins (Fraser-Brunner, 1933). Similarly, male *C. duboulayi* have a more conspicuous white patch behind the eye than do females (Steene, 1978). It seems likely that other members of this genus will eventually be proved dichromatic based on the occurrence of within-species variation in color pattern. *C. personifer*, for example, is commonly found in pairs consisting of a large individual that has a broad white stripe behind its head and a slightly smaller fish that lacks such a stripe.

SPAWNING SEASONS

Although relatively little data are available about spawning seasons of angelfishes, what little there is almost invariably suggests a strong seasonal change in activity. In temperate regions spawning characteristically occurs in the warmest months of the year: *Holacanthus passer* spawns in the summer and fall in the Gulf of California (Moyer, et al., in press) and

Centropyge interruptus spawns in the summer when water temperatures exceed 27°C off southern Japan (Moyer & Nakazono, 1978). The only exception to summer spawning in temperate areas is reported by Suzuki, et al. (1978), who gave fall through spring as the spawning period for two species of *Genicanthus;* all of their observations were made in aquaria, however, and the reported spawning seasons may well be an artifact induced by artificially maintained high water temperatures in captivity. Deeper in the tropics, spawning by angelfishes is also seasonal but seems to peak while water temperatures are still rising. Spring and early summer spawning is the rule for a number of species of *Centropyge* at One Tree Island, Great Barrier Reef, for *Centropyge potteri* off Hawaii (Lobel, 1978), and for several western Atlantic species of *Holacanthus* (Munro, et al., 1973). Spawning seasons of the Atlantic species, however, may vary with locale, since Feddern (1968) found year-round spawning of two species off Florida and Moyer, et al. (in press) found winter and early spring spawning to be the rule off Puerto Rico.

Possible lunar rhythms to spawning activity in angelfishes have been specifically looked for in two species of *Centropyge*, with opposite results. According to Lobel (1978), *Centropyge potteri* spawns only during the week prior to the full moon; *C. interruptus* off Japan, however, spawns daily and shows no conspicuous lunar cycle to its activities (Moyer & Nakazono, 1978). Daily spawning throughout the lunar cycle is also the rule for *Genicanthus lamarck* and *G. semifasciatus* (Suzuki, et al., 1978) and *Holacanthus passer*. In general, lunar spawning cycles do not appear to be common in angelfishes, and daily spawnings throughout all or most of the lunar month are likely to prove to be the rule.

REPRODUCTIVE BEHAVIOR

Spawning by angelfishes varies in detail from genus to genus and to a lesser extent between species within genera, but it incorporates several general features. Spawning, so far as is known, always occurs at dusk and is apparently stimulated by falling light levels. It also occurs slightly off the bottom and ends with a spiraling or straight, relatively slow ascent into the water column to release eggs and sperm. Finally, spawning always involves a single pair of fish. Although males in many species spawn successively with several females and may compete for access to these females, there are no reports in the family of group spawning or even "cheating" by smaller males.

The only pomacanthid genus for which a sizable literature on spawning behavior exists is *Centropyge*. Spawnings in aquaria have been described by Ballard (1970), Bauer & Klaij (1974), Lobel (1974, 1975), Lubbock (cited in Steene, 1978), and J.A. Bauer & S.E. Bauer (1980); spawning in the field has been described by Lobel (1978) and Moyer & Nakazono (1978), both of which go into great detail.

The basic spawning unit for most species of *Centropyge* seems to be a male-dominated harem, with the male larger than any of the females and derived from the previously dominant female by sex change. Pairing is also common in some species, such as *Centropyge argi* and, in some areas, *C. vroliki*, with pairs often composed of one large and one small individual. Lobel (1978) examined in detail the social structure of *C. potteri* and found that the size of the social unit varies with habitat: on patch reefs females outnumber males and each male spawns with several females (suggesting a harem system, though Lobel did not demonstrate such), while on more extensive reefs the sex ratio was closer to one to one, paired fish were commonly seen, and each male spawned with only one partner. The basis for such different spawning systems was not determined, but Lobel suggested that on more extensive reefs there were a greater number of spawning sites available, which permitted spawning by several males rather than just a few. Moyer & Nakazono (1978) expanded on this hypothesis, suggesting other limiting factors that may be involved and further suggested that such flexibility may be characteristic of *Centropyge*.

Spawning by *Centropyge*, like that of other pomacanthids, occurs at dusk. Moyer & Nakazono (1978) reported that courtship by *C. interruptus* begins about 30 minutes before sunset, with spawning itself typically occurring between ten minutes before and five minutes after sunset. On overcast evenings or when the water was murky spawning invariably occurred earlier than normal, suggesting that light levels are used by the fish to time their periods of spawning activity. Spawning does not, however, occur at dawn.

Moyer & Nakazono divided the spawning sequence of *Centropyge interruptus* into six action patterns; these patterns are similar, if not identical, to courtship described by other workers. The first two patterns, rushing and circling, are basically aggressive and are the means by which the male maintains his dominance over the harem. Spawning itself begins with "soaring." The male approaches a female and swims one or two meters off the bottom directly above her and

Male *Axoclinus carminalis* showing courtship behavior and colors (red with black tail) in front of female. Photo by Dr. R. E. Thresher.

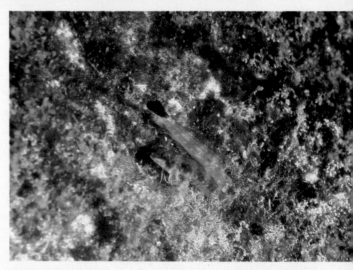

Male *A. carminalis* moving into position to spawn with the female. Photo by Dr. R. E. Thresher.

Actual spawning of *A. carminalis.* Photo by Dr. R. E. Thresher.

Malacoctenus margarite spawning. The male is the darker of the two fish. Photo by Dr. R. E. Thresher.

Salarias fasciatus male tending eggs laid in a clam shell. Photo by Dr. R. E. Thresher.

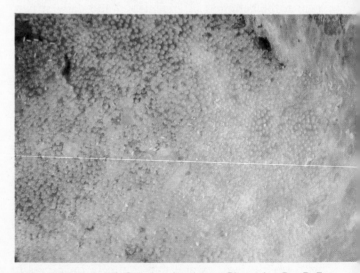

Newly laid eggs of *Salarias fasciatus.* Photo by Dr. R. E. Thresher.

250

Male *Tripterygion nasus* with fins expanded and in bright red courtship coloration. Photo by Stanislav Frank.

Unidentified clingfish tending eggs. Photo by Wade Doak.

Courtship and spawning of the eastern Pacific angelfish, *Holacanthus passer.* (a) The males, which are larger than the females, patrol individual territories near the outer edge of the reef. When a female approaches, the males swim above her, each with his brilliantly white pelvic fins erected, and attempt to lead her up off the bottom. Once the female has ascended, the male "dances" back and forth ahead of and slightly above her (b) with his fins fully spread. If the female is ready to spawn, she ascends toward the surface, following the "dancing" male, who eventually moves behind and below the female and places his head alongside her abdomen (c). Gametes are shed in this position as the pair spirals toward the surface. After gametes are shed, the pair split apart and dash for the bottom (d).

hovers there, making only slight stabilizing movements with his pectoral and caudal fins. He slowly leans over toward the female until he lies at an angle of 45 to 90° with his fins fully spread. If the female is ready, she ascends off the bottom toward him and also begins to soar, either in front of or beside him (a stage referred to as "mutual soaring"). The male circles behind and below the female and "nuzzles" her, placing his snout against her abdomen. The female frequently seems hesitant at this point and may move off slightly, forcing the male to attempt nuzzling over and over again until she is ready. Finally the position is achieved and is held for 2 to 18 seconds, after which the male opens and closes his mouth and flutters his pectoral fins quickly. Spawning follows immediately, with both fish shedding their gametes simultaneously in a single burst. Afterward both dash for the bottom, with the male occasionally chasing the female before going off to spawn with another.

Spawning by other species of *Centropyge* is either identical to that described above (e.g., *C. tibicen, C. bicolor*) or varies from it only slightly. *C. bispinosus, C. vroliki,* and *C. potteri,* for example, perform the same motor patterns but also change color slightly during spawning. *C. fisheri,* instead of hovering, spirals slightly toward the surface while nuzzling (and so is similar to *Holacanthus,* described below) (Ballard, 1970). Lobel (1978) reported hearing various sounds ("clicks and grunts") during the courtship of *C. potteri,* while *C. tibicen* "rattles" during its descent (J. Moyer, pers. comm.). The precise nature and possible role of these sounds are not known.

Spawning by *Holacanthus* has thus far been described only by Moyer et al. (in press), who deal with three American species: *Holacanthus ciliaris* and *H. tricolor* in the Caribbean and *H. passer* in the Gulf of California. Spawning behavior is virtually identical in all three and is similar to that of *Centropyge,* although

"Inciting" behavior during courtship of *Genicanthus semifasciatus.* The male is in front of the female. Photo by Katsumi Suzuki.

"Nuzzling" during courtship of *Genicanthus semifasciatus.* The male is pushing against the abdomen of the female. Photo by Katsumi Suzuki.

Illustrations of male and female Richardson's dragonet, *Callionymus richardsoni,* showing sexual dimorphism. Illustrations by Arita.

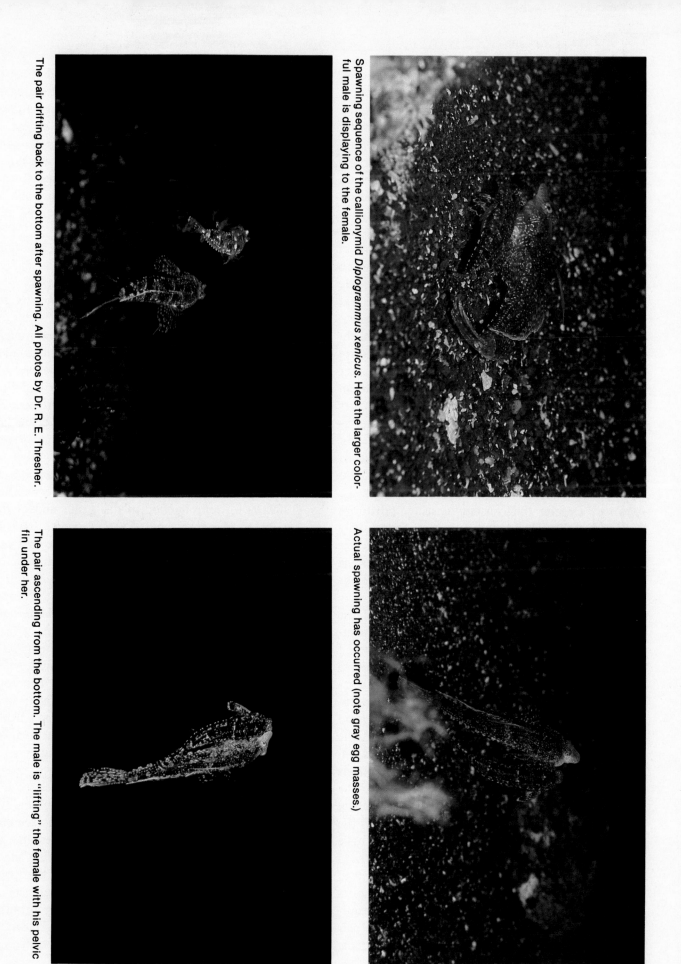

Spawning sequence of the callionymid *Diplogrammus xenicus*. Here the larger colorful male is displaying to the female.

The pair drifting back to the bottom after spawning. All photos by Dr. R. E. Thresher.

Actual spawning has occurred (note gray egg masses.)

The pair ascending from the bottom. The male is "lifting" the female with his pelvic fin under her.

the social systems appear different. *H. ciliaris* and *H. tricolor* are often found in triplets or small groups, suggesting that spawning may normally be haremic. During spawning periods males court as many females as are present and spawn repeatedly. *Holacanthus passer* carries this a step further, with pairs occurring only in marginal, low density areas (Strand, 1978). In preferred habitats, the sexes are found on different parts of the reef and come together only for spawning. The males, larger than the females and bearing brilliant white pelvic fins, forage for plankton in the water column and move about the reef in groups of up to 100 fish; the females, with their pale yellow pelvic fins, remain in a smaller part of the reef and are more bottom-oriented than the males. Spawning occurs at dusk and is preceded by the establishment by the males of temporary spawning territories around a prominent coral outcrop along the outer edge of the reef. The overall system is similar to an avian "lek" (see Loiselle & Barlow, 1978), with males arriving early and vigorously chasing one another as each maneuvers to establish and maintain the best spawning area. Even after spawning begins chases occur sporadically and territories are not abandoned until well after dark and long after the last female has spawned. Despite the constant chasing, territories seem very stable. During three weeks of almost daily observations the same males (recognized based on size and minor differences in markings) occupied the same territories, with the largest male defending the center area and doing the most spawning.

After territorial disputes are more or less settled, the males ascend into the water column and await the arrival of females. These usually approach the spawning area in groups of three or four, beginning about 30 minutes after sunset (about 40 minutes before complete darkness). Courtship begins with a male swimming toward a female then turning away from her and ascending ahead of her into the water column. As he swims, he spreads fully his vivid white pelvic fins and "flutters" with short quivering motions of his body as he swims back and forth in front of the female. If she is ready she follows behind him with her pale yellow pelvic fins spread, and the pair slowly approach the surface. Five to six meters below the surface the male circles behind and below the female, places his head on her abdomen (in a position virtually identical to "nuzzling" in *Centropyge*), and the pair together spiral slowly toward the surface. Gametes are simultaneously shed at the top of the spiral, after which the pair split apart and head for the bottom. Each female

spawns only once and after spawning heads back to the reef; the male, in contrast, begins courting again immediately and continues to do so as long as there are females present. A successful male may spawn six to eight times in an evening. A small male near the periphery of the group may spawn only once or not at all. By ten minutes before complete darkness the action is over for the evening, and after five or six minutes of chasing one another the males, one at a time, depart for the reef.

Spawning by *Genicanthus*, based on aquarium observations, is basically similar to that of the previous two genera and involves "soaring" and "nuzzling" by the courting male and a slow straight swim toward the surface where gametes are shed (Suzuki, et al., 1978). The male "incites" spawning in the female by swimming ahead of her and placing his folded caudal fin immediately in front of her head (a behavior which sounds similar to the fluttering swimming of courting male *H. passer*). If ready, the female ascends, while the male swims behind and below her to assume the spawning position. Each male spawns several times, while females spawn once nightly.

Finally, despite an early report of spawning of *Pomacanthus arcuatus* in captivity by Straughan (1959a & b), reproduction in this genus and in *Euxiphipops*, *Pygoplites*, and *Chaetodontoplus* has not been examined in detail. Individuals in the first two genera are commonly seen in pairs, which may indicate monogamy. Such pairs often consist of similar-sized individuals and do not appear to be territorial. Spawning by the western Atlantic *Pomacanthus paru* appears to confirm monogamous mating in at least one of these genera. As described in Thresher (1980), pair spawning by *P. paru* occurs at dusk, with numerous pairs visible scattered about the reef. Members of each pair were vigorously aggressive toward stray individuals and other pairs such that a regular spacing between pairs was maintained. Periodically, the members of a pair brought their venters close together and ascended in a broad, shallow arc off the bottom. No gamete release was visible, but it would have been difficult to observe in any case due to the dim lighting. Recent observations by J. Moyer (pers. comm.), however, suggest that an assumption of monogamy for typically pairing angelfishes may be premature. Moyer observed that pairs of *Pomacanthus arcuatus* off Panama aggregate at dusk, with members regularly exchanging between pairs. Actual spawning was not observed. Spawning by *Chaetodontoplus* has also not been observed, although Yasuda (cited in Moyer & Nakazono, 1978) reported a harem-like social system in the fish. Finally, J. Aldenhoven (pers. comm.) has recently witnessed spawning by *Euxiphipops sexstriatus* and *Pygoplites diacanthus* at dusk off Lizard Island, Great Barrier Reef. The *Euxiphipops* pair spawned in a manner similar to *Holacanthus passer,* with the pair spiraling to the surface before dashing downward and releasing a cloud of gametes. In *P. diacanthus,* one male and two females (based on size and behavior) were witnessed nuzzling, fluttering, and soaring in a manner very similar to various species of *Centropyge.* No gamete release was visible.

The only estimate of fecundity in the family was provided by S.E. Bauer & J.A. Bauer (1980), who estimated that female *Centropyge* release 300 to 500 eggs each evening.

EGGS AND LARVAE

Based on a sample of only a few species, pomacanthid eggs are small, spherical, nearly transparent, and contain from one to several oil droplets. Egg diameter seems to vary among genera, though there is clearly broad overlap. Known egg diameters are as follows: *Centropyge*—0.6 to 0.68 mm (Bauer & Klaij, 1974; Bauer, 1975); *Genicanthus*—0.75 to 0.80 mm (Suzuki, et al., 1978); *Holacanthus*—0.65 to 0.85 mm (Moe, 1976); *Chaetodontoplus*—0.82 to 0.88 mm (Fujita & Mito, 1960); and *Pomacanthus*—approximately 0.90 mm (Moe, 1976, 1977). Details of egg development are provided by Fujita & Mito (1960), Moe (1977), Suzuki, et al. (1978), and S.E. Bauer & J.A. Bauer (1980).

Hatching for most species occurs in 15 to 20 hours at 28°C (for *Chaetodontoplus septentrionalis* in approximately 23 hours at 24 to 26°C). Newly hatched prolar-

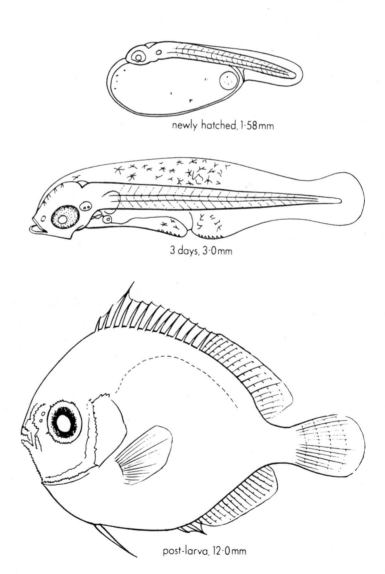

newly hatched, 1·58mm

3 days, 3·0mm

post-larva, 12·0mm

Larval development of an angelfish. Newly hatched and three-day-old larvae are of *Genicanthus semifasciatus*, based on Fujita & Mito (1960). The generalized post-larva is based on Burgess (1978).

257

vae are long and slender, lack functional jaws and pigment in the eyes, and have scattered melanophores and a moderate sized oval yolk containing a single oil droplet (at the anterior end of the yolk in *Chaetodontoplus* and at the posterior end in *Genicanthus*). As with egg diameter, size of the newly hatched prolarvae varies among genera, ranging from 1.58 mm for *Genicanthus* and 1.9 to 2.0 mm in *Chaetodontoplus* to 2.5 mm for an unspecified western Atlantic species of either *Holacanthus* or *Pomacanthus* (see Moe, 1976, 1977). Feeding begins at approximately 48 hours for *Holacanthus tricolor* and at about 72 hours for *Genicanthus* (at 26°C), by which time the eyes are pigmented, the fins have developed, and only remnants of the oil droplet remain. Each scale of the increasingly rounder and more laterally compressed larvae and small juveniles bears thin, elongate teeth, which give them a hairy appearance when viewed under a microscope (e.g., Fourmanoir, 1976). Moe (1977) reported that in aquaria pomacanthid larvae are not very active during early stages of development but become aggressive shortly before metamorphosis. Settlement to the bottom occurs at an age of three to four weeks.

Juveniles of many species differ dramatically in color pattern from the adults, which in some species relates to their activity as cleaners. Juveniles of several species of *Holacanthus* and *Pomacanthus* maintain regular long-term cleaning stations on and near the reef, to which larger fishes come to have external parasites removed (e.g., Brockmann & Hailman, 1976). The color patterns of such juveniles usually involve combinations of blue, black, and yellow, colors that are particularly conspicuous to other fishes (Thresher, 1978). Small juveniles of one species of *Holacanthus, H. tricolor,* have not been observed cleaning but may feed on the cutaneous mucus of larger cave-dwelling fishes (Thresher, 1979b). Fricke (1980) has suggested, based on field experiments, that juvenile specific coloration may also serve to reduce aggression by adults toward juveniles. Among the apparent noncleaning juveniles, young *Genicanthus* resemble the females, young *Pygoplites* and (in general) *Centropyge* resemble the adults, and juvenile *Euxiphipops* are similar to the cleaning species of *Pomacanthus*, although they are apparently not cleaners.

Natural hybrid between *Holacanthus ciliaris* and *H. bermudensis* showing some intermediacy in color features. Photo courtesy New York Zoological Society.

Natural hybrids occur in at least three pomacanthid genera: *Holacanthus*, *Centropye*, and *Euxiphipops*. Indeed, interspecific spawnings in *Centropyge* at Guam have recently been described by Moyer (1981). In the most detailed work on such hybrids, Feddern (1968) examined those between the western Atlantic sibling species *Holacanthus ciliaris* and *H. bermudensis*. The hybrids, in the past erroneously recognized as a third species, *H. townsendi*, are common off Florida (where the ranges of the two parent species overlap) and are intermediate in almost every color pattern feature between the two parents. The hybrids are also fertile and back-cross with the parent species regularly, producing a cline of hybrid individuals ranging from nearly pure *H. ciliaris* at one end to nearly pure *H. bermudensis* at the other. A similar hybridization, involving both an intermediately colored hybrid and apparent back-crossing with the parents, also occurs in *Centropyge flavissimus* and *C. vroliki* (Takeshita, 1976). Such hybrids also have a number of blue spots and lines on the unpaired fins that are not characteristic of either parent species. Another hybrid, also involving *C. flavissimus* but this time with *C. eibli*, has been collected by Allen & Steene (1979) at Christmas Island, Indian Ocean. Regarding *Euxiphipops*, little is known about hybridization in the genus, but one apparent hybrid between *E. sexstriatus* and *E. xanthometopon* has been collected off the Great Barrier Reef. The specimen is illustrated in Steene (1978). Finally, Moe (1976, 1977) produced artificial hybrids of several western Atlantic species of *Holacanthus* and *Pomacanthus* by stripping the adults. The hybrid between *Pomacanthus paru* and *P. arcuatus* proved viable and may occur occasionally in nature (mixed pairs of the adults are occasionally seen).

REPRODUCTION IN CAPTIVITY

Representatives of three pomacanthid genera have spawned in aquaria: *Centropyge*, *Genicanthus*, and *Pomacanthus*. Spawnings by any angelfishes are uncommon, however, although based on the small size of the fish at maturity and the behavior involved, it is likely that various species of *Centropyge* could be spawned regularly with only moderate effort.

A number of species of *Centropyge* have thus far been spawned: *C. fisheri* (Ballard, 1970; Lobel, 1974), *C. bispinosus* (Bauer & Klaij, 1974; Lobel, 1975—apparently describing the same spawnings), *C. flavissimus* (Lubbock, cited in Steene, 1978), and six species by Bauer & Bauer (1980). Only Bauer & Klaij, unfortunately, provided detailed information about

aquarium conditions and maintenance. The aquarium in which spawning occurred was a community tank containing three individuals of *Centropyge bispinosus* (one male and two females), a small *Chaetodontoplus mesoleucus*, several gobies and blennies, and a juvenile batfish (*Platax* sp.). A heavy coating of green algae had been allowed to develop on the walls and corals. The tank (approximately 454 liters) was filled with natural sea water (specific gravity 1.020 to 1.022) and filtered by means of an undergravel filter under a bed of crushed coral. A diatomaceous earth filter was also used on the tank occasionally and 5 to 10% water changes were made every two to three weeks. Water temperatures ranged between 23 and 26°C and, aside from fluorescent illumination each evening, no regular photoperiod was followed. The fish were fed a mixture of flake food, frozen brine shrimp, and vitamin supplements. Spawning began in the spring, when water temperatures presumably rose slightly, and continued daily until late summer.

Comparable results should be attainable with many species of *Centropyge*. Given the probable widespread occurrence of protogyny in the genus, obtaining a spawnable pair should require little more than selecting a large and a small individual. Best results might also be obtained by providing several small fish for each large one and, if more than one large fish is to be kept, by ensuring there is enough space for each to establish a territory. *Centropyge* usually do well in captivity, so getting them into spawning condition should not be a problem. Spawning, when it occurs, may be preceded by increased agonistic activity between the male and the females, which will culminate with soaring and spawning at dusk. Since decreasing light levels apparently trigger spawning, the use of a dimmer switch or a series of lights that can be turned off in sequence, duplicating the normal decrease in light intensity in the evening, might prove an effective way to stimulate spawning.

The only report of spawning by *Genicanthus* in captivity is Suzuki, et al. (1978), who spawned two species in a 5000-liter display aquarium. Water temperature ranged between 21 and 27°C, salinity averaged 33.5 ppm, and the pH averaged 7.8. The adults had been collected from the field five to 20 months prior to spawning. Finally, Straughan (1959a & b) reported spawning *Pomacanthus arcuatus* in aquaria but provided no details. Despite this report, it is unlikely that any of the larger angelfishes, such as *Pomacanthus*, *Holacanthus*, *Euxiphipops*, and *Pygoplites*, will prove amenable to regular spawning in anything smaller than large public aquaria.

Moe (1976, 1977) successfully reared western Atlantic *Pomacanthus* in aquaria but provided no details of the techniques used. Fertile eggs were obtained by stripping ripe adults and mixing the obtained gametes in small bowls of natural sea water at a temperature of 28°C and a specific gravity of 1.026. After 30 to 45 minutes the eggs were transferred to a rearing tank. Subsequent details are lacking, but given his previously described techniques and his location in the Florida Keys, it is likely that Moe reared the larvae in barren all-glass aquaria filled with natural sea water, that the system was partially open, i.e., there was a continuous slow inflow of clean salt water, and that the larvae were fed a combination of sorted wild-caught plankton and cultured rotifers. Moe reported that compared to anemonefishes, angelfishes are very difficult to rear.

Literature Cited

Allen, G.R. and R.C. Steene. 1979. The fishes of Christmas Island, Indian Ocean. *Aust. Nat. Parks Wildl. Serv., Spec. Pub.* (2):81 pp.

Ballard, J. 1970. Aquarium briefs—mating in captivity. *Bull. South Afr. Assoc. Mar. Biol. Res.*, (8):31-33.

Bauer, J.A., Jr. and S.E. Bauer. 1980. A new reproductive strategy found in *Centropyge* angelfish. *Abst. 2nd Intl. Symp. Biol. Management Mangroves and Trop. Shal. Water Comm.*

Bauer, J.A., Jr. and S.E. Bauer. 1981. Reproductive biology of pigmy angelfishes of the genus *Centropyge* (Pomacanthidae). *Bull. Mar. Sci.*, 31:495-513.

Bauer, J.A., Jr. and G. Klaij. 1974. Pigmy angels spawn. *Octopus*, 1(5):7-18.

Bauer, S. 1975. Reef fish eggs. *Octopus*, 2(1):16-17.

Bauer, S.E. and J.A. Bauer, Jr. 1980. Fecundity of *Centropyge* angelfishes. *Abst. 2nd Intl. Symp. Biol. Management Mangroves and Trop. Shal. Water Comm.*

Breder, C.M., Jr. and D.E. Rosen. 1966. *Modes of Reproduction in Fishes.* Natural History Press, Garden City, N.Y. 941 pp.

Brockmann, H.J. and J.P. Hailman. 1976. Fish cleaning symbiosis: notes on juvenile angelfishes (*Pomacanthus*, Chaetodontidae) and comparisons with other species. *Zeit. für Tierpsychol.*, 42:129-138.

Bruce, R.W. 1980. Protogynous hermaphroditism in two marine angelfishes. *Copeia*, 1980:353-355.

Burgess, W.E. 1974. Evidence for the elevation to family status of the angelfishes (Pomacanthidae), previously considered to be a subfamily of the butterflyfish family Chaetodontidae. *Pac. Sci.*, 28:57-71.

Debelius, H. 1977. The African angelfish kept in the aquarium for the very first time. *Trop. Fish Hobbyist*, 26(3):43-61.

Debelius, H. 1978. Neue Erkenntnisse uber die Kaiserfische der Gattung *Genicanthus*. *Seewasser*:31-34.

Feddern, H.A. 1968. Hybridization between the western Atlantic angelfishes, *Holacanthus isabelita* and *H. ciliaris*. *Bull. Mar. Sci.*, 18:351-382.

Fourmanoir, P. 1976. Formes post-larvaires et juveniles de poissons cotiers pris au chalut pelagique dans le sud-ouest Pacifique. *Cahiers du Pacifique*, (19):47-88.

Fraser-Brunner, A. 1933. A revision of the chaetodont fishes of the subfamily Pomacanthinae. *Proc. Zool. Soc. London*:543-599.

Freihofer, W.C. 1963. Patterns of the ramus lateralis accessorius and their systematic significance in teleostean fishes. *Stanford Ichthyol. Bull.*, 8:81-189.

Fricke, H.W. 1980. Juvenile-adult colour patterns and co-existence in the territorial coral reef fish *Pomacanthus imperator*. *Mar. Ecol., P.S.Z.N.*, 1:135-141.

Fujita, S. and S. Mito. 1960. Egg development and hatched larvae of a chaetodontid fish, *Chaetodontoplus septentrionalis*. *Bull. Japan. Soc. Sci. Fish.*, 26:227-229.

Lobel, P.S. 1974. Another pygmy angel spawns. *Octopus*, 1(7):24.

Lobel, P.S. 1975. Hawaiian angelfishes. *Marine Aquarist*, 6(4):30-41.

Lobel, P.S. 1978. Diel, lunar, and seasonal periodicity in the reproductive behavior of a pomacanthid fish, *Centropyge potteri*, and some other reef fishes in Hawaii. *Pac. Sci.*, 32:193-207.

Loiselle, P.V. and G.W. Barlow. 1978. Do fishes lek like birds? Pp. 31-76. *In: Contrasts in Behavior*, (E.S. Reese & F. Lighter, Eds.). Wiley Interscience, N.Y.

Masuda, H., C. Araga, and T. Yoshino. 1975. *Coastal Fishes of Southern Japan.* Tokai Univ. Press, Tokyo, Japan.

Moe, M.A., Jr. 1976. Rearing Atlantic angelfish. *Marine Aquarist*, 7(7):17-26.

Moe, M.A., Jr. 1977. Inside the eggs of an angelfish. *Marine Aquarist*, 8(3):5-13.

Moyer, J.T. 1981. Interspecific spawning of the Pygmy angelfishes *Centropyge shepardi* and *Cen-*

tropyge bispinosus at Guam. *Micronesica*, 17:119-124.

Moyer, J.T. and A. Nakazono. 1978. Population structure, reproductive behavior and protogynous hermaphroditism in the angelfish *Centropyge interruptus* at Miyake-jima, Japan. *Japan. J. Ichthyol.*, 25:25-39.

Moyer, J.T., R.E. Thresher, and P.L. Colin. In Press. Courtship, spawning and inferred social organization of American angelfishes (Genera *Pomacanthus*, *Holacanthus*, and *Centropyge*: Pomacanthidae). *Env. Biol. Fish.*

Munro, J.L., V.C. Gaut, R. Thompson, and P.H. Reeson. 1973. The spawning seasons of Caribbean reef fishes. *J. Fish. Biol.*, 5:69-84.

Neudecker, S. and P.S. Lobel. 1982. Mating systems of chaetodontid and pomacanthid fishes at St. Croix. *Z. Tierpsychol.*, 59:299-318.

Randall, J.E. 1975. A revision of the Indo-Pacific angelfish genus *Genicanthus*, with descriptions of three new species. *Bull. Mar. Sci.*, 25:393-421.

Randall, J.E. and W. Klausewitz. 1976. *Centropyge flavipectoralis*, a new angelfish from Sri Lanka (Ceylon). *Senckenberg. Biol.*, 57:235-240.

Randall, J.E. and J. Yasuda. 1979. *Centropyge shepardi*, a new angelfish from the Mariana and Ogasawara Islands. *Japan. J. Ichthyol.*, 26:55-61.

Shen, S-C. and C-H. Liu. 1976. Ecological and morphological study of the fish-fauna from the waters around Taiwan and its adjacent islands. 17. A study of sex reversal in a pomacanthid fish, *Genicanthus semifasciatus* (Kamohara). *Sci. Repts. Nat. Taiwan Univ.*, (6):140-150.

Smith, J.L.B. 1955. The fishes of the family Pomacanthidae in the western Indian Ocean. *Ann. Mag. Nat. Hist., Ser. 12*, 6:377-384.

Smith-Vaniz, W.F. and J.E. Randall. 1974. Two new species of angelfishes (*Centropyge*) from the Cocos-Keeling Islands. *Proc. Acad. Nat. Sci. Philadelphia*, 126:105-113.

Strand, S. 1978. Community structure among reef fish in the Gulf of California: The use of space and interspecific foraging association. Ph. D. Diss., Univ. California, Davis.

Straughan, R.P.L. 1959a. Salt water black angelfish spawned. *The Aquarium, Philadelphia*, 28(7):211-212.

Straughan, R.P.L. 1959b. Salt water black angelfish spawned. *Aquarium J., San Francisco*, 30(9):338, 340.

Suzuki, K., S. Hioki, and K. Iwasa. 1978. Spawning behavior, eggs, larvae, and sex reversal of two pomacanthine fish, *Genicanthus lamarck* and *G. semifasciatus*, in an aquarium. *Publ. Mar. Sci. Mus., Tokai Univ.*, 12:149-165.

Takeshita, G.Y. 1976. An angel hybrid. *Mar. Aquar.*, 7:27-34.

Thresher, R.E. 1978. Eye ornamentation of Caribbean reef fishes. *Z. Tierpsychol.*, 43:152-158.

Thresher, R.E. 1979. Possible mucophagy by juvenile *Holacanthus tricolor* (Pisces: Pomacanthidae). *Copeia*, 1979:160-162.

Thresher, R.E. 1980. *Reef Fish*. Palmetto Publ., St. Petersburg, Fla. 172 pp.

Thresher, R.E. 1982. Courtship and spawning in the emperor angelfish, *Pomacanthus imperator*, with comments on reproduction by other pomacanthid fishes. *Mar. Biol.*, 70:149-156.

Batfishes and Spadefishes
(Ephippidae)

Batfishes and spadefishes are laterally flattened fishes with elongate dorsal and anal fins. They are usually seen swimming in schools in relatively shallow water. Though often found on the reef, they are equally abundant in other inshore habitats, often being seen around mangroves, docks, and jetties, where they feed on sessile benthic invertebrates. The family is considered closely related to the angelfishes (Pomacanthidae), which they clearly resemble in overall appearance. Members of the two families can be readily distinguished, however, by the heavier body and drabber coloration of the spadefishes and by their pronounced tendency to school (angelfishes are usually solitary or found in pairs). Adult ephippids are various shades of gray, brown, black, and silver, often in patterns involving vertical bands or spots. Juveniles are generally darker colored and often have red or orange areas outlining the fins. This strikingly contrasting color pattern, combined with their very long dorsal and anal fins, makes such juveniles extremely attractive. Not surprisingly, they bring high prices in the retail trade.

The ephippids are a small family, with less than two dozen species in six or seven genera. The best known of these are *Chaetodipterus* in tropical American waters, *Parapsettus* in the eastern Pacific, and *Platax* in the Indo-West Pacific. Despite their relative abundance and wide distribution, little is known about their reproduction. Spawning has been observed in only one species, and little information is available about eggs or larval development.

SEXUAL DIMORPHISM

There are no reports of sexual dimorphism in the ephippids. Schools of such fishes appear to consist of essentially identical individuals, suggesting that any dimorphism present is either inconspicuous (e.g., a slight size difference) or temporary (e.g., temporary spawning colors). There is no information available on ephippid sexuality.

SPAWNING SEASON

The western Atlantic species *Chaetodipterus faber* appears to spawn only in the summer along the coast of the United States (a warm temperate area) (Smith, 1907); farther into the tropics spawning occurs in summer and, perhaps, spring (Munro, et al., 1973). No information is available on frequency of spawning or possible lunar cycles. Spawning in the spotted batfish, *Drepane punctata*, off India occurs in late summer and early fall (September to perhaps November) based on the ripeness of individuals obtained by fishermen (Mahadevan Pillai & Devadoss, 1975). Ova diameters in the ovaries form a single sharp peak in distribution, suggesting a short spawning period for each individual.

REPRODUCTIVE BEHAVIOR

Thus far, presumed spawning has been observed only for the western Atlantic spadefish *Chaetodipterus faber* (Chapman, 1978). On July 23, 1976, at 1100 and 1430 hrs., respectively, two small schools (10 and 20 fish) were discovered milling about near the surface near floating objects about 40 m offshore. All activities occurred within 10 m of the floating objects.

Apparent spawning was preceded by individuals pairing off and engaging in "mouth pushing." Two fish would swim toward one another, join mouths, and then each would apparently attempt to dislodge the other by vigorously shaking the forward half of the body. The fish's grips on one another were frequently broken, after which the pair separated, refaced one another, and again locked jaws. After several such bouts each pair broke apart, assumed a head-to-tail orientation about 0.3 to 0.5 m apart, and slowly spiraled toward the surface. Immediately upon reaching the surface the pair swam forward 1 to 6 m, with one, presumably the male, parallel to and slightly behind the other. During this parallel swim both fish shook vigorously, presumably releasing gametes. Although no eggs were sighted and none were collected, the similarity of the reported behavior to the spawning actions of the related angelfishes supports the idea that spawning was actually occurring.

Based on this report and on the general biology of the family, it seems likely that spawning by most, if not all, ephippids involves an offshore migration.

Fecundity of *Drepane punctata* ranges from approximately 150,000 to 800,000 eggs, depending upon the size of the female (Mahadevan Pillai & Devadoss, 1975).

EGGS AND LARVAE

Infertile mature ova of *Drepane punctata* are 0.8 to 0.9 mm in diameter (Mahadevan Pillai and Devadoss, 1975). Fertile eggs and larval development, however, have been described thus far only for the western Atlantic *Chaetodipterus faber* (Ryder, 1887; Hildebrand & Cable, 1938). Spadefish eggs are spherical, buoyant, about one mm in diameter, and contain a single oil droplet. Hatching occurs in 24

hours at 27°C, resulting in a 2.5 mm long prolarva. These prolarvae are characterized by a single ventrally located oil droplet, only weakly developed fin precursors, and numerous scattered pigment cells. By an age of 53 hours after hatching the eyes are pigmented, the mouth and intestines are open and apparently functional, the yolk sac is largely absorbed, and the pigment cells are concentrated in a vertical band across the center of the larva. Older larvae are laterally compressed with a deep head and prominent preopercular spines. Fin rays are well developed by the time a length of 4.2 mm is reached, and by a length of 10 mm the larva is recognizably a spadefish. Settlement of *C. faber* occurs at about this size and presumably at a similar size in other species.

An apparently newly settled juvenile *Platax pinnatus* measured 16 mm SL, lacked scales, and had small slender canine teeth (larger fish have tricuspid teeth) (Randall & Emery, 1971). The authors considered the lack of scales and presence of canines postlarval features. At this small size juvenile batfish are extremely laterally compressed with very large, broad, and conspicuous dorsal and anal fins. Because of these fins the fish is as tall, or even taller, than it is long. With growth, these broad rounded fins become

Postlarval ephippids collected near a submerged light at night off Papua New Guinea. The uppermost individual is 11.9 mm total length. Photo by Dr. R. E. Thresher.

even more elongate and eventually develop a point and sweep to the rear. At maturity the relative lengths of the fins decrease markedly as the schooling adults develop a more streamlined shape.

Juvenile ephippids have long been considered mimics of detritus based on their dark colors, broad flattened shape, and behavior. Juveniles of *Platax* and *Chaetodipterus* have been observed near wave-washed shores floating among leaves and other detritus, slowly rocking back and forth with this material (Willey, 1904; Uchida, 1951; Randall & Randall, 1961). Such mimicry is thought to camouflage them from potential predators. Randall & Emery (1971) observed a very small (16 mm SL) *Platax pinnatus* on the reef drifting on its side while fluttering its long dorsal and anal fins. Because of this behavior and its color (black, with a bright orange outline), the juvenile was originally mistaken for a swimming flatworm. The authors consequently suggested that the fish was mimicking such a worm, perhaps because the latter is distasteful and avoided by predators. However, this *Platax* was observed only after the area was treated with an ichthyocide (it died shortly after being observed). Since many fishes float on their sides and vibrate when poisoned, the case for mimicry by *Platax pinnatus* of a flatworm would be strengthened significantly by observations of such juveniles in an unstressed condition.

REPRODUCTION IN CAPTIVITY

The large size of ephippids at maturity (most are apparently over 30 cm) and their schooling behavior make it unlikely that any will be spawned in captivity except possibly in extremely large public aquaria. They generally do well given this much space and are popular oceanarium fishes, so perhaps spawning activity should be watched for.

Literature Cited

Chapman, R.W. 1978. Observations of spawning behavior in Atlantic spadefish, *Chaetodipterus faber*. *Copeia*, 1978:336.

Hildebrand, S.F. and L.E. Cable. 1938. Further notes on the development and life history of some teleosts at Beaufort, N.C. *Fish. Bull. (U.S.)*, 24:505-642.

Mahadevan Pillai, P.K. and P. Devadoss. 1975. A note on the fecundity and spawning period of *Drepane punctata* (Linnaeus). *Indian J. Fish.*, 22:262-264.

Munro, R.L., V.C. Gaut, R. Thompson, and P.H. Reeson. 1973. The spawning seasons of Caribbean reef fishes. *J. Fish Biol.*, 5:69-84.

Randall, J.E. and A.R. Emery. 1971. On the resemblance of the young of the fishes *Platax pinnatus* and *Plectorhynchus chaetodontoides* to flatworms and nudibranchs. *Zoologica, N.Y.*, 56:115-119.

Randall, J.E. and H.A. Randall. 1961. Examples of mimicry and protective resemblance in tropical marine fishes. *Bull. Mar. Sci.*, 10:444-480.

Ryder, J.A. 1887. On the development of osseus fishes, including marine and freshwater forms. *U.S. Fish. Comm. Rpt.*, 1885:488-604.

Smith, H.M. 1907. The fishes of North Carolina. *North Carolina Geol. Econ. Surv.*, (2).

Uchida, K. 1951. Notes on a few cases of mimicry in fishes. *Sci. Bull. Fac. Agric., Kyushu Univ.*, 13:294-296.

Willey, A. 1904. Leaf-mimicry. *Spoila Zeylan.*, 2:51-55.

Butterflyfishes
(Chaetodontidae)

Butterflyfishes are at the same time one of the most characteristic and conspicuous groups of reef fishes and also one of the least well known behaviorally. Laterally compressed, with a prominent spiny dorsal fin and in many instances even more prominent elongated snout, butterflyfishes are ubiquitous on the reef and, perhaps for that reason, are often "overlooked" in favor of studying the more "exotic" species. It is also a common observation that butterflyfishes "don't do anything." Damselfishes are spectacularly territorial or at least have complex, interesting, and very visible spawning systems, and wrasses change sex and color and spawn in groups in the middle of the day when everyone can see them. The equally abundant butterflyfishes, however, seem to be content just swimming around in the background being ignored. This is, of course, an oversimplification. Several biologists have looked at the family in considerable detail, yet it emphasizes the remarkable sparsity of even basic information on the family, especially on their reproductive biology.

In a recent systematic review of the family, Burgess (1978) recognized 114 species of butterflyfishes in ten genera. The family is distributed circumtropically, with a peak in abundance and diversity in the Indo-Pacific; a few species, however, inhabit temperate areas, most notably off the southern coasts of Australia. With few exceptions, butterflyfishes are benthic foragers (e.g., Birkeland & Neudecker, 1981). Many, in fact, subsist entirely on the polyps of live coral. This mode of feeding is reflected in the somewhat protruding set of jaws characteristic of the butterflyfishes, a feature that reaches its zenith in the various "longsnout" butterflyfishes of the genera *Chelmon, Forcipiger,* and *Chaetodon (Prognathodes).*

SEXUAL DIMORPHISM

There are no reports of permanent external sexual dimorphism in the butterflyfishes, and the family as a whole appears to be sexually monomorphic. It is, in fact, a common observation that for at least the pair-forming species, pairs consist of roughly same-sized and similar appearing individuals. Aiken (1975), for example, found no significant differences in male and female sizes of three species of *Chaetodon* collected in fish traps off Jamaica. E. Reese (pers. comm.) similarly found only a slight size difference between pair members of the Indo-Pacific *C. trifasciatus*. In contrast, Yamsonrat (1980) examined pairs (and occasional triplets) of 15 Pacific species of *Chaetodon* and *Heniochus* and found first, that about 90% of the pairs were heterosexual and, second, that in such pairs males were larger than females 84% of the time. Within the population as a whole males and females do not differ significantly in size, so using size differences to sex an individual is nearly impossible without precise measurements (M. Sutton, pers. comm.). Immediaely prior to spawning, however, it is possible to sex at least some species, since the female becomes visibly distended with eggs (Lobel, 1979; Suzuki, et al., 1980; pers. obs.).

The gonads of butterflyfishes are paired and lie in the upper posterior portion of the visceral cavity. Those of males are elongate and slender; those of females are heavier, rounder, and bilobed, joining posteriorly (M. Sutton, pers. comm.). Nalbant (1974) indicated that the gonads of *Chaetodon ephippium* occupy ⅓ to ⅔ of the visceral cavity. As yet such gonads have not been examined histologically, so the mode of sexuality characteristic of the family has not yet been determined. Both sexes of *Chaetodon miliaris*, however, reach sexual maturity at the same size (approximately 90 mm), suggesting either separate sexes or, at least, no conspicuous size-related hermaphroditism in this schooling species (Ralston, 1977). The characteristic pairing of some species, along with the occasional observation of a triad consisting of an apparently adult pair and a small subadult individual, suggests that juvenile ambisexuality along the lines suggested for the pair-forming anemonefishes may occur.

SPAWNING SEASON

Only scattered information is available on spawning seasons of butterflyfishes, almost all based on changes in some form of gonad index. In tropical areas available data indicate that levels of spawning activity peak in winter and early spring (Ralston, 1976, 1977, 1981; Lobel, 1979), though at least some species continue to spawn at lower levels of activity throughout the year (Aiken, 1975; E. Reese, unpubl. data). Two studies conducted near the temperate edges of butterflyfish distribution, however, suggest midsummer spawning, when water temperatures are near their annual highs (Suzuki, et al., 1980; M. Sutton, pers. comm.).

Lobel (1979) observed apparent spawning (see below) by three Hawaiian species of *Chaetodon* during the week preceding the full moon, implying that this was the only time spawning occurred. Insufficient data are available to confirm this lunar rhythm of spawning activity in the field, but at least one species, *C. nippon*, shows no such rhythm in captivity (Suzuki, et al., 1980). Moreover, neither Ralston (1981), working with *Chaetodon miliaris* off Hawaii, nor M. Sutton (pers. comm.), studying two species of *Chaetodon* on the Great Barrier Reef, found any correlation between a gonadal index and moon phase for their respective species.

REPRODUCTIVE BEHAVIOR

Butterflyfishes, like most other reef-associated fishes, exhibit a broad spectrum of social systems ranging from solitary home-ranging and/or territorial species (e.g., *Chaetodon trifascialis*) to those that routinely form large foraging groups and even school (e.g., *Chaetodon miliaris*—Reese, 1973, 1975; Hobson, 1974). Nonetheless, the characteristic social unit for most species is a stable heterosexual pair (Bardach, 1958; Burgess, 1978; Yamsonrat, 1980). Thirteen of nineteen species of *Chaetodon* at Enewetak, for example, are typically, if not invariably, found in such pairs (Reese, 1975), as are all four common western Atlantic species. Such pairs have been reported as stable for at least three years (Fricke, 1973) and are probably lifelong.

Fricke (1973) suggested that permanent pair formation in the butterflyfishes is a response to their particularly high within-habitat diversity and has evolved as a means of reducing the frequency of interspecific mating and subsequent wastage of gametes. In support of this argument, he noted that pairing is often most common in low density populations, where the probability of encountering a conspecific mate might otherwise be very low. The hypothesis, in general, seems logical but can be questioned on several points. First, many other families of reef fishes, such as labrids and damselfishes, also have numerous broadly sympatric species and yet do not form conspicuous pairs. Second, there does not seem to be a clear-cut correlation between sympatric within-genus diversity and the percentage of species found in conspicuous pairs. There are, for example, only four common species of *Chaetodon* in the western Atlantic, all of which pair regularly, whereas of 20 common species of *Chaetodon* at Enewetak, only about 75% form pairs (Reese, 1975). Finally, pairing by itself does not necessarily prevent hybridization, since such hybrids have been reported for several species of butterflyfish (e.g., Randall, et al., 1977; Burgess, 1978).

Courtship and spawning have thus far been observed in only a handful of species, most in the genus *Chaetodon*, so that generalities about the reproductive behavior of the family are difficult to make. Three potential generalities are offered, however: first, butterflyfishes typically, if not always, spawn at dusk; second, gamete release typically takes place well up into the water column following what is generally a slow ascent; and third, at least in the genus *Chaetodon*, courtship is usually in the form of the male swimming behind and below the female, nudging her abdomen with his snout and forehead. Spawning usually takes place with the fish in this position.

Complete and clear-cut spawning sequences in the field have been observed for only three species of *Chaetodon* (and also one species of *Heniochus*, described below)—*C. striatus*, a western Atlantic species (P. Colin, pers. comm.); *C. rainfordi*, off One Tree Island, Great Barrier Reef (pers. obs.); and *C. (Prognathodes) aculeatus*, another western Atlantic species (A. Gronell, pers. comm.). The spawning behavior of the first two species was almost identical. In both cases pairs of the fish were observed at dusk, beginning just about the time of sunset. The male courts by swimming just ahead of the female (which, in *C. rainfordi*, was conspicuously heavy with eggs) and "fluttering" his body back-and-forth in front of her. The female ascends slightly and the male moves behind and below her, placing his snout against her abdomen, often while angled slightly head-down (this head-

Possible courtship activity in *Chaetodon (Megaprotodon) trifascialis.* Photo by P. Motta.

Spawning of *Chaetodon rainfordi.* Note the heavy female in front and how high the fish are in the water column. Photo by M. Sutton.

down posturing at the female's abdomen has also been observed in three species of *Chaetodon* off Hawaii (Lobel, 1978) and in various species in the Red Sea and Indo-Pacific (Fricke, pers. comm.; Reese & Mota, pers. comm., respectively)). In this position the pair ascend into the water column. Prior to actual spawning there are usually several "false starts" in which the pair ascend only a few meters and then break apart and return to the bottom before moving into position again. Finally, however, the pair remain together and ascend 10 to 15 m off the bottom in a slow, even ascent. At the peak of the ascent a clearly visible white cloud of gametes is released, after which the pair dash toward the bottom. Each female produces about 20,000 eggs (Aiken, 1975). After spawning the male may court the female weakly, though a second spawning in the same evening by a pair has not yet been observed.

Spawning by *C. (Prognathodes) aculeatus* is similar to that described above but takes place much closer to the bottom. At dusk the pair began to chase one another around a large sponge, a prominent local landmark, with the lead fish swimming with its fins fully spread. During spawning, the pair chased one another to the top of the sponge, assumed a side-by-side position, and then made a short, slow "dash" into the water column to release gametes.

Apparent spawning has also been observed in the field and described by Lobel (1978) for three species of Hawaiian *Chaetodon*. In all three species the female, heavily laden with eggs, was chased across the reef by one, and sometimes two, males. Presumed spawning occurred when the female tilted forward, the male approached her and placed his snout by her abdomen, and the pair ascended from 0.5 to 2 m off the bottom while quivering slightly. No gamete release was visible, however, suggesting that the observed behavior may not have been actual spawning, but rather was "false starts" like those described for *C. striatus* and *C. rainfordi* prior to actual spawning. (E. Reese, pers. comm., reports witnessing similar "false starts" in *C. trifascialis*.) The short distance which the pairs ascended into the water column seems to support this interpretation.

Finally, spawning also has been observed for a species of *Chaetodon* in captivity. In a detailed description of the behavior of *C. nippon*, Suzuki, et al. (1980) reported that courtship and spawning began approximately an hour before dark, at which time the males present ascended off the bottom and began to chase one another. About 20 minutes later a gravid female ascended to a point just below the surface and began to swim in broad circles. She was joined by the largest male, who "pecked" at her abdomen while the pair circled together. Attempts to join the pair by other males were vigorously contested by the dominant. Eventually, shortly before dark, the circling became more rapid, the male pushed the female toward the surface, and gametes were shed. At the point of gamete release, other males often rushed to the pair and shed sperm, apparently trying to fertilize eggs at the dominant male's expense. Whether this behavior is an artifact of crowded aquarium conditions or results from the school-based social system of this species (in contrast to previously described species, all of which are found in pairs) is not known.

The only observation of apparent spawning by a butterflyfish other than a species of *Chaetodon* was made at dusk off One Tree Island (pers. obs.). A pair of *Heniochus acuminatus* was observed to ascend slowly off the bottom until about 8 meters up and then move close together, side-by-side, with one slightly behind the other. There was no further ascent as the pair moved forward slightly and then suddenly dashed for the bottom. Gamete release was not seen but was not likely to be visible given the low light conditions and the distance from which the fish were observed (about 17 m).

Lastly, there are two additional reports of chaetodontids spawning, both in aquaria and both sounding somewhat unlikely. Frank (1975) reported that a pair of *Chelmon rostratus* produced demersal eggs that they tended for two weeks. The eggs ultimately hatched into demersal larvae that were large enough to immediately take a commercially available liquid "fry food." Although the distantly related monodactylids, a primarily brackish water group, produce demersal eggs, it is unlikely that *Chelmon* does so. Finally, Lorenz (cited in Burgess, 1978) observed *Chaetodon kleini* to spawn in a cichlid-like fashion, with the female laying eggs on a flat surface while being followed by the male. The eggs did not adhere to the substrate, but rather floated to the surface. Lorenz interpreted the spawning behavior as a holdover from an ancestral substrate-spawning method. While possible, it seems more likely that the observed sequence was a captivity-induced artifact in this normally water column foraging species.

EGGS AND LARVAE

Chaetodontid eggs are buoyant, transparent, and spherical. Those of *Chaetodon ocellatus*, a western

Atlantic species, are 0.6 to 0.7 mm in diameter (Moe, 1976); those of *C. nippon* from off Japan range from 0.70 to 0.74 mm and contain a single oil droplet measuring 0.18 to 0.19 mm in diameter (Suzuki, et al., 1980).

The eggs of *C. nippon*, the only species for which early development has been described, hatch in 28 to 30 hours at water temperatures of 22.2 to 23.7°C. The newly hatched larvae are 1.43 to 1.53 mm long and have a large yolk sac that extends forward to a point beyond the head of the larvae. A single oil droplet is located at the posterior edge of the yolk, with the result that the newly hatched butterflyfish float upside-down just below the water's surface. The yolk is largely absorbed by 48 hours after hatching, and by 72 hours the mouth and anus open, the eyes

are pigmented, and pectoral fin precursors are present (Suzuki, et al., 1980). Butterflyfish larvae larger (and presumably older) than this have been described by Leis & Miller (1976) based on specimens obtained in plankton tows. The 4.5 mm larva of an unidentified species of *Chaetodon* was smooth-skinned and elongate, with a row of pigment spots along its back, rudimentary vertical fin folds, and a single large preopercular spine. By 5.5 mm total length the fins are well developed (including several short spines leading the dorsal fin) and the head of the rather deep-bodied larva is encased in a bony sheath. From the lower rear edge of this sheath project a pair of spines that extend below the pectoral fins. The next stage in larval development, known as the tholichthys, is one that is unique to the butterflyfishes among reef fishes.

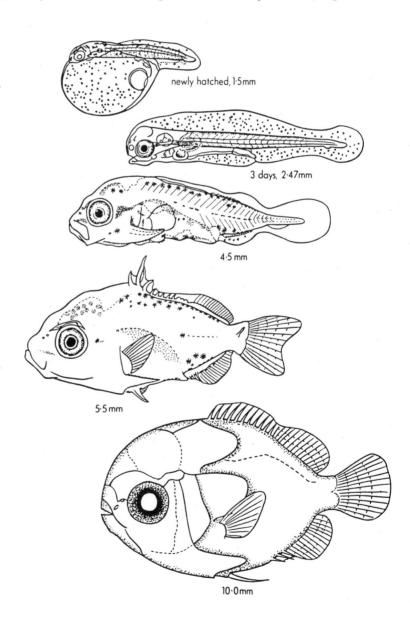

newly hatched, 1·5mm

3 days, 2·47mm

4·5 mm

5·5 mm

10·0mm

Larval development of *Chaetodon*. Newly hatched and three-day-old larvae based on Suzuki, et al. (1980), reproduced with permission of the authors. The 4.5 and 5.5 mm larvae based on Leis & Miller (1976). 10 mm larva based on Burgess (1978).

Tholichthys larvae are deep-bodied, laterally compressed, and silver-colored. Each is characterized by a series of thin transparent bony plates that completely encase the head of the larvae and that extend dorsally and ventrally to form bony spines. These plates remain until after the fish have settled to the bottom as juveniles but are difficult to see and are absorbed within a few weeks (Burgess, 1978). All genera of butterflyfish thus far studied have proved to have tholichthys larvae, most of which are indistinguishable. Those of *Forcipiger*, however, are more elongate than others and have a ragged edge to the dorsal plate. Those of *Chaetodon meyeri* and *C. ornatissimus* are also unusual in that they possess a forward-directed spine projecting over each eye (Burgess, 1978). In the early stage larvae these spines are thick and straight, while those of older larvae are more slender and curved. Such larvae in the past have been considered those of various species of *Heniochus* (e.g., Fourmanoir, 1976), which as adults have "horns" on their heads.

The duration of the planktonic stage is not known, but Burgess (1978) suggested it is likely to be at least several months. Settlement onto the reef occurs at night, followed by a rapid transformation into the juvenile stage of development. Many, though not all, chaetodontids have a distinctive juvenile color pattern often characterized by the presence of one or more "eyespots," circular dark spots outlined by either white or pale blue located below the soft dorsal fin (e.g., Böhlke & Chaplin, 1966; Fricke, 1973; Burgess, 1978). Fricke suggested that such juvenile-specific color patterns "mask" the juveniles' "species-membership," which permits them to occupy the territories of adults without eliciting attack by the adults.

Juvenile butterflyfishes often occupy a part of the reef different from that occupied by the adults (Fricke, 1973; Thresher, 1980). When this occurs, the juveniles are usually found in shallower areas often devoid of coral. Whether such individuals represent a surplus population that eventually dies or whether they migrate individually into deeper water and to the reef as they mature is not known. In the few species in which it has been examined, sexual maturity occurs at a length of roughly 80 to 100 mm (Aiken, 1975; Ralston, 1976), at which point the fish are probably a year or so old. Presumably the large species, such as *Chaetodon lineolatus*, mature more slowly and at a larger size.

REPRODUCTION IN CAPTIVITY

There is only one confirmed report of spawning by a butterflyfish in captivity—that of *Chaetodon nippon* by Suzuki, et al. (1980). Over a several month period a captive group of *C. nippon* held in a large (1.3 m deep) and crowded community aquarium spawned repeatedly. The water temperature was approximately 23°C, the tank was illuminated by fluorescent lights for ten hours daily, and the fish were fed once daily with cut or chopped fish, shrimp, and clams. Courtship always began shortly before the lights were turned off, and spawning always took place after "dark" (the room containing the aquarium was always dimly lit, so that darkness was never total).

Based on this report, spawning by other species of butterflyfish should be possible but will probably require a large aquarium, the presence of at least four or five mature individuals to ensure at least one pair (an alternative approach is to collect both members of a naturally occurring pair), and a gradual reduction of light levels at night. Likelihood of spawning will doubtless vary with the size of the species; large species such as *Chaetodon ephippium* and *C. lineolatus* are likely to prove difficult to spawn, while species that mature at a relatively small size, such as *C. (Prognathodes) aculeatus*, should be more amenable to spawning in captivity.

Suzuki, et al. (1980) attempted to rear the larvae of *C. nippon* by feeding rotifers (*Brachionus plicatilis*), larval oysters (*Crassostrea gigas*), and larval sea urchins (*Temnopleurus reevesi*). Although at least a few larvae fed, none lived more than eight days. If a several month estimate of the duration of the planktonic stage is correct, rearing of the larvae may prove to be a long and arduous task.

Literature Cited

Aiken, K.A. 1975. The biology, ecology and bionomics of Caribbean reef fishes: Chaetodontidae (Butterfly and Angelfishes). *Res. Rpt. Zool. Dept., Univ. West Indies*, (3):57 pp.

Bardach, J.E. 1958. On the movement of certain Bermuda reef fishes. *Ecology*, 39:139-146.

Birkeland, C. and S. Neudecker. 1981. Foraging behavior of two Caribbean chaetodontids: *Chaetodon capistratus* and *C. aculeatus*. *Copeia*, 1981:169-178.

Böhlke, J. and C.C.G. Chaplin. 1968. *Fishes of the Bahamas and adjacent tropical waters*. Livingston Press, Wynnewood, Pa. 771 pp.

Burgess, W.E. 1978. *Butterflyfishes of the World*. T.F.H. Publ., Inc., Neptune City, New Jersey. 832 pp.

Fourmanoir, P. 1976. Formes post-larvaires et

juveniles de poissons cotiers pris au chalut pelagique dans le sud-ouest Pacifique. *Cahiers du Pacifique,* (19):47-88.

Frank, J. 1975. Copperband butterfly spawns. *Octopus,* 2:29-31.

Fricke, H.W. 1973. Behaviour as part of ecological adaptation. *Helgoländer Wiss. Meeresunters.,* 24:120-144.

Hobson, E.S. 1974. Feeding relationships of teleostean fishes on coral reefs in Kona, Hawaii. *Fish. Bull. (U.S.),* 72:915-1031.

Leis, J.M. and J.M. Miller. 1976. Offshore distributional patterns of Hawaiian fish larvae. *Mar. Biol.,* 36:359-367.

Lobel, P.S. 1978. Diel, lunar, and seasonal periodicity in the reproductive behavior of the pomacanthid fish, *Centropyge potteri,* and some other reef fishes in Hawaii. *Pac. Sci.,* 32:193-207.

Moe, M.A., Jr. 1976. Rearing Atlantic angelfish. *Mar. Aquar.,* 7:17-26.

Nalbant, M. 1974. Butterflyfishes. Pp. 117-120. *In: Animal Life Encyclopedia* (H.C.B. Grzimek, ed.-in-chief), Volume 5 (English ed.). Van Nostrand-Reinhold Co., N.Y.

Ralston, S. 1976. Age determination of a tropical reef butterflyfish utilizing daily growth rings of otoliths. *Fish. Bull. (U.S.),* 74:990-994.

Ralston, S. 1977. Anomalous growth and reproductive patterns in populations of *Chaetodon miliaris* (Pisces: Chaetodontidae) from Kaneohe Bay, Oahu, Hawaiian Islands. *Pac. Sci.,* 30:395-403.

Randall, J.E., G.R. Allen, and R.C. Steene. 1977. Five probable hybrid butterflyfishes of the genus *Chaetodon* from the central and western Pacific. *Rec. West. Aust. Mus.,* 6:3-26.

Reese, E.S. 1973. Duration of residence by coral reef fishes on "home" reefs. *Copeia,* 1973:145-149.

Reese, E.S. 1975. A comparative field study of the social behavior and related ecology of reef fishes of the family Chaetodontidae. *Z. Tierpsychol.,* 37:37-61.

Suzuki, K., Y. Tanaka, and S. Hioki. 1980. Spawning behavior, eggs, and larvae of the butterflyfish, *Chaetodon nippon,* in an aquarium. *Japan J. Ichthyol.,* 26:334-341.

Thresher, R.E. 1980. *Reef Fish.* Palmetto Publ. Co., St. Petersburg, Fla. 172 pp.

Yamsonrat, S. 1980. Sex Determination of Butterflyfishes (Chaetodontidae). Master's Thesis, Univ. Hawaii. 15 pp.

Chaetodontid-Related Fishes (Scorpididae, Pentacerotidae, Enoplosidae, and Scatophagidae)

There are a number of nominal families apparently closely related to the butterflyfishes but whose exact systematic positions are not precisely clear. Most generally resemble butterflyfishes in their overall appearance, i.e., laterally compressed, prominent terminal mouth, long dorsal fins, and a deep oval-shaped body. The limits of the group are not well established, but it is generally taken to include the scats (Scatophagidae), a group of brackish water schooling fishes popular in marine aquaria; the stripeys (Scorpididae), a marine group occasionally found on the reef (e.g., *Microcanthus strigatus*) but more common in temperate areas; the boarfishes (Pentacerotidae), another marine temperate group; and the old wife, *Enoplosus armatus* (Enoplosidae), a striking black-and-white banded fish found off the coast of Australia. In toto, there are about 30 species in the group, in about ten genera.

Reproduction by these fishes is not well known. At least one group, the scats, produces demersal eggs which are apparently deposited in nests on the bottom and tended until hatching (e.g., Breder & Rosen, 1966). The other marine groups presumably produce pelagic eggs, though these have not yet been described. According to J. Bell (pers. comm.), spawning by *Enoplosus armatus* occurs in the spring, at which time the fish leave their normal schools to form closely associated pairs. There is no evidence of sexual dimorphism, nor are the fishes hermaphroditic. Juveniles are typically found in sea grass beds and migrate offshore to rocky reefs as they mature. At least one group, the scats, has a tholichthys-type larval stage in its development (Burgess, 1978), whereas another, the scorpidids, apparently does not.

Literature Cited

Breder, C.M., Jr. and D.E. Rosen. 1966. *Modes of Reproduction in Fishes*. Natural History Press, Garden City, N.Y. 941 pp.

Burgess, W.E. 1978. *Butterflyfishes of the world*. T.F.H. Publ., Inc., Neptune City, N.J. 832 pp.

The scat, *Scatophagus argus,* has a tholichthys-type larval stage that transforms into a typical juvenile like that shown here. Photo by Milan Chvojka.

Surgeonfishes and Tangs
(Acanthuridae)

The majority of reef fishes are predators, either large and active piscivores like the various basses or small planktivores and omnivores like the pseudochromids and damselfishes. Given the diversity of the fauna, however, it is not surprising that several families are primarily herbivorous, feeding on the algae that are abundant on the shallow reef. The best known, most colorful, and often most abundant of these herbivores are the acanthurids.

The surgeonfishes and tangs, family Acanthuridae, are laterally compressed fishes with long, low, and even dorsal and anal fins, leathery skin, a small but conspicuous terminal mouth, and usually a bluntly rounded head. In profile, they range from oval and almost circular (most species) to a few elongate fishes. The family is characterized by a distinctive spine or series of spines on each side of the caudal peduncle. Known as the scalpel, these spines are both defensive and offensive weapons: during intraspecific combat the fish circle one another warily, each with its spine angled toward the other. Many species have color marks that emphasize the scalpels, apparently facilitating their role in intraspecific communication.

There are six acanthurid genera (*Acanthurus, Ctenochaetus, Naso, Paracanthurus, Prionurus,* and *Zebrasoma*) and about 70 species (Randal, 1955a, b & c, 1956; Smith, 1966). Genera are separated on the basis of shape of the teeth, number of dorsal spines, and shape and number of spines on the caudal peduncle (Randall, 1955a; Tyler, 1970). All are restricted to the tropics, and all but five species (in *Acanthurus*) are Indo-Pacific.

SEXUAL DIMORPHISM
Surgeonfishes vary widely in the type and degree of sexual dimorphism they exhibit. Permanent dimorphism is relatively uncommon in the family and takes two forms: differences in morphology and size differences between the sexes. The former is, so far as is known, characteristic of only one genus—*Naso*. In some species, such as *N. brachycentron* and *N. unicornis*, males have a well developed "horn" that projects anteriorly from their foreheads (hence their common name of "unicornfish"); females either lack such a horn altogether (e.g., *N. brachycentron*) or have a relatively small horn (e.g., *N. unicornis*) (Smith, 1951, 1966; Barlow, 1974). Females also tend to have less well developed spines or "keels" on the caudal peduncle and may have shorter filaments trailing off the edges of the caudal fin.

Size differences between the sexes have been reported for various species of *Naso* (males larger than females) (Smith, 1966; Barlow, 1974), for two western Atlantic *Acanthurus* (Reeson, 1975), and for several species in *Acanthurus, Ctenochaetus,* and *Zebrasoma* at Aldabra (Robertson, et al., 1979). For the last group, the magnitude of the size difference ranges from none at all (*A. lineatus*) to over 200% (*A. leucosternon*), with either the male or the female the larger depending upon the species (male larger in, for example, *A. nigricaudus (= A. gahhm,), C. strigosus,* and *Z. scopas;* female larger in *A. leucosternon* and *Z. veliferum*). Robertson, et al. (1979) discussed in detail the patterns of size dimorphism in Aldabran surgeonfishes but came to few conclusions. Rather, they suggested only that such dimorphism is the result of a complex interaction between: 1) a positive correlation between female size and fecundity; 2) the food type eaten by a species; 3) territory size and defendability; and 4) size and distribution of patches of suitable habitat, complicated further by historical constraints and the influence of selective factors operating on the fishes outside of the study area.

Temporary color differences between the sexes during courtship and spawning appear to be widespread in the surgeonfishes. Typically, the male darkens all or part of his body during courtship displays. In *Acanthurus coeruleus,* for example, the male pales anteriorly and darkens posteriorly, forming a horizontal bicolor; in *Prionurus microlepidotus* the male develops a series of dark stripes on his otherwise gray abdomen (J. Moyer, pers. comm.); and in a variety of large *Acanthurus* and *Naso* off One Tree Island, Great Barrier Reef, males develop a conspicuous white bar behind the head during courtship. Such temporary colors have been reported for various species of *Acanthurus* (Randall, 1961a; Robertson, et al., 1979; Thresher, 1980; Colin, in prep.), for *Ctenochaetus striatus* (Randall, 1961b), and are suggested for *Naso* (Barlow, 1974). Both Barlow (1974) and Moyer (pers. comm.) noted that the temporary spawning colors they observed were identical to those developed by the males during agonistic encounters. The colors seem to be, therefore, less specific courtship colors as much as indicators of a general state of arousal.

According to Reeson (1975), the paired gonads of *Acanthurus* lie in the upper portion of the coelom. Testes are white, elongate, and separated dorsally, whereas ovaries are pink, round, and hollow. Otherwise there is no detailed information regarding gonad structure or histology in acanthurids, nor has anyone specifically examined these fishes for evidence of hermaphroditism. That several species have a harem-based social system and also show consistent size differences between the sexes suggests that at least some acanthurids may be sequential hermaphrodites.

SPAWNING SEASONS

Acanthurids, as a rule, seem to have a peak in spawning activity in late winter and early spring based on gonad samples (e.g., Randall, 1961a; Munro, et al., 1973), observations of spawning (Colin, in prep.), and recruitment of young (Randall, 1961a; Russell, et al., 1977). Spawning of *Acanthurus triostegus* in Hawaii occurs primarily from December to June, and for several species in the Caribbean it is from December to about April. Some spawning does occur year-round, however, and Randall (1961a) suggested that there may not be any specific spawning season in the deep tropics where seawater temperatures change only slightly seasonally. In that regard, Robertson, et al. (1979) and Johannes (1981) suggested year-round spawning by such tropical fishes, though Johannes seems to feel that spawning is usually greatest in early summer. It is also noteworthy that all areas for which good data on seasonal peaks are available (the Caribbean, Hawaii, and the southern Great Barrier Reef) also exhibit marked seasonal changes in water temperature.

Lunar cycles to spawning by acanthurids are suggested by several works (e.g. Randall, 1961a & b; Bagnis, et al., 1972; Johannes, 1978, 1981) and has been well documented for Caribbean species of *Acanthurus* by Colin (in prep.). Randall, for example, found that twice as many females of *A. triostegus* in Hawaii were running ripe during the first half of the lunar month than during the second half. For those species examined (several species of *Acanthurus* and *Ctenochaetus striatus*), spawning occurs primarily near the new moon, near the full moon, or both. All such data, however, are based on group spawning species; data available to date on pair spawners suggest spawning throughout the lunar month (Robertson, et al., 1979; P. Atkins, pers. comm.; pers. obs.).

REPRODUCTIVE BEHAVIOR

Surgeonfishes produce pelagic eggs which, in most species, are spawned in the water column at the peak of a short spawning ascent. In most cases spawning takes place at dusk. It can involve either individual pairs, groups spawning as a unit, or, in some species, both.

Pair spawning has been documented in *Zebrasoma veliferum* and *Ctenochaetus strigosus* and suggested for *Acanthurus leucosternon*, *A. nigricaudus* (= *A. gahhm*), and *Naso lituratus* by Robertson, et al. (1979), in *Prionurus microlepidotus* by J. Moyer (pers. comm.), in *Acanthurus dussumieri* at One Tree Island, and in *Naso brevirostris* at Enewetak Atoll (pers. obs.). Barlow (1974) also suggested that several species were likely to be pair spawners, largely because he commonly observed the fish moving about the reef in pairs. Pair spawning occurs in three situations: 1) between members of a stable pair or harem which defend a common territory (*Z. veliferum* and *C. strigosus*); 2) at the edge of the outer reef, where males establish temporary spawning territories, court passing females, and spawn pair-wise with them (*Z. flavescens* — P. Atkins, pers. comm.); and 3) between members of a school or foraging group that spawn at the periphery of the group (*P. microlepidotus*, *N. unicornis*, and *A. dussumieri*). Though details are still largely lacking, general spawning behavior appears to be similar throughout these fishes. The male approaches the female, may develop a distinctive spawning color, and then moves parallel to her. In *A. dussumieri* the male also rolls dorsally toward the female while lying on the bottom before her and vibrates his body vigorously while moving across in front of her. Together and side-by-side, the pair then dash a few meters off the bottom, release gametes at the peak of the ascent, and then return to the bottom. These spawning ascents usually do not seem to be as fast as those performed by other groups of reef fishes, such as scarids and mullids. Pair spawning in *Prionurus microlepidotus* is even simpler than that of the other species described. According to the observation, the male suddenly developed dark stripes, a female moved half a meter or so from the school to join him, and the pair touched ventral surfaces and spawned. There was no upward dash, and the entire procedure was quite slow. *Naso brevirostris* behaved comparably, though spawning involved a rapid forward dash of about 2 m while these planktivorous fishes were several meters off the bottom. Similar behavior has been observed at dusk for various species of *Prionurus* and *Naso* off One Tree Island, though no gamete release was conspicuous (pers. obs.).

Group spawning has been more commonly reported in acanthurids (e.g., Randall, 1961a & b; Bagnis, et al., 1972; Lobel, 1978; Robertson, et al., 1979) and is similar to that described elsewhere for scarids and labrids. Prior to spawning, the fish aggregate, often in the tens of thousands, at specific and possibly "traditional" spawning sites along the edge of the outer reef. In several species, for example *Acanthurus triostegus* and *Ctenochaetus striatus,* these sites are near the mouths of channels in the reef, in areas of strong offshore currents (e.g., Randall, 1961a; Johannes, 1981) and are presumably "chosen" because they facilitate transport of the fertilized eggs offshore. In most species spawning is preceded by an agitated milling of the fishes about the spawning site and the coalescing of the fish into tight groups. At first these groups tend to form and break up rapidly, but as darkness approaches they last increasingly longer and the movements of the fish increase in speed. Finally, one group will suddenly dash toward the surface, the fish in a tight parallel formation. At the peak of the ascent the fish turn back, leaving behind a cloud of gametes, while they disperse into the milling school below. Spawning ascents often occur throughout the school more or less simultaneously, resulting in such a mass of eggs and sperm being released that visibility near the water's surface may be momentarily reduced. Based on the color patterns of the participants, Robertson, et al. (1979) suggested that each spawning group of *A. triostegus* consisted of a single female and several attendant males.

A somewhat different group spawning behavior has recently been reported for the blue-lined tang, *Acanthurus lineatus* (Johannes, 1981). Repeated spawnings were observed early in the morning, at peak high tide, on the new moon (this is the only report, thus far, of anything other than dusk spawning in the family). Groups consisted of five to 15 individuals that slowly spiraled upward to about 3 meters over the reef edge, released gametes at the top of the spiral, and then slowly returned to the bottom. Based on their general analysis of acanthurid social behavior, Robertson, et. al. (1979) accurately predicted that *A. lineatus* would prove to be a group spawner, but their data does not indicate why this species should differ so fundamentally from other acanthurids with respect to reproductive behavior or timing of spawning. Robertson, et. al. also did not see *A. lineatus* spawning, perhaps because of few early morning dives.

In the labrids and scarids the difference between pair spawning and group spawning, in terms of sperm competition, has resulted in significant differences in testes size between males in each category. Pair spawning males do not compete for fertilizations (though they do compete for females), shed relatively little sperm during each spawning rush, and consequently have relatively small testes. Group spawning males, however, must compete for fertilization even after attracting a female, so they produce copious amounts of sperm and have large and muscular testes. Robertson, et al. (1979) examined the gonads of males of ten species of acanthurids for which they either observed spawning or could make a reasonable guess at its form, and found a similar correlation. Mean testes weight of pair spawning species ranged from 0.3 to 0.7% of body weight, while that of group spawning species ranged from 1.7 to 3.7%.

Finally, both pair spawning and group spawning in the same species have been reported for several western Atlantic *Acanthurus* (Colin, in prep.), for *Zebrasoma flavescens* (Lobel, 1978, group spawning; P. Atkins, pers. comm., pair spawning), and for *Z. scopas* (Randall, 1961b, group spawning; Robertson, et al., 1979, pair spawning). The factors underlying such diversity of spawning systems within a given species are not clearly understood, but chief among them must be population density. Barlow (1974), for example, noted that in different areas the social units within various species ranged from pairs to large aggregations and suggested that reproductive systems would vary in a parallel manner. No data to test this hypothesis have been collected, though a similar variation in reproductive behavior of some labrids and scarids has been documented. Given the demonstrated lability of spawning system by at least these few species of acanthurids, it would not be surprising if numerous other species showed similar flexibility in response to changes in local conditions.

No hard data are available concerning frequency of spawning by acanthurids. Randall (1961a), however, found two size classes of eggs in the ovaries of *Acanthurus triostegus*, which he interpreted as suggesting multiple spawnings by each female. Indeed, based on the number of females he collected at any given time, he suspected that each female spawned once monthly. In the only estimate of fecundity in the family, Randall estimated that an average size female (123 mm) released 40,000 eggs during each spawning bout.

There is only one report of hybridization in surgeonfishes, that between *Acanthurus achilles* and *A. glaucopareius* in Hawaii (Randall, 1956b). The latter is apparently uncommon in the area, and Randall suggested that hybridization was the result of an inability of the fish to find a conspecific mate. Why *A. achilles*, which should have abundant potential mates, should participate in such spawnings is not clear.

EGGS AND LARVAE

So far as is known, the eggs of all surgeonfishes are spherical, transparent, nonadhesive, and positively buoyant. Egg diameters are available for only two species: the eggs of *Acanthurus triostegus* at Hawaii range from 0.66 to 0.70 mm in diameter and contain a single oil droplet 0.165 mm in diameter (Randall, 1961a); the egg of an unidentified acanthurid, also at Hawaii, was similar but smaller—only 0.575 to 0.625 mm in diameter (Watson & Leis, 1974). Egg development has been described only for *A. triostegus* (Randall, 1961a); hatching occurs in about 26 hours at 24°C.

The newly hatched prolarva of *A. triostegus* is approximately 1.7 mm long and has an immense spherical yolk sac that contains an oil droplet at its most posterior point. Such early larvae float head-down near the surface and are in such a rudimentary state that the heart doesn't begin beating until five hours after hatching. As the yolk is absorbed and the prolarva develops, the fish gradually loses buoyancy and begins to make short duration swimming movements to remain near the surface. The yolk is completely absorbed within one to two days of hatching, and by the second day the larva has begun to develop eye pigments, vertical fin folds, and pelvic fin rudiments (anlagen). After three days the jaws are developed, and after four, the gut. Feeding

presumably begins at this point.

Later stage acanthurid larvae are frequently collected in plankton tows and have been described for various genera by numerous workers, including Lütken (1880), Poey (1875), Randall (1955a & b, 1956, 1961a), Aboussouan (1965), Smith (1966), Fourmanoir (1976), and, in great detail, Burgess (1965). Such postlarvae are diamond-shaped and strongly laterally compressed, with a prominent serrated dorsal spine, two equally large and serrated pelvic fin spines, and a single smoother spine at the leading edge of the anal fin. Randall (1961a) suggested that these spines are poisonous and protect the larvae from at least some predators, but there is no direct evidence for such toxicity. Late-stage postlarvae are finely scaled with well-developed fins, prominent dorsal and anal spines, and still have a diamond-shaped profile. Immediately before settlement, however, the larvae undergo a striking transformation into a characteristic "acronurus" stage. These postlarvae are immediately recognizable as surgeonfishes because of their oval bodies, rounded foreheads, long and even dorsal and anal fins that lack prominent spines, and even small scalpels on the caudal peduncles. Acronurus larvae roughly 20 to 25 mm long are transparent except for a silvery abdomen and gut. At night they actively swim inshore, seek shelter in the reef, and begin final transformation into juvenile surgeonfish, a process that involves growth of scales and intestines (e.g., Clavijo, 1974) and takes five to seven days (Randall, 1961a; Burgess, 1965). Size of the settling acronurus is inversely proportional to water temperature, at least in *Acanthurus triostegus* (Randall, 1961a). Their development into juveniles is apparently stimulated by light levels; Randall (1961a), for example, found that acronurus of *A. triostegus* held for 12 hours in a darkened aquarium showed little development, while those normally illuminated were clearly developing pigment and juvenile characteristics.

Surgeonfish larvae are primarily found well offshore and are most abundant in the surface 100 m (Burgess, 1965; Sale, 1970; Miller, 1974). Duration of the planktonic stage has been estimated only for *A. triostegus*, at two and a half months (Randall, 1961a). The factors that affect larval settlement and habitat selection have been examined by Randall (1961a) and Sale (1968, 1969a & b). Juvenile *A. triostegus*, like juveniles of several other species (e.g., Robertson, et al., 1979) are found in shallower water than the adults. Sale found that such juveniles make a continuous active choice of their habitats and determine its suitabili-

newly hatched, 1·7mm

5 days, 2·6mm

2·8mm

5·4 mm

9·5 mm

26·5mm

Larval development of acanthurids. Newly hatched and 5-day-old larvae are of *Acanthurus triostegus,* based on Randall (1961). Older larvae are of Western Atlantic species of *Acanthurus,* based on Burgess (1965).

ty based on the amount and kind of cover available, the depth (the larger the juvenile, the deeper the water it preferred), and the presence of conspecifics. The amount of exploratory activity performed by a newly settled individual before settling into one place is an inverse function of the combined effect of these characters.

Juveniles of many acanthurids are differently colored than the adults and differ from them in behavior. The juvenile of at least one species, *Acanthurus pyroferus,* appears to be a mimic of an angelfish (*Centropyge flavissimus*), though the reasons for the mimicry are not clear (Randall & Randall, 1960). Juveniles of many species are more strongly territorial than the adults, which in many cases form large feeding groups (e.g., Robertson, et al., 1979; Thresher, 1980). Color changes between juveniles and adults may well reflect such striking changes in social behavior. In the western Atlantic, for example, juveniles of *Acanthurus coeruleus* are bright yellow and territorial, whereas adults are blue and gregarious; juveniles of *A. chirurgus* are colored much like the adults and are as gregarious as the adults. Juveniles of all species eat constantly and grow rapidly. *Acanthurus triostegus* reaches sexual maturity in one to two years at a length of approximately 10 cm (Randall, 1961a). Females of two western Atlantic species, *A. coeruleus* and *A. bahianus,* mature at about 15 cm, whereas males mature at approximately 19.5 and 17.5 cm, respectively (Reeson, 1975).

REPRODUCTION IN CAPTIVITY

There are no reports of spawning of surgeonfishes in aquaria, nor have the larvae been successfully reared to metamorphosis. Indeed, the only attempt thus far to rear the larvae resulted in nearly total mortality even before the yolk sac was completely absorbed (Randall, 1961a); there have been, however, no recent reports of attempts to rear such fish. In contrast to rearing, which will likely prove difficult because of the rudimentary nature of the newly hatched larvae and the long planktonic stage, spawning of at least some of the smaller species should be possible. Spawning of *Zebrasoma flavescens,* for example, may be possible given use of a large aquarium (probably no less than 500 liters), heavy feedings with live food and plant material, and four or five mature fish kept by themselves.

Literature Cited

Aboussouan, A. 1965. Oeufs et larves de Teleosteens de l'Ouest Africain. I. *Acanthurus monroviae* Steind. *Bull. de Inst. Fr. Afr. Noire, Ser. A,* 28:1183-1187.

Bagnis, R., P. Mazellier, J. Bennett, and E. Christian. 1972. *Fishes of Polynesia.* Les Editions du Pacifique, Papeete, Tahiti. 368 pp.

Barlow, G.W. 1974. Contrasts in social behavior between Central American cichlid fishes and coral-reef surgeon fishes. *Amer. Zool.,* 14:9-34.

Breder, C.M., Jr. 1949. On the taxonomy and post larval stages of the surgeon fish, *Acanthurus hepatus. Copeia,* 1949:296.

Burgess, W.E. 1965. A description of the larvae of the surgeonfish genus *Acanthurus* Forskal of the western North Atlantic. Master's Thesis, Univ. of Miami, Coral Gables, Fla. 59 pp.

Clavijo, I. 1974. A contribution on feeding habits of three species of Acanthurids (Pisces) from the West Indies. Master's Thesis, Florida Atlantic Univ., Boca Raton, Fla.

Fourmanoir, P. 1976. Formes post-larvaires et juveniles de poissons cotiers pris au chalut pelagique dans le Sud-Ouest Pacifique. *Cahiers du Pacifique,* (19):47-88.

Johannes, R.E. 1978. Reproductive strategies of coastal marine fishes in the tropics. *Env. Biol. Fishes,* 3:65-84.

Johannes, R.E. 1981. *Words of the Lagoon.* Univ. Calif. Press, Los Angeles, Calif. 320 pp.

Lobel, P.S. 1978. Diel, lunar, and seasonal periodicity in the reproductive behavior of the pomacanthid fish, *Centropyge potteri,* and some other reef fishes in Hawaii. *Pac. Sci.,* 32:193-207.

Lütken, C.F. 1880. Spolia Atlantica. Bidrag til kundskab om formforandringer hos fiske under deres vaext or udvikling, saerligt hos noble af Atlanterhavets Hojsofiske. *K. Danske. Vidensk. Selsk.,* 5:409-613

Miller, J.M. 1974. Nearshore distribution of Hawaiian marine fish larvae: effects of water quality, turbidity and currents. Pp. 217-231. *In: The Early Life History of Fish,* (J.H.S. Blaxter, Ed.). Springer-Verlag, N.Y. 765 pp.

Munro, J.L., V.C. Gaut, R. Thompson, and P.H. Reeson. 1973. The spawning seasons of Caribbean reef fishes. *J. Fish Biol.,* 5:69-84.

Poey, F. 1875. Enumeratio piscium cubensium, Part 1. *Soc. Esp. Hist. Nat.,* 4:75-161.

Randall, J.E. 1955a. A revision of the surgeon fish genus *Ctenochaetus,* family Acanthuridae, with descriptions of five new species. *Zoologica, N.Y.,* 40:149-168.

Randall, J.E. 1955b. A revision of the surgeon fish genera *Zebrasoma* and *Paracanthurus*. *Pac. Sci.,* 9:396-412.

Randall, J.E. 1955c. An analysis of the genera of surgeon fishes (family Acanthuridae). *Pac. Sci.,* 9:359-367.

Randall, J.E. 1956a. A revision of the surgeon fish genus *Acanthurus*. *Pac. Sci.,* 10:159-235.

Randall, J.E. 1956b. *Acanthurus rackliffei,* a possible hybrid surgeon fish (*A. achilles X A. glaucopareius*) from the Phoenix Islands. *Copeia,* 1956:21-25.

Randall, J.E. 1961a. A contribution to the biology of the convict surgeon fish of the Hawaiian Islands, *Acanthurus triostegus sandvicensis*. *Pac. Sci.,* 15:215-272.

Randall, J.E. 1961b. Observations on the spawning of surgeon fishes (Acanthuridae) in the Society Islands. *Copeia,* 1961:237-238.

Randall, J.E. and H.A. Randall. 1960. Examples of mimicry and protective resemblance in tropical marine fishes. *Bull. Mar. Sci.,* 10:444-480.

Reeson, P.H. 1975. The biology, ecology and bionomics of Caribbean reef fishes: Acanthuridae (surgeonfishes). *Res. Rpt. Zool. Dept. Univ. West Indies,* (3):61 pp.

Robertson, D.R., N.V.C. Polunin, and K. Leighton. 1979. The behavioral ecology of the Indian Ocean surgeonfishes (*Acanthurus lineatus, A. leucosternon* and *Zebrasoma scopas*):their feeding strategies, and social and mating systems. *Env. Biol. Fishes,* 4:125-170.

Russell, B.C., G.R.V. Anderson, and F.H. Talbot. 1977. Seasonality and recruitment of coral reef fishes. *Aust. J. Mar. Freshwater Res.,* 28:521-528.

Sale, P.F., 1968. Influence of cover availability on depth preference of the juvenile manini, *Acanthurus triostegus sandvicensis. Copeia,* 1968:802-807.

Sale, P.F. 1969a. Pertinent stimuli for habitat selection by the juvenile manini, *Acanthurus triostegus sandvicensis. Ecology,* 50:616-623.

Sale, P.F. 1969b. A suggested mechanism for habitat selection by the juvenile manini, *Acanthurus triostegus sandvicensis* Streets. *Behav.,* 35:9-44.

Sale, P.F. 1970. Distribution of larval Acanthuridae off Hawaii. *Copeia,* 1970:765-766.

Smith, J.L.B. 1951. Sexual dimorphism in the genus *Naso* Lacepede, 1802, with a description of a new species, and new records. *Ann. Mag. Nat. Hist.,* 12:1126-1132.

Smith, J.L.B. 1966. Fishes of the sub-family Nasinae with a synopsis of the Prionurinae. *Ichthyol. Bull. Rhodes Univ.,* (32):635-682.

Thresher, R.E. 1980. *Reef Fish.* Palmetto Publ. Co., St. Petersburg, Fla. 172 pp.

Tyler, J.C. 1970. Osteological aspects of interrelationships of surgeonfish genera (Acanthuridae). *Proc. Acad. Nat. Sci. Phila.,* 122:87-124.

Watson, W. and J.M. Leis. 1974. Ichthyoloplankton of Kaneohe Bay, Hawaii. *Sea Grant. Tech. Rpt. TR-75-01.*

Moorish Idol
(Zanclidae)

The Moorish idol, *Zanclus canescens,* is the sole member of this family. Closely related to the surgeon-fishes (Acanthuridae), to the point where some authors consider it in the same family, it shares many acanthurid features—a deep, strongly compressed body; long, low, and same-sized dorsal and anal fins; and a similar overall appearance. Unlike the surgeon-fishes, the Moorish idol is characterized by an elongate snout (used to pick out food items from within the interstices of coral) and a long flowing extension of the dorsal fin arising at its anterior base.

Despite its prominence in the aquarium trade and its conspicuousness, abundance, and wide distribution on the reef, almost nothing is known about spawning or reproduction in this species. Given its systematic affinities, it is likely, however, that it produces pelagic eggs, spawns at dusk, and does so on the outer reef slopes. Moorish idols are commonly observed in pairs or in small groups, suggesting that spawning occurs in pairs rather than as a group ascent. About all that is known for sure about their reproduction is that the late stage larvae are very similar to the acronurus stage of acanthurids. Strasburg (1962) described in detail two such larvae, 13.4 and 16.0 mm long (SL), which he distinguished from acanthurid larvae on the basis of two features. Zanclid larvae have an elongate third dorsal spine (the precursor of the adult dorsal fin streamer) more than twice the length of the larva's body, and a prominent spine just above the corner of the mouth lacking in acanthurid larvae. This spine disappears after the larva settles to the bottom at a length of about 7.5 cm. The duration of the larval stage, like most aspects of zanclid reproduction, is not known. The distribution of the adult, from the Indo-West Pacific across to the eastern Pacific coast of North America, suggests a long planktonic existence.

Literature Cited

Strasburg, D.W. 1962. Pelagic stages of *Zanclus canescens* from Hawaii. *Copeia,* 1962:844-845.

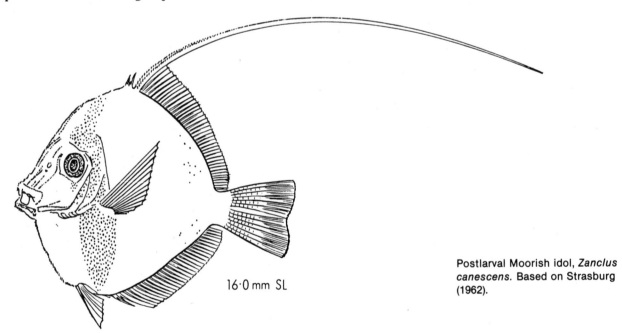

16·0 mm SL

Postlarval Moorish idol, *Zanclus canescens.* Based on Strasburg (1962).

Rabbitfishes
(Siganidae)

Rabbitfishes are fairly large, laterally compressed schooling fishes similar in shape and behavior to the related surgeonfishes. They are common around reefs and inshore areas of the tropical Indo-Pacific. Most have a slightly pronounced snout and many are deep-bodied. Color patterns range from bright yellow with black markings, as in *Siganus (Lo) vulpinus,* to various shades of gray, brown, or pale green with fine blue markings, as in most *Siganus.* As in most groups, the taxonomy of the siganids is not settled, but one recent estimate suggests there are probably 20 to 30 species in the family, all basically similar in most respects, with most, if not all, in the genus *Siganus* (Lam, 1974).

Like the related surgeonfishes, siganids are browsing herbivores feeding in small to large schools on a variety of benthic algae. In captivity they quickly adapt to a meat-based diet and thrive on almost anything. In the field they are common on reef flats, around scattered small coral heads, and near grass flats. A few species are primarily estuarine and have even been successfully introduced into freshwater lakes and ponds.

Breder & Rosen (1966) listed the siganids as one of the Recent families about whose spawning nothing was known. While there is still much to be done, recent interest in mariculture, for which the siganids are well suited (they are considered an excellent food fish), has lead to considerable work on their spawning and rearing.

SEXUAL DIMORPHISM

There is little in the way of definitive differences between the sexes of siganids, but in at least a few species, and probably all, females tend to be larger at maturity than males: the former range in size from 13 to 21 cm SL, the latter only 11 to 14 cm (Manacop, 1937; Lam, 1974). Drew (1971, cited in Lam, 1974) reported the average minimum size of a mature female *S. lineatus* as approximately 22 cm.

Other than size, Bryan, et al. (1975) reported that females of *S. canaliculatus* are slightly deeper-bodied than the males. Other differences become apparent during the spawning season when females are conspicuously heavier-bodied than males, tend to be less active, and have a more enlarged genital opening.

SPAWNING SEASON

Numerous studies (reviewed by Lam, 1974, and Johannes, 1978) report peak siganid spawning in spring and early summer, with a possible smaller peak late in the summer. Like many other coastal marine fishes, there is also a pronounced lunar rhythm to their spawning activity. Popper & Gunderman (1976) found that *S. vermiculatus* most often spawn during the night of the moon's first quarter. Spawning not occurring on this night took place within a few nights on either side of it (Popper, et al., 1976). Similar peaks in activity are reported for *S. canaliculatus* and *S. lineatus,* although *S. argenteus* and *S. spinus* apparently spawn a few days later in the lunar month (reviewed by Johannes, 1978, 1981). Popper, et al. (1976) also reported that siganid larvae follow a lunar rhythm in appearing, with most arriving inshore three to five days after the new moon.

REPRODUCTIVE BEHAVIOR

Siganids are characteristically seen in small to large schools or, in a few species, in pairs (e.g., Woodland & Allen, 1977; Johannes, 1981). So far as is known, however, all species, whether or not normally found in groups, aggregate to spawn. Spawning characteristically occurs at night or early morning, often on an outgoing tide (e.g., Manacop, 1937; Popper & Gundermann, 1975; Popper, et al., 1976; Bryan & Madraisau, 1977; Hasse, et al., 1977—though see also Johannes, 1981, for an account of spawning in mid to late afternoon and evening in *Siganus canaliculatus*), and is preceded by a migration to specific and ap-

Pair of *Siganus virgatus* prior to spawning. The female is above the male. Photo by Jack T. Moyer.

Female of *Siganus virgatus.* Note her rounded, egg-filled abdomen. Photo by Jack T. Moyer.

parently traditional sites. The location of such sites varies widely among species, ranging from near mangrove stands (*S. lineatus,* Drew, 1971), to shallow reef flats (*S. canaliculatus,* Manacop, 1937; Johannes, 1981), the outer reef crest (several species at Palau, McVey, 1972; Johannes, 1978), and even the deeper reef (*S. lineatus,* Johannes, 1981). At Palau, *S. canaliculatus* migrates from all around the island to spawn at one major site and a few minor ones, all inshore and characterized by easy access to the ocean via channels or major breaks in the barrier reef, large areas of sea grass flats nearby, and extensive mangrove stands inshore.

Apparent spawning in the field has thus far been described only for a pair of Red Sea species (Popper & Gundermann, 1975), although schools of *S. canaliculatus* moving about on reef flats just before low tide have been observed in what was probably spawning behavior (Hasse, et al., 1977; Johannes, 1981). In the Red Sea, *S. rivulatus* and *S. luridus* spawn early in the morning in the summer. During such periods both species break up their normal schools to form pairs or small groups. Each group or pair acts "aggressively" toward others, with the result that they are widely spaced over the reef. The apparent male (or males) in each group begins to "nibble" at the flanks of the female(s), resulting in the pair swimming in a circling motion. The males, meanwhile, also develop a marbled pattern (in contrast to their normal uniform gray) and the circling continues for 10 to 15 minutes. Popper and Gundermann assumed that eggs were released during the circling, though none were seen. Given the small size of the eggs, a visible release of gametes might not be expected. A similar circling during apparent spawning occurs in *S. canaliculatus* at Palau (Johannes, 1981), suggesting such behavior may be characteristic of spawning in the family.

Pair spawning has also been reported for *S. canaliculatus* and, apparently, *S. lineatus* in captivity (McVey, 1972, and Bryan & Madrausau, 1977, respectively). In the former, the female nudges the abdomen of the male, rather than vice versa, which McVey suggests might stimulate him to release milt. The female then follows behind the male, releasing her eggs. There does not seem to be any special provision for ensuring complete fertilization.

EGGS AND LARVAE

Siganid eggs are spherical, adhesive, demersal, transparent, colorless, and small, ranging in diameter from 0.42 mm (*S. canaliculatus*–Soh & Lam, 1973) to 0.66 mm (*S. fuscescens*–Fujita & Ueno, 1954). They also contain a variable number of different sized oil droplets. Lam (1974) suggested that a large female may release up to 500,000 eggs at a time, simply scattering them over the bottom. Despite their weight, Hasse, et al. (1977) suggested that the eggs of at least *S. canaliculatus* may be pelagic. Based on the location of the spawning sites, i.e., areas that seem to optimize offshore water movement, and the lack of visible eggs on the bottom after a night of spawning, they hypothesize that the adhesive eggs stick to floating pieces of plant material and other flotsam and so remain in the water column until hatching. If the eggs are, in fact, demersal, there are no reports of parental care in the group.

Egg development is rapid, and hatching occurs in 25 to 32 hours at 27 to 29°C (McVey, 1972; Popper, et al., 1973; Westernhagen & Rosenthal, 1975). Newly hatched larvae range in size from 0.76 mm (*S. canaliculatus*–McVey, 1972) to 2.60 mm (*S. fuscescens*–Fujita & Ueno, 1954), have a small yolk sac containing one to several oil droplets near the anteriormost point, and lack pectoral fin folds, a functional mouth, or pigment in the eyes. Such larvae are photopositive and drift head-down at the surface (Bryan & Madraisau, 1977).

Time to first feeding may vary among species. The yolk sac persists for two to three days, and Popper, et al. (1977) suggested that first feeding by *S. rivulatus* occurs on day four or five, after the mouth has developed and the anus is open. Bryan & Madraisau (1977), however, reported phytoplankton in the guts of larval *S. canaliculatus* within a day of hatching. Duration of the planktonic stage is about 25 days (23 days for *S. canaliculatus*–May, et al., 1974; 29 days for *S. lineatus*–Bryan & Madraissau, 1977), with newly metamorphosed fish between 20 and 26 mm long. In *S. lineatus,* imminent metamorphosis is signaled by an increased interest in the bottom by the larvae (they begin to pick at the substrate), a change in head color from gray to brown, and a marked increase in the length and complexity of the digestive tract (reflecting the shift from feeding on animal material, i.e., zooplankton, to algae).

Juvenile siganids are usually mottled gray and brown, form small schools, and graze along the bottom for benthic algae. Lam (1974) reported that juvenile *S. canaliculatus* appear in late spring, reach sexual maturity by the following winter, and spawn that spring, after which most fish apparently die.

Males apparently reach sexual maturity earlier than females, which would not be surprising given the differences in sizes of the sexes at maturity.

REPRODUCTION IN CAPTIVITY

Natural (not hormonally induced) spawning in captivity has been detailed for three species: *S. canaliculatus* (McVey, 1972), *S. rivulatus* (Popper, et al., 1973), and *S. lineatus* (Bryan & Madraisau, 1977). Westernhagen & Rosenthal (1975) also reported spawning in captivity by *S. oramin* or *S. concatenata*. In each case very large aquaria were used (the smallest, used by Popper, et al., was 2 m³). Bryan & Madraisau (1977) provided few details about spawning of *S. lineatus,* only that it occurred in 7,000-liter concrete tanks, that five fish were present, and that only one pair spawned. McVey induced *S. canaliculatus* to spawn by transferring them from a circular tank 0.9 m deep to a broad and flat one only 18 to 23 cm deep, after which spawning began immediately. Similarly, Popper, et al., induced spawning in ten hours by completely changing the water in the holding tanks. Combined with the field observations, these suggest that an environmental change, probably a drop in water levels, stimulates spawning in these fishes.

Larvae are most frequently raised by feeding them a combination of cultured rotifers (*Brachionis plicatilis*) and naturally occurring copepods (though Popper, et al., 1976, used only the latter, rearing the larvae in "green," fertilized 80 m² concrete pools). Bryan & Madraisau (1977) examined the diet of larval *S. lineatus* in detail and found that for the first few days they ate an unidentified phytoplankter, then took rotifers from day two to day 14, and thereafter ate *Artemia* nauplii and the nauplii and adults of the marine copepod *Oithono* sp. May, et al. (1974) reported increased larval mortality with heavy feedings of *Artemia,* but Bryan & Madraisau had no such problem. The latter authors did find, however, that strong light and high phytoplankton density adversely affected survival of *S. lineatus* and *S. canaliculatus* during the first few days after hatching and recommended the use of plastic filters to reduce light levels.

No information is yet available regarding spawning and rearing of the foxface, *Siganus (Lo) vulpinus,* a brightly colored and popular aquarium fish. Based on work with confamilials, however, inducing *S. vulpinus* to spawn in captivity should not be difficult, given an aquarium large enough to comfortably hold five or six

No information is as yet available on the spawning and rearing of the foxface, *Lo vulpinus,* but induced spawning should not prove difficult given adequate conditions. Photo by Gerhard Marcuse.

fish and some provision for either reducing water level or wholly or partially changing the water in the aquarium. Efforts would probably be most successful a few days after the new moon, with actual spawning probably occurring early in the morning. Despite the probable small size of eggs and larvae, there should be few major problems in rearing the latter given a good algal culture and adequate feedings with rotifers and *Artemia*.

Literature Cited

Breder, C.M., Jr. and D.E. Rosen. 1966. *Modes of Reproduction in Fishes*. Natural History Press, Garden City, N.Y. 941 pp.

Bryan, P.G. and B.B. Madraisau. 1977. Larval rearing and development of *Siganus lineatus* (Pisces: Siganidae) from hatching through metamorphosis. *Aquacult.*, 10:243-252.

Bryan, P.G., B.B. Madraisau, and J.P. McVey. 1975. Hormone-induced and natural spawning of captive *Siganus canaliculatus* (Pisces: Siganidae) year round. *Micronesica*, 11:199-204.

Fujita, S. and M. Ueno. 1954. On the development of the egg and prelarval stages of *Siganus fuscescens* (Houttuyn) by artificial insemination. *Japan. J. Ichthyol.*, 3:129-132.

Hasse, J.J., B.B. Madraisau, and J.P. McVey. 1977. Some aspects of the life history of *Siganus canaliculatus* (Park) (Pisces: Siganidae) in Palau. *Micronesica*, 13(2):297-312.

Johannes, R.E. 1978. Reproductive strategies of coastal marine fishes in the tropics. *Env. Biol. Fishes*, 3:65-84.

Johannes, R.E. 1981. *Words of the Lagoon*. Univ.Calif. Press, Los Angeles, Calif. 320 pp.

Lam, T.J. 1974. Siganids: their biology and mariculture potential. *Aquacult.*, 3:325-354.

Manacop, P.R. 1937. The artificial fertilization of dangit, *Amphacanthus oramin* (Bloch and Schneider). *Philippine J. Sci.*, 62:229-237.

May, R.C., D. Popper, and J.P. McVey. 1974. Rearing and larval development of *Siganus canaliculatus* (Park) (Pisces: Siganidae). *Micronesica*, 10:285-298.

McVey, J.P. 1972. Observations on the early-stage formation of rabbit fish, *Siganus fuscescens*, at Palau Mariculture Demonstration Centre. *South Pacific Isl. Fish. Newsletter*, 6:11-12.

Popper, D., H. Gordin, and G.W. Kissil. 1973. Fertilization and hatching of rabbitfish *Siganus rivulatus*. *Aquacult.*, 2:37-44.

Popper, D. and N. Gundermann. 1975. Some ecological and behavioral aspects of siganid populations in the Red Sea and Mediterranean coasts of Israel in relation to their suitability for aquaculture. *Aquacult.*, 6:127-141.

Popper, D. and N. Gundermann. 1976. A successful spawning and hatching of *Siganus vermiculatus* under field conditions. *Aquacult.*, 7:291-292.

Popper, D., R.C. May, and T. Lichatowich. 1976. An experiment in rearing larval *Siganus vermiculatus* (Valenciennes) and some observations on its spawning cycle. *Aquacult.*, 7:281-290.

Soh, C.L. and T.J. Lam. 1973. Induced breeding and early development of the rabbitfish, *Siganus oramin* (Schneider). *Proc. Symp. Biol. Res. Nat. Dev.*:49-56.

Westernhagen, H. von and H. Rosenthal. 1975. Rearing and spawning siganids (Pisces: Teleostei) in a closed seawater system. *Helgo. Wissenschaft. Meeresunters.*, 27:1-18.

Woodland, D.J. and G.R. Allen. 1977. *Siganus trispilos*, a new species of Siganidae from the Eastern Indian Ocean. *Copeia*, 1977:617-620.

Tetraodontiformes

The fishes of the order Tetraodontiformes, or Plectognathi, are conspicuous, if somewhat bizarre, elements of the coral reef fish fauna. Winterbottom (1974) recognized eleven extant families in the order, four of which are common on or around coral reefs: Balistidae (triggerfishes and filefishes), Ostraciidae (trunkfishes and cowfishes), Tetraodontidae (puffers and sharpnose puffers), and Diodontidae (burrfishes and porcupinefishes). All are related to the giant ocean sunfish, *Mola mola,* and resemble it about as much as they resemble each other; that is, not very much. The entire group, about 300 species strong, is believed to be derived from some surgeonfish-like ancestor, to which the triggerfishes bear a vague resemblance.

Information on spawning by members of the order has until recently been sparse, restricted mainly to a few notes on sexual dimorphism (e.g., Fraser-Brunner, 1940) and early stages of development (e.g., Welsh & Breder, 1922). Spawning by some temperate water tetraodontids, however, had been described many years ago (Welsh & Breder, 1922; Schreitmüller, 1930) based mainly on spawnings in large aquaria.

Literature Cited

Fraser-Brunner, A. 1940. Notes on the plectognath fishes IV. Sexual dimorphism in the family Ostraciontidae. *Ann. Mag. Nat. Hist., London, Ser. 11,* 6:390-392.

Schreitmüller, W. 1930. Kugelfische. *Das Aquarium, Berlin,* Jan.:12-16; Feb.:20-26.

Welsh, W.W. and C.M. Breder, Jr. 1922. A contribution to the life history of the puffer, *Spheroides maculatus* (Schneider). *Zoologica, N.Y.,* 2:261-276.

Winterbottom, R. 1974. The familial phylogeny of the Tetraodontiformes (Acanthopterygii: Pisces) as evidenced by their comparative myology. *Smithsonian Contrib. Zool.,* (155):201 pp.

Triggerfishes and Filefishes (Balistidae)

The merging of the filefishes and triggerfishes into a single family is a recent development, though their close relationship has long been recognized. Both are characterized by strong lateral compression, a high profile, prominent beak-like jaws, and two dorsal fins. The second dorsal, which is long and low, is directly above an almost identical anal fin, and the primary mode of swimming is by means of slow, steady flapping of these fins. Differences between the two groups involve size, body shape, and the structure of the first dorsal fin. Triggerfishes tend to be large and often colorful and are only slightly elongate when viewed in profile. The first dorsal fin consists of one large and two smaller spines. The large anterior spine can be locked in an erect position by means of a "trigger" mechanism involving the two smaller spines. When frightened, triggerfishes typically dash head-first into a crevice or hole and, by means of the locking dorsal spines, wedge themselves in until the danger has passed. In contrast, filefishes tend to be small, drab, and often more elongate. The first dorsal fin usually consists of a single spine that has developed into a slender, finger-like "file," often followed by a second very small spine.

SEXUAL DIMORPHISM

Sexual dimorphism is widespread, though not universal in the triggerfishes. Two of the five recognized species of *Xanthichthys*, for example, are permanently dichromatic, a third is temporarily dichromatic, and two are monochromatic. In *X. mento* (Pacific) and *X. auromarginatus* (Indo-West Pacific), females are overall more drab than the males and differ in color of the second dorsal, anal, and caudal fins (Berry & Baldwin, 1966; Randall, et al., 1978). Females of *X. mento* have broad maroon margins on the dorsal and anal fins and a bright yellow-orange margin on the caudal fin; males have yellow margins on the dorsal and anal fins and a red margin on the caudal fin. Similarly, the fins of female *X. auromarginatus* are brown; those of the male are bright yellow. The remaining three species of the genus are normally monochromatic, but at least one, *X. ringens*, is dichromatic during spawning, when male colors intensify and the unpaired fins, in particular, become more colorful than those of the female (C. Arneson, pers. comm.).

Similar differences have been reported for other triggerfishes. Several, and perhaps all, species of *Sufflamen* are sexually dimorphic, with the males larger and more colorful than females (e.g., Berry & Baldwin, 1966). Males of *Balistapus undulatus* are larger than females and are banded with orange on the body only (females are also banded on the head) (Matsuura, 1976). Males of *Balistes vetula* are somewhat brighter colored than females and have more pronounced filaments trailing from the dorsal, anal, and caudal fins (Breder & Rosen, 1966). Males also tend to be larger than females, though the size ranges overlap considerably (Aiken, 1975). Similarly, the males of *Odonus niger, Pseudobalistes fuscus,* and *Hemibalistes chrysopterus* are conspicuously larger than the females (Fricke, 1980). Finally, the sexes of *Melichthys niger*, a spectacular black triggerfish distributed circumtropically, differ in the shape of the lower jaw. Males have rounded, blunt lower jaws; females have a conspicuous notch below the jaw (Berry & Baldwin, 1966; Moore, 1967). The male also reaches a larger maximum size, 18.3 to 30.6 cm SL as compared to 19.2 to 23.0 cm SL for females. *M. niger* is apparently the only dimorphic species in the genus (Randall & Klausewitz, 1973).

Sexual dimorphism is less widely reported in the filefishes, suggesting it is less common and, where present, less conspicuous. The males of two western Atlantic species, *Monacanthus tuckeri* and *M. ciliatus*, have strongly pronounced ventral flaps that are lacking altogether in the females. The sexes of the latter species also differ in size (males average about 8 mm longer than females), in spination on the caudal peduncle (males have three to eight well developed recurved spines on each side, whereas females lack such spines), and in courtship colors (both sexes develop a spotted pattern, and the male further develops a bright green area on the membrane of his first dorsal fin and the ventral flap becomes peacock blue) (Clark, 1950). Sexual dimorphism has also been reported for several species of Red Sea filefishes (Clark & Gohar, 1953). Females of *Paramonacanthus oblongus* and, apparently, *P. barnardi* are smaller, more deep-bodied than the males, and have a more concave upper profile. Females also have more narrow dorsal and anal fins, while the males frequently have an exaggerated second dorsal ray. The sexes also differ in appearance in the reef-dwelling species of *Oxymonacanthus*. Females of *O. halli* have colorless dorsal and anal fins and a dusky olive-gray ventral flap with black margins; males have pale orange rays in the fins and a bright red, black-edged ventral flap. Color differences are similar in *O. longirostris*. Finally, Randall (1964) reported that males in the monospecific filefish genus *Amanses* have a group of five or six long stout spines on the side of the body between the soft dorsal and anal fins; females, in contrast, have a short brush-like mass of setae in the same position.

There has been, thus far, little work done on balistid gonad structure. Chiba, et al. (1976) reported the testes of the triggerfish *Canthidermis rotundatus* to be paired, flat, and elongate, tapering anteriorly; ovaries are paired, round, and in contact with each other. The gonads of both sexes are located just dorsally of the rectum. The testes of *Sufflamen verres* are gray-white, small, rounded, and paired; the ovaries are similar but larger and more elongate (pers. obs.). Both Chiba, et al. (1976) and Matsuura (1976) sought, but did not find, histological evidence of hermaphroditism in, respectively, *Canthidermis rotundatus* and *Balistapus undulatus*.

SPAWNING SEASON

Little is known about balistid spawning seasons, with little more than scattered notes about a few species available. Most of these suggest a limited spawning season but are often based on samples obtained near the edges of the tropics. Hildebrand & Cable (1930), for example, reported that *Monacanthus hispidus* spawns off South Carolina from May to November, with a peak in activity in July. *Odonus niger* and *Pseudobalistes fuscus* in the Red Sea and *Sufflamen verres* in the Gulf of California similarly spawn

only during the warmer parts of the year (Fricke, 1980, and pers. obs., respectively). More detailed data are available for a number of western Atlantic species (Randall, 1964; Munro, et al., 1973; and especially Aiken, 1975). Data for six species in an equal number of genera all suggest long spawning seasons, with peaks of activity in the spring, the fall, or both.

Thus far only limited data are available concerning lunar rhythms in balistids. Lobel & Johannes (1980) found nests of *Pseudobalistes flavimarginatus* only within one or two days of the new and full moons, suggesting a semilunar cycle for the species. Colin (pers. comm.) has also seen a cycle to nest-guarding activity of *Rhinecanthus aculeatus* at Enewetak Atoll; it is, however, a month-long one, with peaks of activity within a day or so of the full moon. In contrast, Fricke (1980) reported no conspicuous cycles in the spawning of two Red Sea species. Nonetheless, it seems likely that some form of lunar cycle will prove to be the case with most benthic-spawning balistids.

REPRODUCTIVE BEHAVIOR

The basic balistid spawning system involves production of demersal eggs that may or may not be tended. Among triggerfishes, demersal eggs have been reported for several species of *Balistes* (e.g., Garnaud, 1960; Lythgoe & Lythgoe, 1971; Strand, 1977; and probably Cousteau & Diolé, 1971 — they do not specifically identify the balistid involved, but from their photograph it appears to be *B. conspicillum*), *Sufflamen* (e.g., Ballard, 1970; Sweatman, pers. comm.; pers. obs.), *Odonus niger*, *Pseudobalistes fuscus* and *Hemibalistes chrysopterus* (Fricke, 1980), *Pseudobalistes flavimarginatus* and *Balistapus undulatus* (Lobel & Johannes, 1980), and *Rhinecanthus aculeatus* (P. Colin, pers. comm.). Demersal eggs for filefishes are reported for *Alutera schoepfi* (Clark, 1950) and *Monacanthus cirrhifer* (Fujita, 1955) and suggested by Hildebrand & Cable (1930) for *M. hispidus*.

Not all triggerfishes, however, apparently produce demersal eggs. What appears to be spawning by the sargassum triggerfish, *Xanthichthys ringens*, takes place well off the bottom over relatively deep water. According to C. Arneson (pers. comm.), the male and female rush horizontally at one another, cross bodies in an "X" (heads up), and spiral around one another into the water column. Similar behavior has been observed for the western Atlantic species *Balistes vetula* (P. Colin, pers. comm.). In neither case was release of gametes observed, but both observers are reasonably sure that the behavior observed was indeed

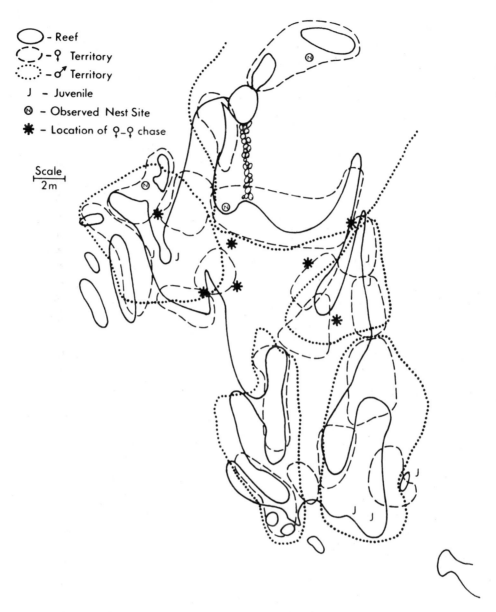

Territory configurations for males and females of the eastern Pacific triggerfish *Sufflamen verres* on a reef in the Gulf of California. The outlines are based on following each individual for a period of one hour.

spawning and, consequently, that the eggs are pelagic. The spiraling behavior is similar to that described for *Balistes bursa* by Salmon, et al. (1968) and considered by them to be an aggressive interaction, so that the behavior alone cannot rule out possibilities other than spawning.

The only detailed work thus far done on balistid social systems is by Fricke (1980), who worked for several years on the Red Sea species *Odonus niger* and *Pseudobalistes fuscus*. The overall system for both is harem-based and is strikingly similar to that of the eastern Pacific species *Sufflamen verres* (pers. obs.). In all three species adults of both sexes are strongly territorial, with the larger male defending an area that encompasses the territories of up to ten females in *O. niger* and up to five females in *S. verres* and *P. fuscus*. In the latter species juveniles and subadults are also territorial, vigorously defending small areas against one another that are sandwiched between the larger territories of the females. Spawning takes place in the female territories, apparently before dawn, and the eggs are tended and guarded principally, or exclusively, by the female. Fricke (1980) reported that in *O. niger* females aggregate before spawning, spawn synchronously, and then are simultaneously defended by the dominant male during brood-tending. The occasional observation of a large male vigorously chasing a smaller individual from his territory suggests the presence of surplus males in the population, corroborating the few histological studies on balistids that indicate gonochorism. Fricke (1980) noted in passing that the social system of *Hemibalistes chrysopterus* is similar to that of *P. fuscus* and *S. verres*.

Spawning behavior of demersal-spawning balistids has not yet been described. Lobel & Johannes (1980) saw *Pseudobalistes flavimarginatus* aggregating in loose associations over the spawning area just prior to the appearance of eggs. The fish frequently circled one another, occasionally nudging one another, a behavior they referred to as "carouseling" and which seems similar to the spiraling behavior of the apparently pelagic-spawning species. Carouseling was not seen during interspawning periods. Fricke (1980) reported courtship in *Pseudobalistes fuscus* to begin at dusk the day prior to spawning, with the male and female swimming back and forth between the nest site dug by the female and her sleeping site. Spawning itself occurs so early in the morning that detailed observations were not possible, though in *Odonus niger* the male was seen "shivering" briefly over the nest site of each female, presumably fertilizing her eggs.

After spawning, the demersal eggs are tended and guarded principally by the female in all cases where the sex of the defender has been determined (e.g., Ballard, 1970; Cousteau & Diolé, 1971; Strand, 1977; Fricke, 1980; Sweatman, pers. comm.; pers. obs.). Especially in the larger species, such defense can be extremely vigorous and can even involve attacks on divers. Triggerfishes are the only known group of reef fishes that have evolved extensive maternal care; in all other groups the eggs are cared for and guarded by the male or, in a few cases, both parents. Female parental care in balistids may be a consequence of the harem-based social structure of the fishes (though one could equally well argue that the haremic system results from the female parental care) and the short time that such care is required. In order to maintain a harem a male must vigorously defend it from other males, something he could not do if he were required to tend demersal eggs. Female territories, however, are much smaller than those of the male, so that even while tending eggs the female could still maintain her position in the social structure. Moreover, the eggs are clearly the fastest hatching eggs yet reported for a demersal-spawning reef fish, so that even the female is not "burdened" with parental duties for long. Given this apparent relationship between a harem-based social system and maternal care, it would be interesting to determine if non-harem-forming trigger-fishes defend demersal eggs and, if so, the identity of the defender.

EGGS AND LARVAE

The eggs and larval development of balistids are poorly known. All, except possibly those western Atlantic species that may spawn in the water column, produce demersal and adhesive eggs. Those of file-fishes have been described for only two species: the eggs of *Monacanthus cirrhifer* are approximately 0.66 mm in diameter and spherical (Fujita, 1955); those apparently of *Alutera schoepfi* are spherical, adhesive, and bright green (Clark, 1950). Triggerfish eggs have been described for species in five genera, *Balistes, Odonus, Pseudobalistes, Balistapus,* and *Sufflamen* (Garnaud, 1960; Fricke, 1980; Lobel and Johannes, 1980; and Thresher, unpubl. data). All are spherical, slightly over 0.5 mm in diameter, and translucent. The adhesive eggs are deposited, as a flattened circular mass containing large amounts of sand and rubble, in shallow pits excavated by the parents.

The eggs hatch the first night after being laid, indicating an incubation time of perhaps as little as 12 hours and no more than 24 hours, one of the shortest of any demersal spawning reef fish. Filefish eggs apparently take longer to hatch: Fujita (1955) reported 58 hours for the eggs of *Monacanthus cirrhifer* at 20°C.

Lobel and Johannes describe the newly hatched larvae of *Pseudobalistes flavimarginatus* as "free-swimming and active," about 1 to 2 mm long, and carrying a small yolk sac. Larval development of balistids has been described in greatest detail for the western Atlantic species *Monacanthus hispidus* by Hildebrand and Cable (1930) and by Aboussouan (1966), who based their descriptions on larvae collected in plankton tows. The smallest specimens of *M. hispidus* obtained were 1.7 mm long and were collected close to the bottom, suggesting they had recently hatched from demersal eggs. Such larvae are robust and elongate, with a long and slender pointed tail, a large and blunt head, small terminal jaws, and a single prominent spine dorsally. The eye is large, well developed, and already pigmented. By a length of 2.8 mm, the soft dorsal, anal, and caudal fin folds are distinct and the jaws have become very blunt. By 15 mm the larvae are virtually identical to the adults in shape and finnage and are easily recognizable as filefishes. *M. hispidus* is pelagic as an adult (living in drifting mats of *Sargassum*), so there is no conspicuous metamorphosis or settlement to the bottom. The development of larval *Balistes forcipatus* is

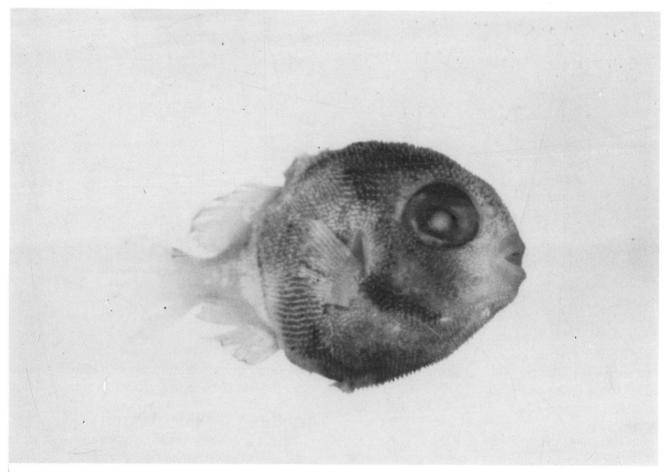

Larval balistid, 7.0 mm total length, collected in a plankton tow off Miami, Florida. Photo by Dr. R. E. Thresher.

Larval filefish, 4.4 mm total length, collected in a plankton tow off Miami, Florida. Photo by Dr. R. E. Thresher.

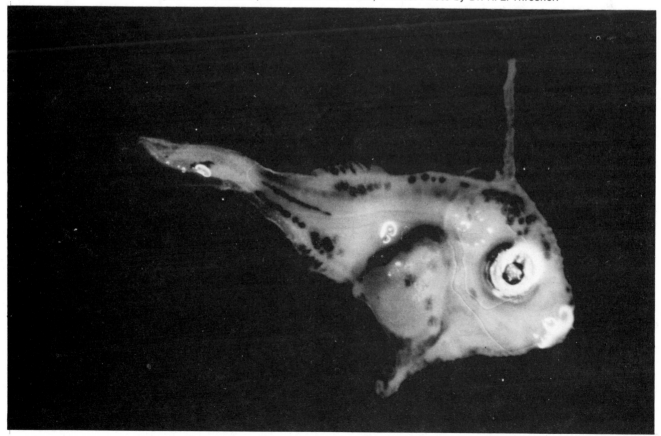

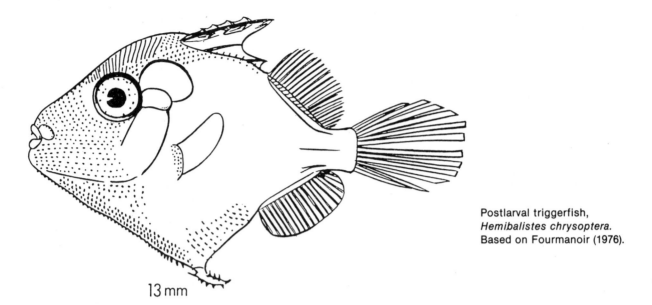

Postlarval triggerfish,
Hemibalistes chrysoptera.
Based on Fourmanoir (1976).

13 mm

similar. By the time they reach a length of 2.1 mm, such larvae have well developed pectoral fin rudiments, the beginnings of a caudal fin, and fully pigmented eyes.

Several triggerfishes have extremely prolonged planktonic stages and can reach a large size before settling to the bottom. This prejuvenile stage is normally reached at a length of 10 to 15 mm (Fourmanoir, 1976), by which time the fish are easily recognizable as triggerfishes. Prejuveniles of several species of *Melichthys*, however, can reach as much as 144 mm before settling and in the past have been described as distinct pelagic species (Randall, 1971; Randall & Klausewitz, 1973). The prejuvenile of *M. niger*, for example, is brass-colored with black-edged fins, a striking contrast to the jet black adult. The prejuveniles of one triggerfish, *Xanthichthys ringens,* are so commonly collected in drifting *Sargassum* patches that the species derives its common name, sargassum triggerfish, from the mottled green and brown prejuveniles, even though the adults are found entirely on the reef. Prejuveniles of some other species are also often associated with floating algae (e.g., Behrstock, 1979). Juvenile triggerfishes, as a rule, are not so brilliantly colored as the adults, and many are spotted or otherwise cryptically colored (e.g., Jonklaas, 1973). Small juvenile filefishes are cryptic and frequently resemble bits of plant material (e.g., Randall & Randall, 1960).

Berry & Baldwin (1966) suggested that maturation of *Sufflamen verres* and *Melichthys niger* occurs at approximately half maximum size. Such fish are probably a year or more old. The smaller filefishes prob-ably mature within a few months after hatching, based on their often phenomenal growth rates in captivity.

REPRODUCTION IN CAPTIVITY

There is only one report of triggerfish spawning in aquaria, that for *Sufflamen capistratus* by Ballard (1970), who observed nest preparation and defense and care of the eggs in a large public aquarium. Given the large size of most adult triggerfishes (often more than 20 cm), this is likely to be the only way these fishes will be spawned. Unfortunately the smaller species, such as those in *Xanthichthys*, may spawn in the water column; if so, they will probably never be spawned without great difficulty in captivity.

Spawning filefishes, however, should be relatively easy, even though no such spawnings have been reported. Various species of *Monacanthus*, for example, adapt quickly to captivity, eat virtually anything, and grow at a phenomenal rate. Given a reasonable sized tank, 100 liters or so (the fish are occasionally aggressive and will need room to avoid one another), well provided with shelter, temperatures around 24°C, and heavy feedings of live and prepared foods, spawning such filefishes should prove simple once a pair is obtained.

Literature Cited

Aiken, K.A. 1975. The biology, ecology and bionomics of Caribbean reef fishes: Balistidae (Triggerfishes). *Res. Rpt. Zool. Dept. Univ. West Indies,* (3):57 pp.

Aboussouan, A. 1966. Oeufs et larves de Téléostéens de l'Ouest africain. III. Larves de *Monacanthus hispidus* (L.) et de *Balistes forcipatus* Gm. *Bull. Inst. Fr. Afr. noire, Ser. A., Sci. Nat.,* 28:276-282.

Ballard, J. 1970. Aquarium briefs—mating in captivity. *Bull. S. Afr. Assoc. Mar. Biol. Res.,* (8):31-33.

Behrstock, R.A. 1979. A juvenile ocean triggerfish, *Canthidermis maculatus* (Bloch) (Pisces, Balistidae) from the Gulf of California. *Calif. Fish and Game,* 65:169-170.

Berry, F.H. & W.J. Baldwin. 1966. Triggerfishes (Balistidae) of the eastern Pacific. *Proc. Calif. Acad. Sci., 4th Ser.,* 34:429-474.

Breder, C.M. Jr., and D.E. Rosen. 1966. *Modes of Reproduction in Fishes.* Natural History Press, Garden City, N.Y. 941 pp.

Chiba, A., S. Yoshie, and Y. Honma. 1976. Histological observations of some organs in the triggerfish, *Canthidermis rotundatus,* stranded on the coast of Niigata facing the Japan Sea. *Japan J. Ichthyol.,* 22:212-220.

Clarke, E. 1950. Notes on the behavior and morphology of some West Indian plectognath fishes. *Zoologica, N.Y.,* 35:159-168.

Clark, E. and H.A.F. Gohar. 1953. The fishes of the Red Sea: Order Plectognathi. *Publ. Mar. Biol. Stat. Al Ghardaqa (Red Sea),* (8):1-80.

Cousteau, J.-Y. and P. Diolé. 1971. *Life and Death in a Coral Sea.* A & W Visual Library, N.Y. 302 pp.

Fourmanoir, P. 1976. Formes post-larvaires et juveniles de poissons cotiers au chalut pelagique dans le sud-oest Pacifique. *Cahiers du Pacifique,* (19):47-88.

Fricke, H.-W. 1980. Mating systems, maternal and biparental care in triggerfish (Balistidae). *Z. Tierpsychol.,* 53:105-122.

Fujita, S. 1955. On the development and prelarval stages of the file fish, *Monacanthus cirrhifer* Temminck et Schlegel. *Sci. Bull. Fac. Agr., Kyushu Univ., Fukuoka,* 15:229-234.

Garnaud, J. 1960. La ponte, l'eclosion, la larve du baliste *Balistes capriscus* Linné 1758. *Bull. Inst. Ocean. Monaco,* 1169:1-6.

Hildebrand, S.F. and L.E. Cable. 1930. Development and life history of fourteen teleostean fishes at Beaufort, North Carolina. *Bull. Bur. Fish.,* 46:383-488.

Jonklaas, R. 1973. The meanest trigger of them all. *Tropical Fish Hobbyist,* 22:50-64.

Lobel, P.S. and R.E. Johannes. 1980. Nesting, eggs, and larvae of triggerfishes (Balistidae). *Env. Biol. Fishes.,* 5:251-252.

Lythgoe, J. and G. Lythgoe. 1971. *Fishes of the Sea* (1975 Edition). Anchor Press. Garden City, New York. 320 pp.

Matsuura, K. 1976. Sexual dimorphism in a triggerfish, *Balistipus undulatus. Japan. J. Ichthyol.,* 23:171-174.

Moore, D. 1967. Triggerfishes (Balistidae) of the western Atlantic. *Bull. Mar. Sci.,* 17:689-722.

Munro, J.L., V.C. Gaut, R. Thompson, and P.H. Reeson. 1973. The spawning seasons of Caribbean reef fishes. *J. Fish Biol.,* 5:69-84.

Nolan, R. 1975. The ecology of patch reef fishes. Ph. D. Diss., Univ. Calif., San Diego. 230 pp.

Randall, J.E. 1964. A revision of the filefish genera *Amanses* and *Cantherhines. Copeia,* 1964:331-361.

Randall, J.E. 1971. The nominal triggerfishes (Balistidae) *Pachynathus nycteris* and *Oncobalistes erythropterus,* junior synonyms of *Melichthys vidua. Copeia,* 1971:462-469.

Randall, J.E. and W. Klausewitz. 1973. A review of the triggerfish genus *Melichthys,* with description of a new species from the Indian Ocean. *Senckenberg. biol.,* 54:57-69.

Randall, J.E., K. Matsuura, and A. Zama. 1978. A revision of the triggerfish genus *Xanthichthys,* with description of a new species. *Bull. Mar. Sci.,* 28:688-706.

Randall, J.E. and H.A. Randall. 1960. Examples of mimicry and protective resemblance in tropical marine fishes. *Bull. Mar. Sci.,* 10:444-480.

Salmon, M., H.E. Winn, and N. Sorgente. 1968. Sound protection and associated behavior in triggerfishes. *Pac. Sci.,* 22:11-20.

Strand, S. 1978. Community structure among reef fishes in the Gulf of California: The use of reef space and interspecific foraging associations. Ph. D. Diss., Univ. Calif., Davis.

Trunkfishes
(Ostraciidae)

Trunkfishes and cowfishes are the familiar box-like fishes encased in living armor-plate. Like the related balistids, they swim by flapping their anal and dorsal fins back and forth, a slow but steady means of propulsion. Speed is apparently of little concern for these fishes, since they are protected from most predators by their rigid armor, by spines and ridges that adorn the armor of many, and possibly by a toxin most, if not all, secrete from glands in the skin (Thomson, 1964, 1969; but see also Moyer, 1979).

Members of the family Ostraciidae are distributed circumtropically, most inhabiting shallow inshore areas such as grass flats and rubble beds. Many species, especially the more colorful ones, are routinely found on coral reefs, however. All are apparently opportunistic predators, feeding on a wide variety of soft-bodied invertebrates picked from the bottom. The systematics of the family has not been examined in some time. Currently there are about a dozen genera recognized and about twice that number of species.

SEXUAL DIMORPHISM

The degree and type of sexual dimorphism vary widely among the ostraciids. On one extreme, the species of the genus *Ostracion*, found in the Indo-Pacific, are typically dichromatic. Females tend to be dark with many small pale spots; males tend to be larger and brightly colored, often yellow on each side and dark blue dorsally (e.g., Randall, 1972, 1975; Moyer, 1979). As in other families of sexually dichromatic reef fishes, such color differences resulted in males and females of some species being treated as separate species in the past, a situation which has now been largely corrected (e.g., Fraser-Brunner, 1940). On the other extreme, most Caribbean ostraciids (*Lactophrys* and *Acanthostracion*) seem to be only slightly dimorphic. Breder and Clark (1947) suggested that males of *A. quadricornis* are smaller than the females,

narrower across the carapace, and more brightly colored, but not distinctively so. At least one species of ostraciid, *Lactoria fornasini*, is temporarily dichromatic; males assume a distinctive black and yellow pattern during courtship (Moyer, 1979). In one of the few reports of morphometric differences between the sexes, Clark & Gohar (1953) reported that males of *Ostracion cyaneus* have a heavier, more convex snout than females.

Based on the similar color patterns of juveniles and females and the strikingly different color patterns of the males, Randall (1975) suggested that the species of *Ostracion* were probably protogynous hermaphrodites. Males and females, however, are nearly equal in size and, in other genera, males may even be the smaller sex, which would seem to make protogyny unlikely. Moyer (1979) histologically examined the gonads of *Lactoria fornasini* and found no evidence of sex reversal; he further suggested that both *L. diaphanus* and *Ostracion cubicus* are gonochoristic as well.

SPAWNING SEASON

Little work has been done on reproduction in ostraciids, so, not surprisingly, little is known about their spawning seasons. Munro, et al. (1973) suggested, based on gonad samples obtained at scattered times throughout the year, that the Caribbean species *Lactophrys triqueter* spawns from January to March; Breder & Clark (1947), however, collected eggs of this species throughout the summer off the coast of Florida. Munro, et al. also suggested that *Acanthostracion quadricornis* spawns throughout the year, with a peak in activity in the early spring. The only other data available are for three species (two *Lactoria* and *Ostracion cubicus*) off southern Japan, a warm temperate area. All three spawn from May to October at water temperatures between 21.5° and 29.5°C (Moyer, 1979; also below). All spawn daily, and there was no apparent relationship with either tide or phase of the moon.

REPRODUCTIVE BEHAVIOR

Ostraciid social behavior has thus far been described only for three species off the southern coast of Japan, *Lactoria fornasini*, *L. diaphanus*, and *Ostracion cubicus* (Moyer, 1979, 1980). They are all haremic, with each male defending a large (approximately 500 m²) territory within which three or four females and often a subordinate immature male forage. The females roam about vaguely defined home ranges within the male's territory and are themselves not territorial; in fact, they often form temporary aggregations. In the center of each male's territory there is usually a conspicuous large rock or outcrop, referred to by Moyer as the "rendezvous" rock, at which more than 90% of the spawnings observed took place. *L. fornasini* spawns shortly after sunset near the peak of this rock and not far above it; *L. diaphanus* and *Ostracion cubicus* spawn earlier in the day (in late afternoon) and go much higher in the water column, a difference Moyer attributes to the larger size of the latter two species and hence their relative immunity from predation. Overall spawning behavior is similar in the three species. In *L. fornasini*, the best studied of the three, spawning begins at dusk with the male searching about his territory for the females and returning to the rendezvous rock about once every three to five minutes, a site at which the ready females congregate within approximately 20 minutes after sunset. Upon locating such a female, the male approaches her, stops motionless about 60 cm from her, and develops a vividly contrasting black-and-yellow pattern. If the female is ready to spawn she rises slowly off the bottom, to be followed closely by the male. Several meters up she stops and turns to one side with her caudal fin curled tightly against her body. The male, whose colors have returned to normal, rises to her level and does the same, but facing in the opposite direction. Their caudal peduncles about 2 cm apart (at no time do the fish ever come into physical contact), the pair hover in place for a few seconds. Spawning occurs shortly thereafter and is synchronized by a sound produced by the male. He emits a high-pitched hum, at the end of which both fish release their gametes and rush to the bottom. Afterward the female departs and seeks shelter for the night, while the male continues to search for other females.

Spawning by *L. diaphanus* is virtually identical, except that the male does not develop a distinctive courting color. Rather, he courts by swimming rapidly in circles around the female. Spawning by *O. cubicus* was not observed closely but is apparently similar to that of *Lactoria*. Generally similar spawning behavior has also been recently observed in the western Atlantic species *Lactophrys triqueter* by Colin (in prep.).

Recent work by Moyer has focused on the central role in the spawning system of *Lactoria fornasini* of the rendezvous rock and on strategies between neighboring males concerning access to the rock and to the females. Large successful males apparently spend a substantial amount of time attempting to "steal" spawnings and females from their neighbors. Such a male adopts "female" colors and approaches the spawning rock of his rival. Upon finding a female, he quickly develops his courting colors and apparently attempts to solicit spawning, usually unsuccessfully. In most cases the resident male detects the intruder and vigorously chases after it. Moyer has also observed "ritualistic" spawning by males and females in midwinter, in which courtship and spawning proceed as normal, but in which no gametes are released. He interprets this as a means of strengthening the pair-bonds between a male and his females and improving the speed and efficiency of their spawning behavior (hence minimizing the likelihood of their being interrupted while spawning). One could also argue, however, that as an essentially tropical species at the temperate limits of its range, the species is "programmed" for year-round spawning and the lack of viable gametes in midwinter is an artifact of the, for it, abnormally low temperatures. If the Japanese population is not genetically isolated from more tropical populations, then selection for more precise locally adapted controls on spawning may be too slight to be effective.

EGGS AND LARVAE

Little is known about either ostraciid eggs or larvae. Leis (1978) described the eggs of Hawaiian ostraciids (*Lactoria* and *Ostracion*) as slightly oblong, containing less than ten oil droplets, and possessing a patch of "bumps" at one end. According to Mito (1962), the eggs of "Ostraciontina" are 1.62 to 1.96 mm in diameter and contain a single oil droplet. The eggs of *Acanthostracion quadricornis*, a western Atlantic species, are spherical, nonadhesive, and 1.4 to 1.6 mm in diameter. Each contains a single light amber colored oil globule about 0.15 mm in diameter and begins to float shortly after spawning. *A. quadricornis* eggs hatch in about 48 hours at 27.5°C into short chunky larvae that apparently begin life floating upside down (Breder & Clark, 1947; Palko & Richards, 1969). Early ostraciid larvae, 2.20 to 2.40 mm long at hatching, quickly develop the rudiments of their armor plating in the form of a thick pliable shell that completely encases them except for the caudal fin (Mito, 1962). About 114 hours after hatching *A. quadricornis* reaches a distinctive postlarval stage that is square-bodied, heavily armor-plated, and looks like nothing else so much as a swimming bean. These postlarvae and the juveniles are dark and spotted, resembling the females in color pattern. They are commonly collected in grass beds and other shallow areas and are rarely seen on the reef. In contrast, the young of some species of *Ostracion*, though similar to *A. quadricornis* in shape, are brighter colored (yellow with black spots) and are commonly seen on shallow reefs, especially in late summer. The duration of the planktonic stage, in general, seems to be short in the trunkfishes, in contrast to other tetraodontids. It is therefore of some interest that Moyer (1980) reported collecting quite large specimens, up to 90 mm long, still in a pelagic color pattern (dark blue above, transparent below), suggesting that some species may be facultatively able to prolong the planktonic stages. Color changes associated with growth in the western Atlantic species *Lactophrys trigonus* were detailed by Tyler (1967).

REPRODUCTION IN CAPTIVITY

There are no reports of spawning trunkfishes in aquaria, nor are any likely in anything smaller than a large public aquarium, given the relatively large size of the adults at maturity (most greater than 20 cm) and the often great heights off the bottom such fishes ascend to spawn. Palko & Richards (1969) reared the larvae of *Acanthostracion quadricornis*, collected as eggs from the plankton, by feeding them wild-caught and size-sorted plankton.

Literature Cited

Breder, C.M., Jr. and E. Clark. 1947. A contribution to the visceral anatomy, development, and relationships of the Plectognathi. *Bull. Amer. Mus. Nat. Hist.*, 88:287-310.

Clark, E. and H.A.F. Gohar. 1953. The fishes of the Red Sea: Order Plectognathi. *Publ. Mar. Biol. Stat. Al Ghardaqa (Red Sea)*, (8):1-80.

Fraser-Brunner, A. 1940. Notes on the plectognath fishes. IV. Sexual dimorphism in the family Ostraciontidae. *Ann. Mag. Nat. Hist.*, London, Ser. 11, 6:390-392.

Leis, J.M. 1978. Systematics and zoogeography of the porcupinefishes (*Diodon*, Diodontidae, Tetraodontiformes), with comments on egg and larval development. *Fish. Bull (U.S.)*, 76:535-567.

Mito, S. 1962. Pelagic fish eggs from Japanese waters. 8. Chaetodontina, Balistina and Ostraciontina. *Sci. Bull., Fac. Agric. Kyushu Univ.*, 19:503-506.

Moyer, J.T. 1979. Mating strategies and reproductive behavior of ostraciid fishes at Miyake-jima, Japan. *Japan. J. Ichthyol.* 26:148-160.

Moyer, J.T. 1980. The mating strategy of the thornback cowfish ("Shimaumisuzume") *Lactoria fornasini*. *ANIMA*, Oct. 1980:50-55.

Palko, B.J. and W.J. Richards. 1969. Rearing of cowfish. *Salt Water Aquarium*, 5(3):67-70.

Randall, J.E. 1972. The Hawaiian trunkfishes of the genus *Ostracion*. *Copeia*, 1972:756-768.

Randall, J.E. 1973. *Ostracion trachys*, a new species of trunkfish from Mauritius (Ostraciontidae). *Matsya*, 1:59-62.

Thomson, D.A. 1964. Ostracitoxin: an ichthyotoxic stress secretion of the boxfish, *Ostracion lentiginosus*. *Science, N.Y.*, 146:244-245.

Thomson, D.A. 1969. Toxic stress secretions of the boxfish *Ostracion meleagris*. *Copeia*, 1969:335-352.

Tyler, J.C. 1967. Color pattern changes with increasing size in the western Atlantic trunkfish *Lactophrys trigonus*. *Copeia*, 1967:250-251.

Puffers
(Tetraodontidae)

The puffers share many features with the ostraciids, for example a heavy rotund body, slow and deliberate swimming that involves mainly the dorsal and anal fins, and a prominent beak-like mouth. They lack the armor plating characteristic of the ostraciids, however, and have instead developed the ability to inflate themselves into a sphere, apparently to deter would-be predators. The ability to inflate is also characteristic of the related porcupinefishes, family Diodontidae, but members of the two families can readily be distinguished by their appearance. Puffers are covered by only minute spines, which in many species are so small that the fish appear smooth; porcupinefishes, not surprisingly, have long and prominent spines.

Puffers are found in both temperate and tropical seas and, as a rule, are characteristic of inshore areas such as grass flats and mangrove swamps. Many are found on the reef only sporadically, though a few of the larger species, such as those in the genus *Arothron*, are typically found on or near reefs. One subfamily of small puffers is also typical of the reef—the Canthigasterinae, sharpnose puffers, are colorful fishes that rarely exceed 12 centimeters in length and that are often seen poking about the base of coral heads and rocky crevices. Like most puffers, though, they seem to prefer mixed zones of coral, rock, and sand or the edges of the reef rather than the reef itself.

Virtually no work has been done on reproduction of reef-dwelling puffers, but a fair amount has been done on the inshore temperate species. Based on the latter, some reasonable assumptions about spawning by the more tropical species are possible.

SEXUAL DIMORPHISM

Among the tetraodontin puffers sexual dimorphism seems to most often involve size: males are characteristically smaller and more brightly colored than females (e.g., Welsh & Breder, 1922; Abe, 1954; Uno,

1955). Whether such is also the case for the reef-dwelling species is not known. Among the sharpnose puffers, most appear to be monochromatic (e.g., Fraser-Brunner, 1940; Allen and Randall, 1977), though personal observations suggest that in at least some species the male may be somewhat larger than the female. Clark (1950) suggested that males of *Canthigaster rostrata* have a prominent point on the dorsal midline, immediately above the gill opening, that is lacking in the female. The sexes of *Canthigaster punctatissima*, an eastern Pacific species, differ in both size and color. Males are larger and green-spotted (over black); females have white spots.

SPAWNING SEASON

Nothing concrete is known about the spawning seasons of reef-dwelling species; temperate species, however, appear to spawn mainly in the warmer part of the year (e.g., Welsh & Breder, 1922; Kusakabe, et al., 1962; Habib, 1979). Uno (1955) reported a strong lunar cycle in the spawning of *Sphoeroides niphobles* (as *Fugu niphobles*) off Japan, but the spawning behavior of this species, involving deposition of eggs on a beach at the peak of high tide, differs so dramatically from that of other tetraodontids that it seems unwise to generalize by imparting such a lunar pattern to other members of the family.

REPRODUCTIVE BEHAVIOR

Temperate puffers typically lay demersal adhesive eggs which may or may not be tended by the male (e.g., Welsh & Breder, 1922; Munro, 1945), as do a number of small freshwater species (e.g., Schreitmüller, 1930; Krapp, 1974). The eggs are deposited either on a rock or in a shallow depression in the sand and are laid and fertilized by the parents while in a side-by-side position. Very different spawning has been described for two species, however. *Tetraodon schoutedeni*, an African freshwater puffer, spawns near the surface with two or more males biting and hanging on to the ventral surface of the female. The eggs are scattered about while the fish gyrate in this position and apparently fall to the bottom (Feigs, 1955; Mercken, 1958). *Sphoeroides niphobles* spawns on beaches, swimming ashore in large numbers at high tide and leaving its eggs partially buried in sand and rubble (Uno, 1955). Again, spawning is stimulated by two to four males that bite and hang on to the female as they lie together on the beach.

Given this range of spawning behavior, it is difficult to predict that of the larger reef-dwelling species, except to say that it probably involves demersal eggs. Nothing is known yet about the spawning of large puffers such as those in the genera *Sphoeroides* and *Arothron*, nor has conspicuous egg tending been reported. The absence of the latter suggests either egg-scattering, deposition of demersal eggs that are not tended by the adults, or egg tending that is not conspicuous. Males, for example, may simply lie on the eggs when approached, such that one would mistake egg tending for simply an individual lying on or partially buried in the sand (a common observation for many species of puffer).

Spawning by one of the smaller species, however, in the subfamily Canthigasterinae has recently been examined by Gladstone (in prep.). Preliminary results indicate that the common western Pacific species *Canthigaster valentini* has a social system based on three types of social groupings: harems, pairs, and solitary individuals. Harems consist of a single dominant male and from four to seven females, each of which defends its own territory within the larger area defended by the male. Pairs, in turn, jointly defend a much smaller territory and spend most of the day together. Finally, solitary individuals appear to be excess males that are attempting to gain an area within established territories. Spawning takes place in mid-morning and, in the haremic fish, apparently involves a different female each day. Courtship takes between five and 35 minutes and consists of the male following closely behind the female, displaying to her and nudging her. It is apparently the female that chooses the spawning site, a tuft of algae growing on a piece of coral rubble, at which she begins to pick while the male continues to display and nudge her and also begins to go through spawning motions. The female initiates spawning by pressing her ventral surface onto the algae, whereupon the male lies across her and, during a period of five to eight seconds, approximately 350 eggs are extruded into the algae. After spawning, the

male departs immediately, while the female remains at the nest site for an additional two to three minutes, pressing into the algae and apparently fanning the eggs deeper into the algal tuft. Although spawning sites are always within the boundaries of a female's territory, she apparently does not re-visit the nest after spawning has ended. Interestingly, considering their apparently close relationship, a very similar spawning behavior has recently been described for a freshwater puffer, *Carinotetraodon somphongsi*, by Richter (1982). Paired spawning, in the aquarium, occurred in tufts of a stringy green aquatic plant, the spawners shedding large, spherical eggs throughout. Unlike its marine relative, however, the male of *C. somphongsi* apparently tends the eggs after spawning.

Social organizations basically similar to that of *C. valentini* appear to be common for other canthigasterin puffers, including the western Atlantic *C. rostrata* and eastern Pacific *C. punctatissima*, both of which tend to be strongly territorial and often seen in pairs (Thresher, 1980; pers. obs.). In the latter species, individuals invariably bear numerous scars and torn fins by the end of the summer. On three occasions, pairs were observed to dash into the water column, with one hanging onto and apparently wrestling with the other. Such dashes, all of which were observed in midmorning, took the pair 3 to 5 m off the bottom. While such interactions looked more like the climax of a particularly vigorous fight than spawning, the latter is possible, especially considering the behavior of the related *Tetraodon schoutedeni*. In the only decent photograph obtained of such a pair, the female is biting the male, holding him near, if not on, the operculum.

EGGS AND LARVAE

Most of the available literature on eggs and larvae concerns temperate species but is so consistent that the reef-dwelling puffers are likely to be similar. Puffer eggs are spherical, approximately 0.9 to 1.4 mm in diameter, contain a "frothy" mass of tiny oil globules, and are covered with an adhesive coating (Welsh & Breder, 1922; Breder & Clark, 1947; Munro, 1945; Fujita & Ueno, 1956; Uchida, et al., 1958). The egg of the tropical inshore species *Lagocephalus lunaris* from Japan is similar but smaller, only 0.61 to 0.70 mm in diameter (Fujita, 1966). Incubation periods given range from four to 40 days at temperatures from 26° down to 11°C but are likely to be much shorter at the higher temperatures characteristic of the reef. The eggs of *Canthigaster valentini*, for example, hatch in about 105 hours (Gladstone, in prep.).

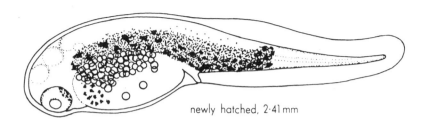

newly hatched, 2·41 mm

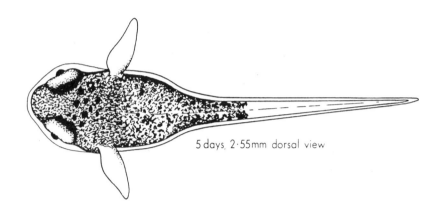

5 days, 2·55mm dorsal view

Larval development in the puffer
Spheroides maculatus. Based on
Welsh & Breder (1922).

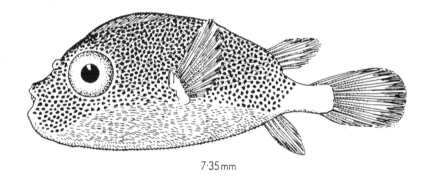

7·35 mm

Development of eggs and larvae for various temperate species of *Sphoeroides* is described by Welsh & Breder (1922), Munro (1945), Uno (1955), Fujita (1956a & b), Fujita & Ueno (1956), and Uchida, et al. (1958). That for a tropical species, *Lagocephalus lunaris,* is given by Fujita (1966). All are similar. Newly hatched larvae range in size from 1.9 to 2.4 mm and have a moderate-sized elongate yolk sac that contains several small oil droplets. Such larvae are active and swim freely. The eye becomes pigmented on the second or third day and the animal begins to feed on the fourth. In *L. lunaris,* dorsal, anal, and caudal fins are formed by day 18; by day 30 the animal is 6.6 mm long, pigmented, and can inflate itself. Settlement to the bottom apparently occurs shortly thereafter. In at least a few species, e.g., *Arothron stellatus* and *Canthigaster callisterna,* color patterns change as the animal matures (Kuthalingham, et al., 1973; Allen & Randall, 1977).

REPRODUCTION IN CAPTIVITY

The only reports of tetraodontids spawning in aquaria are for temperate species and occurred in large public aquaria (e.g., Cohn, 1912; Welsh & Breder, 1922). Spawning of the larger tropical species seems unlikely in aquaria much smaller than these. Spawning of *Canthigaster,* however, is far more likely, given the small size of the fish at maturity and their ready adaptability to aquarium conditions. Minimum tank size should probably be around 250 liters, and abundant cover should be provided. Many species of *Canthigaster* are strongly territorial, and the biggest problem likely to be encountered in spawning them is to get a pair to tolerate each other in the same tank.

299

Rearing of larvae in captivity has been reported for two species. Fujita (1966) reared *Lagocephalus lunaris* from eggs by feeding newly hatched larvae boiled egg yolk and, after nine days, newly hatched *Artemia*. The survival rate was not given but was probably not very high. Valenti (1975) used a more sophisticated approach to rear *Sphoeroides maculatus*. The rearing tanks were 160 liters, contained an undergravel filter, and were fertilized and illuminated to produce a bloom of planktonic algae (*Chlorella* spp.). The first food was cultured rotifers, *Brachionis plicatilis*, newly hatched *Artemia* were provided at ten days, and commercial fish food given at 30 days. Twenty- to 40-day-old juveniles were extremely aggressive and some were killed.

Literature Cited

Abe, T. 1954. Taxonomic studies on the puffers from Japan and adjacent regions—Corrigenda and addenda Pt. 1. *Japan. J. Ichthyol.*, 3:121-128.

Allen, G.R. and J.E. Randall. 1977. Review of the sharpnose pufferfishes (subfamily Canthigasterinae) of the Indo-Pacific. *Rec. Aust. Mus.*, 30:475-517.

Breder, C.M., Jr. and E. Clark. 1947. A contribution to the visceral anatomy, development, and relationships of the Plectognathi. *Bull. Amer. Mus. Nat. Hist.*, 88:287-310.

Clark, E. 1950. Notes on the behavior and morphology of some West Indian plectognath fishes. *Zoologica, N.Y.*, 35:159-168.

Cohn, F. 1912. *Tetraodon, cf. T. cutcutia*, seine Plege und Zucht. *Blatt. Aquar. Terrar.*, 23:582-585.

Feigs, G. 1955. Freshwater puffers spawn. *The Aquarium, Philadelphia*, 24(12):373-375.

Fraser-Brunner, A. 1940. Notes on the plectognath fishes. IV. Sexual dimorphism in the family Ostraciontidae. *Ann. Mag. Nat. Hist., London, Ser. 11*, 6:390-392.

Fujita, S. 1956a. On the development of the egg and prelarval stages of the puffer, *Fugu (Shosaifugu) stictonotus* (Temminck et Schlegel). *Sci. Bull. Fac. Agri., Kyushu Univ.*, 15:525-530.

Fujita, S. 1956b. On the development of the egg and prelarval stages of the puffer, *Fugu (Shosaifugu) poecilonotus* (Temminck et Schlegel). *Sci. Bull. Fac. Agri., Kyushu Univ.*, 15:531-536.

Fujita, S. 1966. Egg development, larval stages and rearing of the puffer *Lagocephalus lunaris spadiceus* (Richardson). *Japan. J. Ichthyol.*, 13:162-168.

Fujita, S. and M. Ueno. 1956. On the egg development and prelarval stages of *Fugu (Torafugu) rubripes rubripes* (Temminck et Schlegel). *Sci. Bull. Fac. Agric., Kyushu Univ.*, 15:519-524.

Gladstone, W. (In Prep.). Aspects of the behavior and ecology of the puffer, *Canthigaster valentini*, at Lizard Island (Great Barrier Reef). Ph. D. Diss., MacQuarie Univ., Australia.

Habib, G. 1979. Reproductive biology of the pufferfish, *Uranostoma richei* (Plectognathi:Lagocephalidae), from Lyttelton Harbour. *New Zeal. J. Mar. Fresh. Res.*, 13:71-78.

Kuthalingham, M.D.K., G. Luther, and J.J. Joel. 1973. On some growth stages and food of *Arothron stellatus* (Bloch) (Tetraodontidae:Pisces). *Indian J. Zool.*, 20:240-243.

Krapp, F. 1974. Puffers. Pp. 245-260. *In: Animal Life Encyclopedia* (English Ed.), (Grzimek, H.C.B., Ed.-in-chief.). Van Nostrand Reinhold, New York.

Kusakabe, D., Y. Murakami, and T. Onbe. 1962. Fecundity and spawning of a puffer, *Fugu rubripes* (T. & S.) in the central waters of the Indian Sea of Japan. *J. Fac. Fish. Anim. Husband., Hiroshima Univ.*, 4:47-79.

Merckens, P.J. 1958. Zoetwater-Kogelvissen. 2. *Het Aquarium*, 29:106-109.

Munro, I.S. 1945. Postlarval stages of Australian fishes. No. 1. *Mem. Queensland Mus., Brisbane*, 12:136-153.

Richter, H.J. 1982. Spawning Somphong's puffer, *Carinotetraodon somphongsi*. *Trop. Fish Hobbyist*, 31:8-25.

Schreitmüller, W. 1930. Kugelfische. *Das Aquarium, Berlin, Jan.*:12-16; *Feb.*:20-26.

Thresher, R.E. 1980. *Reef Fish.* Palmetto Publ. Co., St. Petersburg, Fla. 171 pp.

Uchida, K., S. Imai, S. Mito, S. Fujita, M. Ueno, Y. Shojima, T. Senta, M. Tahuhu, and Y. Dotu. 1958. Studies on the eggs, larvae and juveniles of Japanese Fishes, Ser. 1.

Uno, Y. 1955. Spawning habit and early development of a puffer, *Fugu (Torafugu) niphobles* (Jordan et Snyder). *J. Tokyo Univ. Fish.*, 41:169-183.

Valenti, R.J. 1975. Semi-closed system culture of the northern puffer, *Sphoeroides maculatus*. *Proc. World Maricult. Soc.*, 6:479-485.

Welsh, W.W. and C.M. Breder, Jr. 1922. A contribution to the life history of the puffer *Spheroides maculatus* (Schneider). *Zoologica, N.Y.*, 2:261-276.

Porcupinefishes
(Diodontidae)

The porcupinefishes, balloonfishes, or spiny puffers are unique in appearance and therefore readily identifiable. These short, rotund fishes are characterized by the presence of many prominent spines and, like the closely related tetraodontids, an ability to inflate themselves into a ball-like shape. Upon inflation the spines become erect (if not already so), forming a singularly unpalatable mouthful for a prospective predator. Given this protection, it is not surprising that diodontids are relatively poor swimmers and, like other members of the order, propel themselves mainly by means of sculling with their soft dorsal and anal fins.

Diodontids are opportunistic predators that feed upon a wide variety of benthic organisms. Many are nocturnal and by day retire to caves and crevices or hover quietly in some sheltered area awaiting dusk. Almost no work has been done on their social behavior, but in general they seem to coexist with little or no aggression and have broadly overlapping home ranges. Most diodontids are inshore fishes found on grass flats, around mangroves, and in estuaries. Only the species in the genus *Diodon* are routinely found on the reef.

SEXUAL DIMORPHISM

Little is known about external sexual dimorphism (if any) in most diodontids. Sakamoto & Suzuki (1978) reported that male *Diodon holacanthus* off Japan are generally larger than females (25.0 cm maximum versus 23.1 cm). Size ranges of the two sexes overlap broadly, however, suggesting gonochorism. *D. holacanthus* in the Gulf of California (eastern Pacific) are temporarily dichromatic during courtship and spawning: the male retains his normal black-spots-on-brown pattern, while the female develops a broad white band down each side of her body and darkens almost to black dorsally (pers. obs.).

SPAWNING SEASON

Scattered reports are available about spawning of *Diodon holacanthus*, not all of which are in full agreement. Munro, et al. (1973) suggested that peak spawning occurs in the Caribbean in late spring; Leis (1978) found the same to be true off Hawaii, with some spawning from January to September. In Japan, however, *D. holacanthus* in captivity did not begin to spawn until May, when water temperatures reached approximately 24.4° C, and continued spawning until August (Sakamoto & Suzuki, 1978). In the Gulf of California *D. holacanthus* spawns at least in January (the coldest month of the year) at approximately 15° C and in October-November (the warmest time) at approximately 27° C, which suggests it occurs year-round. Finally, Munro, et al., collected ripe *D. hystrix* in February and March.

REPRODUCTIVE BEHAVIOR

Courtship and spawning have thus far been observed only in the balloonfish, *Diodon holacanthus*. Sakamoto & Suzuki (1978) repeatedly witnessed spawning by this species in a large public aquarium. Spawning occurred at dusk or at night and was initiated by the males, who approached a female and repeatedly nudged her underside. Several males courted a single female simultaneously and together pushed her toward the surface. Just below the surface she shed her eggs and the four or five males pushing her simultaneously shed their milt. Afterward, all rushed back to the bottom.

Spawning by this same species in the Gulf of California is similar (pers. obs.) but differs in several important respects: spawning involves pairs only, it occurs at dawn (just after first light) and at dusk, and the female assumes a distinctive white-sided pattern during courtship and spawning. As above, spawning

Spawning by *Diodon holocanthus*. Above: Two males approach the female at the bottom and incite her by nuzzling her. Below: The female (upper fish) is slowly pushed toward the surface by the males. Photos by Katsumi Suzuki.

The female *Diodon holocanthus* goes up to just below the surface of the water. The eggs then are shed, to be fertilized immediately. Photo by Katsumi Suzuki.

begins with a male approaching a white-sided female, who hovers quietly a few meters off the bottom. In one bay six such pairs were observed within a few meters of one another, all courting and all vigorously driving off conspecifics that approached too closely. The male courts the female by following behind her and gently nudging her abdomen whenever she pauses. If receptive, she ascends farther off the bottom, with the male close behind. Pressing his snout against her abdomen, he begins to push the female toward the surface. Initially she is passive, but as they approach the surface she begins to move forward, vigorously and noisily splashing across the surface with the male close behind. Gametes are apparently shed during this splashing. At the same time other pairs are spawning, and all activity is usually over only three or four minutes after it started.

EGGS AND LARVAE

Diodontid eggs are spherical. Those of *Chilomycterus schoepfi* are approximately 1.8 mm in diameter, nonadhesive, and demersal (Nichols & Breder, 1927). *Diodon* eggs, in contrast, are pelagic and buoyant, containing numerous small yellowish oil droplets. Eggs tentatively identified as those of *D. hystrix* are 1.9 to 2.1 mm in diameter; those of *D. holacanthus* are 1.62 to 1.86 mm in diameter (Leis, 1978; Sakamoto & Suzuki, 1978). According to the latter, hatching occurs in four to five days at 24° to 25° C.

Characteristics of newly hatched larvae and larval development have thus far been described only for *Diodon* (Leis, 1978; Sakamoto & Suzuki, 1978). The accounts describe similar schemes of development but differ in the degree of larval development at hatching and in the time scale of development. Regarding the latter, Leis reported slower growth and smaller animals at metamorphosis, but since he did not succeed in rearing the larvae and the oldest appeared to be emaciated, it is likely that his figures are incorrect. According to Sakamoto & Suzuki, newly hatched *D. holacanthus* are approximately 2.6 mm long, have a large yolk sac, unpigmented eyes, a nonfunctional

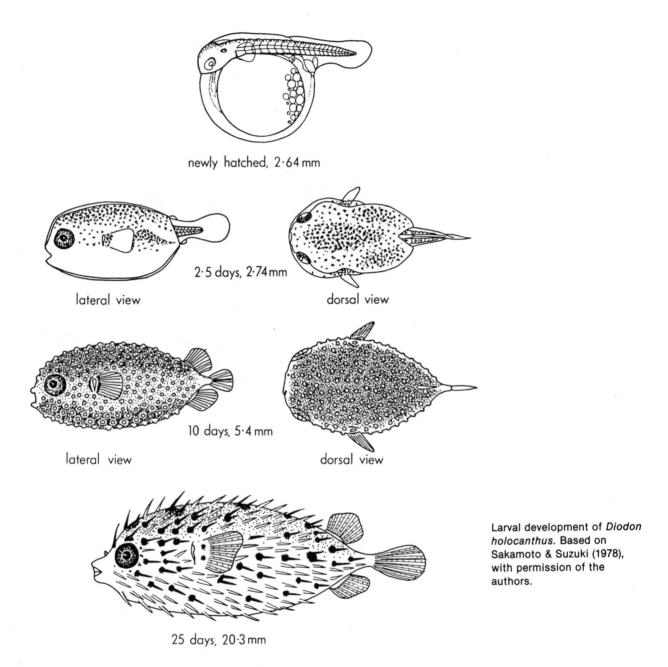

newly hatched, 2·64 mm

2·5 days, 2·74 mm

lateral view

dorsal view

10 days, 5·4 mm

lateral view

dorsal view

Larval development of *Diodon holocanthus.* Based on Sakamoto & Suzuki (1978), with permission of the authors.

25 days, 20·3 mm

mouth, and float upside down at the surface (newly hatched larvae of *D. hystrix* are similar according to Leis). Each is encased in a thick pliable shell similar to that of comparably aged ostraciid larvae. On day two the eyes develop pigment, and on day three the mouth is functional and the yolk sac gone. At this point the larvae are square-bodied with prominent rudiments of the pectoral and caudal fins. According to Leis, *D. holocanthus* from Hawaii hatch at this stage, undergoing earlier development while still in the egg membrane. Leis further reported metamorphosis at a

length of 3 mm and an age of three weeks to a postlarval stage similar in appearance to the adult. Sakamoto & Suzuki reported development to this stage (characterized by numerous short tubercles on the body, rudiments of spines) at 5.4 mm and an age of ten days.

The duration of the larval stage in the field is not known but appears to last at least several months. Very large prejuveniles (up to 86 mm for *D. holocanthus* and 180 mm for *D. hystrix*) are commonly collected in plankton tows.

Böhlke & Chaplin (1968) illustrated a postlarval *Chilomycterus*. The specimen is rotund with short well-developed fins and a number of short thick papillae extending from an otherwise smooth body. Colorwise it is pale with dark ring-like spots. This strange looking stage has in the past been treated as a separate genus, *Lyosphaera*. Heck & Weinstein (1978) suggested that the generally similar postlarvae of *Chilomycterus antennatus* are mimics of an unpalatable sea hare, *Aplysia dactylomela*.

REPRODUCTION IN CAPTIVITY

Aside from Sakamoto & Suzuki's report of spawning of *D. holacanthus* in a large public aquarium, there has been only one other report of spawning in captivity. Wolfsheimer (1957) reported that a pair of *D. holacanthus* (erroneously called *D. hystrix*) spawned in a 30-gallon (112-liter) aquarium in a pet shop. The tank was bare except for an undergravel filter and sand bottom, water temperature was 25.5°C, and the fish had received no special treatment other than heavy feedings. The fish had been in captivity for at least six months. Fecundity was estimated at 3,000 to 4,000 eggs. The eggs sank in seawater and may not have been fertile.

Sakamoto & Suzuki reared *D. holacanthus* to metamorphosis in captivity by feeding them with rotifers (*Brachionis plicatilis*) after hatching, with subsequent feedings of *Artemia nauplii* and nauplii and adults of the marine copepod *Tigriopus japonicus*.

Literature Cited

Böhlke, J.E. and C.C.G. Chaplin. 1968. *Fishes of the Bahamas and adjacent tropical waters.* Livingston Publ., Wynnewood, Pa. 771 pp.

Heck, K.L., Jr. and M.P. Weinstein. 1978. Mimetic relationships between tropical burrfishes and opisthobranchs. *Biotropika,* 10:78-79.

Leis, J.M. 1978. Systematics and zoogeography of the porcupinefishes (*Diodon,* Diodontidae, Tetraodontiformes), with comments on egg and larval development. *Fish. Bull.,* 76:535-567.

Munro, J.L., V.C. Gaut, R. Thompson, and P.H. Reeson. 1973. The spawning seasons of Caribbean reef fishes. *J. Fish Biol.,* 5:69-84.

Nichols, J.T. and C.M. Breder, Jr. 1927. The marine fishes of New York and southern New England. *Zoologica, N.Y.,* 9:1-192.

Sakamoto, T. and K. Suzuki. 1978. Spawning behavior and early life history of the porcupine puffer, *Diodon holacanthus,* in aquaria. *Japan. J. Ichthyol.,* 24:261-270.

Wolfsheimer, F.A.I. 1957. A spawning of porcupine puffers. *The Aquarium, Philadelphia,* 26(9):288-290.

Gobies
(Gobioidei)

Gobies are typically elongate, small, and often cryptic fishes that abound in mangrove areas, sand and mud flats, tropical freshwater lakes and streams, and in many shallow temperate marine areas. The suborder Gobioidei reaches its peak in diversity and abundance, however, on coral reefs. Easily the largest single group of living fishes (perhaps 200 genera and ten times that many species), gobies alone constitute a major segment of the entire reef-fish fauna.

A number of families are variously recognized in the suborder, only two of which, the Eleotridae and Gobiidae, are common on the reef (a third, the Microdesmidae, is found in shallow sand and mud areas, but the species are very cryptic and rarely seen). The Eleotridae and Gobiidae are separated on the basis of a number of morphometric features, the most commonly cited being the degree of pelvic fin fusion. Most gobies have evolved a ventral suction disc (with which they attach to the substrate) formed by the fusion of the pelvic fins into a cup-like structure. Eleotrids as usually defined lack fused pelvic fins. The degree of fusion, however, varies widely even within closely related groups, suggesting that a division based on this character alone is an artificial one. Currently many workers simply treat all reef-associated gobies as members of a single all-inclusive family Gobiidae. All are characterized by a blunt or slightly pointed snout, an elongate body, a small mouth, and, in most, two clearly separated dorsal fins. Most species are benthic and are typically seen sitting on or scooting about the bases of corals and rocks. Even those species that have evolved a hovering life-style, e.g., *Ioglossus, Nemateleotris,* rarely move far from shelter.

Considerable work has been done on goby reproduction, but little of it deals with the reef-associated species. Several major themes pervade the literature, and generalizations that apply to the reef-dwelling fishes are clearly possible.

SEXUAL DIMORPHISM

Most species of reef-associated gobies appear to be sexually monomorphic in terms of permanent coloration and gross morphology. For such species the sexes can be distinguished by only two means. First, shortly before spawning the female typically swells with eggs (very often assuming a "squarish" shape when viewed from above) and may be lighter ventrally than the more slender male. Second, the shapes of the urogenital papillae are diagnostic. That of the male is long and conical, whereas that of the female is short, truncate, and often lobed. It is, however, often necessary to capture and even anesthetize a fish before its papilla can even be seen. Even when visible the use of differences in papilla morphology to correctly identify the sexes takes considerable experience. Examples of such essentially monomorphic gobies include the various neon gobies (*Gobiosoma*, subgenus *Elacatinus*), the obligate coral-dwellers in *Paragobiodon* and *Gobiodon*, and the various hovering gobies in *Ioglossus* and *Nemateleotris* (e.g., Colin, 1975; Lassig, 1977; Randall, 1968; Davis, et al., 1977). One species of *Elacatinus, Gobiosoma puncticulatus,* found in the eastern Pacific, develops a temporary dichromatism during courtship and spawning, the only known reef-associated goby to do so thus far; the snout of the male turns almost black (A. Kerstitch, pers. comm.). Temporary sexual dichromatism associated with courtship and spawning has been commonly reported for gobies inhabiting other habitats (e.g., Tavolga, 1954).

Sexual dimorphism, where it does occur, frequently varies in extent even within genera. Males of *Lythrypnus dalli* and *L. zebra*, two brightly colored eastern Pacific species, and *L. heterochroma* from the western Atlantic, for example, have conspicuously longer first dorsal spines than the females; most other members of the genus, however, are sexually monomorphic (Böhlke & Robins, 1960a). Similarly, males, but not females, of some species of *Coryphopterus* have dorsal and anal spines that reach beyond the base of the caudal fin; other members of the genus lack this sexual difference (Böhlke & Robins, 1960b). In a few species, e.g., *Evermannichthys metzelaari* (Böhlke & Robins, 1969), several species of *Microgobius* (Böhlke & Chaplin, 1968), and *Lythrypnus dalli* in captivity (Bauer & Bauer, 1974), the sexes appear to differ permanently in color pattern, but again there appear to be no general trends. Such permanent color differences between the sexes is a common feature for gobies in other habitats (e.g., Egami, 1960; Brothers, 1975). Why there should be such a high degree of within-genus variation in the type and extent of sexual dimorphism in reef-associated gobies has not been examined. It would appear to be a fertile area, however, for someone interested in the evolution of such features, especially since most gobies are likely to prove easy to spawn in captivity.

Protogynous hermaphroditism has recently been documented in the Indo-Pacific coral-associated gobies of the genus *Paragobiodon* (Lassig, 1977). Variously found in pairs or in small male-dominated harems, the largest individual present is always male, the second largest always the functional female, and additional individuals, if any, non-spawning females. Gonads, like those of other gobies, are paired, uniting posteriorly to run into the urogenital papilla. Ovaries are basically normal for teleosts, except that they contain conspicuous traces of testicular tissue visible during non-spawning periods. The testes are basically ovarian in structure, containing a lumen, lamellae, and undeveloped and regressed oocytes. Sex change is socially controlled, with the largest female turning male if the previous male is removed. Presumably the previously second-ranked female becomes active, reconstituting the breeding pair. Similar hermaphroditism might logically be expected in many of the small pair-forming and strongly site-attached reef-dwelling gobies.

SPAWNING SEASON

There is little information available on spawning seasons of reef-dwelling gobies. Feddern (1967) and Colin (1975) reported that the neon goby *Gobiosoma (Elacatinus) oceanops* spawns from January to May (the coldest part of the year) off Florida. At least some other members of the subgenus, however, living in thermally more stable areas, are ripe nearly every month of the year and several clearly spawn in the summer (Colin, 1975). All probably have extended spawning seasons.

Information on other groups is more fragmentary. According to Tavolga (1954), the intertidal goby *Bathygobius soporator* spawns in the summer off Florida. *Lythrypnus dalli* and *L. zebra*, in the seasonally cool Gulf of California, also have a summer peak to spawning activity, though some spawning occurs from March to October (Wiley, 1976).

Blanco (1958) and Manacop (1953) reported a lunar periodicity to larval recruitment, and presumably spawning, for a number of estuarine gobies. No comparable data are available for reef-dwelling species.

REPRODUCTIVE BEHAVIOR

In most cases gobies appear to spawn promiscuously with either many individuals loosely organized into a social hierarchy, e.g., *Coryphopterus personatus*, or individuals maintaining small contiguous territories, e.g., *Coryphopterus glaucofrenum* and *Lythrypnus dalli*. Heterosexual pairing and apparent monogamy, however, have been documented for a number of gobies, including *Ioglossus* spp. (Colin, 1972), *Gobiodon* spp. (Tyler, 1971), *Valencienna* spp. (Hiatt & Strasburg, 1960; Paulson, 1978), *Gobiosoma (Elacatinus)* spp. (Colin, 1975; Schmale, MS), *Paragobiodon* spp. (Lassig, 1976), *Ptereleotris heteropterus* (Davis, et al., 1977), *Signigobius biocellatus* (Hudson, 1977), and *Amblygobius phalaena* (Paulson, 1978). Pairing has also been documented for a number of temperate species, including the blind goby *Typhlogobius californiensis* (MacGinitie, 1983, 1939). In the latter species pairing is permanent but is not the result of any specific "bond" between pair members. Rather, each individual is strongly territorial toward other members of its own sex, thereby denying its partner the opportunity to breed outside the pair. Wickler & Seibt (1970) referred to such a system as "anonymous monogamy." In contrast, Schmale (MS) has demonstrated that *Gobiosoma oceanops* recognize mates as individuals, even after being separated for two days. He suggests that olfactory cues may be involved in such recognition in this sexually monomorphic species. Sexual, if not individual, recognition in gobies has also been reported for *Bathygobius soporator* (Tavolga, 1955, 1956).

Gobies produce demersal adhesive eggs that are characteristically and nearly universally attached to the roof of a small cavity—either a burrow, the underside of a rock or shell, or the lumen of a sponge. Courtship and spawning have been described for *Lythrypnus dalli* (Turner, et al. 1969; Bauer & Bauer, 1974), *Signigobius biocellatus* (Hudson, 1977),

Paragobiodon spp. (Lassig, 1976), the intertidal *Bathygobius soporator* (Tavolga, 1954, 1955, 1956a & b, 1958), and *Gobiosoma oceanops,* by far the most intensively studied species (Feddern, 1967; Valenti, 1972; Colin, 1975; and numerous popular articles, including those by Boyd, 1956; Straughan, 1956; Moe, 1975, and Musgrave, 1976). Courtship and spawning, in general, are similar for all and also similar to that described for temperate species. Imminent spawning is preceded by an obvious swelling of the female's ventral area and by nest preparation behavior by the male. Such preparation usually consists of cleaning and clearing an area, typically on the roof of a small cave or on the underside of a shell, by means of biting at its surface and rubbing the anal fin and venter against it, while the male holds fast in place by means of his ventral suction disc. The species in *Paragobiodon* differ from the norm in that the male kills a small elliptical spot of coral on the underside of a coral branch and removes the dead tissue from it. A dense mat of algae quickly grows on the spot, and eggs are subsequently laid in the algal mat.

Spawning can occur at any time of the day but is most frequent in the early morning (Moe, 1975; Valenti, 1972). The male characteristically "entices" the female to the nest by swimming back and forth between her and the nest-site he has prepared and may nudge her on the abdomen with his snout or make exaggerated swimming motions in front of her while attached to the bottom by means of the pelvic disc. In the later stages of courtship both fish may participate in this exaggerated swimming, often while sitting side-by-side attached to the bottom. In at least one species (*Bathygobius soporator*) courtship and pre-spawning activities involve a complex suite of visual, chemical, and olfactory cues and were the subject of considerable research by Tavolga in the 1950's. Males court approaching females by vibrating vigorously and by producing a single-pulsed, low-pitched sound, a "grunt," to which the female clearly reponds. The female, in turn, releases a fluid from her ovaries which stimulates the male to full courtship activities, even if the female is not visible. Work in comparable detail has not yet been conducted with any reef-dwelling species, but it is likely that some of these elements will also be found to play a role in their courtship.

One-half to one hour after the beginning of courtship the female follows the courting male to the nest site and begins to lay the eggs one at a time. Spawning occurs with both fish quivering side-by-side over the nest site, the male fertilizing each egg as it is pro-duced. The female may exit and return several times before spawning is completed. The number of eggs laid ranges from as few as 10 to 20 to well over 100,000, depending upon the species and the size and condition of the female (e.g., Smith & Tyler, 1972). Three to five hundred eggs in a clutch is not uncommon. Spawning usually recurs at two- to three-week intervals, though Feddern (1967) reported a pair of *Gobiosoma oceanops* that spawned nine times in two weeks.

Following spawning, the male tends and guards the eggs. Female participation is usually minimal, even in those species that form permanent pairs, though she can usually approach the nest site and may even sleep in it without being attacked by the guarding male. Thus far, joint parental care has been reported only for *Quisquilius hipoliti* in captivity (Bauer & Bauer, 1975).

A different and thus far unique spawning behavior has been reported for the burrowing goby *Signigobius biocellatus* (Hudson, 1977). Prior to spawning each pair constructs a burrow, during which activity the female (sex indicated by a swollen abdomen) frequently "nibbles" at the body of the male. Following spawning in the burrow, the male is sealed by the female into the burrow with the eggs. During incubation the burrow is periodically opened, the nest cleaned by both parents, and after five to ten minutes male and eggs are resealed. After four to five days of such activity the male remains outside the burrow, apparently leaving the eggs to develop on their own. Finally, the adults reopen the burrow, after which Hudson reported seeing a single very small benthic individual present near the tending adults. Hudson interpreted this entire sequence as indicating that after spawning the male was sealed into the burrow to fertilize and tend the eggs. The eggs and larvae subsequently develop entirely within the burrow, deriving nourishment from "food reserves, their surroundings, and possibly by cannibalism." In the end a single juvenile, "the sole survivor of the brood," exits the burrow and immediately begins a benthic existence. If true, reproduction by this goby differs dramatically from that of any other reef fish.

EGGS AND LARVAE

Goby eggs are demersal and attached to the substrate by means of a tuft of adhesive filaments at one end. Most are elongate, smoothly round-ended cylinders ranging in maximum length from 1.1 to 3.3 mm and with a diameter from 0.5 to 1.0 mm. Several, however, have distinctive protuberances, usually at

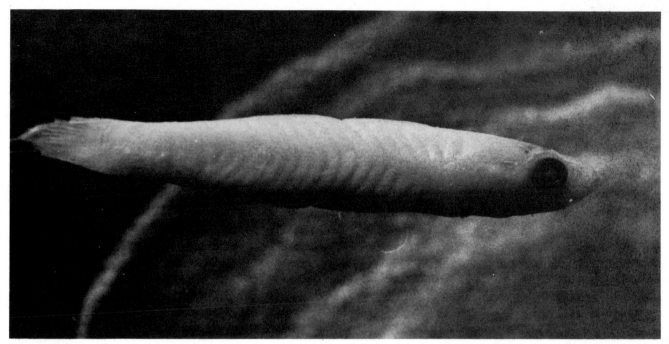

A postlarval goby, 31.0 mm total length, collected near a submerged light at night off the coast of Papua New Guinea. Photo by Dr. R. E. Thresher.

the distal end (farthest from the substrate), and others depart significantly from the generalized cylindrical shape (see discussion in Breder, 1943). All species of the subgenus *Elacatinus* for which data are available, for example, have elliptical eggs that have five or six protuberances at the distal end; such eggs may be characteristic of the subgenus. Within the subgenus, however, egg size varies widely, ranging in longest dimension from 1.7 to 3.3 mm (Feddern, 1967; Valenti, 1972; Colin, 1975; Moe, 1975). Böhlke & Robins (1968) described the prefertilization eggs of other members of the genus *Gobiosoma* as slightly distorted spheres ranging in diameter from 0.45 to 0.8 mm. Prefertilization egg sizes for other Caribbean gobies range from 0.1 to 0.5 mm (Smith & Tyler, 1972).

Eggs range in color from clear and transparent to yellow, depending on yolk color. Development of the egg has been described in detail for only one reef-dwelling goby, *Gobiosoma oceanops* (Wickler, 1962; Feddern, 1967; Valenti, 1972; Colin, 1975; Moe, 1975). Colin indicated that development of other members of the subgenus is similar. Tavolga (1950) described the development of the intertidal goby *Bathygobius soporator*, while Bauer & Bauer briefly described that of *Lythrypnus dalli*. Limited information is also provided for *Paragobiodon* spp. (Lassig, 1976) and for *Quisquilius hipoliti* (Bauer & Bauer, 1975).

In general, incubation lasts from five to six days (though Tavolga reported that *Bathygobius soporator*

eggs hatch in three days) depending on temperature. In at least some species, and perhaps in all, the larvae begin development with the head pointed toward the proximal (attached) end of the egg and reverse position at some point during development. Valenti (1972) indicated that *G. oceanops* that do not reverse position (about 4%) do not hatch, but this is disputed by Moe (1975), who stated that such larvae have no difficulty hatching. The time of rotation seems to vary with the genus: *Coryphopterus personatus* rotates by 48 hours (Brandt, 1974), *G. oceanops* and its near relatives rotate on the third day (at about 50 hours), and *Lythrypnus dalli* reverses position at about 145 hours, within the last 24 hours before hatching.

The size of the newly hatched larva, like that of the egg, varies widely. *G. oceanops* eggs and larvae are quite large, the latter 3 to 4 mm long at hatching, which accounts for the relative ease of rearing them in captivity. In contrast, newly hatched larvae of *Lythrypnus dalli* are only 2.5 mm long, those of *Quisquilius hipoliti* less than 2 mm, and those of *Paragobiodon* spp. only 1.7 mm. In all cases the larvae are relatively advanced at hatching, with pigmented eyes, well developed jaws and digestive tract, well developed vertical fin folds and pectoral fin folds, and a small to completely absorbed yolk sac. The large swim bladder inflates within a few hours of hatching and is located roughly midway along the gut. Such young larvae are also apparently photopositive. Feeding begins within

12 to 24 hours, the larvae striking out at prey from a flexed "S" shape. Older larvae are transparent and elongate and are characterized by the presence of two dorsal fins (the second located immediately over the anal fin) and often by fused pelvic fins.

The duration of the planktonic larval stage is known for only a few species. Based on the interval between first observation of spawning and first appearance of newly settled juveniles, Lassig (1976) estimated the planktonic stage of *Paragobiodon* at Heron Island was about six weeks. Newly settled juveniles were 6 to 7 mm long. Moe (1975) reported metamorphosis of *G. oceanops* larvae in as little as 18 days and as much as 40 days; Colin (1975) reported metamorphosis at 26 days. Imminent settlement in neon gobies is indicated by the development of a dark horizontal line, the precursor of the neon line of the adult, along each side. Newly settled juveniles are about 8 mm long. Post-settlement growth is rapid, and subadult size can be reached in three months; spawning can occur in as little as five months. Colin suggested that neon gobies, and perhaps other small gobies, are essentially "annual" fishes, maturing quickly and probably living for only one to two years.

The atypical larval development reported for *Signigobius biocellatus* (Hudson, 1977) has already been described. Although direct development from egg to benthic juvenile, i.e., development without a planktonic larval stage, occurs in some freshwater gobies (e.g., Kishi, 1979), it is unlikely that there would be sufficient oxygen and nutrients available for the development of even a single juvenile in a sealed burrow.

REPRODUCTION IN CAPTIVITY

Several species of reef-dwelling gobies have spawned in aquaria, including *Signigobius biocellatus* (Hudson, 1977), *Quisquilius hipoliti* (Bauer & Bauer, 1975), *Coryphopterus personatus* (Brandt, 1974), *Lythrypnus dalli* (Turner, et al., 1969; Bauer & Bauer, 1974), *Chriolepis minutillus* (Kerstitch, pers. comm.), and *Gobiosoma oceanops* (Boyd, 1956; Straughan, 1956; Feddern, 1967; Valenti, 1972; Colin, 1975; Moe, 1975; Musgrave, 1976; Walker, 1977). Colin also spawned several other neon gobies, including three sponge-dwelling species and a hybrid pair (*G. oceanops X G. evelynae*). Commercial breeders, in fact, have now spawned and reared several generations of neon gobies in captivity.

The ease with which these species were spawned suggests that other reef-dwelling gobies will prove equally simple. In most cases stimulating spawning required only the presence of a heterosexual pair, reasonably heavy feedings (ideally including some live food), and the presence of a suitable spawning site (if no other suitable sites are available many species, including neon gobies, will spawn inside of short pieces of opaque PVC tubing partially buried in the sand). At least *G. oceanops* and *Lythrypnus dalli* will readily spawn in community tanks; the former can be spawned so easily, in fact, that it commonly does so while still in pet shop aquaria.

Although most species will spawn readily, rearing the young can be much more difficult, the degree of difficulty depending in large part on the size of the newly hatched larvae. As noted, those of the neon goby are relatively large and consequently will accept large food items, facilitating their commercial rearing. In contrast, the newly hatched larvae of *Lythrypnus dalli*, *Coryphopterus personatus*, *Paragobiodon* sp., and probably many other species are much smaller than those of the neons, in some cases less than half their size. None have yet been successfully reared even though several workers have made attempts to rear the colorful *L. dalli*.

Colin (1975) reared neon gobies in 120-liter all-glass aquaria with constant illumination. The larvae were fed wild plankton, ranging in size from 56 to 200 microns, added to the aquarium every eight to 12 hours. Moe (1975) provided no details on commercial rearing other than ". . .the larvae were reared under a carefully simulated pelagic environment . . ." and were fed ". . . small living organisms . . . ." Commercial rearers, however, frequently use 80- to 120-liter all-glass aquaria, constant illumination, bare tanks until metamorphosis and tanks with gravel and undergravel filters after metamorphosis, and feedings of mixtures of rotifers (*Brachionis plicatilis*) and wild plankton. Musgrave (1976) also raised neons, though only to 21 days, by feeding them newly hatched copepod nauplii cultured in "green" (algae-filled) aquaria; the copepods occurred naturally in his community tank. Delmonte, et al (1968) reared two tropical gobies (*Lophogobius cyprinoides*, an estuarine species, and *Bathygobius andrei*, an intertidal one) through use of large (386 to 681 liters) outdoor pools. Several weeks prior to their planned use the pools were enriched with 8-12-4 fertilizer (0.65 grams/liter) and B_{12} (1 mg/pool) to stimulate growth of naturally occurring algae and copepods. The larvae were transferred to these pools after hatching and their diet was supplemented for the first five days with thrice-daily feedings of Liquifry #1 and then similar feedings with Liquifry #2 (both commercially available liquid foods designed for use with juvenile freshwater fishes). At day 20 the larvae began to accept brine shrimp nauplii. Survival rate was good with *L. cyprinoides* but poor with *B. andrei*.

Literature Cited

Bauer, J. and S. Bauer. 1975. First report of spawning rusty gobies. *Octopus*, 2(5):25.

Bauer, S. and J. Bauer. 1974. Spawning the Catalina goby. *Octopus*, 1(8):6-11.

Blanco, G.J. 1958. Assay of the goby fry (ipon) fisheries of the Laong River and its adjacent marine shores, Ilocos Norte Province. *Philippine J. Fish.*, 4(1):31-72.

Böhlke, J.E. and C.C.G. Chaplin. 1968. *Fishes of the Bahamas and adjacent tropical waters.* Livingston Publ., Wynnewood, Pa. 771 pp.

Böhlke, J.E. and C.R. Robins. 1960a. Western Atlantic gobioid fishes of the genus *Lythrypnus*, with notes on *Quisquilius hipoliti* and *Garmannia pallens*. *Proc. Acad. Nat. Sci., Philadelphia*, 112(4):73-98.

Böhlke, J.E. and C.R. Robins. 1960b. A revision of the gobioid fish genus *Coryphopterus*. *Proc. Acad. Nat. Sci., Philadelphia*, 112(5):103-128.

Böhlke, J.E. and C.R. Robins. 1968. Western Atlantic seven-spined gobies, with descriptions of ten new species and a new genus, and comments on Pacific relatives. *Proc. Acad. Nat. Sci., Philadelphia*, 120(3):45-174.

Böhlke, J.E. and C.R. Robins. 1969. Western Atlantic sponge-dwelling gobies of the genus *Evermannichthys*: their taxonomy, habits, and relationships. *Proc. Acad. Nat. Sci., Philadelphia*, 121(1):1-24.

Boyd, D. 1956. Spawning the neon goby, *Elacatinus oceanops*. *Aquarium*, 25:391-396.

Brandt, C. 1974. First spawning of *Coryphopterus personatus*. *Octopus*, 1(5):21-23.

Breder, C.M., Jr. 1943. The eggs of *Bathygobius soporator* (Cuvier & Valenciennes) with a discussion of other non-spherical teleost eggs. *Bull. Bingham Oceanogr. Coll.*, 8(3):1-49.

Brothers, E.B. 1975. The comparative ecology and behavior of three sympatric California gobies. Ph. D. Dissertation, Univ. California, San Diego.

Colin, P.L. 1972. Daily activity patterns and effects of environmental conditions on the behavior of the yellowhead jawfish, *Opistognathus aurifrons*, with notes on its ecology. *Zoologica, N.Y.*, 57:137-169.

Colin, P.L. 1975. *Neon Gobies.* T.F.H. Publ., Neptune City, New Jersey. 304 pp.

Davis, W.P., J.E. Randall, and D.O. French. 1977. The systematics, biology, and zoogeography of *Ptereleotris heteropterus*, Pisces:Gobiidae. *Proc. Third Internat. Coral Reef Symp.*, 1:261-266.

Delmonte, P.J., I. Rubinoff, and R.W. Rubinoff. 1968. Laboratory rearing through metamorphosis of some Panamanian gobies. *Copeia*, 1968(2):411-412.

Egami, N. 1960. Comparative morphology of the sex characters in several species of Japanese gobies, with reference to the effects of sex steroids on the characters. *J. Fac. Sci. Univ. Tokyo, Sect. IV*, 9:67-100.

Feddern, H.A. 1967. Larval development of the neon goby, *Elacatinus oceanops*, in Florida. *Bull. Mar. Sci.*, 17(2):367-375.

Hiatt, R.W. and D.W. Strasburg. 1960. Ecological relationships of the fish fauna on coral reefs of the Marshall Islands. *Ecol. Monogr.*, 30:65-127.

Hudson, R.C.L. 1977. Preliminary observations on the behaviour of the gobiid fish, *Signigobius biocellatus* Hoese and Allen, with particular reference to its burrowing behaviour. *Z. Tierpsychol.*, 43:214-220.

Kishi, Y. 1979. A graphical model of disruptive selection on offspring size and a possible case of speciation in freshwater gobies characterized by egg-size difference. *Res. Popul. Ecol.*, 20:211-215.

Lassig, B. 1976. Field observations on the reproductive behaviour of *Paragobiodon* spp. (Osteichthys:Gobiidae) at Heron Island, Great Barrier Reef. *Mar. Behav. Physiol.*, 3:283-293.

Lassig, B. 1977. Socioecological strategies adopted by obligate coral-dwelling fishes. *Proc. Third Internat. Coral Reef Symp.*, 1:565-570.

MacGinitie, G.E. 1938. Notes on the natural history of some marine animals. *Amer. Midl. Natur.*, 9(1):207-217.

MacGinitie, G.E. 1939. The natural history of the blind goby, *Typhlogobius californiensis* Steindachner. *Amer. Midl. Natur.*, 21(2):489-505.

Manacop, P.R. 1953. The life history and habits of the goby *Sicyopterus extraneus* Herre (anga) Gobiidae with an account of goby fry fishery of Cagayan River, Oriental Misamis. *Philippine J. Fish.*, 2(1):1-58.

Moe, M.A., Jr. 1975. Propagating the Atlantic neon goby. *Marine Aquarist*, 6(2):4-10.

Musgrave, G. 1976. Neon goby spawning. *Marine Hobbyist News*, 4(7):1,6.

Paulson, A.C. 1978. On the commensal habits of *Ptereleotris*, *Acanthurus* and *Zebrasoma* with fossorial *Valencienna* and *Amblygobius*. *Copeia*, 1978:168-169.

Randall, J.E. 1968. *Ioglossus helenae*, a new gobiid fish from the West Indies. *Ichthyologica*, 39(3 & 4): 107-116.

Schmale, M. (MS). Individual recognition in the neon goby, *Gobiosoma oceanops* (Jordan).

Smith, C.L. and J.C. Tyler. 1972. Space resource sharing in a coral reef fish community. Pp. 125-170. *In:* Results of the Tektite Program:Ecology of Coral Reef Fishes. (B. Collette & S.A. Earle, Eds.). *Sci. Bull.* (14), *Natural History Mus., Los Angeles County.*

Straughan, R.P.L. 1956. Salt water fishes are spawned. *The Aquarium, Philadelphia,* 25(5):157-159.

Tavolga, W.N. 1950. Development of the gobiid fish, *Bathygobius soporator*. *J. Morph.*, 87(3):467-492.

Tavolga, W.N. 1954. Reproductive behavior in the gobiid fish, *Bathygobius soporator*. *Bull. Amer. Mus. Nat. Hist.*, 104(5):427-460.

Tavolga, W.N. 1955. Ovarian fluids as stimuli to courtship behavior in the gobiid fish, *Bathygobius soporator* (C. & V.). *Anat. Rec.*, 122(3):425.

Tavolga, W.N. 1956a. Pre-spawning behavior in the gobiid fish, *Bathygobius soporator*. *Behaviour,* 9(1):53-74.

Tavolga, W.N. 1956b. Visual, chemical and sound stimuli as cues in the sex discriminatory behavior of the gobiid fish, *Bathygobius soporator*. *Zoologica, N.Y.,* 41(7):49-64.

Tavolga, W.N. 1958. A tidal zone resident. *Nat. Hist.,* 67(3):156-162.

Turner, C.H., E. Ebert, and R.R. Given. 1969. Man-made reef ecology. *Calif. Dept. Fish and Game, Fish. Bull.,* (146):1-221.

Tyler, J.C. 1971. Habitat preferences of the fishes that dwell in shrub corals on the Great Barrier Reef. *Proc. Acad. Nat. Sci., Philadelphia,* 123:1-26.

Valenti, R.J. 1972. The embryology of the neon goby, *Gobiosoma oceanops*. *Copeia,* 1972(3):477-481.

Walker, S. 1977. Walker successfully spawns five species. *Marine Hobbyist News,* 5(9):1, 4.

Wickler, W. 1962. Ei und Larve von *Elacatinus oceanops* Jordan (Pisces; Gobiiformes). *Senckenb. biol.,* 43(3):201-205.

Wickler, W. and U. Seibt. 1970. Das Verhalter von *Hymenocera picta* (clown shrimp) Dana, einer See-steine fressenden garnele (Decapoda, Natantia, Gnathophyllidae). *Z. Tierpsychol.,* 27:352-368.

Wiley, J.W. 1976. Life histories and systematics of the western North American gobies *Lythrypnus dalli* (Gilbert) and *Lythrypnus zebra* (Gilbert). *Trans. San Diego Soc. Nat. Hist.,* 18(10):169-183.

Blennies
(Blennioidei)

Blennies are elongate, generally robust fishes characterized by a very flexible body (as opposed to the stiffer-bodied gobies) and a fluid swimming motion, pelvic fins inserted ahead of the pectorals, a long anal fin, and usually a single long dorsal fin. Most are cryptically colored, which, combined with their small size, renders them relatively inconspicuous. Many are also quite timid and are rarely seen, even where common.

The blennies (suborder Blennioidei) are commonly divided into eight families (some of which may not be good), four of which are common on the reef: Blenniidae, Tripterygiidae, Clinidae, and Chaenopsidae (the last is sometimes considered a subfamily of the Clinidae). As a group their reproductive biology is well studied. Wirtz (1977), for example, reviewed aspects of the reproductive biology of 110 species of blennioids in 53 genera. The majority of the work, however, has centered on temperate subtidal species (e.g., Mito, 1954; Gibson, 1969; Dotsu & Oota, 1973), with relatively little known thus far about the more numerous reef-associated species. Overall, however,

blenny reproduction appears to exhibit four nearly, if not truly, universal traits: 1) with few exceptions they produce relatively large demersal eggs; 2) the eggs are usually deposited in a cave or shelter-hole; 3) the eggs are tended and/or defended after spawning; and 4) in all known cases the male alone tends the eggs.

Literature Cited

Dotsu, Y. and T. Oota. 1973. The life history of the blenniid fish, *Omobranchus loxozonus. Bull. Fac. Fish. Nagasaki Univ.*, (36):13-22.

Gibson, R.N. 1969. The biology and behaviour of littoral fish. *Ann. Rev. Oceanogr. Mar. Biol.*, 7:367-410.

Mito, S. 1954. Breeding habits of a blenniid fish, *Salarias enosimae. Japan. J. Ichthyol.*, 3:144-152.

Wirtz, P. 1977. Zum Verhalten blennioider Fische, inbesondere der mediterranen *Tripterygion* Arten. Ph.D. Diss., Ludwig-Maximilian Universitat, München, Germany. 91 pp.

Combtooth Blennies
(Blenniidae)

The combtooth blennies are the "common" blennies. They are characterized by a blunt bulbous head, prominent eyes and lips, and no scales on the body. The common name of the family is derived from the arrangement of teeth in each jaw—a single close-set row that resembles the teeth of a comb. Though many species, including the most intensively studied ones, are found in temperate areas such as the coast of Europe and the Mediterranean Sea, the family is primarily a tropical one, with representatives in many shallow-water habitats. On the reef most are benthic, often cryptic, omnivores that feed on a variety of small invertebrates and algae; a few (the saber blennies) have evolved into highly specialized predators that launch high-speed attacks against the sides of larger fishes to feed on mucus and pieces of flesh (see, for example, Wickler, 1960). Some of these ectoparasites are involved in complex mimicries, including one well known species, *Aspidontus taeniatus*, that mimics the colors and swimming pattern of the cleaner wrasse *Labroides dimidiatus* (Hobson, 1969; Springer & Smith-Vaniz, 1972).

The systematics of the Blenniidae is still somewhat unsettled, though considerable progress has been made in the last few years. There are thought to be about 300 species in perhaps 40 to 50 genera.

SEXUAL DIMORPHISM

As in most groups, male blennies tend to be larger than females (e.g., Springer, 1971; Smith-Vaniz, 1976), and females, especially during the spawning season, tend to be more robust. In a few temperate species, and perhaps some tropical ones, the cirri over the eyes of the males enlarge during the spawning season (Zander, 1975). Permanent sexual differences in details of color patterns and morphology, such as a larger head on the males in the genus *Omobranchus* (Springer & Gomon, 1975) or a more pronounced median flap on the head of males, may also be widespread in the family, though not generally recognized. Otherwise, sexual dimorphism is of two types: temporary spawning colors developed by the males, and differences in anal fin morphology.

Regarding the first, it is widespread, though certainly not universal, that during the spawning season the male develops a brighter or at least distinctive color pattern. In many species the underside of the male's jaw and his throat region develop bright colors (e.g., *Hypsoblennius gentilis,* see Losey, 1976), a location that probably accentuates the "head-bobbing" courtship movements typical of the family. Other blennies undergo complete changes in body coloration. *Ecsenius bicolor,* for example, is normally dark blue anteriorly and pale orange posteriorly. During courtship, however, the male develops a dark red ground color with a series of pale white bars posteriorly; the female also assumes a distinctive spawning color—gray-brown anteriorly, yellow-orange posteriorly (Wickler, 1965). After spawning, during egg care, the male develops a third color pattern, this one dark blue with a series of irregular light gray to pink blotches down the center of each side. Similar though usually less complex changes in color pattern are typical of the family.

Regarding anal fin morphology, several authors (e.g., Wickler, 1957; Eggert, 1931; Porter, 1979) have noted the presence of fleshy pads or swellings on the leading edge of the male's anal fin or on the rays of the dorsal and anal fins of several species of blennies (Zander, 1975). Smith (1974) examined these swellings in a large number of blennies and found that: a) they are characteristic of males only; b) their size, at least in some species, varies seasonally, peaking during the spawning season; and c) morphology of the swellings varied widely among the various species. In male *Scartella* (formerly *Blennius) cristata*, a western Atlantic species, two flattened rugose pads are located immediately behind the urogenital papilla, one on each

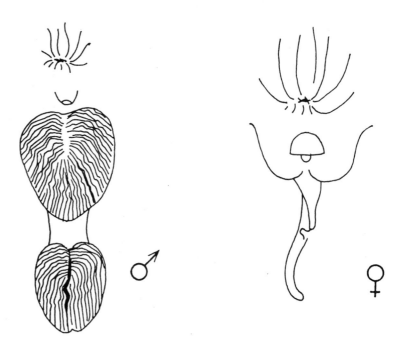

Sexual dimorphism of the anal regions of *Scartella (Blennius) cristata*. Based on Smith (1974), with permission of the *Bulletin of Marine Science*.

of the first two anal spines. Similar pads are reported on males of several species of *Blennius, Hypsoblennius, Hypleurochilus, Cirripectes, Chasmodes,* and *Ophioblennius*. Broad but non-rugose anal pads are characteristic of at least some species of *Petroscirtes, Meiacanthus, Plagiotremus, Aspidontus,* and *Blenniolus*.

The function(s) of such pads are not completely known, but two have been suggested. Smith noted that in most of the blennies he studied, lateral and ventral extensions of the pads develop during the spawning seasons, and he suggested that such extensions might facilitate cleaning or perhaps disinfecting eggs that the male tends. This hypothesis has not yet been tested. Another more widely explored hypothesis is that at least the rugose type of anal pads may be secreting sexual pheromones that facilitate spawning. Eggert (1931) first suggested this possibility when he noted an aromatic secretion emanating from the pads of *Blennius zvonimiri*. Wickler (1957) similarly suggested that some olfactory cues were involved in the courtship of *Blennius fluviatilis*. Following up on these suggestions, Losey (1969) conducted a series of experiments with three eastern Pacific species of *Hypsoblennius (H. jenkinsi, H. robustus,* and *H. gentilis*) to determine if communication by sexual pheromones occurred in these fishes. Individual ripe, but not egg-guarding, males of each species were isolated in 45-liter aquaria into which water from other visually isolated tanks could be slowly added. Test water used ranged from "male-water" and "female-water" drawn from aquaria containing isolated ripe individuals of

each sex, to water drawn off during periods of different intensities of courtship, to "mating-water" drawn off during egglaying. The criterion for measuring response was whether or not the test individuals exited their shelter holes and approached the water inlets within a few minutes of mixing the test water into the normal inflowing water. Losey found that non-egg-guarding ripe males were clearly attracted by water drawn from the fish involved in "high-courtship" and egglaying, but that neither egg-guarding males or females responded. The species also differed in their responsiveness: *H. gentilis* males did not respond to any test water added, while males of *H. robustus* and *H. jenkinsi* both responded strongly to "high-courtship" and "mating" water from conspecifics, and less strongly to "mating" water from *H. gentilis*.

These results clearly demonstrate that an olfactory cue is produced during spawning to which males of at least some species are sensitive and toward the source of which they will move. It was not established, however, whether such chemicals are actively secreted at the anal fin pads or whether they are an incidental, or even undesired, byproduct of, for example, sperm release. Losey suggested that the chemical is a pheromone, presumably actively secreted by the male, and that it serves to promote spawning synchrony and, by attracting many males to a single area, results in social facilitation that possibly enhances a male's sexual receptivity. It is not clear, however, whether these benefits would outweigh the apparently great disadvantage to a courting male of attracting competitors to his area. This apparent disadvantage suggests that the chemical produced is, in fact, incidental.

SPAWNING SEASON

Little is known about the spawning seasons of reef-dwelling blennies, though it seems probable that many spawn year-round. *Ecsenius bicolor*, for example, apparently does so, at least in captivity (Wickler, 1965), as does *Exallias brevis* at Hawaii (B. Carlson, pers. comm.). Limited spawning seasons, however, have been documented for two species at One Tree Island, Great Barrier Reef (Dybdahl, 1978). In both cases spawning occurs only in the warmer half of the year.

There is no information specifically dealing with possible lunar rhythms in blenniid spawning. Wickler (1965), however, found that *Ecsenius bicolor* in captivity spawned at almost exact two-week intervals, which may indicate a semi-lunar rhythm. B. Carlson (pers. comm.) reports spawning by individual females of *Exallias brevis* at Hawaii to occur at three- to four-day intervals not conspicuously synchronized to any specific lunar periodicity. Spawning activity for several species at One Tree Island, Great Barrier Reef, was conspicuously periodic and, when it occurred, widely synchronized. No data specifically relating such spawning periods to lunar cycles was obtained, but general field notes indicate that newly laid eggs and observations of blenny spawning behavior were invariably collected a week or so preceding the new and full moons.

REPRODUCTIVE BEHAVIOR

Although a number of studies have dealt with the reproductive behavior of temperate blennies (e.g., Abel, 1964; Wickler, 1957; Gibson, 1969), only four detailed studies have been made on reef-associated species: Wickler's (1965) study on *Ecsenius bicolor,* Losey's (1968) study of *Hypsoblennius gentilis* and two related species, Fishelson's (1975) study of *Meiacanthus nigrolineatus,* and Dybdahl's (1978) study of *Meiacanthus lineatus* and *Petroscirtes fallax.* Limited information is also available on *Exallias brevis* (Carlson, 1978), with a more detailed report in preparation. Despite the relatively small data base, overall reproductive behavior is similar throughout the family, such that a sequence of behavior followed by a "typical" blenny can be constructed.

Spawning invariably takes place in a narrow crevice or shelter hole, typically at or near the center of the male's territory, with the female depositing her eggs on the sides and roof of the cave. In most cases courtship is initiated by the female, who approaches the territory of the male, occasionally while in a female-specific courtship color. At the approach of the female the male assumes his courting colors (if he is not already in them), leaves his shelter hole, and swims to the female. While swimming, the male frequently travels in an up-and-down sinusoidal motion ("Wippsprungen," Wickler, 1965). Other species pose at the mouth of the shelter hole head-up or head "bob" (shaking their heads up and down vigorously) as the female approaches. The male then either leads the female to the shelter hole (e.g., *Scartella cristata*), again while swimming in a sinusoidal motion, or he sits in front of it and points his head toward it. Males of many species (e.g., *Ophioblennius steindachneri*) also perform an undulating exaggerated swimming motion while leading the female to the nest. If the female is interested she approaches and inspects the male's shelter and then backs into it to begin laying her eggs. If the shelter hole is large enough the male may also enter and the two will sit side-by-side, undulating in unison while the female lays her eggs and the male fertilizes them. If the shelter hole is small the male usually enters it only briefly at irregular intervals to fertilize the eggs. Otherwise he remains near its mouth, driving away intruders and performing sporadic courtship movements. The duration of spawning lasts anywhere from a few minutes or hours (e.g., Wickler, 1965) to more than a day (Fishelson, 1975). In the latter case spawning is intermittent, with the female leaving only a few eggs at a time in the nest then exiting and remaining near it while the male fertilizes them. Fishelson reported that one female *M. nigrolineatus* may continue spawning, with interruptions, for one and a half days and lay between 100 and 160 eggs. Typical blenny clutch sizes are on the order of only a few hundred eggs per female, though successful males often brood eggs produced by several females (e.g., Dybdahl, 1978). Female *Exallias brevis,* which spawn regularly every three to four days, produce, on the average, 200,000 to 300,000 eggs per year.

This "typical" sequence is followed more or less closely by most benthic blennies and a few hovering ones, e.g., *Meiacanthus nigrolineatus* and *Petroscirtes fallax.* The eastern Pacific saber blenny, *Plagiotremus azaleus,* differs in some respects. Courtship is, as usual, initiated by the female but occurs a meter or so off the bottom above the male's shelter hole (pers. obs.). At the approach of a ripe female, who swims back and forth in front of the male's shelter hole, the male, slightly larger than the female but otherwise identical to her, swims toward her and then stops a half meter or so from her, hovering. He then launches himself at her and at lightning speed spins two or three times vertically and tightly around her body as she is swimming; he then flashes back to his starting position. He repeats his "attack" on her five or six times at intervals of five to ten seconds and then swims in an exaggerated motion slowly back to his shelter hole and backs into it. Pulling almost entirely into it, he then "pops" his head in and out of the hole several times, rears about a third of the way out, and begins to "bob" up and down while facing the female. If she does not respond he exits the hole, "attacks" her again, and then releads her back to the hole. If she approaches he exits and she enters the shelter hole. Apparently only a few eggs are laid at a time, as in *M. nigrolineatus*, as she remains in the hole for only a few minutes before exiting and leaving the male's area. That clutch sizes are small in this and similar species would not be surprising, since a female that carried many eggs would seem to be at a substantial handicap in making the high speed attacks necessary to obtain food.

Few of the studies on reproduction of reef-dwelling blennies provide normal spawning times, but it seems likely that spawning in these visually-oriented fishes continues throughout the day. *Plagiotremus azaleus* and *Exallias brevis* characteristically spawn about midday, whereas *Ophioblennius steindachneri*, *Petro-*

scirtes fallax, and *Meiacanthus lineatus* spawn in early morning, within a few hours of sunrise and often at first light.

EGGS AND LARVAE

As is the case with reproductive behavior, most work on blenniid eggs and larvae has involved temperate species (e.g., Ford, 1922; Cipria, 1934, 1936; Mito, 1954; Wickler, 1957; Fishelson, 1963b) or intertidal species (e.g., Dotsu & Oota, 1973). Only two detailed studies on the development of reef-dwelling blennies have been published, but the similarity of their results with each other and with available information on temperate species suggests that the general pattern of development is similar for all blennies.

Blenny eggs are characteristically flattened ovals, usually brightly colored, and about a millimeter in the longest dimension. Those of *Ecsenius bicolor* are about 0.75 mm long and 0.5 mm high and are attached to the sides and roof of the nest hole by means of sticky threads along one side of the egg (Wickler, 1965). At 27° C they hatch in approximately nine days, the male fanning the eggs vigorously during hatching but otherwise ignoring the larvae. Newly hatched larvae are approximately 3.5 mm long and have well developed jaws (lacking teeth), functional pectoral fin rudiments, large black eyes, black pigment spots along the lower edge of the peritoneum, and only a small

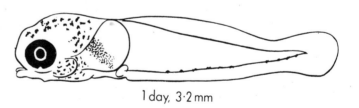

1 day, 3·2 mm

Larval development of the blenny *Meiacanthus nigrolineatus*. Based on Fishelson (1976).

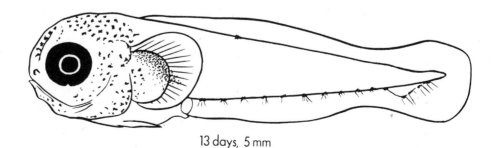

13 days, 5 mm

yolk sac. Upon hatching they immediately swim toward light.

Few data are yet available, but it appears that blenny eggs characteristically hatch at dusk (e.g., Carlson, 1978; Dybdahl, 1978; pers. obs.).

The development of *Meiacanthus nigrolineatus* is similar (Fishelson, 1976). Eggs are orange hemispheres about 1.0 mm across attached to the walls of the shelter hole by means of sticky threads. At 24° C hatching occurs on days 10 and 11 (according to Dybdahl, 1978, *M. lineatus* has a similar incubation period at an average temperature of 27°C). Newly hatched larvae lack a yolk sac and actively swim about feeding, remaining close to the water surface. At one day old they are approximately 3.7 mm long and, except for some scattered pigment spots on the head and gill covers, are nearly identical to larval *E. bicolor*. Later stage blenny larvae, in general, are elongate, with pelvic fins inserted ahead of the pectorals, and have long dorsal and anal fins and well developed (and in some species large) pectoral fins (e.g., Fourmanoir, 1971, 1976). The late-stage larvae of *M. nigrolineatus* develop typical juvenile colors (dark anteriorly, pale posteriorly, with a short horizontal line on each side) by day 20 and begin to stay close to the bottom. By day 30 they are largely benthic.

Miller, et al. (1979) described the planktonic larvae of several Hawaiian blenniids. They reported that larvae of three tribes (Salariini, Omobranchini, and Blenniini) are elongate and slender with a short gut, no apparent gas bladder, long dorsal and anal fins, jugular pelvic fins, and 30-odd myomeres. Salariin larvae are further characterized as having extremely long pectoral fin rays during the later stages of development and large hooked canine teeth in either the lower or both jaws. Larvae of the other two tribes have shorter guts, more rounded heads, and longer preopercular spines than the salariins. Larvae of *Enchelyurus brunneolus* and *Istablennius zebra* are illustrated in Miller, et al. (1979) as representatives of the latter two groups.

REPRODUCTION IN CAPTIVITY

Spawning blennies in aquaria requires little beyond enough space, shelter, and food for at least four or five fish. Many blennies are territorial (especially the males) so that the biggest problem to overcome is likely to be crowding. As with most groups, best results will probably be obtained if the fish are maintained in an aquarium of their own and females outnumber males by at least two to one. In place of natural shelter holes, most blennies will readily accept short pieces of tubular polyvinyl chloride (PVC) partially buried in the gravel.

After spawning, the male will vigorously tend and defend the eggs, and in most cases it is probably best to let him do so. Fishelson (1976), however, hatched eggs of *Meiacanthus nigrolineatus* in the absence of a male by transferring them into a well-aerated 20-liter aquarium that had been filled with millepore-filtered sea water. He reported the newly hatched larvae began feeding immediately and readily took rotifers (*Brachionis plicatilis*). After several days the larvae began to take brine shrimp nauplii.

Literature Cited

Abel, E.F. 1964. Freiwasserstudien zur Fortpflanzungsethologie zweier Mittelmeerfische, *Blennius canevae* Vinc. und *Blennius inaequalis* C.V. *Z. Tierpsychol.*, 21:205-222.

Carlson, B.A. 1978. A contribution to the biology of the spotted blenny, *Exallias brevis* (Pisces; Blenniidae). *Pac. Sci.*, 32:96.

Cipria, G. 1934. Uova, stadi embrionali e postembrionali de *Blennius palmicornis* Cuv. *Mem. Roy. Com. Talassogr. Ital.*, (218):1-12.

Cipria, G. 1936. Uova, stadi embrionali e postembrionali nei Blennidi. 1. *Blennius pavo* Risso. 2. *Blennius inaequalis* C.V. *Mem. Roy. Com. Talassogr. Ital.*, (231):3-8.

Dotsu, Y. and T. Oota. 1973. The life history of the blenniid fish, *Omobranchus loxozonus*. *Bull. Fac. Fish. Nagasaki Univ.*, (36):13-22.

Dutt, S. and V. Visveswara Rao. 1960. On the breeding habits and early developmental stages of *Petroscirtes bipunctatus* Day. *J. Zool. Soc. India*, 12:158-161.

Dybdahl, R.E. 1978. The ecology and behaviour of a mimetic pair of coral reef fishes, *Meiacanthus lineatus* and *Petroscirtes fallax* (Pisces:Blenniidae). Master's Thesis, Univ. of Sydney, Australia. 121 pp.

Eggert, B. 1931. Die Geschechtsorgane der Gobiiformes und Blenniiformes. *Z. Wiss. Zool., Leipzig*, 139(2/3):249-558.

Fishelson, L. 1963a. Observation on littoral fishes of Israel. I. Behavior of *Blennius pavo* Risso (Teleostei, Blenniidae). *Israel J. Zool.*, 12:67-80.

Fishelson, L. 1963b. Observation on littoral fishes of Israel. II. Larval development and metamorphosis of *Blennius pavo* Risso (Teleostei, Blenniidae). *Israel J. Zool.*, 21:81-91.

Fishelson, L. 1975. Observations on behavior of the

fish *Meiacanthus nigrolineatus* Smith-Vaniz (Blenniidae) in nature (Red Sea) and in captivity. *Aust. J. Mar. Freshwater Res.*, 26:329-341.

Fishelson, L. 1976. Spawning and larval development of the blenniid fish, *Meiacanthus nigrolineatus* from the Red Sea. *Copeia*, 1976:798-800.

Ford, E. 1922. On the young stages of *Blennius ocellaris* L., *Blennius pholis* L., and *Blennius gattorugine* L. *J. Mar. Biol. Assoc. U.K.*, 12:688-692.

Fourmanoir, P. 1971. Notes Ichthyologiques. (IV). *Cah. O.R.S.T.O.M., Ser. Oceanogr.*, 9:491-500.

Fourmanoir, P. 1976. Formes post-larvaires et juveniles de poissons cotiers pris au chalet pelagique dans le sud-ouest Pacifique. *Cah. du Pacifique*, (19):47-88.

Gibson, R.N. 1969. The biology and behaviour of littoral fish. *Ann. Rev. Oceanogr. Mar. Biol.*, 7:367-410.

Hobson, E.S. 1969. Possible advantages to the blenny *Runula azalea* in aggregating with the wrasse *Thalassoma lucasanum* in the tropical eastern Pacific. *Copeia*, 1969:191-193.

Losey, G.S. 1968. The comparative behavior of some Pacific fishes of the genus *Hypsoblennius* Gill (Blenniidae). Ph. D. Diss., Univ. California, San Diego.

Losey, G.S. 1969. Sexual pheromone in some fishes of the genus *Hypsoblennius* Gill. *Science, Wash.*, 163(1):181-183.

Losey, G.S. 1976. The significance of coloration in fishes of the genus *Hypsoblennius* Gill. *Bull. So. Calif. Acad. Sci.*, 75:183-198.

Miller, J.M., W. Watson, and J.M. Leis. 1979. An atlas of common nearshore marine fish larvae of the Hawaiian Islands. *Sea Grant Misc. Rpt.* (UNIHI-SeaGrant-MR-80-02). 179 pp.

Mito, S. 1954. Breeding habits of the blennioid fish, *Salarias enosimae. Japan. J. Ichthyol.*, 3:144-152.

Munro, I.S.R. 1955. Eggs and larvae of the sabre-toothed oyster blenny, *Dasson steadi* (Whitley) (Blenniidae). *Aust. J. Mar. Fresh. Res.*, 6:30-34.

Porter, M.C. 1979. An investigation into the nature of anal fin glands in fishes of the family Blenniidae. Master's Thesis, Fla. Atl. Univ. 47 pp.

Rao, V.V. 1970. Breeding habits and early development of two blenniid fishes *Omobranchus japonicus* (Bleeker) and *Cruantus smithi* Visweswara Rao from Godavari estuary. *J. Mar. Biol. Assoc. India*, 12(182):175-182.

Raja, N.S. 1971. Breeding habits, development and life history of *Blennius steindachneri* Day from Waltari coast. *Proc. Indian Acad. Sci.*, 74:37-45.

Smith, R.L. 1974. On the biology of *Blennius cristatus*, with special reference to anal fin morphology. *Bull. Mar. Sci.*, 24(3):595-605.

Smith-Vaniz, W.F. 1976. The sabre-toothed blennies, tribe Nemophini (Pisces:Blenniidae). *Acad. Natur. Sci., Phila., Monogr.*, (19):1-196.

Springer, V.G. 1971. Revision of the fish genus *Ecsenius* (Bienniidae, Blenniinae, Salariini). *Smithsonian Contrib. Zool.*, (72):1-74.

Springer, V.G. and M. Gomon. 1975. Revision of the blenniid fish genus *Omobranchus* with descriptions of three new species and notes on other species of the tribe Omobranchini. *Smithsonian Contrib. Zool.*, (117):135 pp.

Springer, V.G. and W.F. Smith-Vaniz. 1972. Mimetic relationships involving fishes of the family Blenniidae. *Smithsonian Contrib. Zool.*, (112):1-36.

Wickler, W. 1957. Vergleichende Verhaltensstudien an Grundfischen I. Beitrage zur Biologie, besonders zur Ethologie von *Blennius fluviatilis* Asso im Vergleich zu einigen anderen Bodenfischen. *Z. Tierpsychol.*, 14:393-428.

Wickler, W. 1960. Aquarienbeobachtungen an *Aspidontus*, einem ektoparasitischen Fisch. *Z. Tierpsychol.*, 17:277-292.

Wickler, W. 1965. Zur Biologie und Ethologie von *Ecsenius bicolor* (Pisces, Teleostei, Blenniidae). *Z. Tierpsychol.*, 22:36-49.

Zander, C.D. 1975. Secondary sex characteristics of blennioid fishes (Perciformes). *Publ. Staz. Zool. Napoli, Suppl.* 39:717-727.

Scaled Blennies
(Clinidae)

The typical clinid is a slender, heavily and obviously scaled blenny that sits on the bottom propped up on its pelvic fins, feeds on small benthic organisms, and is cryptically colored. Its mouth is terminal and rather small for a blenny, its snout is often pointed, and its dorsal and anal fins are long and continuous. Like most blennies, its head is adorned with several clusters of short hair-like projections known as cirri.

Clinids are found in tropical and temperate shallow seas worldwide but reach their peak in diversity and abundance off South Africa, Australia, and the tropical Americas. The family contains perhaps 35 genera and 150 to 200 species worldwide, about a third to half of which are American. Until recently the group was considered a subfamily of the Blenniidae, and even now they are rather vaguely defined and include a morphologically wide variety of fishes. One group, the chaenopsids, is often elevated to family status, which will be done here as they have a distinctive and characteristic reproductive biology.

Reproductive biology varies widely in the clinids, from a typical blenniid system to some of the few livebearing fishes found on the reef. Detailed information on the majority of reef-dwelling species is still lacking.

SEXUAL DIMORPHISM

External sexual dimorphism is characteristic of the clinids, although detailed information for reef-dwelling species is available only for the New World genera *Labrisomus*, *Malacoctenus*, *Paraclinus*, and *Starksia*. Differences between the sexes are generally of two types: permanent and temporary color differences, and minor morphological differences.

According to Springer (1958), males of *Labrisomus* and *Malacoctenus* are characteristically more colorful than females and tend toward less spotting and bars in the color pattern. Differences are striking for most *Malacoctenus* and a few *Labrisomus* species, e.g., *L. nigricinctus* (see plates in Böhlke & Chaplin, 1968),

but in many species they are not conspicuous. Males of some of the less dichromatic species of *Labrisomus*, however, develop bright colors similar to those of the blenniids during spawning seasons, e.g., *L. gigas*. Thresher (1980) also reported that the sexes of at least one species, probably *L. kalisherae*, are strongly dichromatic during spawning: the male becomes very dark while the female pales almost to white.

Morphological differences between the sexes of *Labrisomus* and *Malacoctenus* involve the shapes of the urogenital papillae and various minor features. On the male the single common papilla is just posterior to the anus; on the female the genital and urinary openings are separate, the former just posterior to the anus and the latter located on a papillose knob behind both. Males of *Labrisomus* also tend to have larger upper jaws than females; males of *Malacoctenus* are more completely scaled ventrally than are the females. Finally, males of at least four western Atlantic species of *Malacoctenus*—*M. aurolineatus*, *M. erdmani*, *M. gilli*, and *M. macropus*—develop large fleshy pads on the innermost pelvic rays when sexually ripe; the function of these pads is not known and their structure has not been examined in detail.

The sexes of the species of *Paraclinus* are similar in permanent color pattern, but males tend to be larger, have more oblique and larger jaws, and have a larger first dorsal fin than females (Breder, 1939; Hubbs, 1952; Springer, 1954). The sexes of *Paraclinus marmoratus* also differ in color during courtship and spawning. The normally dark male becomes deep purple, almost black, with metallic blue spots; the female fades almost to white and has a few tan markings. Spawning and courtship have so far been observed only for *P. marmoratus*, so it is not known how general these temporary color changes may be.

Sexual dimorphism in *Starksia* primarily involves the structure of the urogenital papilla. The starksiins are one of the few known groups of livebearing reef fishes, the males possessing a prominent intromittent organ that is derived from the urogenital papilla and the first anal spine (Hubbs, 1952; Al-Uthman, 1960; Böhlke & Springer, 1961; Rosenblatt & Taylor, 1971). Structure of the organ and shape and length of the urogenital papilla are species-specific. In most species the elongate tubular papilla is closely attached to the first anal spine, which is separated from the anal fin. In two western Atlantic species, *S. hassi* and *S. lepicoelia*, the papilla is completely separate from the spine, and in the monotypic genus *Xenomedea* both papilla and spine are surrounded by a fleshy fold, the height of which varies seasonally. Both papilla and spine are clearly exposed during the nonspawning season, but during the spawning season the fleshy fold enlarges and blocks both when the fish is viewed from the side. In all starksiins, the female urogenital opening and anal fin are unmodified.

Starksiins are not the only clinids with males that possess intromittent organs. The more advanced members of the subfamily Clininae are also ovoviviparous (e.g., Mooreland & Dell, 1950; Milward, 1967; Christensen, 1978; Veith, 1979a, b & c), but the structure of the intromittent organ in these predominantly southern hemisphere fishes differs markedly from that of the starksiins, suggesting ovoviviparity arose independently in the two groups (Penrith, 1965; Rosenblatt & Taylor, 1971).

SPAWNING SEASON

Information on spawning seasons of reef-dwelling clinids is available for only two species, *Paraclinus marmoratus* and *Xenomedea rhodopyga* (Breder, 1941, and Rosenblatt & Taylor, 1971, respectively). Both spawn in the spring, the former when water temperatures are approximately 20° C. Whether the latter copulates and releases young in the same season or copulates after parturition and stores sperm until the next spawning season is not known. Rosenblatt & Taylor suggested that *X. rhodopyga* and perhaps a few species of *Starksia* spawn several times yearly. Such is likely to be true for other clinids as well.

REPRODUCTIVE BEHAVIOR

So far as is known, all reef-dwelling clinids are territorial and spawning takes place in the male's territory. Information on time of spawning is sparse, but it clearly occurs during the day; i.e., it is not restricted to dusk or dawn.

Information on courtship and spawning is available for only three species, one each of *Labrisomus*,

Malacoctenus, and *Paraclinus.* The first two are only partial, covering apparent courtship in *Labrisomus* and spawning alone in *Malacoctenus.* What appeared to be courtship in *Labrisomus kalisherae* was observed on a shallow patch reef at dusk (Thresher, 1980). The male was deep brown with a broad white bar behind his gill covers; the female was pale, whitish with a few tan marks. The two fish approached one another with short intermittent dashes and then the male erected his dorsal fin and presented it broadside to the female. She approached, but unfortunately was then frightened off by a diver.

Spawning was observed on several occasions in the Gulf of California clinid *M. margaritae*, though courtship was not seen (per. obs.). Spawning takes place at midday and occurs in a narrow groove between two rocks (apparently partially cleared of sand by the male). Egglaying takes place on the two opposing faces of the groove, with the pale green and yellow female alternately spawning on one and then the other face. She moves only slightly while spawning, sporadically quivering violently with the darker male beside her, in what appears to result in the release of one or more eggs. The male, in contrast, is extremely active, spending most of his time patrolling the area around the female and chasing from it damselfishes, wrasses, and other intruders. Periodically he returns to the spawning site briefly, either to move perpendicular to the female and fertilize an egg or to nudge her abdomen if she has stopped spawning. Such nudging apparently stimulates the female to continue, since after a few nudges she invariably resumes spawning.

While these two partial observations were made in the field, the only detailed study of clinid spawning is that of the small western Atlantic species *Paraclinus marmoratus* in captivity (Breder, 1939 & 1941). In the field egg masses, usually guarded by the male, were occasionally found in the broken and partially healed-over lumens of old sponges. In aquaria spawning occurs in late afternoon in a rocky crevice normally occupied by the male. It is initiated by a ripe female, who blanches nearly to white. The male, in turn, darkens to a deep purple. After several false starts the female rather unceremoniously enters the small cave that the male has exited, turns upside down, and begins to lay her eggs on the roof of the cave. The male backs in next to her. Periodically the pair vibrate vigorously, apparently releasing an egg. After several such vibration the female leaves. After spawning the male fans and guards the eggs, pulling at their mass

321

periodically in an apparent attempt to facilitate water flow through the eggs, and he also courts any other females that appear. Nests often contain eggs at different stages of development, suggesting that the males are sequentially polygamous.

EGGS AND LARVAE

The eggs and larvae of only one species, *Paraclinus marmoratus*, have been described. According to Breder (1939) the eggs are spherical, about 1.15 to 1.30 mm in diameter, pale amber, contain several oil droplets, and are attached to one another and often to an algal mat by a single cord of twisted strands that emerges from one pole. After spawning, the eggs of *Malacoctenus margaritae* could not be seen in the field, suggesting that they are either much smaller than the eggs of *P. marmoratus* or were deeply buried in algae that covered the spawning site.

P. marmoratus eggs hatch in ten days at 21° C. Newly hatched larvae are approximately 4.1 mm long, bear a moderate-sized yolk sac containing a single oil droplet, and float upside down at the surface. Development is rapid, with well-developed jaws, eyes, and fins forming within the first 24 hours. The larvae become strongly photonegative and seek the bottom within 24 hours of hatching, suggesting an extremely short planktonic stage. Such is apparently not the case for other clinids. Böhlke & Chaplin (1968), for example, reported that *Labrisomus nuchipinnis* has a prolonged larval stage during which the larva is shallow-bodied and lacks scales.

Little is known about the larvae or development of the livebearing starksiins. After copulation the eggs are retained in the follicles of the ovaries and are apparently not nourished, as opposed to the advanced clinins in which the eggs are free in the lumen and the larvae are born well after the yolk is absorbed (Rosenblatt & Taylor, 1971). The number of embryos in a single ovary ranges from 12, in a 28 mm *Xenomedea rhodopyga*, to well over a hundred (along with developing and unfertilized eggs). The largest embryos measured were approximately 5 mm long. Whether or not these larvae have a brief planktonic stage is not known.

REPRODUCTION IN CAPTIVITY

As noted, Breder (1941) reported the successful spawning of *Paraclinus marmoratus* in small aquaria, using several gravid females and a single male that was guarding eggs when it was captured. Similar efforts to spawn other species of *Paraclinus* and the starksiins should be equally successful given their small size at maturity (rarely more than a few centimeters) and limited home ranges. Spawning these fishes may be complicated only by their shyness, which could make observations difficult.

Spawning *Malacoctenus* in captivity should be easier, as these somewhat larger fish (most about 6 cm at maturity) are bold, still relatively small, and adapt quickly to aquarium conditions. Since they are territorial, a minimum 50- to 60-liter aquarium containing several females and a single male might be the best approach. The species of *Labrisomus* are larger than most *Malacoctenus* (up to around 20 cm), are much shyer, and hence do less well in aquaria. Spawning them will require copious amounts of live food, an aquarium containing few, if any, other fish, and patience. Such an effort may be warranted, however, for the brilliantly colorful *Labrisomus nigricinctus* which, because of its shyness, is rarely seen and even more rarely available in the tropical fish trade.

None of the clinids have yet been reared in captivity, but given the relatively large size and short planktonic stage of at least some, such as *P. marmoratus*, they should be among the easiest groups to rear.

Literature Cited

Al-Uthman, H.S. 1960. Revision of the Pacific forms of the tribe Starksiini. *Texas J. Sci.*, 12:163-175.

Böhlke, J.E. and C.C.G. Chaplin. 1968. *Fishes of the Bahamas and adjacent tropical waters.* Livingston Publ., Wynnewood, Pa. 771 pp.

Böhlke, J.E. and V.G. Springer. 1961. A revision of the Atlantic species of the clinid fish genus *Starksia*. *Proc. Acad. Nat. Sci., Philadelphia*, 113:29-60.

Breder, C.M., Jr. 1939. On the life. history and development of the sponge blenny, *Paraclinus marmoratus* (Steindachner). *Zoologica, N.Y.*, 24:487-496.

Breder, C.M., Jr. 1941. On the reproductive behavior of the sponge blenny, *Paraclinus marmoratus* (Steindachner). *Zoologica, N.Y.*, 26:233-235.

Christensen, M.S. 1978. *Pavoclinus myae*, a new species of clinid fish (Perciformes, Blennioidei) from South Africa, with notes on the identity of *P. grammis* and *P. laurenti*, and a key to the known species of *Pavoclinus*. *Spec. Publ. J.L.B. Smith Inst. Ichthyol.*, (18):1-16.

Hubbs, C.L. 1952. A contribution to the classification of the blennioid fishes of the family Clinidae, with a partial revision of the eastern Pacific forms. *Stanford Ichthyol. Bull.*, 4:41-165.

Milward, N.E. 1967. The Clinidae of Western

Australia. *Proc. Roy. Soc. West. Aust.*, 50:1-9.

Moreland, J. and J. Dell. 1950. Preliminary report on the ovoviviparity in a New Zealand blennioid fish, *Ericentrus rubrus. N.Z. Sci. Rev.*, 8:39-40.

Penrith, M.L. 1965. The systematics and distribution of the fishes of the family Clinidae in South Africa, with notes on the biology of some common species. Ph. D. Diss., Univ. Cape Town, South Africa.

Rosenblatt, R.H. and L.R. Taylor, Jr. 1971. The Pacific species of the clinid fish tribe Starksiini. *Pac. Sci.*, 25:436-463.

Springer, V.G. 1954. Western Atlantic fishes of the genus *Paraclinus. Texas J. Sci.*, 6:422-441.

Springer, V.G. 1958. Systematics and zoogeography of the clinid fishes of the subtribe Labrisomini Hubbs. *Publ. Inst. Mar. Sci., Univ. Texas,* 5:417-492.

Thresher, R.E. 1980. *Reef Fish.* Palmetto Publ. Co., St. Petersburg, Fla. 172 pp.

Veith, W.J. 1979a. Reproduction in the live-bearing teleost *Clinus superciliosus. South Afr. J. Zool.,* 14:208-211.

Veith, W.J. 1979b. Viviparity and embryonic adaptations in the teleost *Clinus superciliosus. Canad. J. Zool.,* 58:1-12.

Veith, W.J. 1979c. The chemical composition of the follicular fluid of the viviparous teleost *Clinus superciliosus. Comp. Biochem. Physiol.,* 63:37-40.

Pike Blennies
(Chaenopsidae)

The pike or tube blennies are an amphi-American group of small elongate fishes closely related to, and sometimes considered a subfamily of, the Clinidae. Morphologically they differ from the clinids in lacking scales, having a more reduced lateral line, and having a significantly different skull structure (Stephens, 1970). Behaviorally the family is characterized by a strong tendency to sit in holes on the bottom—either naturally occurring crevices or the abandoned shells and tubes of molluscs and boring invertebrates. There are nine genera and 53 species of chaenopsids (Lindquist, 1975) ranging from the small reef-dwelling planktivores in the genus *Acanthemblemaria* to the highly predatory species in *Lucayablennius* and *Chaenopsis*. The most specialized of these predators is the wrasse blenny, *Hemiemblemaria simulus,* in the western Atlantic, which has evolved a complex pattern of colors and movements to mimic the common wrasse *Thalassoma bifasciatum* (see Randall & Randall, 1960).

SEXUAL DIMORPHISM

The wrasse mimic *Hemiemblemaria simulus* appears to be sexually monomorphic, but all other chaenopsids exhibit at least some degree of sexual dimorphism. The differences are least in the species of *Acanthemblemaria* and in the arrow blenny, *Lucayablennius zingaro*. In the former, the male has a dorsal fin that is only slightly, if at all, larger than that of the female, but he also tends to be darker and more contrastingly or vividly colored (Smith-Vaniz & Palacio, 1974). Males of a few species develop distinctive colors during the spawning season, usually in the form of a darker head or some color highlights on the dorsal fin (Lindquist, 1975). In the arrow blenny the sexes differ only in that the male tends to have more extensive dark areas on the dorsal and anal fins than does the female.

Sexual differences in the development of the spiny dorsal fin and in color pattern increase steadily

through the genera *Protemblemaria, Ekemblemaria, Emblemariopsis,* and *Coralliozetus,* reaching their zenith in *Chaenopsis* and, especially, *Emblemaria* (Böhlke, 1957; Robins & Randall, 1965; Stephens, 1970). Indeed, the sexes of the last two genera are often so different as to be scarcely recognizable as conspecific. In both, males have developed broad sail-like dorsal fins that are spread during courtship. The breadth of the fin is emphasized by the very dark color patterns characteristic of the male (male *Emblemaria* are jet black; male *Chaenopsis* tend to be black anteriorly) combined with color highlights on the fin, best developed in the eastern Pacific species *E. hypacanthus.* Though males tend to maintain these color patterns throughout the spawning period, when stressed or placed in a subordinate position they can quickly fade to grayish brown. The females in both genera lack the broad dorsal fins of the males and are mottled brown and cryptic. They also tend to be smaller than the males.

SPAWNING SEASON

Little is known about seasonal patterns of reproduction in chaenopsids. All court regularly during the warmer parts of the year, and at least some, such as *Emblemaria hypacanthus,* court at least occasionally year-round (pers. obs.). In aquaria both *Emblemaria* and *Chaenopsis* spawn year-round.

REPRODUCTIVE BEHAVIOR

Little is known about the social systems of most pike blennies aside from scattered observations of agonistic behavior (e.g., Lindquist, 1975), which suggests either territoriality or dominance hierarchies (perhaps based on priority of access to favorable shelter tubes). Sporadic observations of apparently the same individuals of several species, such as *Hemiemblemaria simulus* and *Emblemaria hypacanthus,* in the same area or the same shelter tube day after day suggest that most chaenopsids are more or less permanently site-attached to specific parts of the reef. Site-attachment and territoriality in the eastern Pacific pike blenny, *Chaenopsis alepidota,* vary with population density. At low densities the fish change shelter holes daily and foraging areas every few days; at high densities shelter tubes are defended from conspecifics and the animals appear to remain in permanent feeding areas (Thresher, MS).

Spawning activity is most frequent in the morning (Lindquist, 1975; pers. obs.), usually within a few hours of dawn. Courtship and sporadic spawning occur throughout the day, however. There does not seem to be any increase in activity at dusk.

In the basic chaenopsid courtship display the male rears forward and part way out of his shelter tube and erects his dorsal fin. In *Acanthemblemaria* and *Emblemaria* the dorsal fin is "flicked" open and closed four or five times in rapid succession, while the male comes farther and farther out of his tube with each flick (Stephens, et al., 1966; Wickler, 1967). Indeed, excited males of *Emblemaria* often end a bout of vigorous displaying as much as a meter off the bottom and must dash quickly back to their tubes. In the jet black species these courtship displays are both conspicuous and spectacular. Their effectiveness in communicating the male's location is obvious; many times the first indication that *Emblemaria* are present in an area is the noticing of sporadic flashes of black near the bottom. Courtship by *Chaenopsis* is similar to that of *Emblemaria,* except that the dorsal fin is held fully open throughout the in-and-out bobbing.

Upon approach of a female, the male erects his dorsal fin, darkens to his fullest, and begins to bob back and forth in front of her. He then stops and withdraws completely into his shelter tube; the female backs in on top of him. She only stays for a few seconds and presumably spawns only one or a few eggs at a time. Stephens, et al. (1966) reported that female *Acanthemblemaria macrospilus* frequently spawn with several males in quick succession; at the same time, males usually tend eggs at several different stages of development, suggesting a promiscuous mating system. The same is probably true for other species as well (Luckhurst, in prep.).

After spawning the female quickly flees back to her own shelter tube and the male begins to tend and guard the eggs. Longley & Hildebrand (1941) reported defense of eggs by a female *Chaenopsis limbaughi,* but this observation has not since been confirmed and seems unlikely given invariable male defense elsewhere in the family.

EGGS AND LARVAE

Böhlke & Chaplin (1968) reported the eggs of several chaenopsids as "large" or "very large" but otherwise provided no specific descriptions. Brief comments on larval development were made for *Chaenopsis alepidota* by Böhlke (1957) and for *Acanthemblemaria macrospilus* by Stephens, et al. (1966). Larval *C. alepidota* are elongate, have short jaws (as opposed to the pike-like jaws of the adults), and have emarginate or shallowly forked tails not yet united to the dorsal and anal fins (the adult condition). Newly settled individuals are approximately 15 mm long and are nearly transparent; like the adults, the juveniles immediately settle into a shelter tube. Newly settled *A. macrospilus* are similarly between 15 and 17 mm long and are almost completely unpigmented. The cryptic benthic coloration develops between 17 and 19 mm SL.

REPRODUCTION IN CAPTIVITY

The species of at least *Acanthemblemaria, Emblemaria,* and *Chaenopsis* (and probably all chaenopsids except *Lucayablennius zingaro* and *Hemiemblemaria simulus,* which require copious amounts of live food) spawn readily in aquaria if provided enough space, numerous shelter tubes, and lots of food. All will take thawed brine shrimp and eventually flake food, but they clearly prefer live shrimp and small fishes. Species of *Chaenopsis* are predators and should be fed especially heavily if spawning is desired. Courtship and spawning by *Chaenopsis* and *Emblemaria* are particularly spectacular and are well worth the time and effort.

Shelter tube requirements can be easily satisfied for most species by providing small thoroughly cleaned seashells scattered widely about the tank. Species of *Chaenopsis,* which prefer worm tubes, will readily accept soda straws partially buried in the sand or, for larger individuals, short pieces of plastic tubing similarly buried. In "colony tanks" containing eight to 15 individuals, *Emblemaria* and *Chaenopsis* spawned regularly when there were roughly twice as many shelter holes as there were fish and when the shelter holes were located about 0.5 m apart. If they were much closer than this the fish tended to fight a great deal. Previous work with *Emblemaria pandionis* indicates that even in the presence of a female, male colors fade unless another male is present. Addition of a second male, even though he was quickly dominated and lost control, stimulated more intense colors in the dominant male and increased reproductive activity.

Literature Cited

Böhlke, J.E. 1957. A review of the blenny genus *Chaenopsis* and the description of a related new genus from the Bahamas. *Proc. Acad. Nat. Sci., Philadelphia,* 109:81-103.

Böhlke, J.E. and C.C.G. Chaplin. 1968. *Fishes of the Bahamas and adjacent tropical waters.* Livingston Publ., Wynnewood, Pa. 771 pp.

Lindquist, D.G. 1975. Comparative behavior and ecology of Gulf of California chaenopsid blennies. Ph. D. Diss., Univ. Arizona, Tucson, Arizona. 147 pp.

Longley, W.H. and S.F. Hildebrand. 1941. Systematic catalogue of the fishes of Tortugas, Florida, with observations on color, habits and local distribution. *Papers Tortugas Lab, 34 (Carnegie Inst. Wash., Publ. 535)*:1-331.

Randall, J.E. and H.A. Randall. 1960. Examples of mimicry and protective resemblance in tropical marine fishes. *Bull. Mar. Sci.,* 10:444-480.

Smith-Vaniz, W.F. and F.J. Palacio. 1974. Atlantic fishes of the genus *Acanthemblemaria*, with description of three new species and comments on Pacific species (Clinidae:Chaenopsinae). *Proc. Acad. Nat. Sci., Philadelphia,* 125:197-224.

Stephens, J.S., Jr. 1970. Seven new chaenopsid blennies from the western Atlantic. *Copeia,* 1970:280-309.

Stephens, J.S., Jr., E.S. Hobson, and R.K. Johnson. 1966. Notes on distribution, behavior and morphological variation in some chaenopsid fishes from the tropical eastern Pacific, with descriptions of two new species, *Acanthemblemaria castroi* and *Coralliozetus springeri. Copeia,* 1966:424-438.

Wickler, W. 1967. Specialization of organs having a signal function in some marine fish. *Stud. Trop. Oceanogr.,* 5:539-548.

Triplefin Blennies
(Tripterygiidae)

Triplefins are small heavily scaled blennies similar to and sometimes included in the Clinidae. Members of the two groups can be readily distinguished, however, by the structure of the dorsal fins: clinids usually have only one continuous fin or at most two; tripterygiids have three distinct fins, the first two completely spinuous and the third soft. Tripterygiids are common on rock and reef areas in the tropics and in temperate areas. Except during spawning periods, however, these small blennies are easily overlooked. Most reach a maximum size of only 3 or 4 cm.

Böhlke & Chaplin (1968) reported about 16 genera and 100 species of tripterygiids. Reproduction in the group has thus far been studied only for a few temperate species (e.g., Abel, 1955; Gorlina, et al., 1972; Wirtz, 1978).

SEXUAL DIMORPHISM

The only reported sexual dimorphism in tropical triplefins involves temporary spawning colors on the male. In most species females and nonspawning males are various combinations of gray and brown and are extremely cryptic. During spawning periods, however, males generally develop some combination of black and/or red on the body and caudal fin. The most common pattern, exhibited by some species of *Enneanectes* (Rosenblatt, 1960) and *Axoclinus*, is gray dorsally and bright red ventrally, with a jet black caudal fin. Other species reverse this pattern, developing a dark body and a red or pale caudal fin (e.g., *Tripterygion etheostoma*, Shiogaki & Dotsu, 1973a).

It is probable that the sexes also differ in urogenital morphology, size (males larger than females), and possibly seasonal development by the males of pads on the ventral fins.

SPAWNING SEASON

Nothing concrete is known about the length of the spawning seasons of reef-dwelling triplefins.

Temperate species invariably spawn in the warmer parts of the year (e.g., Gordina, et al., 1972; Shiogaki & Dotsu, 1973a; Wirtz, 1978), as does *Axoclinus carminalis* in the Gulf of California (pers. obs.). In the tropics spawning probably occurs year-round.

REPRODUCTIVE BEHAVIOR

Spawning has thus far been observed for only two tropical species: *Tripterygion etheostoma* (Shiogaki & Dotsu, 1973a) and *Axoclinus carminalis* (pers. obs). In both, spawning behavior is similar to that reported for the more intensively studied temperate species (see Wirtz, 1978) and it is likely that courtship and spawning are similar throughout the family.

Prior to spawning the males establish territories on algae-covered rocks, the orientation and location of the spawning site varying with the species. Within this territory the male courts passing females by a long distance signal referred to as "half-circling" by Shiogaki & Dotsu and "loop-swimming" by Wirtz. The shape and speed of this behavior appear to be species-specific but consist of a rapid looping "jump" by the male up and down off the substrate. In *Axoclinus carminalis*, and probably most similarly colored species, the loop-swim is extremely quick, with a flip at the end that emphasizes the stark black caudal fin. Prior to the loop-swim, males of *A. carminalis* often poise on their pelvic fins, holding the caudal fin into the water and waving it to-and-fro. This display seems to emphasize both the caudal fin color and the male's bright red underside. Similar behavior has been reported for western Atlantic *Enneanectes* (Longley & Hildebrand, 1941). Though most frequent before spawning, both loop-swimming and "waving" occur sporadically throughout a spawning bout.

Spawning itself begins when the female approaches the center of the male's territory. In *T. etheostoma* the male prepares the spawning site by cleaning the algal filaments present without removing them. Several

Tripterygion etheostoma, about 40 mm standard length adult. Spawning has been observed in this species. Photo by Dr. Fujio Yasuda.

studies (reviewed by Wirtz) indicate that spawning is one egg at a time, released when male and female are in close enough contact that egg release and fertilization are simultaneous. In *T. etheostoma* the male lies parallel to and beside the female and fertilizes the egg while moving in a sharp curve away from her. In *A. carminalis* the male lies anti-parallel to the female and fertilization occurs as both quiver slightly. Spawning may continue for up to several hours and frequently involves several females. Shiogaki & Dotsu estimated that as many as 500 eggs may be deposited in one nest.

After spawning the eggs are guarded by the male, but otherwise he does little to care for them. This is in sharp contrast to many temperate species in which the male cleans and aerates the eggs, removes those that are dead, and even may help the young to hatch. Wirtz suggested that this difference may be due to the location of the spawning site: those fishes that nest on open sites do little to care for the eggs, whereas those that nest in enclosed areas, such as crevices and caves, tend them carefully.

Wirtz also reported the presence of "satellite" males (and a few females) around spawning pairs of the Mediterranean species *Tripterygion tripteronotus.* Such males are smaller and probably younger than the territory-holding pair spawners, are cryptically colored, and frequently interfere with pair spawning by rushing in to join the spawning individuals. Each is quickly driven out by the territorial male. These satellite males are invariably ripe with sperm, which suggests that they are essentially "streaking" as described for many labrids; that is, the smaller male is shedding his sperm with that of the dominant male in hopes of fertilizing some eggs at the latter's expense. Similar interference from satellite males occurs in *Axoclinus carminalis.*

EGGS AND LARVAE

Tripterygiid eggs are similar to those of blenniids, roughly 0.9 to 1.1 mm in diameter and hemispherical in shape. Each contains a number of small oil droplets and is covered with numerous sticky threads that anchor it in the algae on the nest site. Shiogaki & Dotsu (1973b) reported that eggs of *T. etheostoma* are light yellow-orange shortly after spawning, orange when older, and red-orange shortly before hatching. Hatching occurs in 16 to 17 days at 15.7 to 18.2°C. Wirtz similarly suggested hatching by the Mediterranean *T. melanurus* occurs in 19 to 20 days at 20°C.

Newly hatched larvae of *T. etheostoma* are slender, 4.57 to 5.0 mm long, and have a well-developed mouth, fin folds, and pigmented eyes. The yolk sac is small. Fin rays begin to develop 18 to 19 days after hatching. Settlement occurs after approximately 40 days, with the juvenile approximately 12 mm long. Basically similar development occurs in two temperate species of *Tripterygion* found off New Zealand (Ruck, 1973).

REPRODUCTION IN CAPTIVITY

Spawning tripterygiids in captivity should require little more than an aquarium containing algae-covered rocks, several males and females (with enough space that each male can establish a spawning territory and the females can find refuge from the males), and heavy feedings with live and prepared foods. Efforts to spawn such fish in captivity have thus far had great success. Shiogaki & Dotsu (1973a), for example, observed 21 spawnings in seven and a half months in a captive group of 11 male and 23 female *T. etheostoma*. Four of the males spawned regularly.

Shiogaki & Dotsu (1973b) successfully reared the larvae of *T. etheostoma* in a 30-liter plastic aquarium. The larvae were initially fed newly hatched barnacle *(Balanus amphitrite)* nauplii and then a mixture of rotifers *(Brachionis plicatilis)* and brine shrimp nauplii.

Literature Cited

Abel, E.F. 1955. Freilandbeobachtungen an *Calliony-mus festivus* Pall. und *Tripterygion tripteronotus* Risso, zwei Mittelmeerfischern, unter besonderer Berücksichtigung des Fortpflanzungsverhaltens. *Sitzber. Osterr. Akad. Wiss. Math. Naturw. Kl. Abt. 1,* 164:817-854.

Böhlke, J.E. and C.C.G. Chaplin. 1968. *Fishes of the Bahamas and adjacent tropical waters.* Livingston Publ., Wynnewood, Pa. 771 pp.

Gordina, A.D., L.A. Duka, and L.S. Oven. 1972. Sexual dimorphism, feeding and spawning in the blackheaded blenny (*Tripterygion tripteronotus* Risso) of the Black Sea. *J. Ichthyol.,* 12:401-407.

Longley, W.H. and S.F. Hildebrand. 1941. Systematic catalogue of the fishes of Tortugas, Florida, with observations on color, habits, and local distribution. *Papers Tortugas Lab., 34 (Carnegie Inst. Wash. Publ. 535):*1-331.

Rosenblatt, R.H. 1960. The Atlantic species of the blennioid fish genus *Enneanectes. Proc. Acad. Nat. Sci., Philadelphia,* 112:1-23.

Ruck, J.G. 1973. Development of *Tripterygion capito* and *T. robustum* (Pisces, Tripterygiidae). *Zool. Publ. Victoria Univ. Wellington,* 64:1-12.

Shiogaki, M. and Y. Dotsu. 1973a. The spawning behavior of the tripterygiid blenny, *Tripterygion etheostoma. Japan. J. Ichthyol.,* 20:36-41.

Shiogaki, M. and Y. Dotsu. 1973b. The egg development and larval rearing of the tripterygiid blenny, *Tripterygion etheostoma. Japan. J. Ichthyol.,* 20:42-46.

Wirtz, P. 1978. The behavior of the Mediterranean *Tripterygion* species (Pisces, Blennioidei). *Z. Tierpsychol.,* 48:142-174.

Clingfishes
(Gobiesocidae)

Clingfishes are small and usually cryptic fishes that are most common in the shallow wave-washed areas immediately below the low tide mark. Like many other fishes that occupy this often turbulent area, the clingfishes have developed a means of firmly attaching themselves to the bottom so as to resist the force of surging currents. For the clingfishes, this mechanism is a muscular ventral suction disc with which the animals can securely clamp to any hard substrate. At least one species has also developed mechanisms that permit it to breathe air and so survive even when exposed by a dropping tide (Gordon, et al., 1970).

The family Gobiesocidae consists of about 130 species in approximately 40 genera and is widely distributed in both temperate and tropical seas (Briggs, 1955, pers. comm.). Related most closely to the dragonets, family Callionymidae, clingfishes are characterized by their suction disc, the lack of a spiny dorsal fin, soft dorsal and anal fins set well back on the body, and usually a broad, flat head and slender body (from above many of them look like flattened tadpoles). A few of the reef-associated species, however, are uniformly slender and, because of their suction disc, can easily be mistaken for gobies (which also have a suction apparatus). The reef-associated species are often closely associated with sea urchins, soft corals, or crinoids, in which they find shelter (e.g., Dix, 1969; Teytaud, 1971; Allen & Starck, 1973), and are probably carnivorous. The inshore species, in contrast, are more likely to be free-living and are generally herbivorous or omnivorous.

Because of their small size at maturity (usually less than 7 cm total length), hardiness, and ease of capture, clingfishes have long been kept in aquaria and have proved easy to spawn, even in small tanks. The often shallow depths they occupy have also facilitated observations of spawning in the field. Consequently, at least the basics of clingfish reproduction have been reasonably well known since the late 1800's. Like

many other groups, however, data have been available mainly for temperate species (mostly European ones), and relatively little is known about the reef-associated fishes.

SEXUAL DIMORPHISM

There is wide variation in the family, but in general sexual dimorphism in the clingfishes takes two forms. First, the sexes often differ in the shape and size of the urogenital papillae in a manner reminiscent of other demersal spawning reef fishes. The extent of these differences varies widely among genera (for detailed descriptions see Briggs, 1955), but in general the papilla of the male is conspicuously longer, larger, and wider at the base than that of the female. In using this particular character to sex the fishes it helps to have several on hand in order to make side-by-side comparisons.

The second widespread form of sexual dimorphism in the family is size differences. Detailed studies of two species, *Aspasma minima* (Shiogaki & Dotsu, 1971a) and *Gobiesox rhessodon* (Wells, 1979), reported males consistently larger than females with only a slight overlap in size range between the sexes. In *A. minima*, for example, females ranged in total length from 36.6 mm to 50.2 mm, whereas males ranged from 47.9 mm to 71. 4 mm. This absence of small males and the limited size overlap of the sexes is reminiscent of similar patterns in protogynous hermaphrodites such as labrids and some pomacanthids. As yet no one has looked at the gonad histology of the clingfishes, so the nature of their sexuality is not known.

Two other types of sexual dimorphism have also been reported in clingfishes. A few early reports on European species (reviewed by Breder & Rosen, 1966) suggested that males are more colorful than females during courtship and spawning; subsequent work with other species has not yet confirmed this observation. Finally, Gould (1965) found that the sexes of a tropical western Atlantic species, *Acyrtops beryllinus*, differ in shape: males are widest at the level of the opercula, while females are generally as wide or wider in the trunk region, apparently due to the eggs they carry. In many species, due to their flattened body shape it is even possible to see the developing eggs through the ventral body wall of the female.

SPAWNING SEASONS

Not many data are available on the spawning seasons of tropical clingfishes. The tropical western Atlantic species *Acyrtops beryllinus* spawns year-round off the coast of Florida at temperatures ranging from 18 to 30° C (Gould, 1965), though spawning may cease in mid-summer when water temperatures reach their highest levels (Jachowski, 1970). In more temperate regions, spawning of various species occurs in the spring (e.g., Holt, 1893; Ruck, 1971, 1973), summer (Graham, 1939; Runyan, 1961; Shiogaki & Dotsu, 1971a; Kalinina, 1976), or early fall (Wells, 1979). The factors resulting in such apparent variation are not known.

REPRODUCTIVE BEHAVIOR

Most, if not all, clingfishes are territorial, probably year-round. Although only limited data are available (e.g., Runyan, 1961), it is probable that in most species spawning takes place in the male's territory. Most species spawn on the underside of a rock or shell, so nest preparation behavior by the male is probably also widespread. Finally, spawning appears to occur throughout the day but may occur more often in late afternoon (Martin & Martin, 1971).

Spawning behavior has thus far been described only for two species. In a brief description of the behavior of *Gobiesox strumosus*, Martin & Martin (1971) found that after preparing a spawning site the male courts a female by swimming toward her, attaching to the substratum alongside her, and undulating his body and caudal fin in unison with her. The male then swims back toward the nest site, followed by the female, and the pair enter the cavity. The female deposits eggs one at a time, at two- to five-second intervals, in vertical rows, beginning at the bottom and working upward, while the male undulates next to her, presumably fertilizing the eggs. Spawning continues for several hours, after which the female departs and the male begins to tend and fan the eggs. Runyan (1961) suggested that clutch size in this species varies from 300 to 2500 eggs; Martin & Martin (1971)

similarly reported a single female produced 7100 eggs in eleven clutches over a two-month period in captivity. In the similar *Gobiesox rhessodon* clutch size correlates with female body size (Wells, 1979).

Spawning by *Acyrtops beryllinus* has been described in detail by Gould (1965) and Jachowski (1970) and, in general, is similar to that described above. The emerald clingfish is a small pale green species found in beds of turtle grass, *Thalassia testudinum*, in the tropical western Atlantic. Spawning takes place on the blades of this grass and is initiated by the male, who approaches the female and nudges her abdomen. If she is receptive the male then moves next to her and quivers, alternating between a head-to-head and head-to-tail orientation. Eventually the female starts to undulate and begins laying her eggs, one at a time, shortly thereafter. The mean number of eggs laid is only 24, so that spawning is usually over in 20 to 40 minutes. After spawning, both parents apparently depart, leaving the eggs untended.

Despite this report, the general pattern in the clingfishes is for the male alone to tend the eggs until hatching (e.g., Graham, 1939; Runyan, 1961; Ruck, 1971). Current theory suggests that the type of parental care exhibited by a species depends to a large extent on the mode of fertilization characteristic of that species. Hence, most clingfishes appear to be external fertilizers, as described above, and have single-parent (male) care of the eggs. Clingfishes, however, appear to be one of the few families of fishes that contain both externally and internally fertilizing species. Such internal fertilization is briefly reported (with no details) for the South African species *Eckloniaichthys scylliorhiniceps* (Smith, 1949) and is suggested in other species by the structure of the male's urogenital papilla (Briggs, pers. comm.). Detailed study of these species to determine whether or not the female performs the bulk of parental care as predicted by theory may well provide an interesting test of current hypotheses of the evolution of parental care.

EGGS AND LARVAE

Clingfish eggs are adhesive flattened ovals deposited in a dense single layer. The size of the egg varies latitudinally: tropical and subtropical species have relatively small eggs (larger dimension ranging from 0.94 to 1.25 mm) (Runyan, 1961; Gould, 1965; Martin & Martin, 1971), whereas the eggs of temperate species are as large as 1.81 mm in longest dimension (Holt, 1893; Ruck, 1973; Kalinina, 1976). Size of the newly hatched larvae varies in a similar fashion, from

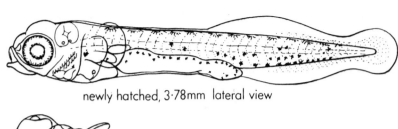

newly hatched, 3·78mm lateral view

newly hatched, dorsal view

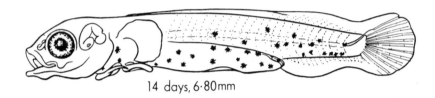

14 days, 6·80mm

Larval development of the clingfish *Aspasma minima*. Based on Shiogaki & Dotsu (1972).

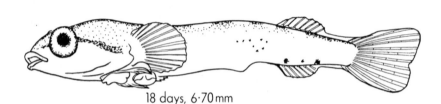

18 days, 6·70mm

2.6 mm for the tropical *Acyrtops beryllinus* (Gould, 1965) to 5.35 to 6.10 mm for the cold temperate *Trachelochismus pinnulatus* (Ruck, 1973). The incubation period ranges from five to 24 days depending on water temperature.

Development of eggs and/or larvae has been described for species in various genera by Runyan (1961), Shiogaki & Dotsu (1971a,b,c & d, 1972), Ruck (1971, 1973), Kalinina (1976), and Allen (1979). Larvae hatch in an advanced state with pigmented eyes, functional jaws, well developed pectoral fin and vertical fin rudiments, and a small yolk sac. Such larvae are characterized by a broad, relatively flat head, large eyes, and an elongate laterally compressed body. Pigment spots are usually present along the gut. By a length of 5 to 12 mm, later stage larvae are easily recognizable as miniature clingfishes. Duration of the planktonic larval stage is not known, but at least for the Japanese species *Aspasma minima* it is probably only a few weeks and may be as little as ten days (Shiogaki & Dotsu, 1972). Newly settled individuals range in size from 7 to 20 mm, in large part depending on the species, grow rapidly, and probably reach sexual maturity in less than a year.

REPRODUCTION IN CAPTIVITY

Spawning in aquaria has been reported for a number of clingfishes, including *Gobiesox strumosus* (Martin, 1968; Martin & Martin, 1971), *Acyrtops beryllinus* (Gould, 1965; Jachowski, 1970), and *Aspasma minima* (Shiogaki & Dotsu, 1971a). In all cases reported spawning occurred readily once the fish had settled in the aquarium, provided members of both sexes were present, the water was relatively warm, and a suitable spawning site was provided (usually in the form of an overturned shell). Despite their hardiness, most clingfishes are shy, so spawning is most likely to occur when only the parents are present in an aquarium. The subtidal species, in particular, are extremely hardy and should require little special equipment or foods to stimulate spawning. The reef-associated species, such as those in *Lepadichthys* and *Derilissus,* are likely to be more delicate.

Based on their advanced state and relatively large size at hatching, larval clingfishes should not prove difficult to rear. Shiogaki & Dotsu (1971a, 1972), for example, readily reared *Aspasma minima* on a diet of rotifers (*Brachionis plicatilis*) and cultured copepods (*Tigriopus japonicus*).

Literature Cited

Allen, G.R. and W.A. Stark II. 1973. Notes on the ecology, zoogeography, and coloration of the gobiesocid clingfishes, *Lepadichthys caritus* Briggs and *Diademichthys lineatus* (Sauvage). *Proc. Linn. Soc. N.S.W.*, 98:95-97.

Allen, L.G. 1979. Larval development of *Gobiesox rhessodon* (Gobiesocidae) with notes on the larva of *Rhimicola miscarum*. *Fish. Bull.*, 77:300-304.

Breder, C.M., Jr. and D.E. Rosen. 1966. *Modes of Reproduction in Fishes*. Natural History Press, Garden City, N.Y. 941 pp.

Briggs, J.C. 1955. A monograph of the clingfishes (Order Xenopterygii). *Stanford Ichthyol. Bull.*, 6:1-224.

Dix, T.G. 1969. Association between the echinoid, *Evechinus chloroticus* (Val.) and the clingfish, *Dellichthys morelandi* Briggs. *Pacif. Sci.*, 23:332-336.

Gordon, M.S., S. Fischer, and E. Tarifeno. 1970. Aspects of the physiology of terrestial life in amphibious fishes. II. The Chilean clingfish, *Sicyases sanguineus. J. Exp. Biol.*, 53:559-572.

Gould, W.R. 1965. The biology and morphology of *Acyrtops beryllinus*, the emerald clingfish. *Bull. Mar. Sci.*, 15:165-188.

Graham, D.H. 1939. Breeding habits of the fishes of Otago Harbour and adjacent seas. *Trans. Proc. Roy. Soc. New Zealand*, 69:361-372.

Holt, E.W.L. 1893. On the eggs and larval and post-larval stages of teleosteans. *Sci. Trans. Roy. Dublin Soc.*, Series 2, 5:5-121.

Jachowski, R.l. 1970. Reproductive behavior of the emerald clingfish, *Acyrtops beryllinus* (Hildebrand and Ginsburg). *Z. Tierpsychol.*, 27:1100-1111.

Kalinina, E.M. 1976. The early ontogeny of two species of Black Sea clingfishes, *Lepadogaster decan-dollei* and *Lepadogaster lepadogaster lepadogaster. J. Ichthyol. (Vopr. Ikhtiol.)*, 16:767-773.

Martin, R.A. 1968. Reproduction of fishes. *Salt Water Aquar.*, 4:34-40.

Martin, R.A. and C.L. Martin. 1971. Reproduction of clingfish, *Gobiesox strumosus. Qt. J. Fla. Acad. Sci.*, 33:275-278.

Ruck, J.G. 1971. Development of the lumpfish, *Trachelochismus melobesia* (Pisces: Gobiesocidae). *Zool. Publ. Vict. Univ. Wellington*, 57:1-9.

Ruck, J.G. 1973. Development of the clingfishes, *Diplocrepis puniceus* and *Trachelochismus pinnulatus* (Pisces, Gobiesocidae). *Zool. Publ. Vict. Univ. Wellington.*, 64:1-12.

Runyan, S. 1961. Early development of the clingfish, *Gobiesox strumosus* Cope. *Chesapeake Sci.*, 2:113-141.

Shiogaki, M. and Y. Dotsu. 1971a. The life history of the clingfish, *Aspasma minima. Japan. J. Ichthyol.*, 18:76-84.

Shiogaki, M. and Y. Dotsu. 1971b. Larvae and juveniles of the clingfishes, *Lepadichthys frenatus* and *Aspasmichthys ciconiae. Japan. J. Ichthyol.*, 18:85-89.

Shiogaki, M. and Y. Dotsu. 1971c. The life history of the clingfish *Conidens laticephalus. Bull. Fac. Fish. Nagasaki Univ.*, 32:7-16.

Shiogaki, M. and Y. Dotsu. 1971d. Egg development and hatched larva of the clingfish, *Lepadichthys frenatus. Bull. Fac. Fish. Nagasaki Univ.*, 32:1-5.

Shiogaki, M. and Y. Dotsu. 1972. The spawning and larva rearing of the clingfish, *Aspasmichthys ciconiae*, and the spawn of the clingfish, *Aspasma minima. Bull. Fac. Fish. Nagasaki Univ.*, 34:29-34.

Smith, J.L.B. 1949 (1961). *The Sea Fishes of Southern Africa*. Central News Agency, Ltd., Cape Town, S.A. 550 pp.

Teytaud, A.R. 1971. Food habits of the goby, *Ginsburgellus novemlineatus*, and the clingfish, *Arcos rubiginosa*, associated with echinoids in the Virgin Islands. *Carib. J. Sci.*, 11:41-45.

Wells, A.W. 1979. Notes on the life history of the California clingfish, *Gobiesox rhessodon* (Gobiesociformes, Gobiesocidae). *Calif. Fish and Game*, 65:106-110.

Dragonets
(Callionymidae)

Dragonets are small (maximum size about 20 cm) bottom-dwelling fishes found in shallow, usually sandy, habitats in tropical and temperate areas worldwide. Thought to be closely related to the clingfishes, family Gobiesocidae (Gosline, 1970), dragonets share many clingfish characteristics: they tend to be small and flattened, have small mouths and long dorsal and anal fins, and are often cryptic in behavior and color pattern. Unlike the clingfishes, which are not often kept in aquaria, two related dragonets, *Synchiropus splendens* and *S. picturatus,* the mandarin fishes, are extremely popular fishes due to their small size, relatively mild manner, and, especially, their magnificent color patterns. There are about eight genera of dragonets and about 40 species. Spawning behavior of the European species has been known since the late 1800's.

Dragonets are characterized by conspicuous sexual dimorphism, apparently a universal feature in the family (e.g., Akazaki, 1957; Böhlke & Chaplin, 1968; Johnson, 1973; Gibson & Ezzi, 1979). Males tend to be larger than females and have longer soft dorsal and anal fins. They are often much more colorful as well, often with complex patterns of spots and lines on the head and dorsal fin that are in striking contrast to the usually drably colored females. Males are also characterized by a much longer first dorsal fin than the females, the fin often bearing one or more colorful markings. This dorsal fin is a key element in courtship displays. To date there is no evidence of hermaphroditism in the family, and sex ratios tend to be close to 1:1 (e.g., Gibson & Ezzi, 1979).

Nothing is known about the spawning seasons of tropical dragonets. Temperate species range in spawning periods from winter and early spring (e.g., *Synchiropus altivelis*–Akazaki, 1957) to late spring and summer (e.g., *Callionymus maculatus*–Gibson & Ezzi, 1979). There is no specific data relating to the presence or absence of lunar cycles, though at least some species spawn daily in captivity (e.g., Takita &

Okamoto, 1979), as does a Japanese species of *Diplogrammus* in the field (Zaiser, in prep.).

There is no detailed information available regarding the spawning behavior of reef-associated dragonets, but spawning has been described for several temperate species, all in the genus *Callionymus: C. lyra* (Holt, 1898; Hardy, 1959; Wilson, 1978); *C. festivus* (Abel, 1955); *C. maculatus* (Hardy, 1959); and two Japanese species (Takita & Okamoto, 1979). Spawning behavior is consistent within at least the genus. Courtship is initiated by the male, who swims in front of the female with his conspicuous dorsal fin fully erected. If the female is ready, she swims forward and the male moves parallel to her, gradually coming closer to her as the pair moves forward. Quickly the fish assume a side-by-side position with the female slightly behind the male and resting one of her pelvic fins on one of his. The pair then prop themselves up on these fins such that they are angled head-up into the water column. From this position they begin to move upward, with the relatively immobile female being "carried" by the male. The ascent is about a meter high, at the peak of which gametes are shed. Holt suggested that the male and female use their anal fins to form a "funnel" into which the gametes are sent, but this was disputed by Takita & Okamoto. After spawning, the fish return to the bottom. Males may spawn several times in a day, while females apparently spawn only once daily. Courtship and spawning take place in late afternoon, before sunset.

Courtship, spawning, and social organization, based on field observations made at Miyake-jima, Japan, will be described in detail for a species of *Diplogrammus* by M.J. Zaiser (in prep.). Spawning in this species is generally similar to that described above and occurs daily at dusk. Social organization is based on male-dominated harems in which one male, the largest, regularly spawns with up to five females nightly.

Smaller males, which are cryptic in color pattern and behavior, inhabit the territory of the dominant male and occasionally attempt to solicit spawning by a female but are usually unsuccessful and are vigorously chased by the dominant male when discovered.

Apparent courtship behavior in *Synchiropus splendens* was reported by Mayland (1975) to consist of a "courtship dance" that occurs most frequently at dawn. In an aquarium containing one male and several females, pairs of individuals would spiral around one another toward the surface, stay there for a short period, and then sink back to the bottom. No gamete release was observed.

There are also vague reports in the literature (e.g., Munro, 1967; Böhlke & Chaplin, 1968) of internal fertilization in the family, followed by release of pelagic eggs in a short spawning ascent. I can find no account specifically describing such behavior, but it might derive from the conspicuous urogenital papillae, similar to those of demersal spawning fishes, that develop prior to spawning.

Callionymid eggs range in size from about 0.55 to 0.75 mm and are spherical, lack an oil droplet, and have a segmented yolk and a mesh-like structure on the egg membrane (e.g., Mito, 1962; Miller, Watson, & Leis, 1979; Takita, 1980). Hatching occurs in about 21 hours at 20° C to produce larvae that range in size from 1.15 to 1.5 mm. Such larvae have a large oval yolk sac, lack fin precursors, and apparently spend their first few days of life floating upside-down in the water column (Takita, 1979). The yolk is absorbed about three days after hatching, producing an elongate large-headed larva that is relatively heavily pigmented. By a length of about 3 mm the larvae have fully developed fins, and by 5 mm they generally resemble the adults. Duration of the planktonic larval stage is not known.

Several species of *Callionymus* have been spawned in captivity, and all of the descriptions above, except Abel (1955), are based on aquarium spawnings. Apparently, little is required to spawn the animals beyond a moderately large aquarium with adequate open space, several females and a male, and heavy feedings of live food. Callionymid larvae have not yet been reared in captivity.

Literature Cited

Abel, E.F. 1955. Freilandbeobachtungen an *Callionymus festivus* Pall. und *Tripterygion tripteronotus* Risso, zwei Mittelmeerfischen unter besondere Berücksichtigung des Fortpflanzungsverhaltens. *Sber. öst. Akad. Wiss. Math.-Naturw. Kl., Abt. 1,* 164:817-854.

Akazaki, M. 1957. Biological studies on a dragonet, *Synchiropus altivelis* (Temminck et Schlegel). *Japan. J. Ichthyol.,* 5:146-152.

Böhlke, J. and C.C.G. Chaplin. 1968. *Fishes of the Bahamas and adjacent tropical waters.* Livingston Press, Wynnewood, Pa. 771 pp.

Gibson, R.N. and I.A. Ezzi. 1979. Aspects of the biology of the spotted dragonet *Callionymus maculatus* Rafinesque-Schmaltz from the west coast of Scotland. *J. Fish Biol.,* 15:555-569.

Gosline, W.A. 1970. A reinterpretation of the teleostean fish order Gobiesociformes. *Proc. Calif. Acad. Sci., Ser. 4,* 37:363-382.

Hardy, A.C. 1959. *The Open Sea. II. Fish and Fisheries.* Houghton Mifflin Co., Boston, Mass. 322 pp.

Holt, E.W.L. 1898. On the breeding of the dragonet (*Callionymus lyra*) in the Marine Biological Association's aquarium at Plymouth; with a preliminary account of the elements and some remarks on the significance of the sexual dimorphism. *Proc. Zool. Soc. Lond.,* (1898):281-315.

Johnson, C.R. 1973. Biology and ecology of three species of Australian dragonets (Pisces: Callionymidae). *Zool. J. Linn. Soc.,* 52:231-260.

Mayland, H.J. 1975. The mandarin. *Mar. Aquar.,* 3:22-23.

Miller, J.M., W. Watson, and J.M. Leis. 1979. An atlas of common nearshore marine fish larvae of the Hawaiian Islands. *SeaGrant Misc. Rpt.* (INIHI-SeaGrant-MR-80-02). 179 pp.

Mito, S. 1962. Pelagic fish eggs from Japanese waters V. Callionyma and Ophidina. *Sci. Bull. Fac. Agric., Kyushu Univ.,* 19:377-380.

Munro, I.S.R. 1967. *The Fishes of New Guinea.* Dept. Agric., Stock & Fish., Port Moresby, N.G. 651 pp.

Takita, T. 1980. Embryonic development and larvae of three dragonets. *Bull. Japan. Soc. Sci. Fish.,* 46:1-7.

Takita, T. and E. Okamoto. 1979. Spawning behavior of the two dragonets, *Callionymus flagris* and *C. richardsoni,* in the aquarium. *Japan. J. Ichthyol.,* 26:282-288.

Wilson, D.P. 1978. Territorial behaviour of male dragonets (*Callionymus lyra*). *J. Mar. Biol. Assoc. U.K.* 58:731-734.

Sandperches
(Mugiloididae)

The sandperches or weeverfishes are elongate cylindrical fishes widely distributed on sand plains and around reefs in the tropical Indo-Pacific and along the coasts of South America and Africa. The family is characterized by a heavy body, long, low, and even dorsal and anal fins, pelvic fins below or slightly in front of the pectoral fins, and a large mouth. Typically they are mottled brown or black on a white background, colors that allow them to blend into the sand on which they typically sit. There are about 50 species in the family in five genera, of which *Parapercis* is the best known.

Until recently little was known about mugiloidid reproduction. Marshall (1950), in a checklist of the fishes from the Cocos-Keeling Islands, suggested *Parapercis hexophthalma* to be sexually dichromatic and a protogynous hermaphrodite. Stroud (1981) recently confirmed both suggestions for several species of *Parapercis* found on the Australian Great Barrier Reef. In both *P. hexophthalma* and *P. cylindrica* males have one or more small dark markings on their heads that are lacking in the females. Such dichromatism is obvious when looked for but is hardly conspicuous. Males, moreover, are larger than females, which is not surprising considering that all species thus far examined have proved to be protogynous and monandric hermaphrodites. Histological evidence for such hermaphroditism included the presence of spermatogenic crypts at the bases of the ovarian lamellae in active females, degenerated oocytes in active males, and conspicuous remnants of the ovarian lumen in the testes.

Spawning in mugiloidids has thus far been described only by Stroud (1981), who examined the social system of *Parapercis cylindrica* in particular detail. Like many other sequentially hermaphroditic fishes, *P. cylindrica* has a social system based on male territoriality and a harem. Each male defends an area, averaging about 17 m², within which two to five females defend smaller territories against one another. Spawning, which always involves harem members, takes place between 18 and 29 minutes after sunset and occurs year-round (though with a peak of activity in the summer). Courtship begins about 40 minutes prior to sunset and starts with the male moving broadside to the female and "bobbing" up and down next to her. The male may also place his head over hers and fan her with his pectoral fins. Eventually the pair make a short (about 60 to 70 cm) ascent off the bottom, at the peak of which gametes are shed. Females release anywhere from 100 to 1500 eggs during each spawning.

Mugiloidid eggs are spherical and pelagic. For tropical species, egg diameters range from 0.63 to 0.99 mm (Mito, 1962, 1965; Stroud, 1981), whereas *P. colias*, a temperate species found off New Zealand, has an egg diameter of about 1.2 mm (Robertson, 1973). According to Stroud (1981), hatching in *P. cylindrica* occurs in 22 to 24 hours at 22.7 to 27° C, producing a larva 1.6 mm long and bearing a large yolk sac. This yolk is fully absorbed in 62 hours. Hatching by other species appears to be slower, i.e., about 70 hours at 12 to 17° C for *P. sexfasciata*, producing a 2.19 mm larva, and 114 hours at 11.5 to 12.5° C for *P. colias*, producing a 3.0 mm larva. Nellen (1973) depicted a 7.3 mm larva as elongate, with well developed eyes, jaws, and caudal fin rays. Duration of the planktonic larval stage is estimated at one to two months.

Parapercis cylindrica males have dark markings (lines) on the head that are absent in females. Photo by Dr. V. G. Springer.

Literature Cited

Marshall, N.B. 1950. Fishes from the Cocos-Keeling Islands. *Bull. Raffles Mus.*, 22:166-205.

Mito, S. 1962. Pelagic fish eggs from Japanese waters IV. Trachina and Uranoscopina. *Sci. Bull. Fac. Agric., Kyushu Univ.*, 19.

Mito, S. 1965. On the development of the eggs and hatched larvae of a trachinoid fish, *Neopercis sexfasciata* (Temminck et Schlegel). *Sci. Bull. Fac. Agric., Kyushu Univ.*, 15:507-512.

Nellen, W. 1973. Fischlarven des Indischen Ozeans. *Meteor Forsch.-Ergebnisse*, 14:1-66.

Robertson, D.A. 1973. Planktonic eggs and larvae of some New Zealand marine teleosts. Ph. D. Diss., Univ. of Otago, New Zealand.

Stroud, G. 1981. Aspects of the biology and ecology of weever-fishes (Mugiloididae) in northern Great Barrier Reef waters. Ph. D. Diss., James Cook Univ., Townsville, Australia.

Rearing Marine Fishes

Ten years ago the possibility of even an advanced aquarist rearing reef fishes from egg to metamorphosed juvenile was remote. Even the largest larvae of marine fishes are only on the order of a few millimeters, far smaller than newly hatched freshwater fishes, and the combination of small size, often long planktonic stages, and pickiness in feeding proved fatal. Within the last decade, and especially within the past few years, however, considerable progress has been made toward unraveling the biology of planktonic larvae. At present, professional laboratories are in a position of routinely rearing larvae of many species. Admittedly there are still some that have not yet been reared successfully despite all efforts, but the number of these is small and grows smaller. Most such laboratories rely on three factors not usually available to aquarists: very large rearing tanks (as much as several thousand liters), unlimited access to clean sea water (usually piped in), and unlimited availability of size-sorted wild-caught plankton (which explains why most such laboratories are located close to the coast and why many are in Florida). Yet the technology of rearing larvae has reached such a position that even land-locked aquarists willing to put some effort into it can reasonably expect to rear some larvae of some species (as testified to throughout the previous chapters).

HATCHING THE EGGS

The first problem encountered in rearing reef fishes is collecting the eggs and/or larvae and transferring them to a special rearing tank. For demersal spawning fishes such as pomacentrids, gobies, and blennies it is usually not possible to transfer the eggs once the parents have attached them to coral, rock, or the side of the aquarium, and, indeed, it is probably not even desirable to do so. Although leaving parent (usually only the male tends the eggs) and eggs together risks their being eaten, this risk can be minimized by

studiously leaving the tending parent alone. In any case, this is more than compensated for by his care of the eggs—aerating and cleaning them, picking out those that are infertile before they affect the healthy brood, and in some cases even helping the young to hatch. The best procedure, then, is to wait until the larvae have hatched and are free-swimming before collecting and transferring them to a rearing tank. Hatching in many groups (e.g., pomacentrids and balistids) occurs at dusk. For such species it is often possible to trigger hatching "on command" by leaving tank lights on past the normal time on the night the eggs are due to hatch (based on information provided in the previous chapters) and then abruptly turning them off. In anemonefishes, and probably many other demersal spawning fishes, hatching occurs within half an hour of lights out (e.g., Sieswerda, 1977). Most, if not all, newly hatched larvae are photopositive and can be collected by illuminating one end of the aquarium (with filters, air stones, and other sources of turbulence minimized) and gently siphoning the larvae into the rearing tank or, somewhat more laboriously, dipping them out with a small cup. **Never use a net;** always keep the delicate larvae fully immersed and treat them gently. The tank to which the larvae are transferred should, of course, contain water as close as possible in composition to that of the parent's tank. Indeed, before hatching the eggs it is often a good idea to drain most of the water out of the latter and into the former (without drawing out so much as to disturb the nest-guarding parent). Both tanks can be subsequently refilled with clean salt water.

For those species that produce pelagic eggs the presence of the adults is not only unnecessary but could result in the eggs being eaten well before hatching. Thus, either the adults can be removed or the eggs can be collected and transferred to a rearing tank. The latter is probably best since aquaria designed to stimulate spawning are not usually best for rearing the eggs. Eggs can be obtained either by swishing a fine net through the water (the eggs are not as delicate as newly hatched larvae and can tolerate some handling) or by siphoning water from the parents' tank and adding the water containing the eggs to a prepared hatching tank. The key feature of the hatching tank is enough turbulence to maintain the positively or neutrally buoyant eggs off the bottom and away from the walls of the tank (either of which can result in fouling). The simplest approach is the use of a barren, commercially available all-glass aquarium filled with filtered sea water (the use of artificial water is probably best, since it minimizes the likelihood of bacterial infection) and agitated by one or two moderately active air stones. Filtration presents obvious problems and, at least until hatching, should not be done. Somewhat more sophisticated units, such as that described by Houde and Ramsay (1971) are also used. Most authors recommend the use of circular tanks to eliminate dead spaces and produce constant circulation. Houde and Ramsay produced this circulation by running a circle of airline tubing around the inside base of the aquarium and securing it with weights. Every few centimeters this tubing has a hole punched in it. When attached to an air pump, the result is a fine curtain of air bubbles around the entire perimeter of the tank, which keeps the water circulating and keeps eggs and larvae away from the walls.

THE REARING TANK

Though cylindrical rearing tanks are sometimes recommended, in general the normal rectangular ones work well and are easier and cheaper to obtain. Most breeders—commercial, scientific, or amateur—now recommend the use of completely bare tanks (no gravel, rocks, etc.) for at least the first week or two of larval existence. Gravel attracts detritus and can trap and damage sensitive larvae. Equally important is the use of large aquaria, essentially the largest compatible with maintaining a suitable density of food items. Several studies (e.g., Houde, 1973, 1975; Walker, 1977) reported significantly higher survival rates of larvae in larger aquaria with lower larval densities. For consistently good results a rearing tank should be no less than 50 liters, and twice or three times that would not be unreasonable. Some scientific rearing tanks exceed 1000 liters in volume. Finally, the rearing tank should be covered on the outside with black construction paper or sheet plastic (polyethylene). Not only does this black background make it easier for the larvae to see food (e.g., Marliave, 1976), but it also shelters the larvae from outside movements and

minimizes possible stress due to such stimulation.

There are two approaches to establishing a rearing tank, both of which have proved successful: a closed system containing an algal culture, and a semi-closed system. The first is probably best suited for amateur breeders since it permits use of relatively small aquaria and does not require large amounts of filtered sea water. The aquarium is established before hatching is anticipated—it should be clean, bare, aerated, and filled with filtered sea water and water from the parent's tank. Nutrients are then added to stimulate algal growth and the tank is inoculated with planktonic algae. The algae act as a "conditioner" of the water, essentially a biological filter, that clears up waste products created by the larvae and by uneaten food organisms (Houde, 1973). Larvae apparently do not eat the algae. The nutrients and algae used vary with different workers. Sieswerda (1977) recommended 1 ml of nutrient stock solution per liter of aquarium water. The stock solution consists of a mixture of sodium nitrate (7.5 gr/100 ml) and sodium hypophosphate (0.5 gr/100 ml). Thomas (1964) recommended a comparable mixture for culturing the alga *Dunaliella*. Less sophisticated approaches include those of Schumann (1969), who dissolved a teaspoon of chemically pure ammonium sulphate in a quart of water and then added one teaspoon of this solution per week to his algal tank, and Schlais (1975), who stimulated growth of *Chlorella* with one to two drops per day of a 5:1 solution of water and an orchid fertilizer named Orchidzyme. All of these techniques work to one extent or another in stimulating algal production. A recently available alternative is the use of recommended dosages of commercially available fertilizer specifically designed for marine planktonic algae. These are often available in local tropical fish outlets or can be ordered through such outlets. Algae that have been used as successful water conditioners include *Dunaliella*, *Chlorella*, and *Anacystis* or some combination thereof. During this period, and indeed throughout rearing, the tank is illuminated 24 hours/-day both to stimulate algal growth and later to facilitate constant feeding by the larvae.

Starter cultures of algae can be obtained from a number of marine laboratories on both coasts that routinely culture the algae for a variety of uses or can be obtained from commercial sources. *Dunaliella* cultures are carried by some larger retail marine fish outlets.

In the "green tank" method, the system is kept closed and unfiltered for about the first week of larval development. After this, the system is either opened (slow addition of clean salt water to the tank with the excess allowed to overflow) or partial water changes (25% to 30%) are made weekly to reduce metabolite levels (e.g., Houde, 1973; Walker, 1977).

The alternative method, which is often preferred by commercial breeders because it frequently results in higher rates of survival and quicker growth, is to eliminate the algal culture and allow clean filtered sea water to flow slowly through one relatively large tank (100 + liters). Flow rate is usually enough to produce a complete water change every two days or so. The overflow outlet, of course, is screened to prevent loss of larvae. The advantage of such a system, aside from the higher success rate, is that little special preparation of a rearing tank is required and that water changes are unnecessary (thus reducing maintenance time and required manpower). Its disadvantage is that it requires the use of large aquaria and the constant availability of clean filtered sea water.

With either method, larvae can be brought through to metamorphosis without the use of additional aquaria. It is a common practice, however, to transfer newly metamorphosed individuals to another tank containing an undergravel filter and gravel bed to help keep the water clean and chemically balanced.

FOODS AND FEEDING

Reef fishes, especially those that produce demersal eggs, have spawned in aquaria for many years. The problem in rearing them has not been getting them to breed, but rather getting the newly hatched larvae to feed. Rarely, workers have been able to rear a few larvae using prepared, non-live foods (e.g., Allen, 1972, fed larval *Amphiprion* a pulverized flake-food suspension; Fujita, 1966, fed newly hatched puffers a boiled egg yolk suspension), but the generally poor survival and high risk of bacterial contamination involved in such procedures made it unreliable. Live foods were clearly needed. Various attempts have been made with mussel larvae (e.g., Houde, 1973; Riley, 1973), barnacle nauplii (Shelbourne, 1971), cultures of various marine organisms (e.g., Schumann, 1969), and, of course, wild-caught size-sorted zooplankton. The "breakthrough" that has permitted success in rearing by aquarists, however, was the discovery that the marine rotifer *Brachionis plicatilis* is not only an acceptable and nutritious food for most newly hatched larvae, but that it is also easily cultured in small containers. *Brachionis* is now the standard first food for reef fishes bred in captivity.

Rotifers are among the smallest of metazoan animals—a mature *B. plicatilis* is approximately half the size of a newly hatched *Artemia* nauplius—and are characterized by an elongate, more or less cylindrical body capped anteriorly by a ciliated organ known as the corona. Different types of rotifers have different kinds of coronas which are correlated with their life styles (some, like *B. plicatilis*, are free-swimming, others are benthic, and some are colonial). The corona of the vase-shaped *B. plicatilis* is broad, heavily covered with hair-like cilia, and is used both to capture food (planktonic algae) and to propel the animal. Rotifers reproduce in two ways. In a normal growing population all rotifers are female and reproduce by parthenogenesis, i.e., without males. Each female carries from one to eight "amictic" eggs which hatch in one day and reach maturity in another. Such a population can double in size almost daily as long as food is available, an obvious advantage for aquarists rearing them for food. Growth rate is also affected by temperature (Theilacker & McMaster, 1971), which permits control of growth rate by the aquarist. Eventually, however, the population reaches the limits of food or nutrients or conditions become stressful, with the result that the females begin to produce males. Normal copulation then results in "mictic" eggs which are extremely hardy and remain viable when desiccated. These "resting" eggs hatch when conditions improve.

Because of their hardiness, rotifers can be cultured with relatively few difficulties and several popular articles describing simple means of doing so have appeared (e.g., McMaster, 1975; Schlais, 1975; Sieswerda, 1977; Kloth & McMaster, 1978). There is also an extensive technical literature on rearing rotifers that was recently reviewed by Solangi and Ogle (1977). Tolerating a wide range of temperatures (approximately 14 to 36° C), salinity (10 to 40 ppt), and lighting (from constant light to constant dark), rotifers require only a suitable container (20-liter aquaria do well), moderate aeration, and food to thrive. The most commonly used food, and the one rotifers do best on, is live planktonic algae, though other foods can also be provided. They will, for example, do well on yeast (Hirayama & Watanabe, 1973) or dried *Chlorella* powder (Hirayama & Nakamura, 1976), either of which might be preferred over live algae if only limited numbers of rotifers are required for only a brief period. Regarding live algae, almost any single-celled planktonic species will do, but based on a survey of a number of species, Theilacker &

McMaster (1971) found best growth resulted from feeding of *Dunaliella* spp. This alga is readily available and is easily cultured following procedures discussed earlier.

Though tolerating wide environmental extremes, rotifers grow best at high temperatures (in excess of 30° C), temperatures too high for the algae, however. The best compromise is between 21 and 25° C and a salinity slightly less than normal (Theilacker & McMaster, 1971). The size of the rotifer tanks and the complexity of the system depend entirely on the aquarist's needs. On one extreme, Schlais (1975) grew *Chlorella* to feed his rotifers in 5-gallon (about 22 liter) buckets and reared the rotifers in gallon (4.5 liter) jars; on the other extreme, Sieswerda (1977) described a fairly complex system of many custom-built rearing tanks for algae and rotifers and recommended regular rotation of stocks to maintain fresh supplies of both. Professional equipment can be even more complex, involving water tables, light racks, numerous filters, and routine schedules of feeding, fertilization, and restarting old cultures (e.g., Theilacker & McMaster, 1971; Houde, 1973; Persoone & Sorgeloos, 1975). If rotifers are needed only for a brief period, establishing the culture requires little more than adding rotifers to a "green" aquarium, i.e., one in which algal growth has been stimulated by fertilization. This tank, about 22 liters, should be illuminated constantly to stimulate algal production. Rotifers can be harvested within three to four days and production prolonged almost indefinitely by regular addition of algae (roughly a liter of "green water" a week). The growth rate of the culture can be controlled by temperature. Under conditions of limited need, or if one only wants to keep the culture viable for future use, a temperature of 14 or 15° C will result in slow growth. When need for them increases, increasing the temperature to 23 to 25° C will result in rapid population growth.

An alternative method for keeping a population of rotifers viable is to collect the "resting" eggs. Egg production increases in constant dark, when food runs low, or when temperature or salinity approaches the limits of tolerance. If mud is allowed to deposit on the bottom of the rotifer tank it can be collected and dried. The eggs contained within will remain viable for up to 12 months and when re-immersed will hatch in seven to 12 days (Ito, 1960).

In feeding rotifers to larvae, the key to successful rearing is to maintain a high enough density of rotifers in the tank so that the larvae do not have difficulty in finding food. Generally, a density of eight to ten rotifers per milliliter is recommended. This density can be measured through the use of a low power microscope or, with experience, by visual inspection of the rearing tank. If enough food is available, the guts of the larvae should be clearly distended and should be whitish (as opposed to transparent) (Walker, 1977). Rotifers can be collected for feeding by filtering them through a 64 micron mesh filter. Such filters are available in some larger fish stores and from scientific supply houses. In lieu of such a filter, Schlais (1975) successfully used plastic rayon towels. Experiments with other materials may result in other usable alternatives to the sometimes difficult-to-obtain plankton mesh.

Brachionis plicatilis starter cultures are available under various brand names from commercial sources such as pet shops and mail order, and occasionally from marine laboratories. Ito (1960) collected *B. plicatilis* from warm and shallow estuarine pools, isolating the rotifer from its relatives by increasing the salinity to nearly normal seawater. The high salinity killed other species. He also obtained *B. plicatilis* by collecting dormant eggs from mud at the bottom of such ponds.

After a week to ten days of feeding with *B. plicatilis*, newly hatched *Artemia* nauplii can be offered to the larvae. Successful feeding on the nauplii will be indicated by a pink color to the gut. Since larvae grow at different rates, rotifers should continue to be offered until it is clear that all larvae are taking *Artemia*. Techniques for rearing brine shrimp in small quantities are well known; for higher volume needs, see Sorgeloos (1973) and Sorgeloos & Persoone (1973).

Literature Cited

Allen, G.R. 1972. *Anemonefishes*. T.F.H. Publ., Neptune City, N.J. 284 pp.

Fujita, S. 1966. Egg development, larval stages and rearing of the puffer *Lagocephalus lunaris spadiceus* (Richardson). *Japan. J. Ichthyol.*, 13:162-168.

Hirayama, K. and K. Watanable. 1973. Fundamental studies on physiology of rotifer for its mass culture. IV. Nutritional effect of yeast on population growth of rotifer. *Bull. Japan. Soc. Sci. Fish.*, 39:1129-1133.

Hirayama, K. and K. Nakamura. 1976. Fundamental studies on the physiology of rotifers in mass culture. V. Dry *Chlorella* powder as food for rotifers. *Aquacult.*, 8:301-307.

Houde, E.D. 1973. Some recent advances and unsolved problems in the culture of marine fish larvae. *Proc. World Maricult. Soc.*, 3:83-112.

Houde, E.D. 1975. Effect of stocking density and food density on survival, growth and yield of laboratory reared larvae of the sea bream *Archosargus rhomboidalis* (Linnaeus) (Sparidae). *J. Fish Biol.*, 7:115-127.

Houde, E.D. and A.J. Ramsay. 1971. A culture system for marine fish larvae. *Prog. Fish Culturist*, 33:156-158.

Howell, B.R. 1973. Marine fish culture in Britain VIII. A marine rotifer, *Brachionus plicatilis* Muller, and the larvae of the mussel, *Mytilus edulus* L., as foods for larval flatfish. *J. du Conseil*, 35:1-6.

Ito, T. 1960. On the culture of mixohaline rotifer *Brachionus plicatilis* O.F. Muller in the sea water. *Rept. Faculty Fish. Prefect. Univ. Mie.*, 3:708-740.

Kloth, T.C. and M.F. McMaster. 1978. The mighty rotifer. *Fresh. and Marine Aquar.*, 1:52-53.

McMaster, M.F. 1975. A first food. *Mar. Hobbyist News*, 3.

Marliave, J.B. 1976. Laboratory rearing of marine fish larvae. *Drum and Croaker*, 16:15-20.

Persoone, G. and P. Sorgeloos. 1975. Technological improvements for the cultivation of invertebrates as food for fishes and crustaceans. I. Devices and Methods. *Aquacult.*, 6:275-289.

Riley, J.D. 1973. Induced spawning of the mussel *Mytilus edulus* L. and its uses in larval fish feeding. *Proc. Challenger Soc.*, 4:116.

Schlais, J. 1975. Put a rotifer in the tank. *Mar. Aquar.*, 6:52-55.

Schumann, G.O. 1969. Rearing marine fish larvae. *Salt Water Aquar.*, 5:131-141.

Shelbourne, J.E. 1971. *The artificial propagation of marine fish*. T.F.H. Publ., Neptune City, N.J. 83 pp.

Sieswerda, P.L. 1977. Grow some rotifers. *Mar. Aquar.*, 8:6-15.

Solangi, M.A. and J.T. Ogle. 1977. A selected bibliography on the mass propagation of rotifers with emphasis on the biology and culture of *Brachionus plicatilis*. *Gulf Res. Rpts.*, 6:59-68.

Sorgeloos, P. 1973. High density culturing of the brine shrimp *Artemia salina* L. *Aquacult.*, 1:385-391.

Sorgeloos, P. and G. Persoone. 1973. A culture system for *Artemia*, *Daphnia*, and other invertebrates, with continuous separation of the larvae. *Arch. Hydrobiol.*, 72:133-138.

Theilacker, G.H. and M.F. McMaster. 1971. Mass culture of the rotifer *Brachionus plicatilis* and its evaluation as a food for larval anchovies. *Mar. Biol.*, 10:183-188.

Thomas. W.H. 1964. An experimental evaluation of the C[14] method of measuring phytoplankton production, using cultures of *Dunaliella primolecta* Butcher. *Fish. Bull. (U.S.)*, 63:273-292.

Walker, S. 1977. Final obstacle overcome. *Mar. Hobbyist News*, 5:1-8.

Patterns in the Reproduction
of Reef Fishes

In terms of their overall reproductive strategies, reef fishes fall naturally into broad classes: 1) demersal spawners, which produce demersal eggs and either deposit them in a preselected and prepared nest or orally brood them; 2) pelagic spawners, which produce pelagic eggs and release them at the peak of an ascent off the bottom made specifically for that purpose; 3) egg-scatterers, which spawn in a manner similar to the pelagic spawners, but which produce demersal eggs that settle back to the bottom in a haphazard fashion; 4) benthic broadcasters, which shed planktonic eggs while remaining on the bottom; and 5) livebearers, which retain fertile eggs *in utero* until after hatching and which subsequently release live free-swimming young. The first four are, in essence, the four possible combinations of two variable characteristics: egg type (pelagic versus demersal) and spawning location (off the substratum versus on it). Of these, the first two groups, the demersal spawners and the pelagic spawners, by far dominate the reef, accounting for all but representatives of one or two families (Table 1). Pelagic spawning is the single most taxonomically widespread mode of reproduction, though in terms of number of individuals the demersal spawning fishes may be more common.

Division of reef fishes into these spawning classes can usually be made at the family and superfamily levels, suggesting that overall reproductive strategies are evolutionarily conservative. Among shallow water tropical groups, within-family differences exist only in the Clinidae, the Antennariidae, and Muraenidae, and possibly in the two tetraodontiform families Balistidae and Tetraodontidae. Latitudinal shifts from pelagic to demersal spawning also occur in three other primarily tropical families—Labridae, Sparidae, and apparently Diodontidae. This shift parallels a comparable poleward

shift in many marine invertebrates (Thorson, 1950) and a trend in both invertebrates and fishes for larger eggs in cooler areas (e.g., Marshall, 1953; Thresher, in prep.).

DEMERSAL VERSUS PELAGIC SPAWNING

The two dominant modes of reproduction in reef fishes differ in many features. Demersal spawning, for example, characteristically involves nest preparation and parental care, whereas pelagic spawning involves neither. Conventional and long-standing knowledge (e.g., Curtis, 1940) also has it, despite reports to the contrary (e.g., Williams, 1959), that the two also differ in egg size and fecundity (e.g., Mann & Wills, 1979; Keenleyside, 1979; Lowe-McConnell, 1979). Demersal spawners purportedly have fewer but larger eggs than pelagic spawners.

Characteristics of the reproduction of reef fishes can be conveniently analyzed at three levels: within species, within families, and over reef fishes as a whole. Data for analysis at the first two levels are available for relatively few taxa but present no conceptual difficulties in analysis. Examination of overall trends in reef fishes, however, requires that within-family data be pooled to produce a family-wide mean value for such things as fecundity and body size. More sophisticated statistical procedures are not yet possible due to the limited data base. Although the approach suggested clearly obscures the variability occurring within families, it has the important consequence of avoiding bias caused by the extensive data base available for a few families, e.g., pomacentrids, and the much more limited data base for others. Instead, each family contributes equally to the patterns that emerge.

Based on analysis of data presented in the previous chapters, the following general statements can be made about demersal and pelagic spawning fishes.

Table 1. Distribution of Reproductive Strategies in Reef Fishes

LIVE BEARERS	Brotulidae	Clinidae*	
	N = 2		

DEMERSAL EGGS	*Demersal Spawners*		*Egg-Scatterers*
	Antennariidae*		Siganidae Tetraodontidae*
	Apogonidae	Opistognathidae	N = 2
	Balistidae*	Plotosidae	
	Blenniidae	Pomacentridae	
	Chaenopsidae	Pseudochromoids	
	Clinidae*	Tetraodontidae*	
	Gobiidae	Tripterygiidae	
	N = 13		

PELAGIC EGGS	*Pelagic Spawners*		*Benthic Broadcasters*
	Acanthuridae	Mugilidae	Muraenidae*
	Antennariidae*	Mugiloididae	N = 1
	Aulostomidae	Mullidae	
	Balistidae*	Muraenidae*	
	Branchiostegidae	Ophichthidae	
	Callionymidae	Ostraciidae	
	Carangidae	Pempheridae(?)	
	Chaetodontidae	Pteroidae	
	Cheilodactylidae	Pomacanthidae	
	Cirrhitidae	Priacanthidae(?)	
	Congridae	Scaridae	
	Diodontidae	Sciaenidae	
	Ephippidae	Scorpaenidae	
	Fistulariidae	Serranidae	
	Grammistidae	Sparidae	
	Haemulidae	Sphyraenidae	
	Holocentridae(?)	Synodontidae	
	Lutjanidae	Tropical Labridae	
	N = 36		

* = families occurring in more than 1 category.

1) Contrary to popular belief, demersal eggs are not significantly larger than pelagic eggs (X^2 = 14.05, df = 9, p < 0.25) (Fig. 1a). They do differ, however, in terms of variance; both the largest and the smallest eggs produced by reef fishes are those of demersal spawning families (plotosids and siganids, respectively). More than half of the pelagic eggs range in size from 0.1 to 0.5 mm^3, whereas the same size range in demersal eggs accounts for less than a third of those for which data are available (Fig. 1b). In general, there appears to be no relationship between egg size and type of parental care in reef fishes.

2) Again, contrary to widespread belief pelagic and demersal spawning fishes do not differ fundamentally in fecundity; i.e., same-sized demersal and pelagic spawners produce similar numbers of eggs (Fig. 2). Because so few estimates of fecundity are available for reef fishes, species level estimates rather than family mean numbers were used for this analysis. Body size was obtained from Böhlke & Chaplin (1968) and Masuda, et al. (1975) if it was not provided by the worker who estimated fecundity. For both pelagic and demersal spawners fecundity was significantly correlated with body length (respectively, r = 0.86, n = 15; and r = 0.89, n = 41), and plotted linearly on a log-log scale, indicating a power function. There was no significant difference between either the slopes of the two lines (F = 0.99, df = 1 and 52, p > 0.25) or the intercepts (F = 1.32, df = 1 and 53, p > 0.25). The pooled relationship between fecundity and

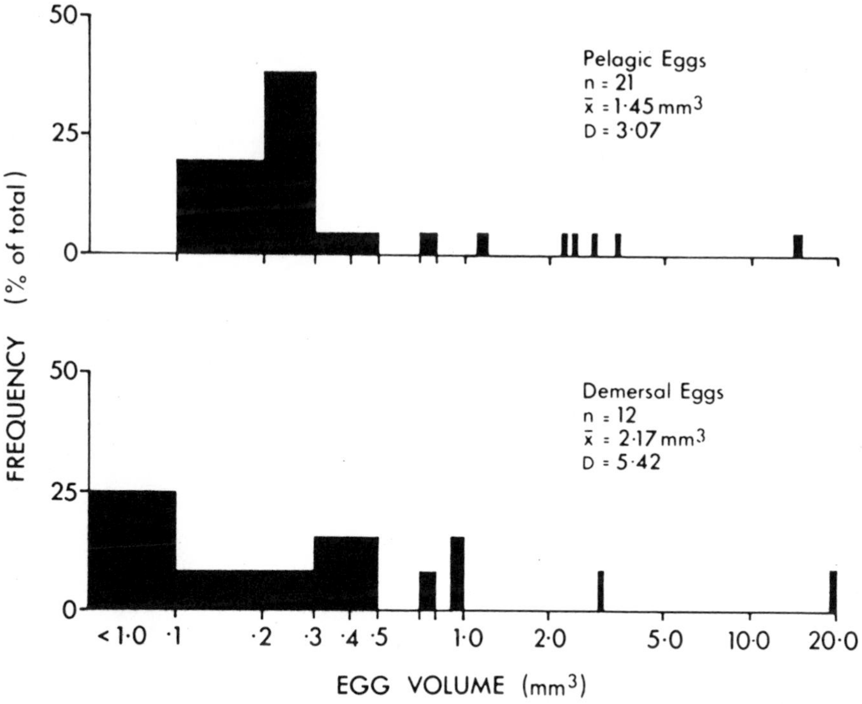

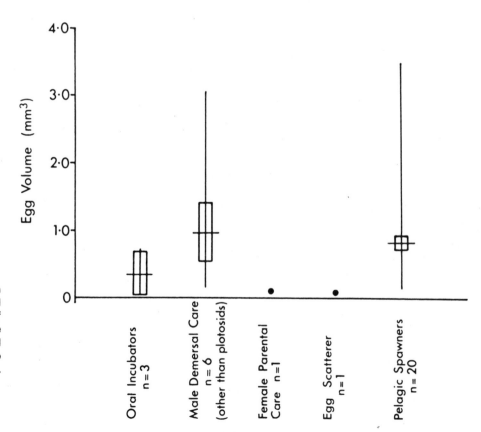

Fig. 1a. Distribution of egg volumes, based on a single mean value per family, for fishes producing pelagic and demersal eggs. In this and the following figures, n = sample size, x = mean, and D = standard deviation.

Pelagic Eggs
n = 21
$\bar{x}$ = 1·45 mm^3
D = 3·07

Demersal Eggs
n = 12
$\bar{x}$ = 2·17 mm^3
D = 5·42

Fig. 1b. The relationship between egg volume and type of parental care, if any, for families of coral reef fishes. Horizontal lines are the means, vertical lines are ranges, and open rectangles represent two standard errors about the mean.

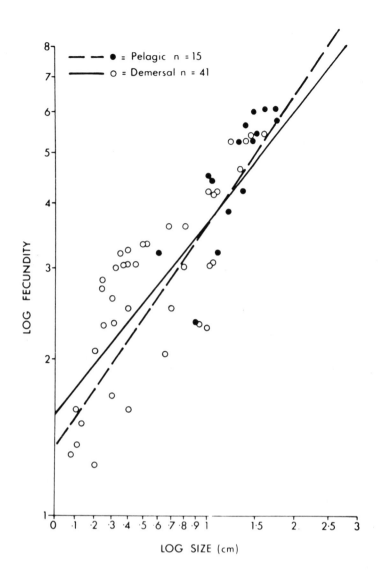

Fig. 2. Effect of body size (SL) on instantaneous fecundity for pelagic and demersal spawning reef fishes. Neither the slopes nor the intercepts of the *Least Squares Lines* differ significantly.

standard length for pelagic and demersal spawning reef fishes is

$$Y = 27.5\ x^{2.37}$$

when Y = instantaneous fecundity
and   X = standard length.

3) Pelagic spawning fishes in general are significantly larger than demersal spawning fishes ($\bar{x}$ pelagic = 44.6 cm, n = 30 families; $\bar{x}$ demersal = 14.4 cm, n = 12 families; X^2 = 28.5, df = 6, p<<0.01) (Fig. 3). Average size for a family was determined from data in Masuda, et al. (1975) by averaging the maximum length provided for all species in each family. As a consequence of this general size difference, pelagic spawners *on average* produce more eggs than demersal spawners despite their common ratio of fecundity to standard length.

4) The primary difference between pelagic and demersal spawning fishes is the incubation period of their respective eggs (Fig. 4). The average incubation period for demersal spawning families is 152.3 hours (n = 10); that for pelagic spawning families is only 43.1 hours (n = 20), all at species-normal temperatures. Despite this three-fold difference in the mean values, incubation periods in the two groups overlap sufficiently that the differences are not quite significant (X^2 = 16.96, df = 10, 0.1 > p > 0.05). Incubation period is a significant function of egg size in pelagic spawning families (r = 0.52, n = 19, p < 0.05), but not for demersal spawners (r = 0.45, n = 11, 0.1 > p > 0.05).

5) Size of the newly hatched larva is a positive func-

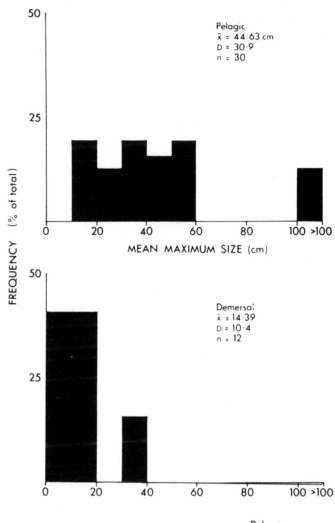

Fig. 3. Size distribution of pelagic and demersal spawning families, based on a single mean maximum size value (see text) per family.

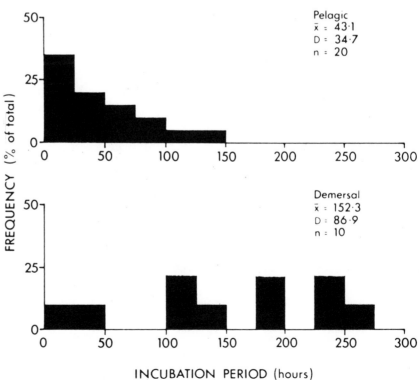

Fig. 4 Distribution of incubation periods, based on a single mean value for each family, for fishes producing pelagic and demersal eggs.

tion of incubation period in both groups (r = 0.67, n = 19, and r = 0.85, n = 10, for pelagic and demersal spawning families, respectively; p in both cases is less than 0.01). Multiple regression analysis was run on both groups to determine the relative importance of incubation period and egg size in determining size of the newly hatched larva. In pelagic spawners both egg volume and incubation period contributed significantly to length of the larva at hatching (F for egg volume alone = 14.46, df = 1 and 18, p < 0.01; F for additional reduction of sum of squares by incubation period = 5.42, df = 1 and 17; p < 0.05, one tailed). The relationship between the three variables is:

$$Y = 1.26 + 0.299x_1 + 0.021x_2$$
$$R^2 = 0.55$$

where Y = length of the newly hatched larva
        x_1 = egg volume
        x_2 = incubation period.

In contrast, length of newly hatched larvae of demersal eggs was entirely a function of incubation period (F = 21.56, df = 1 and 9; p < 0.01); egg volume neither significantly correlated with size of the newly hatched larva (F = 2.46, df = 1 and 9; p > .1) nor contributed significantly to the reduction in variance when added to incubation period in a multiple regression analysis (F = 0.35, df = 1 and 7; p > 0.25).

6) The final difference between the two groups is in degree of larval development at hatching. No statistic was calculated for this difference, but a survey of the previous chapters quickly indicates its nature. At hatching, larvae from pelagic eggs are little more than floating balls of yolk with a sliver of "protofish" attached. Such larvae generally lack pigment in their eyes, lack fins, and are incapable of doing much more than floating. First feeding usually begins four to five days after hatching. In contrast, the newly hatched larvae of demersal eggs are usually strong swimmers with pigmented eyes, well developed fin precursors, and a small yolk sac. Feeding begins within the first 24 hours. There are exceptions, however. The newly hatched larvae of siganids and balistids are more similar to those of pelagic spawners than they are to other demersal spawners.

The above analysis of the differences between demersal and pelagic spawners leads to several ten-tative conclusions. The popular concept of the relative compensating advantages of the pelagic and demersal strategies of reproduction is that demersal spawning fishes "place all their eggs in one basket," so to speak, producing only a few eggs that hatch out into a few advanced and highly competitive larvae. In contrast, pelagic spawners supposedly produce "millions" of tiny eggs which they do not tend, instead simply scattering them to fend for themselves. These alternate strategies were thought to even out in the end since the extra number of larvae produced by pelagic spawners compensates for their increased risk of being preyed upon between spawning and, essentially, first feeding, the point at which the "fewer" demersally produced larvae hatch and until which point they are protected by the adult. This interpretation of reef fish spawning strategies fails primarily on two counts: demersal eggs are not bigger then pelagic ones and consequently do not represent any particularly greater concentration of reproductive effort per female in each offspring, and, on a per unit size basis, pelagic spawning fishes do not have an instantaneous fecundity significantly higher than demersal spawning fishes. The similar sizes of the eggs produced by the two groups also suggests that, at first feeding (i.e., when yolk reserves, which are a function of egg size, are used up), the larvae of pelagic and demersal spawning fishes are about the same size and at a comparable stage of development.

The totality of the analysis thus far is that, if anything, pelagic spawning fishes are at a competitive disadvantage with demersal spawners. By releasing their eggs into the water column, such pelagic spawners start off with the same number of offspring as a demersal spawner, but subject them to longer periods of predation in the water column. Our knowledge of what goes on in the water column is, at best, rudimentary, but it appears that extensive larval mortality regularly occurs there. For example, if one assumes a stable population with a) a one-year generation time, b) an annual fecundity of 100,000 eggs, c) total development period from egg laying to larval settlement of 30 days, d) mortality in the water column density independent, and e) a 90% juvenile mortality between recruitment and spawning, then planktonic mortality must average approximately 30% per day. A pelagic spawner after eight days has had its number of young reduced from 100,000 to only 5,580, a decrease of 94.4%; a demersal spawner incubating its offspring for the same eight days then starts off at day nine with roughly 18 times as many offspring in the water column as its pelagic spawning counterpart. Twenty-one days later the demersal spawner will still have nearly twice as many young settle to the bottom as will the pelagic spawning species. Without a counterbalancing advantage for the pelagic spawner, it would appear to be at a strong competitive disadvantage.

A variety of such counterbalancing factors are possible. Pelagic spawning fishes could: a) increase fecundity, in terms of eggs per clutch, clutches per unit time, duration of spawning season, i.e., clutches per year, or, by increasing life span, net lifetime fecundity; b) reduce planktonic mortality, perhaps by producing a distasteful or spiny larva; c) reduce postsettlement mortality; or d) increase the rate of larval development, shortening the time exposed to predation. Fecundity adjustments require the greatest change (a doubling of egg production) to match our model demersal spawners, whereas the other three options require changes of, respectively, 5.1%, 9.5%, and 6.3%, all of which seem small enough to be feasible. Aside from the data on instantaneous fecundity (cited above) and frequency of spawning which indicate that pelagic spawners do not have twice the fecundity per unit size of demersal spawners, all four options suffer in that all can be tracked by a competing demersal spawning species, i.e., if a pelagic spawning individual can evolve a distasteful larva, there seems no *a priori* reason why a similar demersal spawning species could not do so. Whatever difference between the two groups results in their stable coexistence, it must involve either some benefit of pelagic spawning which is impossible for a demersal spawner or a cost specific to and unavoidable for a demersal spawner.

I suggest three likely hypotheses:

1) Size disadvantage hypothesis—As noted earlier, there is a striking difference in the average sizes of demersal and pelagic spawning reef fishes, with the latter averaging three times the length of the former and presumably several times its mass. There are presumably advantages to large size: large individuals are likely to be relatively immune to predation (see below), facilitating, among other things, migrations to optimal spawning sites. Similarly, large predators may be capable of taking a wider range of prey items than smaller ones. In terms of spawning, however, large size may impart a distinct disadvantage. Small fishes would have little difficulty in finding a secluded and easily defended spot to tend eggs—under a shell, in a burrow, or on a flat wall of rock—and are small enough to be attentive to and respond to small egg

predators. With increasing body size, however, such defendable spawning sites become more difficult to find. Larger caves, for example, often have numerous small holes and entrances through which small egg predators, such as labrids and pomacentrids, could approach an egg mass discretely to eat the eggs despite the parents' guarding of the "front door." Similarly, small fishes are usually more maneuverable than larger ones and are able to utilize topographic features to elude pursuit, again making them difficult to exclude from the nest site. I suggest that this size disadvantage effectively prevents large reef-associated fishes from tending demersal eggs. It seems more than coincidental that the largest demersal spawning fishes are either egg-scatterers and do not tend their eggs (siganids), or prepare shallow nests on open sandy areas where there is little relief to hide approaching egg predators (balistids, tetraodontids). The balistids also have the shortest incubation time of any demersal spawning family, resulting in an absolute minimum period of egg defense and tending. No comparable information is available for tropical tetraodontids, though at least one may be an egg-scatterer. Because of this size disadvantage, all large fishes (mean family maximum size greater than 40 cm) are pelagic spawners. Pelagic spawners, in fact, dominate in all size classes greater than 20 cm mean family maximum (see Fig. 3). Conversely, at very small sizes (less than 10 cm maximum family mean) the advantages of parental care of demersal eggs in terms of overall reduced mortality prior to settlement have resulted in all such families producing demersal eggs. According to this hypothesis, fishes that produce pelagic eggs simply cannot compete effectively in this size range. Families at intermediate sizes (10 to 20 cm) are about evenly split in terms of pelagic and demersal spawning, hypothetically indicating the increasing difficulty of defending demersal eggs as sizes increase in this range.

Based on this hypothesis, one would further predict that at the small size limits of some pelagic spawning families, such as small labrids, one would see the evolution of demersal spawning and egg tending. At least one such case may, in fact, occur: small grammistids (pseudogrammids) appear to be demersal spawners, whereas at least some larger ones are definitely pelagic spawners. Similarly, the pseudochromoids are thought to be small-bodied offshoots of the serranid line; all are demersal spawners, whereas the typically much larger serranids are pelagic spawners. Finally, at least one small anglerfish broods demersal eggs (Pietsch & Grobecker, 1980), in contrast to the pelagic spawning of other species. That such shifts do not occur more often may indicate an overall evolutionary conservatism with respect to basic reproductive strategies.

2) High immediate mortality hypothesis—One fundamental difference between fishes that produce pelagic eggs and those that produce demersal ones is that the fertilized eggs of the former are launched into the water column above the reef, often in areas with currents that carry them quickly away from it, and that the fertilized eggs of the latter stay on the reef and subsequently (after hatching) have to move away from it on their own. The demands of nest tending and associated territorial defense may even preclude launching larvae into particularly favorable currents; rather, the newly hatched larvae emerge from all over the reef. If the reef is a particularly harsh environment for eggs and larvae, because of the filter-feeding and particulate planktivores present, then the gauntlet of such predators run by newly hatched larvae of demersal spawning fishes may be so rigorous that by the time the larvae have reached open water they may have been so reduced in numbers that they equal the young of pelagic spawning fishes that have been in the water column since spawning. The high risk of larval predation by reef-associated predators, then, is postulated to be an unavoidable consequence of demersal spawning that counterbalances their initially reduced mortality.

Several lines of evidence appear to support this hypothesis. The first is that pelagic spawning fishes often shed gametes in areas of strong current that result in their being carried quickly away from the reef. They also spawn at times that result in lowest risks of at least immediate predation by planktivores (i.e., at dusk and/or at the peak of the outgoing tide). Similarly, demersal eggs often, though not always, hatch at dusk and also at the peak tidal flow, again apparently to minimize larval mortality. Most larvae are also photopositive, a mechanism thought to get them off the bottom away from planktivores (e.g., Fricke, 1974). All three observations suggest that predation by reef-based plankitivores is a significant source of larval mortality. The difficulty with this hypothesis is that such features of reproduction and larval behavior are equally well predicted by other, more parsimonious explanations, such as selection for maximum dispersal of fertilized eggs or intraspecific selection. Nor does it obviously lead to any specific predominance of pelagic spawners at large sizes and demersal spawners at small size ranges (see Strathman

& Strathman, 1982, however, for alternative hypotheses regarding this size difference). Finally, given that newly hatched larvae from demersal eggs are relatively effective swimmers capable, at least, of maintaining a position near the water's surface and that hatching occurs typically at dusk, when planktivore activity is lowest (e.g., Hobson, 1972), it seems unlikely that immediate predation on such larvae would be so high as to counterweigh what may be as much as a twenty-fold difference in survivorship of pelagic-derived and demersal-derived larvae.

3) Low mortality of pelagic eggs hypothesis—The underlying assumption of the argument thus far is that pelagic eggs and yolk-sac larvae suffer rates of mortality at least as high as those of post-first-feeding larvae, the state at which demersal egg derived larvae hatch. Almost nothing is known about the ecology of planktonic eggs and larvae of coral reef fishes, so that the validity of this assumption is untested. The eggs of at least one group are unpalatable (Moyer & Zaiser, 1981), but most such eggs appear to be edible. Several workers, for example, report them in the diets of planktivorous fishes (e.g., Randall, 1967; Hobson, 1968) and several others note that such fishes converge on newly released gamete clouds (e.g., Colin, 1976; Nakazono, 1979). Moreover, the swimming ability of prolarvae suggests only limited capability to avoid predation.

Regardless of the factors that select for the maintenance of both pelagic and demersal spawning in reef fishes, one interesting prediction about the two groups derives from their different times spent in the water column. Whatever else happens to planktonic larvae, they are dispersed. Since the young of demersal spawning fishes spend a large part of their developmental period on the bottom, it follows that such young spend, on average, less time in the water column than the young of pelagic spawning fishes and so are likely not to be dispersed as far. Consequently, one would predict that, on average, a demersal spawning species would have a smaller geographic range than a pelagic spawner and show a greater tendency toward development of discrete local populations. Comparative information on geographic ranges to test this prediction is difficult to obtain, but information on the degree of speciation in each family, one apparent consequence of the development of discrete populations, is readily available. Numbers of species in the families of reef-dwelling fishes were obtained from Nelson (1976), a recent review of fish systematics. Since it was also expected that there might be a difference in number of species between pelagic and demersal spawning reef fishes due to their average size differences (big fishes perhaps less speciose than small ones), both mean family maximum size (obtained from Masuda, et al., 1975) and mode of spawning (demersal versus pelagic eggs) were run in a multiple regression against number of species per family. Based on this analysis, there was, as predicted, a significant correlation between number of species and egg type (demersal spawners with more species) ($r = 0.331$, $n = 42$; $F = 4.92$; $df = 1$ & 40, $p_{ONE-TAILED} < 0.05$). The correlation between number of species and body size was not significant ($r = -0.24$; $n = 42$; $F = 2.35$, $df = 1$ & 40, $p_{ONE-TAILED} > 0.1$), nor did adding size into the regression between number of species and egg type significantly improve the regression ($F = 0.01$, $df = 1$ & 39, $p \gg 0.5$).

Spawning Systems of Reef Fishes

There are five spawning systems characteristic of reef fishes:

1) Monogamy and Long-term Pairing:

Direct evidence of monogamy (i.e., exclusive repeated matings between one male and one female) is uncommon for any reef fish. The only documented examples, in fact, are a few anemonefishes (Fricke, 1974; Moyer & Bell, 1976), the brood-tending damselfish *Acanthochromis polyacanthus* (Thresher, in prep.), and a reef-associated syngnathid (Gronell, in press). That monogamy *may* be more common on the reef than this sparse list indicates, however, is suggested by the widespread occurrence of pairing in reef fishes. Pairing is characteristic, though not universal, in two groups, the anemonefishes and the Chaetodontidae. As documented in the previous chapters, however, it has been reported for fishes in at least 14 other families, ranging from blennies and gobies to hawkfishes and siganids. In the vast majority of these cases, there are no data whatsoever regarding the stability of such pairs. To date, evidence of long-term pairing, lasting several years, has been provided only for *Amphiprion* (Fricke, 1974; Moyer & Bell, 1976), for several species of butterflyfish (Reese, 1975; Fricke, 1973a; Sutton, pers. comm.), and an angelfish (Strand, 1978). Schmale (MS) has also documented individual partner recognition in pairs of the goby *Gobiosoma oceanops*, suggesting that it, too, forms long-term pairs. Similar small and relatively isolated pair-forming fishes, such as the hawkfish *Oxycirrhites typus*, are also likely to be eventually documented as permanently paired.

There is, of course, considerable danger in equating pairing with monogamy, especially for fishes like the siganids which aggregate for spawning. Conversely, however, it is equally presumptuous to rule out monogamy in those species that do not form conspicuous pairs. A prime example of the latter is the reef-associated syngnathid *Corythoichthys intestinalis*, which is documented by Gronell (in press) to be monogamous but in which males and females do not associate outside of brief spawning periods. Similarly, a number of apogonids mill about in large aggregations during the day but form conspicuous pairs before dispersing for the night. One gets the distinct impression that monogamy and/or pair-formation is far more common on the reef than is generally assumed.

Several papers have addressed the question of monogamy in reef fishes, each from a different point of view.

1) Fricke (1973a), reviewing the chaetodontids, suggested that permanent pair-formation and monogamy in this particularly speciose family may reduce the frequency of interspecific mating and subsequent gamete wastage, and that such pairing can, in part, account for the co-occurrence of many closely related species on the same reefs. In support of this argument, Fricke noted that pairing is often most common in low-density populations, i.e., where the probability of encountering a conspecific mate is low. Two points regarding this hypothesis, however, seem questionable. First, many reef fishes have numerous broadly sympatric species yet do not form conspicuous pairs; some are demonstrably polygamous (e.g., labrids and some pomacentrids). Even so, such families do not seem any more prone to hybridization than the pair-forming chaetodontids. Second, there is no apparent correlation between sympatric within-genus diversity in butterflyfishes and the number of species found in conspicuous pairs. There is, for example, only one common eastern Pacific species of *Chaetodon* that only occasionally pairs (Strand, pers. comm.; pers. obs.); there are four common species of *Chaetodon* in the Caribbean, all of which pair regularly (Thresher, 1980) and which breed as pairs (Colin & Clavijo, in prep.); and there are 19 common Indo-Pacific species of *Chaetodon* surveyed by Reese (1975), of which only about 75% pair.

2) Low population density combined with high risks of predation between suitable habitat patches could logically lead to pairing, monogamy, and hermaphroditism (Ghiselin, 1969; Wilson, 1975) and has been suggested to account for pairing not only by

several small, brightly colored, and site-restricted reef fishes (Fricke, 1974; Lassig, 1976; Fricke & Fricke, 1977), but also by small reef crustaceans (Knowlton, 1976, 1978). The arguments for and data supporting this hypothesis seem sound but can be applied only to select examples of pairing by reef fishes—mainly anemonefishes, small site-attached gobies, and the like—and can be applied only with difficulty to the many large, abundant, and more motile species, such as the chaetodontids. Moreover, the same arguments that lead to pairing by such site-restricted fishes can lead equally well to harem formation by the same or similar fishes, e.g., various species of *Dascyllus* (Schwarz, pers. comm.) and several smaller pomacanthids (Moyer & Nakazono, 1978a).

3) Robertson, et al. (1979) provided data on the spawning behavior, general ecology, and social organizations of several species of surgeonfishes (Acanthuridae) found at Aldabra Atoll, Indian Ocean, and speculated on the factors underlying manifested patterns of social behavior. Based on these data, the authors suggested that "relatively permanent pairing" (they avoided use of the term monogamy, justifiably so since they provided no direct data regarding mating patterns; such monogamy is implied, however, in their descriptions of permanent pairing, pair defense of a territory, and spawning occurring only within the territory) could evolve as a result of three processes: first, a male that pairs may be able to guarantee itself a mate, may increase the reproductive output of that mate to a level higher than those mates it could find in some other social system, and/or may enjoy reduced levels of predation by avoiding migrations to group spawning areas; second, in a social system that results in extreme variation in male reproductive success, a "lower quality" male may be able to offset its disadvantages by providing a female access to some limiting resource and so inducing it to spawn with it regularly; and third, pairing may be adaptive when the benefits of joint defense of a territory, in terms of efficient sequestering of resources, outweigh the disadvantages of limiting oneself to a single partner. Robertson, et al. (1979) favored the last hypothesis in accounting for the relatively permanent pairing they observed in *Zebrasoma scopas*. They subsequently argued that the absence of pair-formation in a second territorial species, *Acanthurus lineatus*, results from the small territories defended by the fish and the consequent likelihood of male-male interference during spawning. They did not, however, witness spawning in this species and so provided no information as to

where the fish spawn or whether or not such interference occurs. The single report of spawning by *A. lineatus* indicates group-spawning in shallow water.

Subsequently, Pressley (1981) has also suggested the increased effectiveness of joint territorial defense may underlie long-term pairing in, in this case, the serranine *Serranus tigrinus*. As an alternative, he suggested that pairs may also benefit from increased foraging efficiency. In support of this, he noted synchronized prey stalking by pairs on the reef.

4) Gronell (in press) documented in detail the social organization of the reef-associated pipefish *Corythoichthys intestinalis* at One Tree Island, Great Barrier Reef, and reported it to be monogamous, based on repeated observations of spawnings over more than a year. The species, however, is not territorial or site-attached to the extent considered above, nor do pair members even associate during the day. Rather, each pair member roams about a specific home range during the day, but then moves during brief spawning periods to a spawning site uniquely used by that pair. Gronell reviewed hypotheses previously offered to account for monogamy and pairing in reef fishes and found none to be appropriate. Rather, she suggested that such pairing derives from selection for maximal spawning efficiency and minimum time spent searching for a suitable mate.

Each of these hypotheses can account for some of the examples of pair-formation in reef fishes. Of them, however, only the last, i.e., increased reproductive efficiency deriving from minimal time spent searching for a suitable mate, seems likely to be a general principle that can be applied to, for example, pair-forming siganids, chaetodontids, or apogonids. This argument will be developed further below.

2) Harems:

Harems consist of one male dominating (usually), controlling access to, and spawning exclusively with more than one female. Such harems, consisting of anywhere from two to seven females, have thus far been reported for both demersal (e.g., balistids, pomacentrids, tetraodontids) and pelagic (e.g., pomacanthids, mugiloidids, labrids, scarids) spawners and will probably turn out to be the dominant social mode of relatively small site-attached reef fishes. Among large fishes, enhanced mobility, due to perhaps lower risks of predation, often leads to migrations to optimal spawning areas, producing large groupings and apparently promiscuous spawning.

The advantages gained by a male in controlling a harem at first appear obvious; i.e., it spawns several times each cycle and consequently enjoys a level of reproductive success higher than any female or any non-harem-controlling male. In practice, however, harems are an adaptive strategy for a male only so long as control of the harem does not reduce his life span so drastically as to lower his net lifetime reproductive rate; i.e., a male that spawns with one female 50 times is more successful than one that spawns with five females, but only five times each. Similarly, the presence of large numbers of peripheral males "poaching" females may result in harem defense being uneconomical, creating a shift to a promiscuous spawning system. The logic behind such shifts and the adaptive significance of harem formation are treated by, among others, Warner, et al. (1975) and Warner & Hoffman (1980).

The advantages to a female that remains in a harem are less clear than those accruing to the male. At the most basic level, a female may not lose much by being in a harem, since she spawns as frequently as she would otherwise. Complications arise, however, when harem formation is accompanied by female dominance systems in which lower ranking females may be denied access to resources they might be able to obtain elsewhere. Robertson & Hoffman (1977) have argued that in the harem-forming labrid *Labroides dimidiatus*, females are site-attached cleaners that remain at a conspicuous and defended cleaning station and that such site-attachment permits male territoriality over access to them. Thresher (1979b) reached a similar conclusion in examining the social systems of two western Atlantic wrasses, noting that in the harem-forming species high levels of female density and site-attachment appeared to result in male defense of those females, whereas in a sympatric promiscuous species females occurred in low density and roamed over large home ranges. Emlen & Oring (1977) referred to such direct control over access to females as "female defense polygamy." An alternate route to harem-formation in reef fishes, referred to by Emlen & Oring as "resource defense polygyny," has been suggested for pomacanthids (Moyer & Nakazono, 1978a; Lobel, 1978) and for ostraciids (Moyer, 1979). They argued that prime spawning sites are in short supply for their respective species and that males can control access to females by controlling these sites. Similar arguments involving male control of feeding areas have been made for labrids and scarids.

Harem-formation clearly grades into monogamy (a harem of one), and transitions between pairing and harems have been documented for both labrids (various species of *Labroides*) and pomacanthids. Indeed, it is likely that at least some of the reports of pairing in reef fishes are, in fact, such minimum sized harems, emphasizing again the caution that must be exercised in generalizing from pairing to monogamy in the absence of evidence concerning mating preferences.

3) Explosive Breeding Assemblages:

Borrowing a term from Emlen and Oring (1977), explosive breeding assemblages in reef fishes occur when the fishes, usually though not always pelagic spawners, aggregate at "traditional" spawning sites for brief spawning periods. Examples include some epinephelines, some siganids, some acanthurids, some scarids, and some mugilids. Such a breeding assemblage may be the primary mode of reproduction of some larger reef species, e.g., lethrinids, ephippids, carangids, sphyraenids, and lutjanids, but as yet there are not enough observations on spawning by such fishes to test for generality. Johannes (1978) suggested that migrations to spawning grounds are adaptive because they result in gametes being shed in areas that facilitate offshore transport, e.g., at the mouths of channels through the reef or at the end of coral promentories, and presumably minimize larval predation. For the same reasons such aggregations occur most often at times of peak tidal flow offshore. Smaller fishes presumably are prevented from migrating to such optimal spawning areas by the high risk of predation they would incur while traveling.

4) Promiscuity:

Data on the spawning patterns of small site-attached reef fishes are still sparse and thus far available only for the pomacentrids (e.g., Doherty, 1980; M. Schmale, 1981). The long-standing assumption about such fishes, however, is that they are basically promiscuous, with each male courting any female that passes and each female spawning with any one of a number of males available to her. Such an assumption is supported by frequent observation that males of at least the demersal spawning species frequently tend more than one clutch of eggs or many more eggs than could be produced by a single female (e.g., Doherty, 1980; Breder, 1941; Fishelson, 1975a). Evidence of female promiscuity, however, has not yet been obtained; current theories suggest that females should be discriminating with respect to choice of male and would probably spawn with that male repeatedly. Data presented later, in fact, suggest that in such "promiscuous" species both sexes may frequently exhibit high degrees of mate fidelity.

5) Lek-like Spawning Aggregations:

A lek is a "communal display area where males congregate for the sole purpose of attracting and courting females and to which females come for mating" (Emlen and Oring, 1977). The term was originally coined to describe the breeding systems of birds (e.g., Robel & Ballard, 1974; Rippin & Boag, 1974) and other terrestrial vertebrates (see review by Loiselle & Barlow, 1978) but has since been applied to reef fishes by numerous workers (e.g., Sale, 1978; Warner, 1978) with respect to temporary spawning aggregations of both pelagic spawning (e.g., some labrids) and demersal spawning fishes (e.g., some pomacentrids). Loiselle & Barlow (1978) reviewed the occurrence of such "leks" in fishes, compared it to the similar-appearing lekking in terrestrial animals, and concluded that the term legitimately describes reproductive systems in both groups. Moyer and Yogo (in press) reexamine the concept in great detail and test it against the spawning behavior of a Japanese labrid. Whether or not lekking occurs in both terrestrial animals and reef fishes, however, depends on the limits one wishes to set on the phenomenon and, by implication, the precision of its usage. Loiselle & Barlow, for example, expanded the definition of a lek considerably, referring to it as any "temporary aggregation of sexually active males for reproduction," eliminating the key phrase "for the sole purpose of attracting and courting mates." In all cases examined in which males aggregate for spawning, either pelagic spawning itself takes place (e.g., labrids, acanthurids) or demersal spawning occurs and the males subsequently tend eggs in place. Emlen & Oring's definition of a lek specifically states that the selective factor resulting in a lek is the communal display of the males and presumably their heightened attractiveness to females as a consequence of this joint display. This is relatively simple to demonstrate in an internally fertilizing terrestrial animal, since after copulation the female leaves; i.e., the factors that result in lek formation are clearly independent of those involved in rearing young. In a reef fish, however, the distinction is much more difficult to make. Many larger pelagic spawners, for example, are group spawners and thus can hardly be considered to form leks, but as part of

their spawning behavior they migrate to specific spawning sites thought to be optimal locations for off-shore transport of eggs and larvae (e.g., Johannes, 1978). If a smaller pair-spawning species, such as a wrasse, did the same thing on a smaller scale, with individual males competing for the optimal spot to spawn (perhaps the highest coral outcrop or the main axis of an offshore current), the result would be an aggregation of such males in a single area spawning with females which sought the same area for exactly the same reasons, i.e., because it is the best spot for gamete launching.

A similar argument can be made for the demersal spawning "lek" species, such as some damselfishes. Not only does spawning take place in the "leks," but afterward the male defends and tends the eggs until hatching. Several recent studies suggest that similar forms of communal breeding and nesting in birds develop not because of increased male attractiveness in such aggregations, but because they result in reduced mortality of both parents and offspring (e.g., Clark & Robertson, 1979; Wiklund & Anderson, 1980), presumably because of group defense of spawning grounds and/or temporary overloading of the local predators' capability of taking prey. Such a selective regime operating in demersal spawning "lek" species of reef fishes seems entirely reasonable, even though it has yet to be examined in detail.

Temporary spawning aggregations of territorial males in both pelagic and demersal spawning reef fishes, then, can logically be predicted to occur on the basis of underlying factors fundamentally different from those demonstrated to occur in terrestrial lek-forming species. This is not to state that enhancement of male attractiveness through communal display is not occurring in reef fishes, but rather that the evidence for such an effect is, to date, nonexistent. To use the same term "lek" for what may be only superficially similar spawning systems in terrestrial animals and fishes seems unwise since, even with the best of intentions, such common usage can quickly lead to a widespread assumption of identical underlying processes. I suggest that for the moment the more conservative term "lek-like" be employed to describe such fishes, at least until more solid data on the dynamics of such spawning aggregations are obtained.

COSTS AND BENEFITS OF REEF FISH SPAWNING SYSTEMS

Although many factors may ultimately affect the social structure and mating system of a reef fish (e.g.,

Fricke, 1975b; Robertson & Hoffman, 1977), three factors appear to have the strongest influence.

1) *Limited Resources*—As emphasized for reef fishes by Fricke (1975a & b), Thresher (1977a), and Moyer & Nakazono (1978a), resource abundance and distribution directly control the size of the social units, the distribution pattern of individuals in the unit, and the amount of aggression, if any, manifested by such individuals (for a review of the concept see Brown & Orians, 1970, and Wilson, 1975). Obviously, if population size is limited by the scarcity of some resource, e.g., food or shelter, large and dense social units are unlikely and, depending upon the economic defensibility of that resource, the development of broadly overlapping home ranges or individual territories will be favored. Similarly, if a limited resource is defendable but not abundant enough to support a pair of fishes, individual territoriality should be selected for.

2) *High Individual Fecundity*—Currently available estimates of instantaneous fecundity in reef fishes range from a low of only a few hundred eggs, as in small pomacentrids, blennies, and angelfishes, to well over a million eggs, as in the case of large serranids. Given that many of these fishes also spawn repeatedly during the year, annual fecundity of many, if not most, fishes is likely to be measured in the tens of thousands to tens of millions. Yet as a result of immense, largely random, mortality, very few of these eggs ever result in juveniles that return to the reef. Fitness for a reef fish, therefore, is to a large extent a function of the rate at which an individual produces offspring (Maynard Smith, 1977), with the most fecund individuals those most likely to have the highest number of offspring successfully run the "gauntlet" of planktonic and juvenile existence. Under such conditions, there is likely to be strong selection for efficient reproduction, a conclusion also drawn by Johannes (1978), who based his arguments in part on larval mortality and in part on the "lottery" aspect of competition for space on the reef (e.g., Sale, 1974).

3) *Limited Spawning Periods*—Although a few species spawn over long periods during the day (e.g., clinids, some pomacentrids, and some labrids), most—especially those with pelagic eggs—restrict spawning activity to narrow "launch windows" when conditions are most favorable for at least immediate survival of offspring (Jones, 1968; Lobel, 1978). In some cases spawning periods amount to no more than a few minutes each day (e.g., Moyer & Nakazono,

1978a). Spawning in many cases is also restricted to particular tidal or current conditions, often in a lunar rhythm and limited to a few days each month, apparently to ensure that the young are released into the most favorable current patterns for return to the reef at the end of the planktonic phase (e.g., Johannes, 1978; Lobel, 1978). The apparently strong selection for spawning during these brief critical periods has recently been emphasized by Johannes (1978), who even suggested that spawning during these short periods is so critical that normal antipredator behavior is overridden to facilitate spawning. Given this narrow time slot in which to spawn and then, in some cases, seek shelter for the night, prolonged search for a suitable mate and lengthy courtship are likely selected against.

The combined effect of the three factors above is a trade-off between, on the one hand, large social units and consequently the continuous presence of one or more mates and, on the other hand, possible resource limitations. Reef fishes can be classified as belonging to one of four social "types" (Fricke, 1975b), each of which has associated costs and benefits for the individual participants. Of these types, two—pairing and harems—appear to be most effective in terms of guaranteeing both sexes immediate access to a ready partner, minimum courtship time (e.g., Moyer & Bell, 1976), and maximum reproductive synchrony (the significance of the latter for reef fishes is not known and probably varies with the species). Multi-male heterosexual groups can also be an effective spawning system, as evidenced by the widespread occurrence of "group" spawning in reef fishes (e.g., labrids—Randall & Randall, 1963; Robertson & Hoffman, 1977; scarids—Randall & Randall, 1963; Barlow, 1975; acanthurids—Randall, 1961; Colin, in prep.; mullids—Helfrich & Allen, 1975; mugilids—Colin & Clavijo, 1978), but is limited to those species for which resource distribution permits large social units (planktivores and benthic foragers, in particular). Sperm competition in such a system, however, may have led to the development in many such fishes of an alternative spawning system based on pairs of single "superior" males (often territory holders that can successfully expel other males) and single females in sequential polygamy (e.g., Warner, et al., 1975).

As compared with the previous three systems, a mating system based on solitary promiscuous individuals appears to be remarkably inefficient, involving a relatively long and uncertain search for a mate, prolonged courtship, and perhaps imperfect syn-

Table 2. Costs and Benefits Associated with Each of the Four Major Social Systems Manifest by Reef Fishes. References to Fricke are Fricke (1975b).

Social System	Costs	Benefits
Solitary, with individual territories (Fricke Type 1)	1. Finding mate 2. Imperfect synchronization 3. Continuous reassessment of mate quality necessary 4. Potentially prolonged courtship 5. Care of demersal eggs may restrict movement and limit choice of mates	1. Exclusive use of resources 2. Defense of mates not required 3. Individuals potentially free to mate with many partners
Paired (Fricke Type 4)	1. Shared resources 2. Defense of mate required 3. Individuals possibly restricted to single mate (may be mitigated by limited spawning period)	1. Mate always accessible 2. Full synchronization 3. Mate reassessment not necessary 4. Courtship potentially reduced in intensity or duration 5. Joint care of spawn possible 6. Joint defense of resources possible 7. Possibly lowered susceptibility to predation
Harems (Fricke Type 3)	1. Shared resources 2. Defense of mates required	1. Successful male has access to several females 2. Mate always available 3. Full synchronization 4. Mate reassessment not necessary 5. Possible benefits of schooling 6. For male, temporary exclusion of other males from gene pool (may be mitigated by sex change).
Multi-male, heterosexual groups (Fricke Type 2)	1. Shared resources 2. Possible sperm competition and wastage	1. Many mates potentially available for both sexes 2. Full synchronization 3. Mate generally available 4. Defense of mates not required 5. Mate probably capable, but mate selection may not be possible 6. Possible benefits of schooling

356

chronization. Such disadvantages, however, can be ameliorated by two strategies: temporary pairing with joint defense of a territory and preferred mate polygamy. Temporary pairing can occur either during periods of especially intense and frequent reproductive activity, in which members of a pair accept the reduced levels of resources available to each in return for the ready availability of a spawning partner, or during periods of abundant resources, at which times a defendable territory can sustain two individuals. Conspicuous seasonal pairing does not seem to be common on the reef, perhaps in part due to the relative lack of seasonality on the reef (seasonal pairing may be more common near the temperate limits of the reefs, where there are more pronounced seasonal effects) and in part due to the potential penalty incurred by one partner in having to obtain a new territory or having to fight to regain its old one at the end of the spawning period.

An alternate compromise is for individually territorial fishes to exhibit long-term mate fidelity, either as discrete monogamous pairs of fish each of which defends its own territory or as a polygamous system in which each individually territorial fish "knows" where its partners are, is synchronized with them, and needs "waste" little time searching for a suitable mate. Such "preferred mate polygamy" seems to be an effective means of optimizing both rate of production of offspring and foraging ability given relatively constant conditions and limited resources.

Direct evidence for mate fidelity by individually territorial and sexually monomorphic fishes is provided by several recent studies which indicate that spawnings by such fishes are not random "affairs," but rather that individuals demonstrate clear mate preference and mate fidelity. Such mate preference by a reef fish was first described by Wickler (1967), who noted that, in aquaria, "the males of the blenny *E. bicolor* prefer a specific female even when other females are equally close and no male rivals are present" (author's translation). More recent studies of a few systematically very diverse reef fishes indicate similar behavior. The hamlets, small Caribbean serranids in the genus *Hypoplectrus*, often spawn with the same partner over at least several spawning periods, even though this partner is frequently not the nearest available mate (Fischer, 1978, pers. comm.; Colin, pers. comm.). Similarly, individuals of the Caribbean tilefish *Malacanthus plumieri* commonly maintain separate territories but have been observed to spawn in consistent pairs for as long as two months

(Colin, pers. comm.). The yellowhead jawfish, *Opistognathus aurifrons*, a burrow-constructing (and defending) planktivore found in colonies of up to several dozen individuals on sand flats around Caribbean reefs, apparently behaves in a similar manner, spawning with the same partner over at least several spawning periods, based on both aquarium (Leong, 1967) and field (Colin, 1971a & b, 1972) studies. More recent information suggests that its spawning pattern, and those of other jawfishes, may approach exclusive mating between members of a pair (Colin, pers. comm.). Colin (1971b) even reported apparently altruistic behavior between members of such a pair.

Such studies, unfortunately, are rarely conclusive since they characteristically focus on some other aspect of the subject's biology and obtain data on spawning patterns on an opportunistic basis. The strongest evidence to date of significant mate fidelity by solitary territorial reef fishes concerns the western Atlantic damselfish *Eupomacentrus variabilis* and is provided by two independent studies. The cocoa damselfish is a common benthic herbivore on reefs throughout the tropical western Atlantic. Off Florida it is found in two color phases that are not sexually correlated—a dark, or gray, phase and a light, or blue, phase (Emery, 1973). Gronell (1978) found consistent mate fidelity in this species, based in part on female preference for specific males and in part on male aggression toward unfamiliar females. Specifically, males must guard the demersal eggs and, while doing so, are strongly territorial and site-restricted (similar behavior is typical of other damselfishes and other demersal spawning reef fishes). Home ranges of this fish, however, are large and often widely distributed. Such site-restriction provides the males few opportunities to seek out and court females; rather, each must often wait and court those that approach its nest site. Gronell found that in most cases a female regularly returned to spawn with a particular male and routinely "ignored" and bypassed other actively courting males to reach him. The male, meanwhile, courted only those females that were of the same color phase as his last mate, mating with others only if they were persistent and he was completely without eggs. Given a dispersed population, an approximate 1:1 sex ratio, and an even distribution of the two color phases in each sex, all of which are indicated by Gronell (1978), the result is nearly exclusive mating between a given male and female.

Data corroborating such pairing by the cocoa damselfish were provided by Clarke (pers. comm.).

While surveying the distribution of damselfishes off Bimini (see Clarke, 1977), Clarke sampled seven relatively, though not completely, isolated pairs (i.e., two fish holding territories closer to each other than to any other conspecifics—there is no external sexual dimorphism in this species) of *E. variabilis* and found all to be heterosexual. The overall male:female ratio in the area was approximately 2:3, so that the probability that chance distribution alone would result in such consistent pairing was only 0.006. Such fish, therefore, appear to be actively pairing, even though each member of the pair maintains a territory against not only strange conspecifics, but also each other. Nor is there any indication, either behavioral or histological (Fishelson, pers. comm.), for the alternative possibility of socially-controlled sex change, despite the recent demonstration of such a mechanism in other damselfishes. Similar pairing by an individually territorial reef fish also appears to be the case in the butterflyfish *Chaetodon trifascialis* (formerly *Megaprotodon strigangulus*) (Reese, 1973, pers. comm.).

The mating system manifested by the cocoa damselfish, the hamlets, and other individually territorial species I term "preferred mate polygamy," defined as the restriction of all or most spawning activity by members of both sexes to a disproportionately small number of the mates available to it. For such species the line between monogamy and polygamy (mainly polygyny) may be a fine one and individuals may vary in their degree of mate fidelity depending on population density, accessibility of mates, and length of each spawning period (i.e., the amount of time available to search for other mates).

In a similar manner, local variations in resource abundance and distribution, by affecting the size of the social unit, may result in a single species manifesting a spectrum of reproductive systems. While the reproductive biology of most reef fishes is too poorly documented to test this prediction, it does appear to be borne out by one particularly well studied genus. The cleaner wrasses in the genus *Labroides* are dependent upon a fixed, often limited, and often defended resource—a cleaning station. These wrasses are also unusual, though not unique, among labrids in being sexually monomorphic. Both Randall (1958), reviewing the genus, and Potts (1973), working with *L. dimidiatus*, confirmed conspicuous pairing by these fishes and even described pair-defense of the cleaning station. Other work, however, indicates that in some areas at least two members of the genus, *L. dimidiatus* and *L. bicolor*, form male-dominated harems, with the socially-controlled sex change typical of such systems (Robertson, 1972, 1973; Robertson & Hoffman, 1977). At the other end of the spectrum, both *L. phthirophagus* (Youngbluth, 1968) and *L. dimidiatus* (Fricke, 1966; Slobodkin & Fishelson, 1974) have been reported as commonly solitary fishes only occasionally found in pairs. The factors underlying such apparent diversity of social systems manifested by in one case even the same species have not been investigated, but there is some evidence that the limiting resource may differ from group to group. Youngbluth (1968) suggested that food abundance was the limiting resource for his solitary fish, whereas Robertson & Hoffman (1977) suggested that the harem-forming fishes they studied were more directly affected by the limited availability of "foraging sites," i.e., cleaning stations, than by food supply directly. Similar ranges in the size of "normal" social units have been reported for pomacanthids (Feddern, 1968; Moyer & Nakazono, 1978a; Lobel, 1978), and chaetodontids (Reese, 1975; pers. obs.).

REPRODUCTIVE BEHAVIOR

Given the diversity of coral reef fishes and the diversity of their spawning systems, it is hardly surprising that there are few candidates for "universal" courtship patterns. Aside from a number of motor patterns (such as nest skimming by demersal spawners) which appear to be inevitable consequences of spawning, only two movements occur with sufficient regularity to warrant consideration.

The first and more ubiquitous of the two is also the less conspicuous, yet it may well be the most nearly universal motor pattern in the spawning of fishes in general. "Nudging" or "mouthing" the female's abdomen by the male, in one form or another, occurs in the courtship of many demersal spawners (e.g., pomacentrids, pseudochromoids, gobies) and most pelagic spawners (e.g., antennariids, chaetodontids, cirrhitids, diodontids, labrids, ephippids, pomacanthids, pteroids, serranids). The specific form of such nudging varies from a direct and repeated butting of the female's abdomen in diodontids and open-mouthed pushing in some epinepheline serranids to a stereotyped and apparently ritualized head-standing in pomacentrids and similarly ritualized laying of the male's head against the female's abdomen during the spawning ascent of chaetodontids and pomacanthids. In such a ritualized form the evolutionary significance of the motor pattern is not clear. In its apparently more simple forms, however, the repeated butting or shoving by the male against the female's egg-swollen abdomen suggests that the male is attempting to begin movement of the ripe eggs, possibly by stimulating muscle contractions or by squeezing them out of the ovarian lumen. That such a strategy might be effective is suggested by the numerous observations of "ripe" and "running ripe" females from which eggs can be stripped by means of only slight pressure along the abdomen. Along similar lines, Warner, et al. (1975) suggested that repeated tactile stimulation of a ripe female by males in a group-spawning aggregation of bluehead wrasses, *Thalassoma bifasciatum*, spurs the female into spawning with subordinate males.

The second common form of reproductive behavior in coral reef fishes is the spawning ascent, otherwise—if somewhat inaccurately—referred to as a "spawning rush" (Robertson & Hoffman, 1977; Johannes, 1978; Colin, 1982; Smith, 1982) or, prejudging its function somewhat, "gamete launching" (Ehrlich, 1975). Such spawning ascents are characteristic of (by definition) all families of pelagic spawning fishes. In its simplest form it consists of a paired movement by a single male and female into the water column, followed by the release of gametes at the peak of the ascent. More complex variations include a male-head-to-female-tail spiral toward the surface in ephippids and large pomacanthids, side-by-side spiraling in some acanthurids, tail-to-tail hovering off the bottom in ostraciids, and a slow spawning clasp with participants wrapped about one another, as in some serranines and muraenids.

Two hypotheses have been offered to account for the widespread occurrence of spawning ascents in reef fishes. Randall (1961) and Randall and Randall (1963) suggested that such rapid ascents serve a mechanical function. Expansion of each spawning individual's swim bladder during an ascent, caused by the decreasing water pressure as the fish move toward the surface, results in pressure being placed on the gonads. Once the appropriate muscles are relaxed, this pressure forces out the gametes, scattering fertilized eggs into the water column. Randall also suggested that the typical sharp flex of the spawners' bodies at the peak of the spawning ascent serves a similar function, facilitating the squeezing of eggs and sperm from the gonads. This strictly mechanical interpretation, however, has been generally discounted in the literature (e.g., Ehrlich, 1975; Johannes, 1978), largely as the result of several apparent shortcomings. First, not all fishes that spawn in this manner have swim bladders (e.g., cirrhitids—Randall, 1963a). Second,

within a species spawning height does not appear to vary with depth. If swim bladder expansion due to pressure changes is the primary function of the spawning ascent, then the ascent should be higher at greater starting depths in order to produce the same amount of expansion. Although few observations have been made of spawning by the same species at very different depths, these few data suggest a relative constancy of ascent height. At One Tree Island, for example, *Centropyge tibicen* (Pomacanthidae) ascends approximately 0.8 m to spawn both in the lagoon at a depth of about 1.5 m and at the outer reef edge at a depth of nearly 20 m (pers. obs.). Finally, height of the spawning ascent of many fishes, especially the smaller ones, is commonly less than a meter, even at depths of 15 to 20 meters. The change in swim bladder volume produced by a 1 m ascent in 15 m of water is only 4%, which seems hardly likely to put enough pressure on the gonads to significantly affect gamete release.

An alternative hypothesis for the evolution of the spawning ascent was suggested by Jones (1968), Ehrlich (1975), Robertson & Hoffman (1977), and Johannes (1978). Following up on a comment by Randall (1961), all suggested that the upward dash during spawning results from two competing selection pressures. Firstly, the ascent results in the free-floating eggs being released above the bottom. This is thought adaptive in that it minimizes the risk of fertilized eggs being eaten by the numerous filter-feeding invertebrates that make up the reef, e.g., corals and sponges, and by the dense concentrations of site-attached planktivorous fishes, e.g., many pomacentrids and labrids near the bottom. Secondly, the height of the ascent is hypothesized to be constrained by the risk the adults incur while spawning. By ascending into the water column to release gametes, the spawners move away from the shelter of the complex coral and sponge substrate and presumably run a high risk of being attacked due to their increased visibility and vulnerability. Indeed, many of the "predatory events" observed in reef fishes have been attacks on courting or spawning individuals (e.g., Emery, 1973; Robertson & Hoffman, 1977; Moyer & Yogo, in press). Consequently, spawning ascents should be rapid, minimizing time exposed, and just high enough to afford the newly released gametes a relatively low risk of at least immediate predation.

Although the risk-of-predation hypothesis is currently widely accepted to account for the ubiquity of spawning ascents in reef fishes (e.g., Keenleyside, 1979), it has never been adequately tested against the behavior of the fishes involved and against the activity patterns of planktivores and piscivores. Direct experimental evidence of the roles of such factors is likely to be of little value, even if obtainable, since short term effects (such as scaring a spawning individual to see if its spawning behavior can be changed) confuse proximate and ultimate causal factors. Studies involving such things as the behavior of spawning individuals in the conspicuous presence or absence of predators and/or numerous planktivores are likely to be of greater value, but have yet to be done. In the absence of such information, it is nonetheless possible to derive several testable predictions based on the risk-of-predation hypothesis and then compare such predictions to observed behavior. The results of such a comparison provide only a crude measure of the utility of the hypothesis, since they are necessarily based on a number of simplifying assumptions and on estimates of spawning height, rather than actual measurements, but may have value if they indicate potential problem areas.

Four predictions readily come to mind based on the risk-of-predation hypothesis.

1) Daytime spawners should ascend higher into the water column to spawn than dusk spawners. This prediction follows from differences in foraging heights and levels of activity by planktivores at the two times. Specifically, during the day the water column immediately over the reef seems full of actively foraging fishes. As dusk approaches these fishes slowly and steadily descend closer to the bottom and eventually seek shelter in coral crevices for the night, producing an "evening interim period" (Hobson, 1972), during which the water column is virtually empty, before the arrival of the nocturnally active fishes. Species that spawn at dusk, therefore, are likely to face minimal activity from planktivorous fishes, and these planktivores will be concentrated close to the bottom. Hobson (1972), for example, estimated that eight to ten minutes after sunset, roughly the time dusk spawning occurs, the planktivores still active are foraging only 1 to 2 m above the bottom. In contrast, diurnal planktivores characteristically forage up to two to three times this height (Collette & Talbot, 1972; Hobson, 1972; Hobson & Chess, 1978). Emery (1973), for example, provided "normal" and "occasional" foraging heights for the seven common species of planktivorous damselfishes found on coral reefs off Florida (Fig.5). "Normal" heights for most species reach 3 m, while "occasional" foraging heights reach as

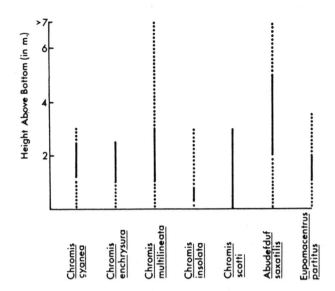

Fig. 5. "Normal" (solid lines) and "occasional" (dashed lines) foraging heights for the seven common species of planktivorous pomacentrids found off the coast of Florida. Based on data from Emery (1973).

much as 7 m off the bottom. If the risk-of-predation hypothesis is correct, therefore, one would expect diurnal spawners to ascend farther off the bottom to spawn than dusk spawners, all else being equal (Fig. 6).

2) Ascent heights, in general, should be positively correlated with adult size. Without question many factors interact to determine the risk of predation run by an individual, such as its habitat preferences, periods of peak activity, and conspicuousness (whatever determines conspicuousness in the eye of the predator; see Endler, 1978). One important factor, however, must be the individual's size. Although there are few hard data regarding the size-frequency distribution of reef-associated piscivores, even casual observations suggest that there are far more small predators than large ones. Consequently, small fishes are likely to run a disproportionately much greater risk of being attacked while spawning than larger ones. Indeed, above a certain size, the chances of being attacked while off the bottom likely approach zero. As an extreme example, it is unlikely that a 3 m jewfish, *Epinephelus itajara*, ever seriously needs to worry about being attacked and eaten at any time, regardless of what it's doing. Ideally, one should be able to estimate this "release-from-predation" size from the relative abundances and sizes of reef-associated piscivores and the size-frequency distribution of fishes in their diets. Again, unfortunately, most such data are not yet available. Some information is available concerning the epinepheline piscivores, at least.

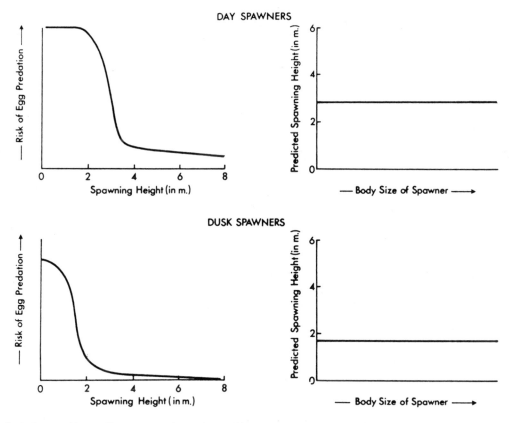

Fig. 6. Theoretical plots and immediate egg survivorship at different spawning heights for day- and dusk-spawning fishes, along with predicted spawning heights.

A size-frequency distribution of the dominant Indo-Pacific piscivore *Plectropomus leopardus* suggests that most large individuals are in the range of 50 to 55 cm SL (Goeden, 1978)(Fig. 7). Generally similar maximum sizes for the dominant piscivorous epinephelines appear to be the case in other major reef areas as well. Maximum sizes for the prey for such fishes, in turn, can be crudely estimated by comments made by Randall (1967), who noted particularly large food items found in several Western Atlantic epinephelines. Specifically, he noted that a deep-bodied

species, an acanthurid, was about 0.29 times the standard length of the fish that ate it; whereas a more slender-bodied species, a synodontid, was just under half (0.48) the length of its captor. Applying these figures to, first, the relative proportions between body depth and standard length for, respectively, the dusk spawners (most of which, like the acanthurid, are relatively deep-bodied; for the 26 dusk-spawning species for which ascent height information is available, average body depth to SL ratio is 0.49) and the day spawners (mainly slender labroids; for 24 species the

Fig. 7. Size distribution of *Plectropomus leopardus,* the dominant Indo-West Pacific epinepheline piscivore at Heron Island, Great Barrier Reef. Figure modified from Goeden, 1978.

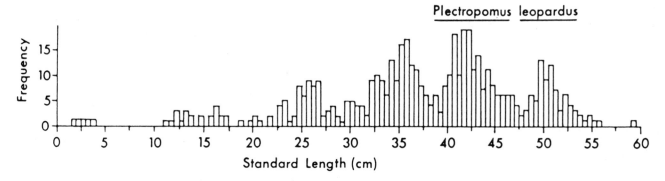

Species of fishes with poisonous spines, such as this *Pterois volitans* is more predator-immune and can ascend as high as necessary to avoid planktivores. Photo by Müller-Schmida.

ratio averages 0.33), and second, an average "large" piscivore of 60 cm SL, one can determine an approximate maximum sized prey of 17 cm for a dusk spawner and 25 cm for a day spawner. These correspond to body areas of roughly 135 cm² and 195 cm², respectively. Body area, rather than either length or depth alone, is likely to be a best indicator of the size of a target, since it is the combination of the two that determines whether a piscivore can choke down an item or not.

Summarizing, the risk-of-predation hypothesis implies, first, that there should be a positive correlation between body size and ascent height, and second, that above some minimum size, roughly estimated above, predator inhibition on spawning adults should be ineffective. The numerous assumptions involved in the latter half of this prediction make it relatively non-robust. Even the first half is not particularly compelling since, even if true, it would hardly be surprising that larger animals take more space to do something than smaller ones.

3) Unpalatable or otherwise predator-immune species, such as cleaners, should spawn higher in the water column than similar-sized "vunerable" species. This prediction follows from the hypothesized inhibitory effect of piscivores on spawning heights. Clearly, if a fish is "immune" to being attacked, it can ascend as high as is necessary to elude planktivores. In practice, only a few groups fit this category: the cleaner wrasses in the genus *Labroides* and various armored and/or poisonous species of ostraciids, pteroids, and scorpaenids. Immunity from attack is

likely to be a relative thing for these fishes, since at least some of them are eaten on occasion (e.g., Randall, 1967; Lobel, 1976).

4) Finally, predictions 2 and 3 above can be equally well cast in terms of spawning speed, since both speed and height determine length of time in the water column and, consequently, exposure to piscivore attack. The fastest moving fishes during spawning, therefore, should be the small ones, whereas large fishes, as well as those "immune" from predators, should spawn notably more slowly.

These predictions are tested against the documented spawning heights of reef fishes (see Figs. 8-10). Only field data are used, with the further stipulations that the observer provided an actual numerical estimate of spawning height and that the species did not "splash" along the surface while spawning, since this suggests that the fish might have ascended higher given the option. Finally, only data for pair spawnings, rather than group spawnings, were used, so as to minimize potentially confounding social effects. Also, if a range of spawning heights was provided, a mean height was calculated to typify the species. Information on fish sizes, if not provided by the observer, was obtained from the literature. In all, data are available for 24 species in six families of day spawners and 26 species in 12 families of dusk spawning fishes.

Because of the imprecision involved in numerous observers independently estimating ascent heights while lying on the reef, one cannot get too excited about slight differences between species. On a broad level, however, several points emerge that are relevant

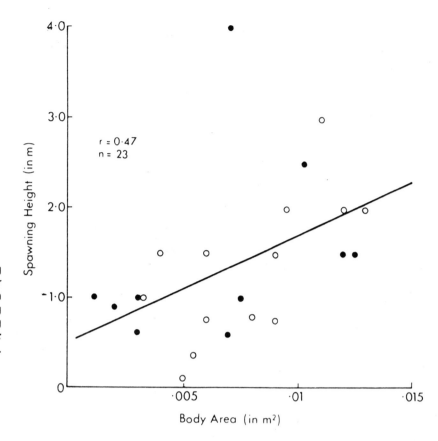

Fig. 8. Correlation between spawning height and body size for species smaller than calculated release-from-predation sizes. Open circles = day-spawning species; solid circles = dusk-spawning species. *Least Squares Regression Line:* $Y = 0.52 + 119.4X$; $r = 0.47$, $p < 0.05$.

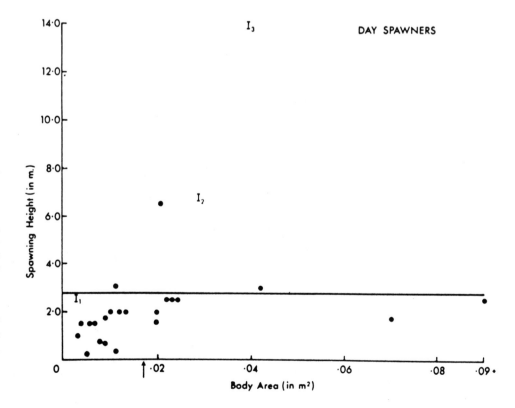

Fig. 9. Spawning heights for 24 species of day-spawners in six families. Solid horizontal line indicates minimum spawning height predicted on the basis of planktivore foraging heights. "Predator-immune" species are indicated by an I. I_1 = *Labroides dimidiatus*; I_2 = *Ostracion meleagris*; I_3 = *Lactoria diaphanus*. Arrow indicates predicted release-from-predation size.

363

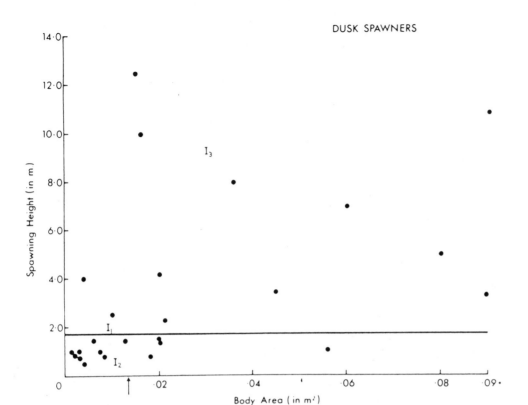

Fig. 10. Spawning heights for 26 species of dusk-spawners in 12 families. Solid horizontal line indicates minimum spawning height predicted on the basis of planktivore foraging heights. "Predator-immune" species are indicated by an I. I₁ = *Lactoria fornasini;* I₂ = *Dendrochirus zebra;* and I₃ = *Diploprion bifasciatum.* Arrow indicates predicted release-from-predation size.

to the predictions above. 1) For fishes smaller than the estimated "release-from-predation" sizes, there is a positive correlation ($r = 0.47$, $p < 0.05$, $n = 23$) between ascent height and body size, as predicted in #2 above (Fig. 8). 2) For the same group of fishes, however, there is no significant difference between ascent heights for dusk and day spawning fishes (for the slope, $F_{1,19} = 0.48$, $p > 0.25$; for the intercepts, $F_{1,20} = 0.57$, $p > 0.25$). 3) Again, contrary to predictions, dusk spawners in general ascended higher to spawn than day spawners ($\bar{x} = 3.62$ m and 1.91 m, respectively) (Figs. 9 & 10). 4) Day spawning "immune" species ascend to heights well above those of predator-vulnerable comparable sized species, whereas at dusk there is no apparent difference between the two groups in terms of ascent height. 5) There appears to be an abrupt increase in ascent height at the estimated "release-from-predation" size for dusk spawning fishes; such an effect is conspicuously absent for spawning fishes after "immune" species are excluded. 6) Finally, there is only one accurate measurement of ascent speed for spawning reef fishes (Colin, 1978), but general observations are consistent with the predictions regarding ascent speed. Among dusk spawning species, very small fishes (e.g., cirrhitids, anthiines, small serranines)

spawn so rapidly they are difficult to follow by eye. In contrast, large species frequently spawn quite slowly, often, like the larger pomacanthids and chaetodontids, courting off the bottom as well. Medium-sized dusk spawners, in general, exhibit a range of spawning speeds. Some, like the acanthurids, move rapidly up and down, whereas others, like the large serranines and small pomacanthids, are relatively slow spawners. Most day spawning species, in contrast, regardless of their size, spawn at extremely high speeds. Colin (1978), based on a frame-by-frame analysis of movie film, calculated that the moderate-sized day spawning parrotfish *Scarus croicensis* reaches a speed of nearly 40 km/hr during its ascent. Other day spawners appear to reach comparable speeds, with two exceptions. The unpalatable trunkfish *Lactoria diaphanus* is probably the slowest of all pelagic spawning fishes, taking several minutes to complete its ascent (J. Moyer, pers. comm.). Similarly, the equally unpalatable porcupine-fish, *Diodon holacanthus,* also ascends slowly and spends more than a minute splashing along the surface while shedding gametes.

In general, these results bear out predictions relating to the effects of piscivore inhibition, but do not support predictions derived from consideration of planktivorous fishes as predators on fish eggs. Regarding the latter, two pieces of evidence seem to be particularly damning: small diurnal and dusk spawners do not differ in ascent heights, although planktivore foraging activities appear to be much more intense during the day and extend to greater heights off the bottom; and day spawners, in general, spawn at heights well within the foraging ranges of most day-active planktivorous fishes, i.e., even the larger species do not appear to be "attempting" to spawn above such planktivores. If planktivorous fishes are not important egg predators, then why should reef fishes ascend at all when spawning, especially given that they are probably exposed to high risks of predation at such times? Two hypotheses seem reasonable. First, although fishes may not pose a significant threat to free-floating fish eggs, benthic invertebrates (especially those that filter feed) may take such eggs. Spawning ascents may be a means of releasing gametes at a height above the reef that is sufficient for such eggs to achieve positive buoyancy before drifting within reach of the many filter-feeding corals, sponges, and echinoderms that abound on the reef. A second hypothesis is that spawning ascents, and also a general tendency in reef fishes to begin their ascent from a particularly high spot on the reef, may be adaptive in that they result in the eggs being transported rapidly offshore (see also, however, Thresher, 1982, for discussion of an alternative hypothesis regarding use of high spots for spawning), not because the reef is such a particularly dangerous place for an egg to be (there is no evidence that offshore waters are any more benign for such eggs), but rather because there may be strong selection to place the eggs where they will be in an area of high food concentrations upon hatching and subsequent first feeding. Eggs trapped in eddies on the reef and transported offshore late in their development may be in relatively food-poor waters during the critical period of first feeding. The assumptions underlying this hypothesis, as well as the former hypothesis, are badly in need of careful field testing.

The data also suggest that, contrary to widespread opinion, risks of predation on reef fishes may be higher during the day than at dusk. Three lines of evidence specifically suggest this conclusion: dusk spawners, in general, ascend higher to spawn than day spawners; dusk spawning, in general, is slower than that during the day; and the predicted effects of predator "immunity" are evident only during the day. Other data also support the idea that day may be a time of particularly intense predator activity. Gut

content analysis of piscivores, for example, often documents feeding at uniform levels throughout the day rather than the dusk peaks in activity commonly assumed to be the case (e.g., Hobson, 1974; Goeden, 1978). Several families of piscivorous fishes, such as the synodontids (Hobson, 1974) and sphyraenids (de Sylva, 1963; Randall, 1967), are wholly or largely active only during the day. Finally, a number of supposedly dusk-active piscivores spawn at dusk (e.g., serranids, aulostomids) and are often preoccupied with their own courtship activities at such times rather than, as widely assumed, hunting other fishes. Overall, such data suggest that the widespread assumption of high levels of piscivore activity at dusk should be more critically examined before it is widely applied to explanations of spawning behavior or timing of reef fishes or their behavior and ecology in general.

Again, it must be emphasized that this aspect of reef fish reproductive behavior is badly in need of well-designed field studies. The analysis above is, unfortunately, a crude one. It is, however, all that is currently possible given the limited data available. I suggest that it casts reasonable doubt on the role of, at least, planktivorous fishes in the evolution of reef fish ascent behavior and consequently that caution be used in applying this as an explanatory principle for the behavior observed.

TIMING OF SPAWNING

Diel

It is a general observation among divers that one rarely sees spawning during the day. To a large extent this explains why reproduction by even the most ubiquitous fishes had not been described until recently, when biologists began diving intensively at dawn and dusk. In order to analyze more precisely the diel patterns of spawning activity in reef fishes, I have plotted the distribution of spawning times by demersal and pelagic egg producing families. Reported spawning times were summed into seven classes (from Pre-dawn to Night). Each family was assigned a value of 1.0 and then divided into the various classes on the basis of the distribution of spawning times reported for the family. (Fig. 11). Damselfishes, for example, are primarily pre-dawn spawners but have also been reported spawning in late morning and early afternoon; the 1.0 value for the family, consequently, is distributed 0.9 to Pre-Dawn, 0.05 to Late Morning and 0.05 to Early Afternoon. The figures were produced by summing such

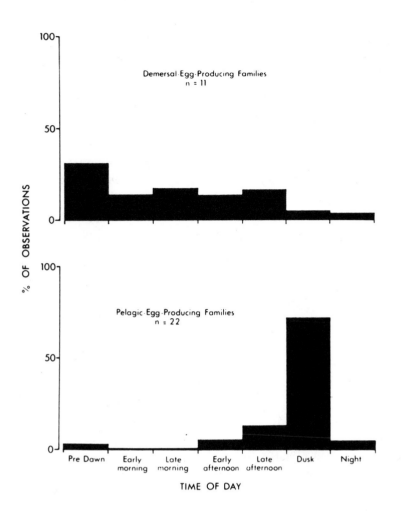

Fig. 11. Distribution of spawning times for demersal and pelagic egg-producing families.

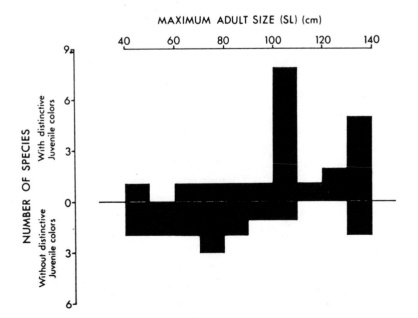

Fig. 12. Effect of maximum adult sizes of 36 benthic, territorial species of damselfishes on the occurrence of conspicuous juvenile-specific color patterns.

values for all families in each of the two broad classes for all seven spawning periods. The only fishes excluded from the analysis are those labrids and scarids that spawn at high tide, apparently irrespective of time of day. These fishes are discussed separately below.

There is a striking and highly significant difference between the distributions of spawning times for demersal egg producing and pelagic egg producing families ($X^2 = 107.6$, df = 6, p<<0.01) (Fig. 11). The former essentially spawns throughout the day with a small peak in activity prior to sunrise and a marked decrease at dusk and at night. The pelagic egg spawners, in contrast, have a pronounced peak in activity at dusk, i.e., after sunset and before complete darkness, though some spawning also occurs at late afternoon, early afternoon, and at night. There are only two reports of spawning by a pelagic egg producing species prior to noon—*Acanthurus lineatus* (Johannes, 1981) and *Diodon holacanthus* (in this volume), both of which spawned before sunrise.

The function of concentrated spawning at dusk for pelagic egg producers has been discussed by Lobel (1978) and Johannes (1978), both of whom argued that such timing is a response to low levels of planktivore activity at dusk and, possibly, low levels of piscivore activity as well. This argument was analyzed above, where it was concluded that, if anything, levels of piscivore activity are lowest at dusk. This being the case, then, it is somewhat puzzling why any pelagic egg producing fishes would spawn during the day. Such daytime spawning has been reported for one species in each of four families (Acanthuridae, Muraenidae, Diodontidae, and Serranidae) and is characteristic, though in at least one case not universal, of three others (Ephippidae, Labridae, and Scaridae). Information on the first four families is too fragmentary to discern any patterns, and, of the second three, only the labrids and scarids deserve detailed discussion. The single report of spawning by an ephippid had it occurring well offshore and far from the reef. The labroid fishes, however, spawn immediately over the reef and in a manner identical to that of many dusk spawning species and so should be subject to a similar selective regime. That such fishes continue to spawn during the day, despite the apparent disadvantages, suggests either an inability of the fishes to respond to selection for dusk spawning or the action of some counterbalancing factor. The first explanation is unlikely, both on general grounds (if everything else can adapt, why not labroids?) and

because a few primitive labrids do spawn at dusk. The species in the genus *Bodianus*, for example, all spawn after sunset, so far as is known. *Bodianus* is a member of the primitive Hypsigenyini (Gomon, 1979), which suggests that dusk spawning may be the primitive condition in the group.

One possible selective advantage that could be utilized by these day spawning fishes is the option of spawning when current patterns are optimal for transport of eggs away from the reef. A dusk spawner, constrained to spawn during the half hour or so between sunset and darkness, has an optimal offshore current at dusk roughly every two weeks (just as the tide begins to ebb). In contrast, a day spawner can track the tide as its time of peak height changes through the day, permitting it to spawn in optimal current conditions almost daily. Such tracking of tidal currents was suggested by numerous workers, including Randall (1961), Choat & Robertson (1975), and Johannes (1978). Data on the overall spawning patterns of labroid fishes, however, indicate that less than half adjust their spawning times to follow the tide. Most scarids are reported spawning in mid-afternoon, while labrids consistently spawn at all times of the day from mid-morning until dusk. On reefs in the Gulf of California, for example, some species spawn at mid-morning, others at dusk, and still others at high tide, even though all are sympatric and subject to the same current regime. If optimization of egg transport is an important counterbalance to the apparent disadvantages of day spawning, then it is not one that is taken advantage of by most day spawning species. Nonetheless, it can not be discounted for those species that do track the tides. In areas of very strong tidal currents, even some fishes that otherwise "always" spawn at dusk have been reported spawning during the day at high tide, e.g., *Parupeneus cyclostomus* and, possibly, *Acanthurus lineatus*.

What other factors day spawners may be taking advantage of can only be speculated upon. Robertson & Hoffman (1977) suggested that various species spawn at different times due to differences in the risk of predation encountered by spawning adults in each, based on an assumption that risk of predation, in general, increases toward dusk. Presumably the more predation-prone a species, the closer to midday it will spawn. This hypothesis has yet to be critically evaluated, but at least superficially does not appear to be borne out.

Spawning times by demersal spawning fishes have

received less attention in the literature than those of pelagic spawning species, perhaps because they are less conspicuous while spawning. The relative evenness of spawning times throughout the day suggests that whatever is selecting for their timing is either weak or is uniformly manifest throughout the day. The decrease in the frequency of spawning at dusk may be due to decreasing light levels that limit the effectiveness of visual signals involved in courtship interactions. The only other trend apparent in the data is for larger fishes, e.g., balistids, siganids, and pomacentrids, to spawn earlier in the day than other families. This trend is hardly significant, however, and may be an artifact of the sample size.

Despite this lack of overall trends, there appear to be substantial differences between families of demersal spawners in terms of their spawning times (e.g., most pomacentrids spawn beginning at first light, whereas most blennioids, so far as is known, spawn at midday); the significance of such differences is not known but may relate to incubation period and selection for some optimal hatching time.

Demersal spawners, in fact, are united by the timing of egg hatching. In all families for which data are available (e.g., blenniids, balistids, pomacentrids, pseudochromoids, siganids), hatching occurs at dusk. Several workers, e.g., Allen (1972), Fricke (1974), Lobel (1978), and Johannes (1978), suggested that dusk hatching results in the larvae being transported offshore when planktivore activity is lowest, i.e., the same argument applied to timing of pelagic spawning. Again, there is no direct evidence to support the hypothesis, but it at least sounds reasonable.

Lunar

The distribution of lunar cycles to spawning activity, if any, in coral reef fishes is given in Table 3. Data for many families are still weak or lacking, but even given the limited data base several trends are evident.

1) All but two groups of fishes that spawn daily or irregularly are pelagic spawners, and most are relatively small and site-attached. Spawning

Table 3. Distribution of lunar spawning cycles in reef fishes.

No Lunar Cycle	Semilunar Cycle	Lunar Cycle
Acanthuridae (non-migratory)	Some Apogonidae	Acanthuridae (migratory)
Anthiinae(?)	Some Balistidae	Some Apogonidae
Antennariidae	Some Blenniidae*	Some Balistidae
Some Blenniidae	Some (probably most)	Carangidae(?)
Chaetodontidae*	Pomacentridae	Some Chaetodontidae(?)
Some Epinephelinae	Pseudochromoids*	Most Epinephelinae
Some (probably most) Labridae	Some Pteroidae	Some Labridae(?)
Ostraciidae	Opistognathidae*	Lutjanidae
Some Pomacentridae		Mugilidae(?)
Pomacanthidae		Mullidae(?)
Most Pteroidae		Some Pomacentridae
Serraninae		Large Scaridae(?)
Scaridae		Siganidae(?)
		Sphyraenidae(?)
		Sparidae(?)

? = circumstantial or unconfirmed evidence

* = based on aquarium observations

migrations have been reported in only two of these families, the epinephelines and scarids. The only demersal spawning families in this class are the Pomacentridae, which is represented by only two species out of the many that have been examined, and the Blenniidae, represented by only one species.

2) All but one group of fishes with a semilunar (i.e., biweekly) cycle of spawning activity are demersal spawners. The only exceptions are the pteroids, based on one species (*Dendrochirus zebra*) which apparently has a weak semilunar cycle (Moyer & Zaiser, 1981); all other pteroids, so far as is known, lack a lunar spawning cycle (Fishelson, 1975).

3) Lunar cycles of spawning activity are characteristic of large migrating pelagic spawners and a few demersal spawners.

A variety of hypotheses have been offered to account for lunar cycles of spawning activity in reef fishes (Allen, 1972; Fricke, 1974; Lobel, 1978; Johannes, 1978; Ross, 1978; and Pressley, 1980). Hypothesized advantages for such cycles include the following. 1) Hatching or spawning at dusk on the strongest outgoing tide, which occurs biweekly, minimizes predation on eggs and larvae and facilitates transport offshore. 2) Such a cycle minimizes larval dispersal. 3) Many invertebrates spawn on a lunar cycle; consequently having your eggs hatch then maximizes the density of food available to the newly hatched larvae. 4) Having eggs hatch on the night of the full moon results in the photopositive larvae swimming toward the brightly illuminated surface and away from the predator-filled reef. 5) Similar lunar illumination at the full moon facilitates care of eggs at night or early morning spawning. 6) The lunar cycle is just a convenient clock; the important thing is that spawning is synchronized, overwhelming predators.

Of these, number five seems immediately unlikely, since: a) nocturnal care of eggs has not been reported in any species; and b) most fishes, either demersal or pelagic spawning, do not spawn at first light. Even early morning spawning damselfishes do so, generally, well after the sun's illumination is dominant. Lobel (1978) made the point that none of these hypotheses are mutually exclusive and any combination of them may result in whatever pattern a given species manifests, if any. Unfortunately, it also makes it difficult to falsify any of them. To date, the overall con-

census seems to lean strongest toward the first hypothesis, that lunar spawning cycles minimize immediate larval and egg predation and facilitate their transport offshore, although both Fricke (1974) and Pressley (1980) presented evidence that such may not be the case in their respective species.

The distribution of lunar cycles in reef fishes bears on the hypotheses above in two ways. First, that semilunar cycles are characteristic of demersal spawners but not site-attached pelagic spawners argues strongly that any hypothesis that applies equally well to both groups is unlikely to account for the evolution of semilunar cycles. The fundamental difference between the two is that one produces a free-swimming, relatively advanced larva, whereas the other launches a passively drifting egg. The adaptive significance of semilunar cycles of spawning activity lies somewhere in the ecology of these newly hatched larvae. Of the hypotheses offered thus far, all but one appear to apply equally well to the two groups of fishes and hence are not likely to be involved (unless one makes additional assumptions about, for example, the relative risk of predation run by an egg and a larva). The only hypothesis that appears to apply only to demersal spawners, unfortunately, predicts a lunar cycle rather than a semilunar one. If surface illumination is an important factor in attracting the positively phototropic larvae to the surface and away from the planktivores below, then it certainly isn't much of a factor on the new moon peaks, when the night is at its darkest each month. Consequently, I suggest that none of the hypotheses currently offered to "explain" semilunar cycles does so adequately and that the significant factors involved have yet to be discovered.

The second point evident from Table 3 is that the occurrence of a lunar spawning cycle in pelagic spawners is very much a function of whether or not the fishes migrate to specific spawning grounds. Virtually every pelagic spawning species reported to have such a lunar cycle migrates (even the demersal spawning siganids), whereas comparable sized species that do not migrate lack a lunar cycle (e.g., acanthurids). Johannes (1978) suggested that this lunar cycle results from the same selective pressures that result in the spawning migration, i.e., situating the eggs and larvae in the optimal location for transport offshore and subsequent return to the reef. The table appears to support at least the linkage of the two behaviors, whatever their combined function may be, and further suggests that lunar spawning cycles, in at least pelagic spawning fishes, are not adaptive without spawning migrations.

Seasonal

Duration of spawning seasons of reef fishes was reviewed by Johannes (1978) and Sale (1978, 1980), both emphasizing, first, that such fishes generally have long spawning seasons, and second, that within these seasons there tend to occur one or more peaks of activity. Such peaks are often synchronous in a given locale across a wide variety of fishes, suggesting the action of some pervasive factor whose timing varies geographically. Johannes (1978) presented a solid case for seasonality to be a function of selection for spawning during periods of minimal currents offshore so that larvae have the highest possibilities of developing and metamorphosing into juveniles without being carried away from suitable substrates. Lobel (1978) made a similar argument for Hawaiian reef fishes, following up upon earlier suggestions by Sale (1970) and Leis & Miller (1976).

Family by family surveys of the spawning seasons of coral reef fishes indicate, first, general agreement with the arguments above, and second, no obvious patterns in the data. Demersal and pelagic spawners, for example, appear to have comparable length spawning seasons, both involving considerable family to family variation, and both appearing to spawn at about the same times of the year.

SEXUAL DIMORPHISM

Sexual dimorphism is a widespread, though hardly universal, characteristic of reef fishes, with three broad categories of fishes readily identifiable: permanently dimorphic fishes, permanently monomorphic species, and temporarily dimorphic fishes, the last typically involving the development by either or both sexes of sex-specific color patterns during courtship and spawning. About equal numbers of fishes are found in each of these major categories (Table 4), reflecting the diverse reproductive strategies evolved by these animals. A number of species also bracket two categories; that is, they are both permanently sexually dimorphic and also develop male or female specific courting colors. Examples of such fishes include a number of labrids, pomacanthids, and acanthurids.

In general terms, sexual dimorphism is widely considered an indication of habitat partitioning between the sexes (especially where such differences relate to feeding mechanisms or size); differential growth rates or age at maturity (usually with respect to sexual size dimorphism); or polygamy, the result of intrasexual selection for mates and/or intersexual (epigamic) selection of mates (see review by Ralls, 1977). Conversely, sexual monomorphism is characteristic, though not

Table 4. Taxonomic Distribution of Sexual Monomorphism and Dimorphism in Reef Fishes

Permanently Monomorphic		Temporary Dichromatism			Permanently Dimorphic
	Male Only	*Female Only*	*Both Sexes*		
Most Amphiprioninae	Some Acanthuridae	Some Diodontidae	Some Pomacentridae		Some Acanthuridae
Antennariidae	Some Apogonidae	Some Pomacentridae(?)	Pteroidae		Some Aulostomidae(?)
Some Apogonidae(?)	Blenniidae	Some Apogonidae	Some Pomacanthidae		Some Balistidae
Branchiostegidae	Some Carangidae				Callionymidae
Chaetodontidae	Cirrhitidae(?)				Some Chaenopsidae
Most Gobiidae	Epinepheline Serranids				Some Clinidae
Some Grammistidae	Some Labridae				Epinepheline
Some Labridae	Some Ostraciidae				Serranids
Lutjanidae	Most Pomacentridae				(Size Only)
Mugilidae	Siganidae				Some Grammistidae
Muraenidae	Some Sparidae				Most Labridae
Most Opistognathidae	Tripterygiidae				Mugiloididae
Many Pomacanthidae					Some Opistognathidae
Most Pseudochromoids					Some Ostraciidae
Some Sparidae(?)					Some Pomacanthidae
Synodontidae					Some Pseudochromoids
					Most Scaridae
					Anthiine Serranids

diagnostic, of monogamy, since with the development of long-term pairs sexual selection is presumably reduced or at least is not a constant feature in the mating systems of the animals.

There is to date no evidence relating sexual dimorphism to habitat partitioning between the sexes for any species of reef-associated fish. It may be occurring in some of the sexually dichromatic pseudochromids (e.g., *Pseudochromis fuscus*) found on the Barrier Reef, where one sex appears to aggressively mimic one species and the other mimics some other species, but the detailed observation to support such speculation has not been done. Sexual bimaturism, however, involving differential ages at maturity or growth rates, has been suggested for a number of fishes, including serranids, lethrinids, and scarids. The greatest interest, however, has been on the relationship between social organization and degree of sexual dimorphism. Perhaps the best example of such a relationship to date involves the anemonefish *Amphiprion clarkii*, studied by Moyer (1976, 1980) and Moyer & Bell (1976). At Miyake-jima, Japan, the species is sexually dichromatic and at least occasionally polygamous (the latter apparently due to relatively low levels of predation, facilitating inter-anemone movements). In contrast, most other anemonefishes, and *Amphiprion clarkii* in other areas, are sexually monochromatic, paired, and widely assumed to be (in a few cases demonstrated to be, e.g., Fricke, 1974) monogamous (Allen, 1972; Moyer, 1976).

The correlation between permanent sexual monomorphism and pairing in reef fishes is complicated by a factor not common in other vertebrate groups, sequential hermaphroditism. As recently emphasized by Lassig (1976), Fricke & Fricke (1977), and Moyer & Nakazono (1978b), scattered, small, and specialized microhabitats combined with high risks of predation between habitats can logically lead to both permanent pair formation and sequential hermaphroditism, the latter usually involving social control of sex and, consequently, sexual dimorphism based on often slight differences in size. A similar situation occurs in many harem-forming species (e.g., Robertson & Hoffman, 1977; Moyer & Nakazono, 1978a). It seems noteworthy in this regard, however, that even though slightly dimorphic with respect to size, conspicuously paired hermaphroditic species show little or no additional sexual dimorphism or dichromatism (e.g., Fricke, 1974; Lassig, 1976; Moyer & Nakazono, 1978b), whereas harem-forming species are frequently, though not always, sexually dichromatic as well

(e.g., Winn, 1960; Robertson & Warner, 1978; Warner & Robertson, 1978; Moyer & Nakazono, 1978a; Thresher, 1982; see, however, the discussion in Robertson & Hoffman, 1977).

Sexual monomorphism is not restricted to pair-forming species, however, but also occurs widely among individually territorial reef fishes. This could result from factors such as predator pressure, and for some species such selection pressures are conspicuous. Cleaner fishes, for example, regardless of their spawning system, may be subject to selection for morphometric and chromatic conservatism since an oddly colored cleaner may not be readily recognized as such by potential hosts. Similarly, for midwater foraging species the advantages of uniform appearance as an antipredator device and the selection against oddly colored individuals in a school (see Davis & Birdsong, 1973) may well outweigh any selection for permanent sexual dimorphism in these demonstrably polygamous species (e.g., Albrecht, 1969; Fishelson, 1970). Indeed, extreme predator pressure on reef fishes in general could result in reduced permanent dimorphism, even among non-schooling species. This, however, seems unlikely. The wide-ranging morphological and chromatic variation between species and the often brilliant and apparently conspicuous colors of even monomorphic species (see Brockmann, 1973; Thresher, 1977b) suggest predator constraints on coloration to be minimal. Similarity in size, color, and shape among members of a species could also facilitate intraspecific recognition and, in the case of individually territorial species, intraspecific territoriality (see Geist, 1975, 1977). The widespread occurrence of interspecific territoriality in reef fishes (e.g., Colin, 1971a; Low, 1971; Myrberg & Thresher, 1974; Thresher, 1976a & b) and the apparent ease with which such fishes recognize and respond appropriately to even very dissimilar appearing fishes (e.g., Thresher, 1979a), however, suggest that such selection is weak.

There are, in fact, a number of permanently strongly sexually dimorphic reef fishes, some of which are territorial against both similar and dissimilar conspecifics (e.g., some scarids, some chaenopsids, some clinids). Such permanent dimorphism includes color (e.g., some labrids—Roede, 1972; Robertson & Hoffman, 1977), size (e.g., some serranids—Smith, 1965), or both (e.g., some scarids—Rosenblatt & Hobson, 1969; Randall, 1963). Where their spawning systems have been examined all such species, with the exception of the paired sequential hermaphrodites noted

Table 5. Strongly Sexually Dimorphic Fishes

Family	Reference	Spawning System	Reference
Most Scaridae	Choat & Robertson (1975); Randall (1963b)	Haremic; Polygamous	Choat & Robertson (1975); Winn & Bardach (1960); Buckman & Ogden (1973)
Most Labridae	Roede (1972); Robertson & Hoffman (1977)	Haremic; Polygamous	Randall & Randall (1963); Robertson (1972); Warner, Robertson & Leigh (1975)
Epinepheline Groupers	Smith (1965); Colin (pers. comm.)	Polygamous	Smith (1972)
Some Acanthuridae e.g., *Naso*	Barlow (1974)	Haremic; Polygamous	Barlow (1974); Robertson, et al. (1979)
Some Chaenopsidae e.g., *Emblemaria*	Stephens (1970)	Polygamous	Wickler (1967a); Stephens, et al. (1966)
Some Clinidae e.g., *Malacoctenus*	Springer (1959)	?	
Some Ostraciidae e.g., *Ostracion meleagris*	Randall (1972); Fraser-Brunner (1940)	Haremic	Moyer (1979)
Some Pomacanthidae e.g., *Holacanthus passer*	Strand (1978)	Polygamous	This volume
Genicanthus spp.	Randall (1975)	Polygamous	Suzuki, et al. (1978) (aquarium)
Some *Centropyge*	Moyer & Nakazono (1978a)	Haremic	Moyer & Nakazono (1978a)
Some Opistognathidae e.g., *Opistognathus gilberti*	Bohlke & Chaplin (1968)	?	
Small Grammistidae e.g., *Pseudogramma polyacanthus*	This volume	?	
Anthiines	Popper & Fishelson (1973)	Polygamous	Popper & Fishelson (1973)
Some Balistidae e.g., *Sufflamen verres*	Berry & Baldwin (1966)	Haremic	This volume
Some Pseudochromoids e.g., *Anisochromis straussi*	Springer, et al. (1977)	Paired ?	Lubbock (1975)
Callionymidae	Akazaki (1957)	Polygamous	Takita & Okamoto (1979) Zaiser (in prep.)

above, have proved to be polygamous or promiscuous (Table 5). Both intrasexual competition and intersexual selection have been cited to account for such dimorphism (Choat & Robertson, 1975; Robertson & Hoffman, 1977; Thresher & Moyer, in press).

Two hypotheses readily come to mind to account for the general lack of permanent sexual dimorphism in individually territorial or home-ranging fishes. First, given that such dimorphism derives from strong sexual selection, it is possible that such selection is, in fact, relatively weak in these fishes. Data to support or refute this hypothesis are, to date, limited. As noted earlier, there has long been a general assumption that individually territorial or home-ranging reef fishes are

fundamentally promiscuous, and should consequently be subject to potentially significant levels of sexual selection. This, however, may not always, or even often, be the case for two reasons. The first was discussed above. At least some of these fishes appear to exhibit preferred mate polygamy, and others, such as some pipefishes, serranines, and opistognathids, appear to form long-term pairs and may be monogamous. The general lack of widespread permanent sexual dimorphism may well reflect the occurrence of a good deal more mate fidelity than these fishes are generally given credit for. The second reason why spawnings may not be entirely promiscuous is limited population sizes. Although it is often true that some species are present in large numbers, in other cases population densities can be quite low. Recent work (e.g., Sale, 1980; Doherty, 1980) suggests that populations of some fishes are, in fact, recruitment-limited, such that not all available habitat space is occupied. Especially on Indo-Pacific reefs, where species diversity is high and absolute density of any single species in a guild may be low, it may often happen that the number of potential mates for an individual is limited. How often this occurs, if at all, is not known, but at least theoretically it could lead to limited mate choice and low levels of sexual selection.

The second hypothesis that can account for the widespread absence of permanent sexual dimorphism in many reef fishes is quite different. Fishes, unlike most "higher" vertebrates, are often capable of rapid and striking temporary color changes, such that the males, females, or in some cases both sexes of many species (e.g., many pomacentrids, pomacanthids, blennioids, and acanthurids) assume temporary "courtship" colors. Such temporary colors may be a means of achieving sexual dimorphism when needed and sexual monomorphism when it is not. Robertson & Hoffman (1977) emphasized this point for western Atlantic labrids (all of which are conspicuously polygamous), pointing out that temporary sexual dichromatism is characteristic of those species found in high predator-risk areas such as grass beds, sand flats, and open water. Similarly, the water column foraging damselfishes in the genera *Chromis* and *Abudefduf* are probably under strong selection to maintain monomorphism during nonspawning periods but often develop conspicuous sexual dichromatism during the typically brief spawning periods (e.g., Fishelson, 1970; Sale, 1971); these fishes are also conspicuously polygamous. For more reef-associated species, however, the role of temporary sexual dichromatism as a direct substitute for permanent dimorphism is less clear. As emphasized by Robertson & Hoffman (1977), the nearby and constant presence of abundant shelters from predators should result in little selective pressure against the development of permanent sexual dimorphism. Indeed, since, as is widely accepted, reef predators are most active during low-light periods (see Hobson, 1968) and least active during daytime periods, when color signals would be most conspicuous (see Thresher, 1977b), predator pressure for sexual monomorphism should be relatively slight. Further, even if such pressure existed selection for crypsis does not preclude sexual dichromatism. As recently pointed out by Endler

(1978), "All of the color pattern types of a given species could be equally cryptic if each is a random sample of the background as the predators see it; they are simply different random samples." The argument that predator pressure generally precludes sexual dichromatism confuses the issues of dichromatism and conspicuousness. The selective factors underlying each are likely to be quite different, given that one relates to sexual identification and the other with long-distance advertisement that may or may not be courtship-related. Given the abundance of physical refugia from predators, low or at most moderate predator pressure when chromatic signals are most conspicuous, and the visual complexity of the reef, facilitating crypsis, there would seem to be little to prevent the development of permanent sexual dichromatism if sexual selection and intrasexual competition, resulting from widespread polygamy, were prevalent. Yet such permanent dimorphism is not the rule. Invoking heavy predator pressure to "explain" temporary dichromatism as an enforced substitute for permanent dichromatism seems unworkable.

I suggest that a more likely role of temporary spawning colors in such species is communication of readiness to spawn. For many such species the temporary courtship color is only one of several color patterns that the males (and often the females) can assume; each pattern is closely associated with specific motor, and occasionally sonic, patterns and specific situations, suggesting that each represents the manifestation of a different motivational system and serves in intraspecific communication (e.g., Myrberg, 1972; Wickler, 1967). Bottom-associated damselfishes, for example, besides a "normal" color pattern and a courtship pattern often have specific and different patterns associated with aggressive activity, submission, and fright (e.g., Myrberg, 1972; Fricke, 1973b; pers. obs.). Similarly, variable color patterns occur in at least blenniids and many gobies, that is, in other fishes closely associated with the substratum where the enhanced risk of predation that may derive from "flashing" suites of different color patterns is likely minimal.

Several lines of evidence support this hypothesis and suggest a non-equivalence of temporary and permanent sexual dimorphism.

First, if temporary dichromatism was simply a substitute for permanent dimorphism and derived from the same selective regime, i.e., strong sexual selection, one would predict it to occur principally, if not exclusively, in promiscuous fishes. In fact, it is not so distributed. Aside from numerous cases of its coincidence with demonstrably promiscuous spawning systems (as in some epinephelines and acanthurids), temporary sexual dichromatism also occurs in many harem-forming species (e.g., labrids – Robertson & Hoffman, 1977; Thresher, 1978b; and pomacanthids – Moyer & Nakazono, 1978a; Thresher, 1982) and even some pair-forming and potentially, if not actually, monogamous species (e.g., some syngnathids – Gronell, in press; at least one pomacentrid – Thresher, in prep.; and possibly numerous apogonids). In these latter cases, sexual selection is presumably weak, or at least limited in extent and duration. Along similar lines, the invocation of strong sexual selection to account for temporary sexual dichromatism also suggests that the latter be characteristic of males, except in some apparently sex-role reversed fishes like apogonids and some syngnathids. Again, although "courtship colors" are often characteristic of males, as in many pomacentrids and blenniids, they have also been widely documented in females, both coincident with and independent of its occurence in the males. Examples include at least some labrids (Robertson & Hoffman, 1977; for a temperate example see Olla, et al., 1981), pomacentrids (Fricke, 1973b), pteroids (Moyer & Zaiser, 1981; Gronell, pers. comm.), pseudochromids (this volume) and numerous pomacanthids (Bauer & Bauer, 1981; Thresher, 1982; Moyer, et al., in press). That female-specific courtship colors are not even more prevalent may well reflect the often conspicuously egg-swollen abdomens of many such females, an immediately recognizable and unmistakable indication of her readiness to spawn.

The second line of evidence suggesting a non-equivalence of temporary and permanent sexual dimorphism is their co-occurrence in a variety of fishes. If temporary dichromatism is simply a predation-enforced replacement for permanent dichromatism, then why should both occur simultaneously? Not only is such co-occurrence redundant, but the addition of unneccesary temporary color changes could well cost both physiologically and in terms of increased conspicuousness to predators. Yet, permanently dimorphic species are often also temporarily dichromatic. Examples include many labroids (Robertson & Hoffman, 1977, where it is referred to as "courtship intensification"), pomacanthids (reviewed by Thresher, 1982), and at least one conspicuously dichromatic damselfish (Thresher & Moyer, in press), as well as at least some aulostomids,

syngnathids, acanthurids and pseudochromids. The list is even longer if one includes sexual size differences as permanent sexual dimorphism. The different but overlapping distributions of temporary and permanent dimorphism suggest that each derives from a different selective regime and serves a different function.

The final line of available evidence that suggests non-equivalence of temporary and permanent sexual dimorphism is a field study by Thresher & Moyer (in press), who examined courtship behavior, social organization and, especially, patterns of male spawning success in a group of related damselfishes that differed in the type of sexual dichromatism manifested. One species is permanently monochromatic, one is temporarily dichromatic only, and the third is both permanently and temporarily dichromatic. Theory predicted that male spawning success in the permanently dichromatic species would be much more skewed in its distribution among males than in the monochromatic species, reflecting very different levels of sexual selection. Data were sought to test this prediction and also to see where the temporarily dichromatic species fit in. The data obtained, in fact, permitted two robust conclusions. First, the patterns of male success in the permanently dichromatic and monochromatic species differed exactly as predicted. The variance of male success in the temporarily dichromatic species, however, was very different from that of the permanently dichromatic one and, in fact, wasn't significantly different from that of the monochromatic species. Limited data for other temporarily dichromatic fishes was consistent with this result. In terms of intensity of consistent sexual selection, at least, temporary and permanent dichromatism appeared to coincide with very different selective regimes. And secondly, there was a consistent and significant association between the extent of sexual dichromatism and the complexity of courtship behavior. The monochromatic species had a very complex courtship, that of the only temporarily dichromatic species was intermediate in complexity, and the species that was both permanently and temporarily dichromatic had a relatively simple courtship. Again, limited data on other benthic damselfishes were consistent with this trend. Based on this, Thresher & Moyer (in press) suggested that temporary dichromatism in damselfishes is primarily a means of intersexual communication, as is courtship behavior in general. It is simply a different channel of communication which, when used, results in a corresponding reduction in the amount of information transmitted through other channels, such as motor patterns, and their parallel simplification.

Additional field studies of this type are clearly needed if the factors underlying the evolution of different types of sexual dimorphism are to be understood or the generality of various hypotheses tested. Various blenniids, gobies, and small angelfishes (*Centropyge* spp.), each involving species that differ widely in the extent of sexual dimorphism manifested, seem particularly likely to be profitable subjects.

Finally, and related to the above, there has been considerable discussion in the literature regarding the specific selective regime underlying permanent dichromatism in, particularly, the labroids. Robertson & Hoffman (1977) surveyed the socio-sexual systems of labroid fishes and concluded that male-specific color patterns are primarily the result of intersexual selection by females. They offered two pieces of evidence in support of this conclusion. First, they noted that species with variably intense color patterns, e.g., *Halichoeres maculipinna,* maintain brilliant colors during courtship and spawning but lose them during aggressive, i.e., inter-male, activities. Second, they further noted that "strictly haremic" labrids (mainly *Labroides dimidiatus*), in which female mate choice and, consequently, intersexual selection are limited, are sexually monochromatic, while non-strictly haremic species are sexually dichromatic. Thresher (1979b) subsequently suggested that neither assertion is supportable, even within the labrids.

First, in two species studied in detail, *Halichoeres maculipinna* and *H. garnoti,* terminal-phase males characteristically intensify color patterns during aggressive interactions. In the former species even subordinate males, which are usually "dully colored," assume intense colors during aggressive interactions. Male color patterns, in fact, were found to be especially intense during the "tournament" for a vacated territory, a strictly male-male interaction. Second, *H. maculipinna* is not monochromatic, even though it appears to fulfill Robertson & Hoffman's requirements of exclusive access by the dominant male to the females in his territory and, with limited female movement, few, if any, extra-haremic spawnings. *Labroides dimidiatus,* in fact, is also not "strictly haremic," even though Robertson & Hoffman (1977) offered it as the prime example of the genre, but rather spawns with extra-haremic females 2 to 3% of the time (Warner, et al., 1975). Thresher concluded that available data do not unambiguously support the primacy of epigamic selec-

tion in the evolution of labrid sexual dichromatism and that it is entirely premature to rule out the effects of intrasexual competition among males.

This question has been further considered by Robertson (1981), who examined the social organizations and spawning behavior of *Halichoeres garnoti* and *H. maculipinna* off Panama. This interesting paper justifiably criticized Thresher (1979) for a lack of quantitative data and then pointed out several important differences in the social behavior of both species between Panama and Florida. First, whereas off Florida neither species migrates to specific spawning grounds, both apparently do so regularly in Panama. Second, possibly as a consequence of this difference, *H. maculipinna* off Panama does not form harems; rather, females individually migrate, often over long distances, to mate at male-established spawning territories along the edge of the reef. This difference is a most critical one since, as a dichromatic promiscuous species, *H. maculipinna* becomes consistent with Robertson & Hoffman's (1977) proposed link between female mate choice and sexual dichromatism in labroids. Finally, Robertson pointed out that at Panama terminal-phase male colors are not only most intense during courtship and spawning, but, in fact, also "do not develop . . . and also tend to fade . . . during an (territorial) interaction." This is a striking contrast to the behavior of the species off Florida, where male colors not only conspicuously intensify during territorial disputes (even those involving subordinate males, which do not exhibit such colors at any other time), but also are particularly vibrant during the "tournament" between males for a vacant territory. This difference between populations testifies to an extreme lability on the part of the species to adjust even relatively minor aspects of its social behavior to local conditions of topography and population density. Overall, Robertson attributed behavioral differences between fishes in the Florida and Panama populations to differences in reef topography and tidal conditions.

Robertson subsequently made several suggestions about labrid dichromatism which are consistent with his female mate-selection-based hypothesis. First, he suggested that the high contrast conspicuous elements in the terminal-phase pattern function in species recognition and long-distance attraction of females. The former, indeed, seems likely, given the often high sympatric diversity of labrids on the reef, many often spawning near one another (as pointed out by Robertson), and the likely low signal-to-noise ratio for any one species among many others. The striking differences between terminal-phase male color patterns in congeneric wrasses, such as the various species of *Halichoeres,* are very different from courtship colors in other groups, such as the pomacentrids, in which, for example, courtship colors of various sympatric species of *Pomacentrus* are quite similar (pers. obs.). Presumably this difference reflects the greater mobility of the labrids and the greater number of congeners with which each male must contend. That these species-specific male colors also function in attracting females, however, is not so readily apparent. Selection for territorial advertisement by males, along the lines of Lorenz's (1962) "poster coloration" hypothesis, can also result in the divergence and exaggeration of male colors. Currently available data do not permit distinguishing between the two hypotheses, which, in any case, are not mutually exclusive. Second, Robertson suggested that the myriad of fine details of the terminal-phase pattern "has a role in allowing the female to distinguish a sexually active TP male that is in control of a spawning territory . . . and to determine the quality of the male . . . ." On strictly theoretical grounds, this hypothesis seems unlikely, given that females would be basing their decisions on an instantly variable characteristic that is likely to provide, at best, an extremely unreliable indication of a male's quality. Such a system would be extremely vulnerable to "cheaters," low quality males that could simply mimic the characteristics of more successful males. Robertson did admit, however, that thus far there is no direct evidence that elements of the male color pattern significantly affect a female's choice of a mate. In fact, those data available suggest that no such effect operates, contrary to his hypothesis. Jones (1981), while studying patterns of male success in the temperate labrid *Pseudolabrus celidotus,* found no correlation between a male's success and the intensity of its color; rather, spawning success correlated most highly with the vigor of male courtship displays (a similar correlation has been reported for a pomacentrid by Schmale, 1981). Similar studies are clearly needed for tropical species, but at .this point an automatic assumption that female choice is based on elements in the color pattern of tropical labrids seems unwise. Finally, Robertson suggested that selection favors a female that chooses brilliantly colored and active males because such males draw the attention of piscivores away from her and consequently reduce the likelihood that she will be attacked during the spawning ascent. There are several difficulties with this

hypothesis, as well. It assumes, first, that piscivores are impatient enough not to defer an attack on a rapidly moving, visually aware target in favor of a subsequent relatively slow-moving target preoccupied with the mechanics of spawning. Second, if one assumes a genetic basis for the intensity of male colors and a correlation, as Robertson suggested, between brightness and likelihood of being attacked, then the optimal selection of a mate by a female would seem to be a male whose colors are, perhaps, at somewhat less than peak brilliance, i.e., one whose colors are bright enough to result in high levels of reproductive success, but not so bright as to invite imminent attack, in order to pass this optimal combination on to her offspring. The optimal brightness for a male will vary depending upon many things, such as rates of predation and the trade-off between brightness and duration of the male's spawning success. Finally, there is as yet no direct evidence that brilliantly colored individuals are any more likely to be attacked than more cryptic ones, although this assumption is widely made and sounds reasonable.

In the absence of direct evidence indicative of the social role, if any, of sexual dichromatism in labroids, Robertson suggested that indirect evidence supports the female-choice hypothesis. Specifically, he predicted that "species in which *typically* there is no continuing free competition for mates at spawning grounds should be the least dichromatic species" and "if terminal phase males can change color, the terminal phase pattern should be more intensely displayed during sexual interactions than during agonistic interactions . . . ." An obvious difficulty with the first prediction is that it is often difficult to readily determine what is "typical" for a species. In the case of *Halichoeres maculipinna,* for example, it does one thing off Florida, something else off Panama. Presumably one would have to survey all of the habitats the species occupies throughout its range, determine what it does in each, and determine the reproductive output of fishes in each social regime before a "typical" behavior could be defined. Despite this problem, Robertson suggested that three points support his predictions. First, monochromatic species of *Labroides* and, at least, *Bodianus rufus* form harems in which free competition by males is minimal, even if matings are not, as Robertson & Hoffman (1977) emphasized, "strictly haremic." Second, many sexually dichromatic labrids are demonstrably not harem-forming. And third, during male-male interactions the intensity of male color patterns fades, clearly demonstrating that such colors are not significant in intra-male competition. Data to support these three points, unfortunately, remain ambiguous, contrary to Robertson's conclusions.

Regarding a correlation between sexual monochromatism and limited mate choice, Moyer & Yogo (in press) abundantly and quantitatively document that the monochromatic species *Halichoeres melanochir* is promiscuous, with males forming lek-like spawning aggregations to which females migrate. Information on other monochromatic labrids is sparse. At least two (*Bodianus axillaris* and the diminutive *Pseudocheilinus hexataenia*) appear to be harem-forming, based on personal observations during spawning periods (the former cleans, however, even as an adult, which Thresher, 1979b, suggested may constrain the range of colors the species might develop); a third species, *Halichoeres prosopeion*, also appears to be lek-like in its spawning behavior (Moyer & Yogo, in press). The relationship between monochromatism and limited mate choice does not appear to be as clear as Robertson & Hoffman (1977) and Robertson (1981) suggested. Similar complexity arises when one attempts to correlate sexual dichromatism with a promiscuous spawning system. Robertson (1981) cited several examples of dichromatic wrasses that are not yet reported to be harem-forming. He rejected as inadequate the data regarding the strikingly dichromatic species *Duymaeria flagellifera*, although in fact Nakazono & Tsukahara (1974) observed the species for four years and reported that all spawnings take place between a male and the females inhabiting his territory. Another species he rejected as inadequately documented is *Cirrhilabrus temminckii*, a spectacularly dichromatic species, although both Moyer & Shepard (1975) and Suzuki, et al. (1981) reported the species to be harem-forming in the field (although see also L. Bell, in prep., for data indicating female mobility). Moyer & Yogo (in press) comment on several species that, based on long-term observations of spawning, are conspicuously dichromatic and also harem-forming. Personal observations strongly indicate that *Hemipteronotus splendens* (observed for several days off Puerto Rico), *Halichoeres chierchiae* (similarly observed in the Gulf of California), and *Gomphosus varius* (observed at Enewetak and One Tree Island) all form harems. All, nonetheless, are conspicuously dichromatic. While it is doubtless true that most promiscuous wrasses are dichromatic (although *H. melanochir* is not), it is also true that most wrasses, in general, are dichromatic.

Finally, Robertson argued that the fading of intense male colors during agonistic encounters falsifies the intrasexual competition hypothesis. Again, the generality of this observation is questionable; even in *H. maculipinna* it apparently varies between Florida and Panama. Males of several other labrid species, at least, intensify color patterns during aggressive activities (pers. obs.; Moyer, in prep.). Robertson further argued that a territorial male that exhibits terminal-phase colors "most continuously and intensely" (as opposed to nonterritorial males that do not) is only trying "to communicate information on his status to a female." Carried to its extreme form, Robertson suggested that the brilliant colors of all males involved in the vigorous and highly aggressive competition for a vacant territory described in Thresher (1979b) are the result of a temporarily unstable dominance relationship in which "each (male) was behaving as if it was the dominant male and was advertising that to resident females." This suggestion is unrealistic in light of the prolonged high intensity, aggressive interactions between these males. Their behavior then and during subsequent bouts of territorial defense was clearly directed at one another, and accounting for their intense colors as simply coincidental routine displays to potential mates outside of normal spawning periods in untenable. A far more parsimonious explanation is the obvious one—that the males were vigorously displaying warning colors to one another as part of their normal aggressive repertoire.

If it is premature to argue that intrasexual competition is not an important evolutionary determinant of sexual dichromatism in labrids, then it is equally premature to rule out the female-choice hypothesis. To date, the only direct evidence relating male color pattern elements to female mate selection is negative, i.e., there was no evident effect. Nonetheless, male colors are brilliant during courtship and spawning in many species, and a sexual role for such colors would not be hard to accept. Such a role would not, nonetheless, rule out an aggressive role for such colors. The female-choice hypothesis and the intrasexual competition hypothesis are not mutually exclusive in labrids or in any other animal. The important question is not which of the two is operating, but rather how much each has contributed to extant patterns of dichromatism.

Although this "controversy" has dealt entirely with labrids, the group in which the most work on sexual dimorphism has thus far been done, the hypotheses involved are likely to be cast into forms applicable to other reef-associated families. Data for these other families are sparse, but that available are not consistent with a strong link between sexual monochromatism and a harem-based social system (although as noted earlier most strongly dichromatic reef fishes appear to be promiscuous). In the pomacanthid genus *Centropyge*, for example, species run the full gamut from permanently dichromatic through various degrees and types of temporary dichromatism to permanent monochromatism. Yet all appear to be harem-forming; none have been documented or even suggested to be as promiscuous as the dichromatic wrasses cited by Robertson (Thresher, 1982). Similarly, many harem-forming balistids are sexually dichromatic. On the other extreme most, if not all, chaetodontids are sexually monochromatic regardless of their social system.

Other observations also suggest that territorial advertisement may often be a principal role of male-specific colors. In the goatfish *Parupeneus trifasciatus*, for example, males develop a brilliant red flush while patrolling their temporary spawning territories at dusk. These colors are maintained at full intensity during territorial fights between males but immediately disappear when a female approaches. During courtship and the spawning ascent the male's colors revert to normal, the red flush developing again only after spawning is completed.

Speculations on the roles of sexual dimorphism in general, and sexual dichromatism in particular, and broad scale surveys documenting its occurrence in various types of spawning systems can provide only limited insights into the evolution of such color patterning and are fraught with the danger of easily over-interpreting limited data. This is an area of reef fish behavioral ecology that seems to be particularly well suited for detailed observational and experimental study. There are a number of groups, such as the gobies or the angelfishes in the genus *Centropyge*, that exhibit within a group of closely related species a range of sexual dimorphism. A detailed quantitative examination of the socio-sexual systems of such fishes would seem to be both simple to carry out and possibly quite informative.

JUVENILE COLORATION

Distinctive juvenile color patterns have evoked considerable interest among behaviorists (e.g., Fricke, 1973a; Patterson, 1975; Thresher, 1979c; Fricke, 1980) and among ecologists (e.g., Sale, et al., 1980; Itzkowitz, 1977). In the broadest terms, three hypo-

theses have been put forward to account for such color patterns:

1) *Species Camouflage*—Fricke (1973a, 1980) and Hamilton (1973) both suggested that juvenile color patterns that are different from those of the adult are a means by which the juveniles mask their species membership, consequently avoiding attack as competitors by territorial adults and being permitted to occupy such territories. Drawing mainly on his work with chaetodontids, Fricke also noted that apparent "eyespots," dark spots outlined by a lighter color, tend to be most prevalent on juveniles which most closely resemble the adults, which he again interpreted as a mechanism of camouflaging species membership.

2) *Intra-juvenile Advertisement*—Many juvenile fishes are strongly territorial (e.g., many pomacentrids), even where the adults are less so or not at all (e.g., some pomacanthids). Lorenz (1962) suggested that in such territorial fishes bright color patterns serve a function similar to that of song in birds; that is, they advertise the location of a territorial individual and warn away conspecific rivals. This "poster color" hypothesis has subsequently been criticized as too simple an explanation for the diversity of color patterns in reef fishes (e.g., Ehrlich, et al., 1977; Peterman, 1971) but has also been supported by evidence that at least some reef fishes are extremely attentive to the color patterns of conspecific individuals (Brockmann, 1973; Fricke, 1973a; Thresher, 1976b, 1979a). The poster coloration hypothesis can also be applied to juveniles: territorial juvenile fishes have conspicuous color patterns primarily due to their mutual intraspecific interactions, with such colors functioning to space out the juveniles. In this regard Patterson (1975), in the only quantitative study of the interactions of such juveniles to date, found evidence that such juveniles respond aggressively at greater distances to brightly colored conspecific intruders than to dully colored ones.

3) *Adult Habituation*—A third hypothesis for the evolution of distinctive juvenile color patterns is that such patterns, in fact, attract the attention of a territorial adult, much in the manner of a red cape attracting a bull. Like the matador, the small and colorful juvenile evades the adult's attack by dashing into a crevice or hole too small for the adult to follow, a widely observed behavior referred to by Sale, et al. (1980) as "topological deception." In such a system the bright colors of the juvenile enhance its visibility, so that it initially draws frequent chases. Such a high

Table 6. Distribution of Patterns of Juvenile Coloration in Reef Fishes.

No Distinct Juvenile Pattern		More Cryptic	Less Cryptic
Some Acanthuridae	Most Pseudochromoids	Some Balistidae	Some Acanthuridae
Antennariidae	Pteroidae	Carangidae	Some Balistidae
Anthiines	Sciaenidae	Most Chaetodontidae	Some Ephippidae (?)
Apogonidae	Scorpaenidae	Ephippidae	Most Labridae
Aulostomidae	Some Serranines	Most Epinephelines	Some Lutjanidae
Blenniidae	Siganidae	Some Grammistidae	Plotosidae
Branchiostegidae	Most Sparoids	Most Pomadaysiidae	Some Pomacentridae
Callionymidae	Synodontidae	Scaridae	Some Pomacanthidae
Chaenopsidae	Tripterygiidae	Some Serranines	
Some Chaetodontidae	Zanclidae	Sphyraenidae	
Cirrhitidae			
Clinidae			
Eels			
Some Epinephelines			
Fistulariidae			
Gobiidae			
Most Grammistidae			
Gobiesocidae			
Gobiidae		*Same*	
Holocentridae			
Some Lutjanidae		Some Balistidae	
Mugilidae		Some Sparoids	
Mugiloididae		Some Pomacanthidae	
Mullidae		Some Pomacentridae	
Nemadactylidae		Some Pomadasyidae	
Opistognathidae		A few Pseudochromoids	
Pempheridae			
Some Pomacanthidae			
Some Pomacentridae			
Priacanthidae			

chase frequency in turn results in a rapid habituation of the adult toward the presence of the juvenile (Thresher, 1979c). Presumably a small fish with adult colors would draw even more frequent attacks, but such attacks might well be so vigorous that habituation would occur more slowly. Detailed work on the process of habituation indicates that a persistent waning of responses occurs most rapidly with frequent repetition of low intensity, rather than high intensity, stimulation (Hinde, 1970).

Despite widespread interest in the phenomenon, juvenile-specific color patterns, in fact, are relatively uncommon in reef fishes (Table 6). Among the minority with such juvenile patterns three different classes can be distinguished: those with juveniles more cryptic than the adults, those with juveniles similar in conspicuousness to the adults, and those with more brightly colored juveniles. Such a separation is impor-

tant since it is unlikely that any single hypothesis will account for all three situations, i.e., the factors that select for cryptic juvenile colors in one group are not likely to be the same ones that select for extremely bright colors in another. There is about equal representation in all three groups (see Table).

Several trends are readily apparent from the table. First, the fishes that lack distinctive juvenile color patterns are extremely diverse systematically, behaviorally, and ecologically, ranging from eels and synodontids to all groups of nocturnally active fishes and to a variety of small territorial species, including gobies, blennies, and most pseudochromoids. Any explanation for the role of distinctive juvenile coloration in territorial species must also account for the lack of such colors in these species. This widespread lack of distinctive juvenile color patterns is, in fact, somewhat surprising. If one assumes that small individuals are more susceptible to predation than the adults, then one would expect most fishes to have juveniles more cryptically colored than the adults. The fact that this general pattern is not the case could result from the adults being cryptic anyway, which is certainly the case for many of these fishes. Alternatively, it suggests either that rates of juvenile mortality are not as high as is commonly assumed or that other factors, such as the need for species identification at such small sizes, outweigh any advantage of crypsis to the juvenile. A second evident trend is that of those groups that have more cryptic juveniles, six out of ten are predators of one sort or another. This suggests that for such fishes the necessity of crypsis in ambush-style predation may overrule all other considerations.

The class of juvenile-specific color patterns that has aroused the most widespread interest is that in which the juveniles are more conspicuously colored than the adults. Only eight such groups occur in reef fishes, one of which, the ephippids, is questionable (although the juveniles of some are rather strikingly colored it has been suggested that the colors of at least one mimic a distasteful nudibranch). The diverse ecologies of the fishes in this class clearly suggest a diversity of selection pressures that favor the evolution of conspicuous juvenile coloration. In some cases, e.g., most labrid and pomacanthid examples, the juveniles are cleaning species (i.e., they pick ectoparasites off larger fishes) and their bright colors have presumably evolved to advertise their roles. Another species, the catfish *Plotosus anguillaris*, forms massive juvenile schools in which bright juvenile color patterns apparently serve in school coordination. Perhaps the most interesting fishes, however, are those in which bright juvenile colors occur in conjunction with territoriality.

The clearest example of this association is the western Atlantic surgeonfish *Acanthurus coeruleus* (Thresher, 1980). Adults are dull blue, forage in large and often heterospecific schools, and feed on benthic algae. Small juveniles, however, are solitary and vigorously territorial. These juveniles are bright yellow with a blue eye-ring, a color combination that is particularly conspicuous to other fishes (Thresher, 1977b). As the juvenile grows it gradually develops a schooling tendency, first schooling with small non-related fishes, then within monospecific groups of only two or three, and ultimately in the normal large foraging units. This change in sociality is paralleled by a change in color from bright yellow, to blue-bodied with a yellow tail, to bright blue, and then to dull blue. Since the adults are not territorial it is unlikely that the bright juvenile colors function in facilitating habituation of adult aggression; rather, this species seems to fit most closely the juvenile poster-coloration hypothesis.

The social role of juvenile color patterns, however, has been discussed in greatest detail for the fishes in the family Pomacentridae (e.g., Itzkowitz, 1977; Patterson, 1975; Thresher, 1977, 1979c), a group in which such coloration is common, though clearly not universal. A survey of the distribution of juvenile-specific colors in the family (based largely on data provided by Allen (1975), personal communications from P.F. Sale and P. Doherty, and personal observation) indicates two broad trends.

1) Planktivorous fishes rarely, if ever, have juvenile-specific color patterns. Examples of entire genera, all or most species of which are planktivorous, that lack such juvenile color patterns include *Chromis, Dascyllus, Neopomacentrus, Amblyglyphidodon, Abudefduf,* and *Lepidozygus.* This general relationship also holds within those few genera in which some species are primarily planktivorous and others are primarily benthic foragers. Among Great Barrier Reef *Pomacentrus,* seven species are primarily planktivorous and none have a distinctive juvenile color pattern, whereas six of the ten benthic foragers for which data are available have such juvenile coloration. The difference between the two guilds in terms of the prevalence of juvenile color patterns is significant at greater than the 0.025 level (Fisher Exact Probability).

2) Within the benthic foraging species there is a strong tendency for large species to have a conspicuous juvenile-specific coloration and for small species to lack such colors. The transition point is between 90 and 100 mm SL, with 11 of 15 species less than 90 mm lacking juvenile color patterns and 16 of 19 species larger than 100 mm SL having them. The difference is significant at greater than the 0.01 level (Fisher Exact Probability).

I interpret these two trends as supporting the last hypothesis offered, that juvenile colors serve to attract attacks from territorial adults, resulting in the rapid habituation of their levels of attack readiness. Specifically, it appears that if the adults are not territorial, then juvenile-specific color patterns have not developed. And also, if the adults are so small that avoiding their attacks by dashing into a hole too small for them to enter is not feasible then the habituation process of attract and flee cannot work and, again, bright juvenile-specific coloration tends not to develop. These rules are clearly not hard-and-fast, however. The ability of a juvenile to flee into shelter, for example, will vary with the structure of the habitat occupied, such that the minimum acceptable adult size will also vary.

The distribution of distinctive juvenile colors in other families of reef fishes is also consistent with these trends. Small territorial fishes, such as gobies and blennies, lack such colors, as noted earlier. This absence is predicted by the hypothesis. In balistids a comparable trend of increasing likelihood of juvenile colors with increasing adult body size also seems to be the case. Smaller species, such as the numerous file-fishes, by and large lack such colors, whereas the larger triggerfishes, e.g., *Sufflamen* and *Pseudobalistes,* often have them. Many balistids are also planktivores, which might provide an independent test for the negative correlation between such a feeding mode and distinctive juvenile colors. Data are limited, but at least some planktivorous species lack these colors (e.g., *Xanthichthys ringens),* though such an analysis is complicated by the often lengthy prejuvenile plank-tonic stages common in this family.

Direct evidence that the presence of brightly colored juvenile pomacentrids reduces the level of attack readiness of a territory-holding adult against such juveniles was obtained during a brief study of the Sea of Cortez damselfish *Eupomacentrus rectifraenum.* Adults of this eastern Pacific species are dull brown benthic foragers that reach a length of about 130 mm (Thompson, et al., 1979); juveniles, in contrast, are bright blue. The social system of the species in the Gulf of California is based on adult territoriality centered, for the most part, on low relief rocky reefs (e.g., Thresher, 1980b). Juveniles apparently prefer to settle in adult territories (Godsey, et al., 1980). Adults are territorial toward juveniles, but at least small ones are capable of avoiding attack by dashing into cover (Godsey, et al., 1980). Consequently, an adult's territory may contain as many as five co-resident small juveniles, though the average is probably closer to only one or two.

Model-bottle experiments (see Myrberg & Thresher, 1974) were performed by placing a single blue juvenile in an all-glass cylindrical container and setting it in the center of the adult's territory, as determined from five minutes of previous observation. The number of attacks (bites and/or ramming movements) specifically directed against the introduced juvenile during three minutes following the first attack (or after two minutes if no attacks occurred) was counted

before removing the model-bottle. A new juvenile was obtained before testing the next adult. Adults were chosen for testing such that there was a wide range in the number of juveniles already residing in their territories.

The results of the experiment (Fig. 13) document a clear inverse relationship between the number of juveniles in a territory and the attack readiness of that adult toward newly intruding juveniles, i.e., exactly what would be predicted if the adult was habituating to the presence of juveniles at a rate proportional to its encountering them. The data also suggest that the adults do not discriminate between juveniles, that is, they do not ignore residents, but attack newcomers. An alternate explanation for the results, that adults vary in their level of attack readiness toward juveniles, for whatever reason, and the number of juveniles present in their territories is a response to that level, is unlikely. A brief experiment, in need of replication, involving addition and removal of juveniles to territories suggested that changing the number of juveniles present resulted in changes in the adult's attack readiness toward introduced juveniles. An adult to whose ter-

Fig. 13. The effect of the number of juveniles residing in a territory on the attack rate of the territory-holding adult against a "strange" juvenile presented in a glass bottle.

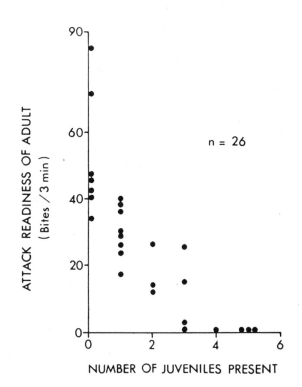

Table 7. Comparison of Adult-colored and Juvenile-colored *E. rectifraenum* of Similar Sizes.
Significance determined by T-test.

Variable Measured	Adult-colored (All Brown) (n=9)	Juvenile-colored (All Blue) (n=13)	Significance Level
Mean Size (mm)	71.1	66.9	N.S.
Home Range Size (m^2)	2.23	2.60	N.S.
Home Range Depth (m)	3.5	3.85	N.S.
% Home Range Exclusive	40.1	27.1	N.S.
Absolute Area of Exclusive Use (m^2)	2.06	0.64	$p < 0.001$
Estimated % Algal Cover	15.6	13.8	N.S.
No. Times Chased/30 min.	1.22	1.31	N.S.
No. Chases/30 min.	1.89	1.54	N.S.
No. Juveniles Co-resident	3.0	2.4	N.S.

ritory five juveniles were added dropped in manifested attack readiness from 45 attacks/three minutes to eight attacks/three minutes in three days; the adult from whose territory the juveniles were removed increased in attack readiness from 0 to 44 attacks/three minutes over the same period.

The size at which juveniles change their color patterns from juvenile to adult varies widely among individuals (e.g., Thresher, 1979a) for reasons that are not well understood. The key stimulus or stimuli probably differ among species depending upon the particular social system involved. Juveniles of *Paraglyphidodon melas* at One Tree Island, Great Barrier Reef, for example, seem to change color all over the lagoon within a few days of each other, suggesting some pervasive environmental cue. In more strongly territorial species, like the Sea of Cortez damselfish, the stimuli involved are likely to be social. To get at least a preliminary indication of what might be involved, data on the social and physical environments of nine adult-colored and 13 juvenile-colored *E. rectifraenum* of similar size were collected at Punta Pescadero in the Gulf of California. Analysis of the data (Table 7) shows no significant differences between the two groups in terms of body size, home range size, estimated amount of algal food present in the home range, or the frequency with which the individual chased or was chased. There was a highly significant difference, however, in the area of the individual's

home range which it had exclusive use of (t = 3.98, df = 20, p < 0.001), based on a 30-minute observation period. These preliminary data suggest that the change from juvenile to adult color pattern is stimulated by the individual's ability to establish itself as an independent territorial individual.

The varying amounts and types of juvenile coloration in several families of reef fishes, such as labrids and balistids as well as pomacentrids, appear to be a fertile ground for detailed observational and experimental work. Some families, such as the chaetodontids, do not readily fit into any of the three hypotheses offered, even though they were the basis for Fricke's species camouflage hypothesis and would be particularly interesting to examine.

Literature Cited

Akazaki, M. 1957. Biological studies on a dragonet, *Synchiropus altivelis* (Temminck et Schlegel). *Japan. J. Ichthyol.*, 5:146-152.

Albrecht, H. 1971. Behaviour of four species of Atlantic damselfishes from Columbia, South America (*Abudefduf saxatilis*, *A. taurus*, *Chromis multilineata*, *C. cyanea*: Pisces, Pomacentridae). *Z. Tierpsychol.*, 26:662-676.

Allen, G.R. 1972. *Anemonefishes: their classification and biology*. T.F.H. Publ., Neptune City, New Jersey. 288 pp.

Allen, G.R. 1975. *Anemonefishes*. Second Ed. T.F.H.

Publ., Neptune City, New Jersey. 352 pp.

Barlow, G.W. 1974. Contrasts in social behavior between Central American cichlid fishes and coral-reef surgeon fishes. *Amer. Zool.*, 14:9-34.

Barlow, G.W. 1975. On the sociobiology of four Puerto Rican parrotfishes (Scaridae). *Mar. Biol.*, 33:281-293.

Bauer, J.A. Jr. and S.E. Bauer. 1981. Reproductive biology of pigmy angelfishes of the genus *Centropyge* (Pomacanthidae). *Bull. Mar. Sci.*, 31:495-513.

Berry, F.H. and W.J. Baldwin. 1966. Triggerfishes (Balistidae) of the eastern Pacific. *Proc. Calif. Acad. Sci., 4th Ser.*, 34:429-474.

Böhlke, J.E. and C.C.G. Chaplin. 1968. *Fishes of the Bahamas and adjacent tropical waters.* Livingston Publ. Co., Wynnewood, Pa. 771 pp.

Breder, C.M., Jr. 1941. On the reproductive behavior of the sponge blenny, *Paraclinus marmoratus* (Steindachner). *Zoologica, N.Y.*, 26:233-235.

Brockmann, H.J. 1973. The function of poster-coloration in the beaugregory, *Eupomacentrus leucostictus* (Pomacentridae, Pisces). *Z. Tierpsychol.*, 33:13-34.

Brown, J.L. and G. Orians. 1970. Spacing patterns of mobile animals. *Ann. Rev. Ecol. Syst.*, 1:239-262.

Buckman, N.S. and J.C. Ogden. 1973. Territorial behavior of the striped parrotfish *Scarus croicensis* Bloch (Scaridae). *Ecol.*, 54:1377-1382.

Choat, J.H. and D.R. Robertson. 1975. Protogynous hermaphroditism in fishes of the family Scaridae. Pp. 263-283. *In: Intersexuality in the Animal Kingdom*, (R. Reinboth, Ed.). Springer-Verlag, Berlin.

Clark, K.L. and R.J. Robertson. 1979. Spatial and temporal multi-species nesting aggregations in birds as anti-parasite and anti-predator defenses. *Behav. Ecol. Sociobiol.*, 5:359-371.

Clarke, R.D. 1977. Habitat distribution and species diversity of chaetodontid and pomacentrid fishes near Bimini, Bahamas. *Mar. Biol.*, 40:277-289.

Colin, P.L. 1971a. Interspecific relationships of the yellowhead jawfish, *Opistognathus aurifrons* (Pisces, Opistognathidae). *Copeia*, 1971:469-473.

Colin, P.L. 1971b. The other reef. *Sea Frontiers*, 17:160-170.

Colin, P.L. 1972. Daily activity patterns and effects of environmental conditions on the behavior of the yellowhead jawfish, *Opistognathus aurifrons*, with notes on its ecology. *Zoologica, N.Y.* 57:137-169.

Colin, P.L. 1976. Filter feeding and predation on the eggs of *Thalassoma* sp. by the scombrid fish *Rastrelliger kanagurta. Copeia*, 1976:596-597.

Colin, P.L. 1982. Aspects of the spawning of western Atlantic reef fishes. *NOAA Tech. Mem.*, NMFS-SEFC-80:69-78.

Colin, P.L. and I. Clavijo. 1978. Mass spawning of the spotted goatfish, *Pseudupeneus maculatus* (Bloch) (Pisces:Mullidae). *Bull. Mar. Sci.*, 28:780-782.

Collette, B.B. and F.H. Talbot. 1972. Activity patterns of coral reef fishes with emphasis on nocturnal-diurnal changeover. *In:* Results of the Tektite Program: Ecology of Coral Reef Fishes, (Ed. by B.B. Collette & S.A. Earle). *Science Bull.* (14), *Natural History Mus., Los Angeles*.

Curtis, B. 1940. *The Life Story of the Fish.* Appleton-Century, New York.

Davis, W.P. and R.S. Birdsong. 1973. Coral reef fishes which forage in the water column. *Helgo. Wissen. Meeresunter.*, 24:292-306.

de Sylva, D.P. 1963. Systematics and life history of the Great Barracuda, *Sphyraena barracuda* (Walbaum). *Stud. Trop. Oceanog.*, 1:179 pp.

Doherty, P.J. 1980. Biological and physical constraints on populations of two sympatric, territorial damselfishes on the southern Great Barrier Reef. Ph. D. Diss., Univ. Sydney, Australia.

Ehrlich, P.R. 1975. The population biology of coral reef fish. *Ann. Rev. Ecol. System.*, 6:211-248.

Ehrlich, P.R., F.H. Talbot, B.C. Russell, and G.R.V. Anderson. 1977. The behaviour of chaetodontid fishes with special reference to Lorenz's "poster colouration" hypothesis. *J. Zool. Lond.*, 183:213-228.

Emery, A.R. 1973. Comparative ecology and functional osteology of fourteen species of damselfish (Pisces:Pomacentridae) at Alligator Reef, Florida Keys. *Bull. Mar. Sci.*, 23:649-770.

Emlen, S.T. and L.W. Oring. 1977. Ecology, sexual selection, and the evolution of mating systems. *Science*, 197:215-223.

Endler, J.A. 1978. A predator's view of animal color patterns. *Evol. Biol.*, 11:319-364.

Feddern, H.A. 1968. Hybridization between the western Atlantic angelfishes *Holacanthus isabelita* and *H. ciliaris. Bull. Mar. Sci.*, 18:351-382.

Fischer, E.A. 1978. On the evolutionary stability of simultaneous hermaphroditism in fishes: reproductive behavioral ecology of *Hypoplectrus nigricans* (Serranidae, Pisces). Ph. D. Diss., Univ. Calif.

Berkeley, Berkeley, Calif.

Fishelson, L. 1970. Behavior and ecology of a population of *Abudefduf saxatilis* (Pomacentridae, Teleostei) at Eilat (Red Sea). *Anim. Behav.,* 18:225-237.

Fishelson, L. 1975a. Observations on behavior of the fish *Meiacanthus nigrolineatus* Smith-Vaniz (Blenniidae) in nature (Red Sea) and in captivity. *Aust. J. Mar. Freshwater Res.,* 26:329-341.

Fishelson, L. 1975b. Ethology and reproduction of the pteroid fishes found in the Gulf of Aqaba (Red Sea), especially *Dendrochirus brachypterus* (Cuvier) (Pteroidae, Teleostei). *Publ. Staz. Zool. Napoli, 39, Suppl.:*635-656.

Fraser-Brunner, A. 1940. Notes on the plectognath fishes IV. Sexual dimorphism in the family Ostraciontidae. *Ann. Mag. Nat. Hist., Ser. 11,* 36:390-392.

Fricke, H.-W. 1966. Zum Verhalten des Putzerfisches *Labroides dimidiatus. Z. Tierpsychol.,* 23:1-3.

Fricke, H.-W. 1973a. Behavior as part of ecological adaptation. *Helgo. Wissen. Meeresunter.,* 24:120-144.

Fricke, H.-W. 1973b. Okologie und Sozialverhalten des Korallenbarsches *Dascyllus trimaculatus* (Pisces, Pomacentridae). *Z. Tierpsychol.,* 32:225-256.

Fricke, H.-W. 1974. Oko-Ethologie des monagemen Anemonenfisches *Amphiprion bicinctus. Z. Tierpsychol.,* 36:429-513.

Fricke, H.-W. 1975a. Evolution of social systems through site attachment in fish. *Z. Tierpsychol.,* 39:206-211.

Fricke, H.-W. 1975b. Sozialstructur und okologische Spezialisierung von verwandten Fischen (Pomacentridae). *Z. Tierpsychol.,* 39:492-520.

Fricke, H.-W. 1980. Juvenile-adult colour patterns and coexistence in the territorial coral reef fish *Pomacanthus imperator. Mar. Ecol. (Naples),* 1:133-141.

Fricke, H.-W. and S. Fricke. 1977. Monogamy and sex change by aggressive dominance in coral reef fish. *Nature (Lond.),* 266:830-832.

Geist, V. 1974. On the relation of social evolution and ecology in ungulates. *Amer. Zool.,* 14:205-220.

Geist, V. 1977. A comparison of social adaptations in relation to ecology in gallinaceous bird and ungulate societies. *Ann. Rev. Ecol. Syst.,* 8:193-207.

Ghiselin, M.T. 1969. The evolution of hermaphroditism among animals. *Q. Rev. Biol.,* 44:189-208.

Godsey, M.S., S. Tanaka, and E.A. Aughley. 1980. Experimental analysis of the factors influencing habitat selection by juvenile Sea of Cortez damselfish, *Eupomacentrus rectifraenum* (Pomacentridae). *Bull. Mar. Sci.,* 30:326.

Goeden, G.B. 1978. A monograph on the coral trout *Plectropomus leopardus* (Lacepede). *Res. Bull. (1) Queensland Fish. Serv.:* 42 pp.

Gomon, M.F. 1979. A revision of the labroid fish genus *Bodianus,* with an analysis of the relationships of other members of the tribe Hypsigenyini. Ph. D. Diss., Univ. Miami, Florida.

Gronell, A. 1978. Home-ranging behavior of the cocoa damselfish, *Eupomacentrus variabilis,* off the coast of Florida. Master's Thesis, Univ. Miami, Miami, Fla.

Gronell, A. (In Press). Courtship, spawning and social organization of the pipefish, *Corythoichthys intestinalis* (Pisces: Syngnathidae) with notes on two congeneric species. *Z. Tierpsychol.*

Hamilton, W.J. 1973. *Life's Color Code.* McGraw-Hill, New York. 238 pp.

Helfrich, P. and P.M. Allen. 1975. Observation on the spawning of mullet, *Crenimugil crenilabris* (Forskal), at Enewetak, Marshall Islands. *Micronesica,* 11:219-225.

Hinde, R.A. 1970. *Animal Behaviour.* 2nd Ed. McGraw-Hill, New York.

Hobson, E.S. 1968. Predatory behavior of some shore fishes in the Gulf of California. *Res. Rept. (73), Bur. Sport Fish. Wildl.:*1-92.

Hobson, E.S. 1972. Activity of Hawaiian reef fishes during the evening and morning transitions between daylight and darkness. *Fish. Bull.,* 70:715-740.

Hobson, E.S. and J.R. Chess. 1978. Trophic relationships among fishes and plankton in the lagoon at Enewetak Atoll, Marshall Islands. *Fish. Bull.,* 76:133-153.

Itzkowitz, M. 1977. Spatial organization of the Jamaican damselfish community. *J. Exp. Mar. Biol. Ecol.,* 28:217-241.

Johannes, R.E. 1978. Reproductive strategies of coastal marine fishes in the tropics. *Environ. Biol. Fishes,* 3:65-84.

Johannes, R.E. 1981. *Words of the Lagoon.* Univ. Calif. Press, Los Angeles. 320 pp.

Jones, G.P. 1981. Spawning-site choice by female *Pseudolabrus celidotus* (Pisces: Labridae) and its influence on the mating system. *Behav. Ecol. Sociobiol.,* 8:129-142.

Jones, R.S. 1968. Ecological relationships in Hawaiian and Johnston Island Acanthuridae (surgeonfishes). *Micronesica,* 4:309-361.

Keenleyside, M.H.A. 1972. The behaviour of *Abudefduf zonatus* (Pisces, Pomacentridae) at Heron

Island, Great Barrier Reef. *Anim. Behav.,* 20:763-774.

Keenleyside, M.H.A. 1979. *Diversity and Adaptation in Fish Behaviour.* Springer-Verlag, Berlin. 208 pp.

Knowlton, N. 1976. Pair bonds in a snapping shrimp commensal with a sea anemone (Abstract). *Amer. Zool.,* 16:100.

Knowlton, N. 1978. The behavior and ecology of the commensal shrimp *Alphaeus armatus,* and a model for relationship between female choice, female synchrony, and male parental care. Ph. D. Diss., Univ. Calif., Berkeley.

Lassig, B. 1976. Field observations on the reproductive behavior of *Paragobiodon* spp. (Osteichthys: Gobiidae) at Heron Island, Great Barrier Reef. *Mar. Behav. Physiol.,* 3:283-293.

Leis, J.M. & J.M. Miller. 1976. Offshore distributional patterns of Hawaiian fish larvae. *Mar. Biol.,* 36:359-367.

Leong, D. 1967. Breeding and territorial behaviour in *Opistognathus aurifrons* (Opistognathidae). *Naturwissenschaften,* 54:97.

Lobel, P.S. 1976. Predation on a cleaner fish *(Labroides)* by a hawkfish *(Cirrhites). Copeia,* 1976:384-385.

Lobel, P.S. 1978. Diel, lunar, and seasonal periodicity in the reproductive behavior of the pomacanthid *Centropyge potteri* and some reef fishes in Hawaii. *Pac. Sci.,* 32:193-207.

Loiselle, P.V. and G.W. Barlow. 1978. Do fishes lek like birds? Pp. 31-75. *In: Contrasts in Behavior,* (E.S. Reese and F. Lighter, Eds.). Wiley-Interscience, N.Y. 406 pp.

Lorenz, K. 1962. The function of color in coral reef fishes. *Proc. Roy. Inst. Great Brit.,* 39:282-296.

Low, R.M. 1971. Interspecific territoriality in a pomacentrid reef fish *Pomacentrus flavicauda* Whitley. *Ecology,* 52:648-654.

Lowe-McConnell, R.H. 1979. Ecological aspects of seasonality in fishes of tropical waters. *Symp. Zool. Soc. Lond.,* (44):219-241. Academic Press, N.Y.

Mann, R.H.K. & C.A. Mills. 1979. Demographic aspects of fish fecundity. *Symp. Zool. Soc. Lond.,* (44):161-177. Academic Press, N.Y.

Marshall, N.B. 1953. Egg size in Arctic, Antarctic and deep-sea fishes. *Evol.,* 7:328-341.

Masuda, H., C. Araga, and T. Yoshino. 1975. *Coastal Fishes of Southern Japan.* Tokai Univ. Press, Tokyo. 378 pp.

Maynard Smith, J. 1977. Parental investment: a prospective analysis. *Anim. Behav.,* 25:1-9.

Moe, M.A., Jr. 1963. A survey of offshore fishing in Florida. *Prof. Pap. Ser., Mar. Lab. Fla.,* (4):1-117.

Moyer, J.T. 1976. Geographical variation and social dominance in Japanese populations of the anemonefish, *Amphiprion clarkii. Japan. J. Ichthyol.,* 23:12-22.

Moyer, J.T. 1979. Mating strategies and reproductive behavior of ostraciid fishes at Miyake-jima, Japan. *Japan. J. Ichthyol.,* 26:148-160.

Moyer, J.T. 1980. Influence of temperate waters on the behavior of the tropical anemonefish *Amphiprion clarkii* at Miyake-jima, Japan. *Bull. Mar. Sci.,* 30:261-272.

Moyer, J.T. and L.J. Bell. 1976. Reproductive behavior of the anemonefish *Amphiprion clarkii* at Miyake-jima, Japan. *Japan. J. Ichthyol.,* 23:23-32.

Moyer, J.T. and A. Nakazono. 1978a. Population structure, reproductive behavior and protogynous hermaphroditism in the angelfish *Centropyge interruptus* at Miyake-jima, Japan. *Japan. J. Ichthyol.,* 25:25-39.

Moyer, J.T. and A. Nakazono. 1978b. Protandrous hermaphrodistism in six species of the anemonefish genus *Amphiprion* in Japan. *Japan. J. Ichthyol.,* 25.

Moyer, J.T. & J.W. Shepard. 1975. Notes on the spawning behavior of the wrasse, *Cirrhilabrus temminckii. Japan. J. Ichthyol.,* 22:40-42.

Moyer, J.T. and Y. Yogo. (In Press). The lek mating system of *Halichoeres melanochir* (Pisces:Labridae) at Miyake-jima, Japan. *Z. Tierpsychol.*

Moyer, J.T., R.E. Thresher, and P.L. Colin. (In Press). Courtship, spawning and inferred social organizations of American angelfishes (genera *Pomacanthus, Holacanthus* and *Centropyge;* Pomacanthidae). *Env. Biol. Fishes,* 8.

Moyer, J.T. & M.J.Zaiser. 1981. Reproductive behavior and social organization of the pygmy lionfish, *Dendrochirus* zebra, at Miyake-jima, Japan. *Japan.J. Ichthyol.,* 28:52-69.

Myrberg, A.A., Jr. 1972. Ethology of the bicolor damselfish, *Eupomacentrus partitus* (Pisces: Pomacentridae): a comparative analysis of laboratory and field behavior. *Anim. Behav. Monogr.,* 5:197-283.

Myrberg, A.A., Jr. and R.E. Thresher. 1974. Interspecific aggression and its relevance to the concept of territoriality in reef fishes. *Amer. Zool.,* 14:81-96.

Nakazono, A. 1979. Studies on the sex reversal and spawning behavior of five species of Japanese labrid fishes. *Rpt. Fish. Res. Lab., Kyushu Univ.,* (4):1-64.

Nakazono, A. & H. Tsukahara. 1974. Underwater observation on the spawning behavior of the wrasse, *Duymaeria flagellifera* (Cuvier et Valenciennes). *Rpt. Fish. Res. Lab., Kyushu Univ.*, (2):1-11.

Nelson, J.S. 1976. *Fishes of the World*. Wiley-Interscience, New York. 416 pp.

Olla, B.L., C. Samet and A.L. Studholme. 1981. Correlates between number of mates, shelter availability and reproductive behavior in the tautog *Tautoga onitis. Mar. Biol.*, 62:239-248.

Patterson, S.A. 1975. The ontogeny of territoriality in juvenile threespot damselfish, *Eupomacentrus planifrons*. Master's Thesis, Univ. Miami, Fla.

Peterman, R.M. 1971. A possible function of coral in coral reef fishes. *Copeia*, 1971:330-331.

Pietsch, T.W. and D.B. Grobecker. 1980. Parental care as an alternative reproductive mode in an antennariid anglerfish. *Copeia*, 1980:551-553.

Popper, D. and L. Fishelson. 1973. Ecology and behavior of *Anthias squamipinnis* (Peters, 1855) (Anthiidae, Teleostei) in the coral habitat of Eilat (Red Sea). *J. Exp. Zool.*, 184:409-424.

Potts, G.W. 1973. The ethology of *Labroides dimidiatus* (Cuv. & Val.) (Labridae, Pisces) on Aldabra. *Anim. Behav.*, 21:250-291.

Pressley, P.H. 1980. Lunar periodicity in the spawning of yellowtail damselfish, *Microspathodon chrysurus. Env. Biol. Fishes*, 5:153-159.

Pressley, P.H. 1981. Pair formation and joint territoriality in a simultaneous hermaphrodite: the coral reef fish *Serranus tigrinus. Z. Tierpsychol.*, 56:33-46.

Ralls, K. 1977. Sexual dimorphism in mammals: avian models and unanswered questions. *Amer. Natur.*, 111:917-938.

Randall, J.E. 1958. A review of the labrid fish genus *Labroides*, with descriptions of two new species and notes on ecology. *Pac. Sci.*, 12:327-347.

Randall, J.E. 1961. Observations on the spawning of surgeonfishes (Acanthuridae) in the Society Islands. *Copeia*, 1961:237-238.

Randall, J.E. 1963a. Review of the hawkfishes (family Cirrhitidae). *Proc. U.S. Nat. Mus.*, 114:389-450.

Randall, J.E. 1963b. Notes on the systematics of the parrotfishes (Scaridae), with emphasis on sexual dichromatism. *Copeia*, 1963:225-237.

Randall, J.E. 1967. Food habits of reef fishes of the West Indies. *Stud. Trop. Oceanogr.*, (5):665-847.

Randall, J.E. 1972. The Hawaiian trunkfishes of the genus *Ostracion. Copeia*, 1972:756-768.

Randall, J.E. 1975. A revision of the Indo-Pacific angelfish genus *Genicanthus*, with descriptions of three new species. *Bull. Mar. Sci.*, 25:393-421.

Randall, J.E. and H.A. Randall. 1963. The spawning and early development of Atlantic parrotfish, *Sparisoma rubripinne*, with notes on other scarid and labrid fishes. *Zoologica, N.Y.*, 48:49-69.

Reese, E.S. 1964. Ethology and marine zoology. *Ann. Rev. Oceanogr. Mar. Biol.*, 1964:455-488.

Reese, E.S. 1973. Duration of residence by coral reef fishes on "home" reefs. *Copeia*, 1973:145-149.

Reese, E.S. 1975. A comparative field study of the social behaviour and related ecology of reef fishes of the family Chaetodontidae. *Z. Tierpsychol.*, 37:37-61.

Rippin, A.B. and D.A. Boag. 1974. Spatial organization among sharp-tailed grouse on arenas. *Can. J. Zool.*, 52:591-597.

Robel, R.J. and W.B. Ballard. 1974. Lek social organization and reproductive success in the greater prairie chicken. *Am. Zool.*, 14:121-128.

Robertson, D.R. 1972. Social control of sex reversal in coral-reef fish. *Science*, 177:1007-1009.

Robertson, D.R. 1973. Sex changes under the waves. *New Sci.*, 31 May:538-539.

Robertson, D.R. 1981. The social and mating systems of two labrid fishes, *Halichoeres maculipinna* and *H. garnoti*, off the Caribbean coast of Panama. *Mar. Biol.* 64:327-340.

Robertson, D.R. and S. Hoffman. 1977. The roles of female mate choice and predation in the mating systems of some tropical labroid fishes. *Z. Tierpsychol.*, 45:298-320.

Robertson, D.R., N.V.C. Polunin, and K. Leighton. 1979. The behavioral ecology of three Indian Ocean surgeonfishes (*Acanthurus lineatus, A. leucosternon* and *Zebrasoma scopas*): their feeding strategies and social and mating systems. *Env. Biol. Fishes*, 4:125-170.

Robertson, D.R. and R. Warner. 1978. Sexual patterns in the labroid fishes of the western Caribbean, II. The parrotfishes (Scaridae). *Smithsonian Contrib. Zool.*, (255):1-26.

Roede, M.J. 1972. Color as related to size, sex, and behavior in seven labrid fish species. *Stud. Fauna Curacao*, 138:1-264.

Rosenblatt, R.H. and E.S. Hobson. 1969. Parrotfishes (Scaridae) of the eastern Pacific, with a generic rearrangement of the Scarinae. *Copeia*, 1969:434-453.

Ross, R.M. 1978. Reproductive behavior of the anemonefish *Amphiprion melanopus* on Guam. *Copeia*, 1978:103-107.

Sale, P.F. 1970. Distribution of larval Acanthuridae off Hawaii. *Copeia*, 1970:765-766.

Sale, P.F. 1971. The reproductive behaviour of the pomacentrid fish, *Chromis caeruleus*. *Z. Tierpsychol.*, 29:156-164.

Sale, P.F. 1974. Mechanisms for co-existence in a guild of territorial fishes at Heron Island. *Proc. 2nd Internat. Coral Reef Symp.*, 1:193-206.

Sale, P.F. 1978. Reef fishes and other vertebrates: a comparison of social structures. Pp. 313-346. *In: Contrasts in Behavior*, (E.S. Reese & F.J. Lighter, Eds). Wiley-Interscience, N.Y. 406 pp.

Sale, P.F. 1980. The ecology of fishes on coral reefs. *Ann. Rev. Oceanog. Mar. Biol.*, 18:367-421.

Sale, P.F., P.J. Doherty, and W. Douglas. 1980. Juvenile recruitment strategies and the coexistence of territorial pomacentrid fishes. *Bull. Mar. Sci.*, 30:147-158.

Schmale, M.C. 1981. Sexual selection and reproductive success in males of the bicolor damselfish, *Eupomacentrus partitus* (Pisces: Pomacentridae). *Anim. Behav.*, 29:1172-1184.

Slobodkin, L.B. and L. Fishelson. 1974. The effect of the cleaner fish *Labroides dimidiatus* on the point diversity of fishes on the reef front at Eilat. *Amer. Natur.*, 108:369-376.

Smith, C.L. 1965. The patterns of sexuality and the classification of serranid fishes. *Amer. Mus. Novitates*, (2207):1-20.

Smith, C.L. 1972. A spawning aggregation of Nassau grouper, *Epinephelus striatus* (Bloch). *Trans. Amer. Fish. Soc.*, 101:256-261.

Smith, C.L. 1982. Patterns of reproduction in coral reef fishes. *NOAA Tech. Mem.*, NMFS-SEFC-80: 49-66.

Springer, V.G. 1959. Systematics and zoogeography of the clinid fishes of the subtribe Labrisomini Hubbs. *Publ. Inst. Mar. Sci. (Univ. Texas)*, 5:417-492.

Springer, V.G., C.L. Smith, and T.F. Fraser. 1977. *Anisochromis straussi*, a new species of protogynous hermaphroditic fish, and synonymy of Anisochromidae, Pseudoplesiopidae, and Pseudochromidae. *Smithsonian Contrib. Zool.*, (252):15 pp.

Stephens, J.S. 1970. Seven new chaenopsid fishes from the western Atlantic. *Copeia*, 1970:280-309.

Stephens, J.S., E.S. Hobson, and R.K. Johnson. 1966. Notes on distribution, behavior, and morphological variation in some chaenopsid fishes from the tropical eastern Pacific, with descriptions of two new species, *Acanthemblemaria castroi* and *Coralliozetus springeri*. *Copeia*, 1966:424-438.

Strand, S. 1978. Community structure among reef fish in the Gulf of California: the use of reef space and interspecific foraging associations. Ph. D. Diss., Univ. Calif. Davis, Davis, Calif.

Strathman, R.R. and M.F. Strathman. 1982. The relationship between adult size and brooding in marine invertebrates. *Amer. Natur.*, 119:91-101.

Suzuki, K., S. Hioki, K. Kobayashi, and Y. Tanaka. 1981. Developing eggs and early larvae of the wrasses, *Cirrhilabrus temminckii* and *Labroides dimidiatus*, with a note on spawning behaviors. *J. Fac. Mar. Sci. Technol., Tokai Univ.*, 14:369-377.

Thomson, D.A. 1969. Toxic stress secretions of the boxfish *Ostracion meleagris*. *Copeia*, 1969:335-352.

Thomson, D.A., L.T. Findley, and A. Kerstitch. 1979. *Reef Fishes of the Sea of Cortez*. Wiley-Interscience, New York. 302 pp.

Thorson, G. 1950. Reproductive and larval ecology of marine bottom invertebrates. *Biol. Rev.*, 25:1-45.

Thresher, R.E. 1976a. Field analysis of the territoriality of the threespot damselfish, *Eupomacentrus planifrons* (Pomacentridae). *Copeia*, 1976: 266-276.

Thresher, R.E. 1976b. Field experiments on species recognition by the threespot damselfish, *Eupomacentrus planifrons* (Pisces: Pomacentridae). *Anim. Behav.*, 24:562-569.

Thresher, R.E. 1977a. Ecological determinants of social organization of reef fishes. *Proc. 3rd Internat. Coral Reef Symp.*, 1:551-557.

Thresher, R.E. 1977b. Eye ornamentation of Caribbean reef fishes. *Z. Tierpsychol.*, 43:152-158.

Thresher, R.E. 1979a. The role of individual recognition in the territorial behaviour of the threespot damselfish, *Eupomacentrus planifrons*. *Mar. Behav. Physiol.*, 6:83-93.

Thresher, R.E. 1979b. Social behavior and ecology of two sympatric wrasses (Labridae: *Halichoeres* spp.) off the coast of Florida. *Mar. Biol.*, 53:161-172.

Thresher, R.E. 1979c. Territoriality and aggression in the threespot damselfish (Pisces; Pomacentridae): an experimental study of causation. *Zeit. Tierpsychol.*, 46:401-434.

Thresher, R.E. 1980a. *Reef Fish*. Palmetto Publ. Co., St. Petersburg, Fla. 172 pp.

Thresher, R.E. 1980b. Clustering: non-agonistic group contact in territorial reef fishes, with special reference to the Sea of Cortez Damselfish, *Eupomacentrus rectifraenum*. *Bull. Mar. Sci.*, 30:252-260.

Thresher, R.E. 1982. Courtship and spawning in the emperor angelfish *Pomacanthus imperator*, with comments on reproduction by other pomacanthid fishes. *Mar. Biol.*, 70:149-156.

Thresher, R.E. and J.T. Moyer. (In Press). Male success, courtship complexity and patterns of sexually monochromatic and dichromatic damselfishes (Pisces: Pomacentridae). *Anim. Behav.*

Warner, R.R. 1978. The evolution of hermaphroditism and unisexuality in aquatic and terrestrial vertebrates. Pp. 77-101. *In: Contrasts in Behavior*, (E.S. Reese & F.J. Lighter, Eds.). Wiley-Interscience, N.Y. 406 pp.

Warner, R.R. & S.G. Hoffman. 1980. Local population size as a determinant of mating system and sexual composition in two tropical marine fishes (*Thalassoma* spp.). *Evol.*, 34:508-518.

Warner, R. and D.R. Robertson. 1978. Sexual patterns in the labroid fishes of the western Caribbean, I. The wrasses (Labridae). *Smithsonian Contrib. Zool.*, (254):1-27.

Warner, R., D.R. Robertson, and E.G. Leigh, Jr. 1975. Sex change and sexual selection. *Science*, 190:633-638.

Wickler, W. 1967. Vergleich des Ablaichverhaltens einiger paarbilden- der sowie nicht paarbildener Pomacentriden und Cichliden (Pisces: Perciformes). *Z. Tierpsychol.*, 24:457-470.

Wiklund, C.G. and M. Andersson. 1980. Nest predation selects for colonial breeding among fieldfares *Turdus pilaris*. *Ibis*, 122:363-366.

Williams, G.C. 1959. Ovary weight of darters: a test of the alleged association of parental care with reduced fecundity in fishes. *Copeia*, 1959:18-24.

Wilson, E.O. 1975. *Sociobiology*. Belknap Press (Harvard Univ.), Cambridge, Mass.

Winn, H.E. and J.E. Bardach. 1960. Some aspects of the comparative biology of parrotfishes at Bermuda. *Zoologica, N.Y.*, 45:24-34.

Youngbluth, M.J. 1968. Aspects of the ecology and ethology of the cleaning fish, *Labroides phthirophagus* Randall. *Z. Tierpsychol.*, 25:915-932.

Glossary

Italic items in the definitions are defined elsewhere
in the glossary.

Atreitic Oocytes—Degenerate *oocytes* found in ovaries and, often, the testes of *secondary males*. In the latter, they are considered good evidence of *hermaphroditism.*

Benthic Broadcaster—A species that releases *pelagic eggs* without ascending off the bottom. The only known reef-dwelling examples are eels.

Demersal Egg—An egg that is heavier than water and consequently sinks to or sits on the bottom after being released. Most, though not all, demersal eggs are elliptically shaped and adhesive.

Demersal Spawner—A species that deposits *demersal eggs* in a pre-prepared nest on or under the bottom. Most demersal spawners exhibit some form of parental care.

Diandry—The condition in which males in a species are of two types: those derived from females by means of sex change and those that are born and remain male. Sometimes referred to as "biandry." See *Monandry.*

Dualistic Reproductive Behavior—The occurrence of both *pair spawning* and *group spawning* in a single species.

Egg Scatterer—A species that ascends off the bottom to release *demersal eggs* which subsequently scatter across the bottom. The only reef-associated examples are belonids, siganids, and possibly some canthigasterin puffers.

Gonochorism—The condition in which the sexes are separate, genetically determined, and do not change throughout the individual's lifetime. See *Secondary Gonochorism, Hermaphroditism.*

Group Spawning—Spawning behavior in which a number of individuals (usually one female and several males) simultaneously release gametes. The term usually refers to *pelagic spawners,* which ascend en masse to release gametes. In contrast to *streaking,* there is usually no overt aggression between males participating in group spawning. See *Pair Spawning.*

Harem—A social system based on the control by one male of access to and spawning with at least two females.

Hermaphroditism—The condition in which an individual can function as both sexes. See *Gonochorism, Simultaneous Hermaphroditism, Sequential Hermaphroditism.*

Initial-phase—In *sequentially hermaphroditic* species, the color pattern characteristic of smaller sexually mature individuals. Usually, though not always, less gaudy and complex than the *terminal-phase* pattern developed in most cases following sex change.

Juvenile—The stage in development following settlement to the bottom and prior to sexual maturation.

Lamellae—Finger-like or branching projections into the ovarian *lumen* on which *oocytes* develop before being released prior to spawning.

Larva—Loosely, the stage of development between hatching and settlement to the bottom. More precisely, the stage following absorption of the *yolk sac* and first feeding and prior to the development of caudal fin flexion. See *Prelarva, Postlarva, Prejuvenile.*

Lek—A reproductive system in which males aggregate for the sole purpose of attracting, courting, and spawning with females. A system that superficially appears to be a lek, but in which other factors may have resulted in male aggregation, such as selection of an optimal position for release of *pelagic eggs* or joint defense of *demersal eggs,* is referred to as "lek-like."

Lumen—The "hollow" fluid-filled space in the center of an ovary into which *lamellae* project and ripe eggs are released and which join posteriorly to form the oviduct.

Lunar Rhythm—A temporal patterning to spawning activity in which spawning either occurs only at certain times of the lunar month or reaches a peak in activity at certain times of the month. Such peaks usually occur near the full and/or new moon. In common usage, a "lunar" rhythm involves spawning once monthly, whereas a "semilunar" rhythm refers to spawning every two weeks.

Milt — Male gametes or sperm.

Monandry — The condition in which all males in a species are either born male and remain male through life or all derive from females through sex change. In common usage the latter condition is usually referred to as monandry, whereas the former is *gonochorism*. See *Diandry*.

Monogamy — A reproductive system in which each member of a heterosexual pair spawns only with the other member of that pair. "Permanent" monogamy indicates pairing is life-long; "temporary" monogamy indicates pairing for long terms, but less than life-long. The latter grades into sequential *polygamy*.

Monomorphism — The condition in which the sexes cannot be distinguished on the basis of external morphology.

Oil Globule (drop, droplet) — A spherical, often amber colored, drop of an oil-like organic liquid located in the egg and *yolk sac* of many *larvae*, which provides buoyancy.

Oocyte — An unfertilized egg still in the *ovary*.

Oral Incubation — A form of parental care in which one parent (usually the male) carries the eggs in his mouth until they hatch. In reef fishes only the cardinalfishes and pseudochromoids are known to brood their eggs in this fashion. Also known as mouthbrooding.

Ovary — The female gonad, usually a Y-shaped organ that is round in cross-section, hollow, and located in the upper rear portion of the abdominal cavity.

Oviduct — A hollow organ formed by the posterior fusion of the paired *ovaries* (the base of the Y described above) that connects the *ovaries* with the pore through which gametes are expelled.

Ovoviviparity — A form of livebearing in which the eggs are retained in the female until hatching, without receiving nutrients from the female other than those stored in the yolk sac.

Pair Spawning — Spawning by a single male with a single female. In common usage, the term usually refers to *pelagic spawners*. See *Group Spawning*.

Pelagic Egg — An egg that is lighter than, or equal in specific gravity to, water and which therefore drifts in the water column after being released. Most, though not all, such eggs are spherical, transparent, and colorless.

Pelagic Spawner — A species that sheds *pelagic eggs*

into the water column at the peak of a conspicuous *spawning ascent*.

Polygamy — A mating system in which each male spawns with several females, who spawn only with him. In sequential polygamy the male forms temporary bonds with each female and remains with her for at least one spawning period; in simultaneous polygamy the male courts and often spawns with several females in a single spawning period. See *Promiscuity, Monogamy, Preferred Mate Polygamy*.

Postlarva — The stage in the development of a planktonic *larva* between development of caudal fin flexion and settlement to the bottom, provided there is no *prejuvenile* stage. See *Larva, Prelarva, Prejuvenile*.

Preferred Mate Polygamy — A form of simultaneous *polygamy* in which individuals restrict all or most of their spawning activities to only a portion of the mates available. See *Monogamy, Polygamy*.

Pre- (or Pro-) larva — The stage in development between hatching and first feeding. Also known as a *yolk sac larva*. Species that produce *demersal eggs* usually lack a prelarval stage or have a very short one. See *Larva*.

Prejuvenile — A specialized *postlarval* stage in which the *larva* has the adult complement of fin spines and rays but otherwise differs conspicuously from either the *postlarva* or benthic juvenile in morphology. Examples include the acanthurid "acronurus" larvae and the chaetodontid "tholichthys" larvae.

Primary Female — A *gonochoristic* female. See *Secondary Female*.

Primary Male — A *gonochoristic* male. See *Secondary Male*.

Promiscuity — A mating system in which each male may mate with more than one female and each female with more than one male. See *Polygamy, Monogamy*.

Protandry — *Sequential hermaphroditism* in which an individual transforms from male to female. See *Protogyny*.

Protogyny — *Sequential hermaphroditism* in which an individual transforms from female to male. See *Protandry*.

Secondary Gonochorism — The condition in which the sexes are separate and genetically fixed but in which this condition is clearly derived from *hermaphroditic* ancestors. See *Gonochorism, Hermaphroditism*.

Secondary Female — A female derived through sex

change from a *protandrous* male. See *Primary Female*.

Secondary Male — A male derived through sex change from a *protogynous* female. See *Primary Male*.

Sequential Hermaphroditism — A form of *hermaphroditism* in which, except for a brief transitional stage, the sexes are separate. See *Simultaneous Hermaphroditism*.

Sexual Dimorphism — The condition in which males and females can be distinguished externally. In common usage, the term excludes behavioral differences between the sexes. Differences between the sexes in coloration, a form of sexual dimorphism, is referred to as sexual dichromatism. See *Monomorphism*.

Simultaneous Hermaphroditism — A form of *hermaphroditism* in which each individual is simultaneously male and female and can release either sperm or eggs while spawning. The only known reef-associated examples are a few sparids and the serranine sea basses. See *Sequential Hermaphroditism*.

Sneaking — Surreptitious *pelagic spawning* between a subordinate male and a female within the territory of a dominant male.

Spawning Ascent — Spawning behavior of *pelagic spawners* in which the individuals rise off the bottom and into the water column before releasing gametes.

Sperm Duct — A hollow tube that runs along the edge of each *testes* and that fuses posteriorly with its counterpart from the other *testes* to connect with the genital pore.

Spermatogonia — Crypts of sperm-producing cells within the *testes*.

Streaking — Spawning behavior in which subordinate males rush in and shed sperm at the climax of spawning between a dominant male and a female.

Terminal-phase — In *sequentially hermaphroditic* species the color pattern developed after sex change, usually, though not always, more complex and more colorful than the *initial-phase* color pattern characteristic before sex change.

Testes — The male gonads; usually a pair of long and slender organs located in the upper rear portion of the abdominal cavity and joined posteriorly.

Urogenital Papilla — A small conical tube located just ahead of the anal fin and usually visible only during, or shortly before, spawning, through which gametes are expelled. Such papillae are characteristic of most *demersal spawning* species.

Viviparity — Loosely, any form of reproduction in which free-swimming young, rather than eggs, are released by the female. Specifically, livebearing in which the young within the female receive nourishment from her other than that stored in the *yolk sac*. The latter is sometimes referred to as "euviviparity. See *Ovoviviparity*.

Yolk-sac — A ball of nutritive material, usually pale yellow or orange, around which newly hatched *larvae* from *pelagic eggs* develop and on which they feed during their early stages of development.

Yolk Sac Larva — A *prelarva*.

General Index

Illustrations Index